Conceptual Academy Chemistry

UNDERSTANDING OUR WORLD OF ATOMS AND MOLECULES

John Suchocki

Conceptual Academy

Conceptual Academy

Editor in Chief: Paul G. Hewitt
Operations Specialist: John Wright
Senior Marketing Manager: Graham Orr
Composition: Tracy Suchocki
Photo Research: Conceptual Productions, LLC
Copyediting: Susan Robeck
Cover Image Credit: Wikanda/Shutterstock

Credits and acknowledgments appear with images.

Library of Congress Cataloging-in-Publication Data available upon request from Publisher.

Printed in the United States
3 4 5 6 7 8 9 10

Conceptual Academy

Conceptual Academy, PBC www.ConceptualAcademy.com

ISBN: 979-8-9862706-3-0

About Conceptual Academy

Conceptual Academy is a public benefit company of science teachers, media producers, and authors known for their conceptual science textbooks, such as *Conceptual Physics*, *Conceptual Chemistry*, *Conceptual Biology*, and others. Conceptual Academy provides a comprehensive library of video lessons that, altogether, create the "video textbook"—just as comprehensive and cohesive as the traditional textbook but featuring video. Conceptual Academy also produces traditional textbooks, such as the one you are now holding, as well as many other learning resources, such as lab activities, worksheets, podcasts, test banks, an automated encouragement-based assessment system, and resources for an instructor's professsional development all now hosted by the PocketLab Notebook learning and lesson plan system. Variety is the key to serving the varying needs of our diverse students and helping to keep each classroom truly interactive and learner-centered. Toward this, our programs follow a double helical model as described here:

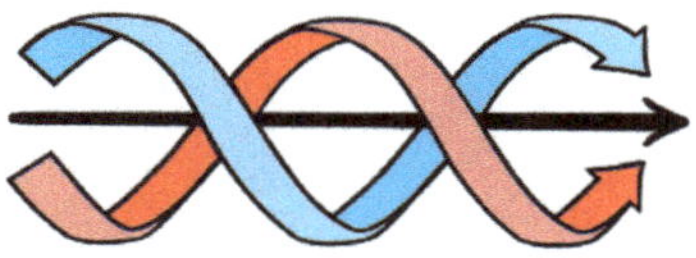

About PocketLab

PocketLab is a hands-on, classroom-ready, learning and lesson plan system that brings STEM labs and lessons to life. This learning system features hand-held data sensors and digital lab software where students collect, visualize, and analyze real-time experimental data. PocketLab hosts hundreds of comprehensive lesson plans available for instructors to customize for their own classroom needs. Lesson plan developers include PhET Interactive Simulations, Kesler Science, TeacherGeek, Green Ninja, Open SciEd, Hooked on Science with Jason Lindsay, StarTalk with Neil deGrasse Tyson, and Conceptual Academy with Paul Hewitt. PocketLab also sponsors the community-building "Science is Cool" virtual unconferences for science teachers across the world. Known for a relaxed and engaging atmosphere, each webinar unconference allows teachers to connect with science celebrities and, just as celebrated, science educators doing amazing things in their classroom.

TEXAS EDITIONS

About the Conceptual Academy Science Series

SBOE approved

The Conceptual Academy authors have revived their physical science and integrated science "Explorations" titles once published for middle and secondary level students. The content has been 1) updated and reorganized into distinct module lessons, 2) integrated with the Conceptual Academy video library and other resources including PocketLab sensor activities, 3) made available through the dynamic Pocket Lab Notebook platform that allows each instructor to organize and implement this library of content according to their classroom needs, and 4) aligned to the TEKS and ELPS Texas state standards. To both teach and reach our modern students effectively, we need not just curricula of high quality but also of great variety and adaptability. Combine this with affordability, then we have the means to pay it forward to a sustainable future built by compassionate, curious, and scientifically literate citizens who love and respect the natural universe for what it truly is. That's good energy!

To the Student

Dear Student,

Welcome to the world of chemistry—a world where everything around you can be traced to these incredibly tiny particles called *atoms*. Chemistry is the study of how atoms combine to form materials. By learning chemistry, you gain a unique perspective on what things are made of and why they behave as they do.

Chemistry is a science with a very practical outlook. By understanding and controlling the behavior of atoms, chemists have been able to produce a broad range of new and useful materials—alloys, fertilizers, pharmaceuticals, polymers, computer chips, recombinant DNA, and more. These materials have raised our standards of living to unprecedented levels. Learning chemistry, therefore, is worthwhile simply because of the impact this field has on society. Understanding chemistry also empowers us to protect the natural environment.

Furthermore, there's value in the insights we gain into everyday phenomena: How does ice float? What do fish "breathe" underwater? (It's not the water.) Why does it take so long for water to boil? Just as important, with a foundation in chemistry, we are in a better position to understand the value and potential impact of the latest research. Along with an awareness of how things work, learning science also strengthens our thinking skills, which is likely to help you in your career, regardless of what career path you follow.

This book presents chemistry conceptually, focusing on the concepts of chemistry with only a minor emphasis on calculations. Though sometimes wildly bizarre, the concepts of chemistry are straightforward and accessible. I enjoy chemistry and I know you can too. So, put on your boots and let's go explore this world from the perspective of its fundamental building blocks. Fantastic discoveries await you.

Good chemistry to you!

John Suchocki

To the Instructor

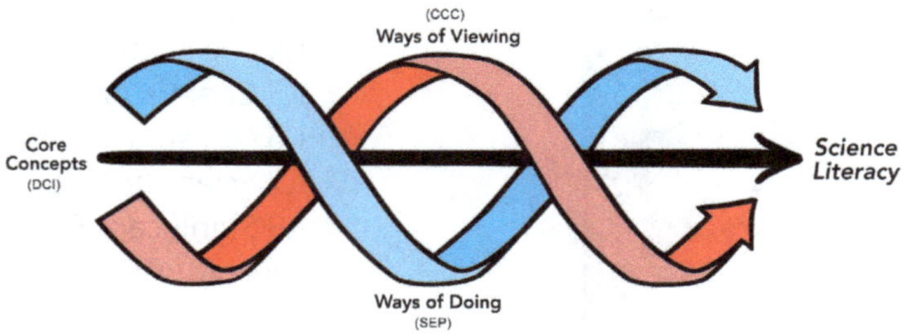

As instructors, we share a common desire for our teaching efforts to have a long-lasting positive impact on our students. We focus, therefore, on what we think is most important for the student to learn. For students taking introductory chemistry courses, certain learning goals are clear. They should become familiar with and even interested in the basic concepts of chemistry, especially the ones that apply to their daily lives. They should understand, for example, how soap works and why ice floats on water. They should be able to distinguish between stratospheric ozone depletion and global climate change and also know what it takes to ensure a safe drinking water supply. Along the way, they should learn how to think about matter from the perspective of atoms and molecules. Furthermore, by studying chemistry, students should come to understand the methods of scientific inquiry and become better equipped to pass this knowledge along to future generations. And perhaps the most pivotal learning goal is nurturing a student's curiosity and passion for the natural world. Students should come to enjoy what they learn, with delight! In short, these students should become citizens of above average science literacy.

These are noble goals, and it is crucial that we do our best to achieve them. I have come to realize, however, that these feats are not what my former students usually cherish most from having taken a course in chemistry. Rather, it is the personal development they experienced through the process.

As all science educators know, chemistry—with its many abstract concepts—is fertile ground for the development of higher thinking skills. Thus, it seems reasonable for us to share this valuable scientific offering—tempered to an appropriate level—with all students. Students come to class not just to learn specific subjects but for personal growth as well. In fact, I would propose that this personal growth is perhaps the most vital commodity of one's education. This growth should include improvements in critical, analytical, and verbal reasoning skills, along with a boost in self-confidence from having successfully met well-placed challenges. The value of our teaching, therefore, rests not only on our ability to help students learn chemistry but also on our ability to help students learn about themselves.

You will find in this program the standard discussions of the applications of chemistry, as shown in the table of contents. True to its title, this textbook also builds a conceptual base from which students may view nature more perceptively by helping them visualize and model the behavior of atoms and molecules and showing them how this behavior gives rise to our macroscopic environment. Numerical problem-solving skills and memorization are not emphasized. Instead, chemistry concepts are developed in a storytelling fashion, with the frequent use of analogies and tightly integrated illustrations and photographs. Follow-up questions are designed to challenge the students' understanding of concepts and their ability to synthesize and articulate conclusions. Concurrent with helping students learn chemistry, *Conceptual Chemistry* aims to be a tool by which students can learn how to become better thinkers and reach their personal goals of self-discovery. That's good chemistry.

Table of Contents

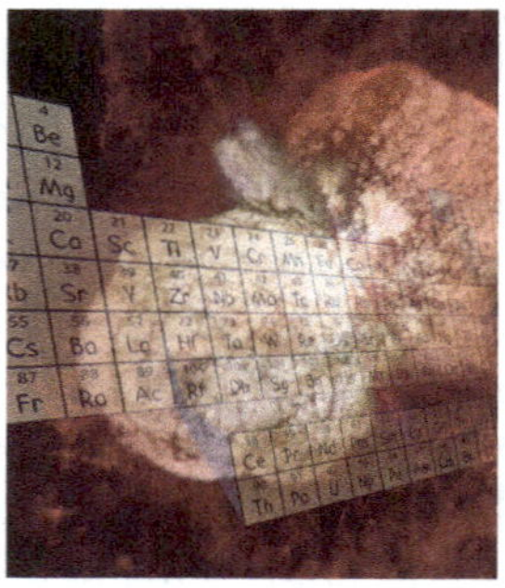

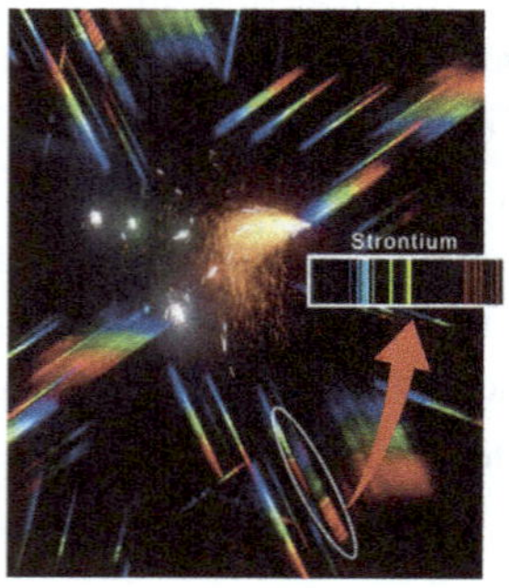

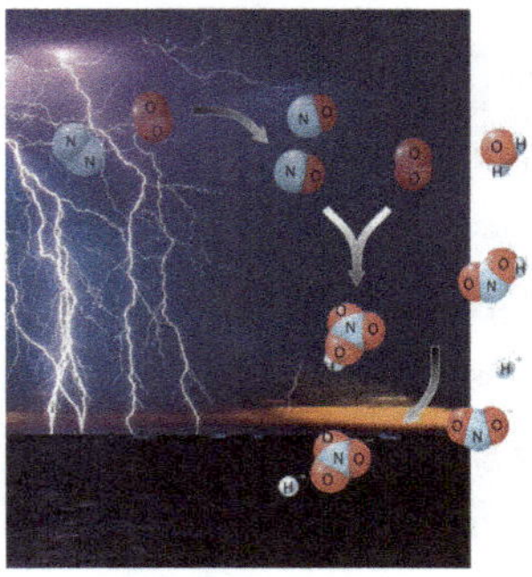

CaCO3

Tetramethyl
pyrazine
Vanillin

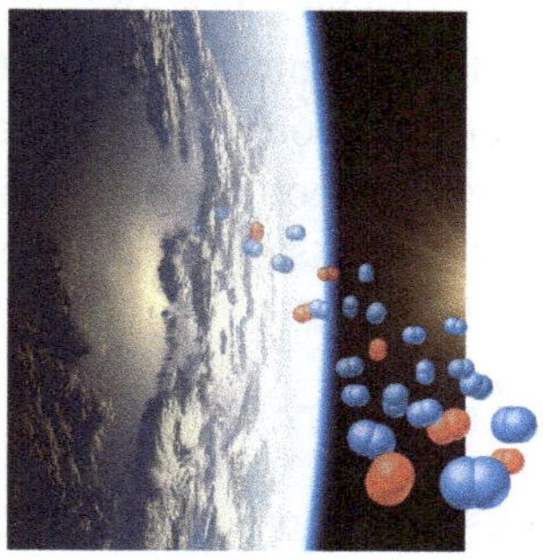

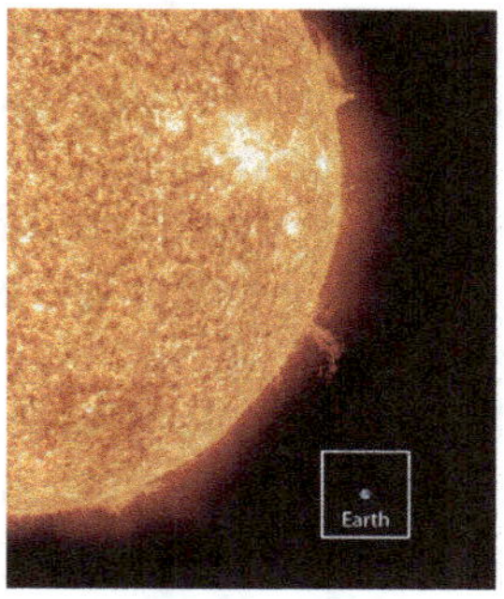

Conceptual Academy

TEXAS EDITION

PocketLab
SCIENCE. EVERYWHERE.

About this Textbook

Conceptual Academy Chemistry is derived from its parent textbook *Conceptual Chemistry*, now in its sixth edition and used primarily by students taking a lower level introductory science course in college. For both titles, all content has been updated to current events including issues related to the Covid pandemic. New content includes discussions of rare-earth metals, chemical equilibrium, nanotechnology, regenerative agriculture, global warming, mind-enhancing drugs, thorium nuclear reactors, the role of water in generating winds, and science denialism. Because automated assessment has been primarily moved to the online PocketLab platform, this printed edition features a slimmed version of end-of-chapter concept review questions along with answers to odd-numbered questions. All content featured within this printed textbook, and much more, is also featured at PocketLab. We have produced this printed textbook for the benefit of students preferring the printed page and also for instrucors to use as a reference for their lesson plans.

Acknowledgements

My great thanks and appreciation to the many instructors using the *Conceptual Chemistry* textbook through our Conceptual Academy platform. Your continued support has truly helped to make the further development of this curriculum possible, which includes the development of *Conceptual Academy Chemistry* as hosted through PocketLab. A special shout-out to Annabelle Anderton, Karson Bader, Stephanie Blake, David Brink, Pat Bunt, Brandon Burnett, Marie Buse, Eddie Case, Mark Cubberley, Jennifer Doyle, Jarden Durden, Eddie Edwards, Jeanette Eggert, Ludmilla Eggleton, Paul Ehrich, Kay Fincher, Sue Geer, Donna Gibson, Rocky Glenn, Salomon Itza, Jerry Jensen, Nadine Jessen, Colleen Jones, Dave Kardelis, Chong Kim, Connie Lewis, Steven Mamanakis, Grant Maru, Jeremy Mason, Jason Moncure, Hai Nguyen, Deanna Pearce, Phil Reedy, John Reilly, Richard Roberts, Carlos Roldan, Lisa Schiller, Nancy Schorschinsky, Richard Synder, John Taylor, Luba Voloshko, Keith Warren, John Weber, Gail Wyant, and the many other instructors also using the Conceptual Academy platform. Much thanks is also extended to numerous past reviewers of the *Conceptual Chemistry* manuscript.

I would also like to thank and acknowledge my earlier publishers who through a process of many mergers include: Pearson; Prentice-Hall; Benjamin-Cummings; Addison-Wesley; Scott-Foresman; Harper Collins; and Little Brown and Company. Much gratitude is extended to all my previous editors, most notably Ben Roberts, Jim Smith, Jessica Moro, Kate Brayton, Hilair Chism, and Irene Nunes. If you are new to Conceptual Chemistry, please know this curriculum has the benefit of a rich history of development courtesy of some of the most talented book-worms within the publishing industry.

There are numerous individuals to whom I am grateful and indebted for their assistance in the development of *Conceptual Chemistry*. Standing at the head of this crowd is my uncle, mentor, and Conceptual Academy co-founder, Paul G. Hewitt. He planted the seed for this book in the early 1980s and has lovingly nurtured its growth ever since. To my parents, thank you for your continued love and support. To my wife, Tracy, I remain deeply thankful for your endless patience and for the love and time you give to me daily. Tracy's assistance in producing the manuscript and her persistent and creative efforts on the test bank were particularly helpful. Most notably, she spearheaded the page layout to this new edition with the assistance of our talented post-college son's Ian and Reece. To Ian, Reece, and Maitreya, who have grown up knowing only parents who pore for hours over their computer, thank you for reminding us of the important things in life.

Special thanks to my very inspirational high school science teachers, Linda Ford (chemistry) and Edward Soldo (biology) of Sycamore High School, Ohio. Their positive impact on me has been life-long.

To the faculty and staff of the chemistry and physics departments of Saint Michael's College I send a grand thank you for your continued support and friendship. Special thanks are extended to Frank L. Lambert to whom the fourth edition was dedicated, Professor Emeritus, Occidental College, for his much-appreciated assistance in the development of *Conceptual Chemistry*'s presentation of the second law of thermodynamics. I send a big mahalo to the crew that helped in the filming of CCAlive! including Michael Reese, Peter Elias, Camden Barruga, Ed Nartatez, Kelly Sato, Sharon Hopwood, Patrick Garcia, Irwin Yamamoto, Stacy Thomas, Kai Dodge, and Maile Ventura. Also, I am grateful for the past support of the faculty and staff of Leeward Community College. Many thanks to Bradley Sieve of Northern Kentucky University for his assistance with the lecture PowerPoint slides, To

Phil Reedy of Delta College of San Joaquin, I send much thanks for revising the Clicker Response System questions as well as for valuable feedback on the development of Conceptual Academy and for coordinating the creation of the complete Conceptual Chemistry video lecture transcripts, which were created courtesy of Delta College.

To Jeff Paradis of CSU Sacramento, I am thankful for his efforts in creating the activities manual, which complements this textbook so well. For developing the *Conceptual Chemistry Laboratory Manual*, I am forever grateful to Donna Gibson of Chabot College. For past work on the Conceptual Chemistry test bank, I am deeply indebted to Bill Centobene of Cypress College as well as Dan Stasko of the University of Southern Maine. Thanks to Kelly Befus of Anoka-Ramsey Community College for checking the accuracy of the *Conceptual Chemistry* test bank as well as the lab manual. For the "Wheel of Scientific Inquiry" shown in Chapter 1, I thank William Harwood and his graduate students at Indiana University.

A special thanks to John Wright, CFO, and Graham Orr, VP of marketing and sales, who have helped to steer Conceptual Academy toward financial sustainability, while maintaining our mission as a science-education public benefit company. To Susan Robeck, thank you for sharing your copyediting eagle eye. Much appreciation to Roger Guest and Donna Stephens for reviewing the chapter on medicinal chemistry. And a super thanks to the talented folks at PocketLab for their amazing support and deep trust in our endeavors: Clifton Roozeboom, Dave Bakker, Robby Douthitt, Jolyn Janis, Jim Schwapach, Scott Kiessig, Susan Hein, and the whole team. And a special shout out to Conceptual Academy's remarkable Loft Team members: Graham Orr, Wendee Payne, Jordan Rouleau, Stephanie Ryan, Greg Simmons, Ian Suchocki, Tracy Suchocki, David Sumida, Jennifer Williams, and John Wright.

To the struggling student, thank you for your learning efforts—you are on the road to making this world a better place.

Much effort has gone into keeping this program error-free and accurate. It is possible, however, that some errors or inaccuracies may have escaped our notice. Your forwarding such errors or inaccuracies to us would be greatly appreciated. Your questions, general comments, and criticisms are also welcome. We look forward to hearing from you.

Good chemistry,

John Suchocki
John@ConceptualAcademy.com

"Curiosity is a delicate little plant that, aside from stimulation, stands mainly in need of freedom."

—Albert Einstein

(1879 - 1955)

Chapter 1

About Science

THE MAIN IDEA

Science is the study of nature's rules

Λ Our home is a blue marble of a planet covered mostly with oceans. Land makes up only 30 percent of its surface. The atmosphere we breathe is quite thin compared to the size of our planet—about as thin as an apple skin is compared to an apple.

Through science we have learned much about the natural world. For example, we have learned that matter is made of very small fundamental particles called *atoms*. These atoms can then join to form larger fundamental structures called *molecules*. This sort of knowledge has allowed us to create some amazing technologies—from agriculture to medicine to space travel.

Yet science is more than just a body of knowledge. It is also a method for exploring nature and discovering the order within it. **Science** is the product of observations, common sense, rational thinking, experimentation, and (sometimes) brilliant insights. It has been built up over many centuries and gathered from places all around the Earth. It is a huge gift to us today from the thinkers and experimenters of the past.

What is so special about science? Why is science such an effective tool for discovery and for solving problems? How is science different from technology? Why is it so important that each of us have an understanding of this eye-opening and creative human endeavor?

1.1 Understanding the Natural World

We humans are good at observing. We are also good at explaining our observations. What we recognize today as modern science, however, began not with our powers of observation, nor with our creative explanations. Rather, modern science began when people first became skeptical of their observations and explanations. They wondered whether their observations were accurate. They wondered whether their explanations were correct. To resolve their doubt, they turned to *experimentation*.

READING CHECK
When did modern science begin?

The greatly respected Greek philosopher Aristotle (384–322 BCE) claimed that an object falls at a speed proportional to its weight. In other words, the heavier the object, the faster it falls. This idea was held to be true for nearly 2000 years, in part because of Aristotle's compelling authority.

Figure 1.1 >

Place a feather on top of a book. Drop the two together to the floor. Which falls faster: the book or the feather?

The Italian physicist Galileo (1564–1642) was doubtful and allegedly showed the falseness of Aristotle's claim with one experiment—demonstrating that heavy and light objects dropped from the Leaning Tower of Pisa fall at nearly equal speeds. You too can refute Aristotle's claim with a simple experiment, as shown in **Figure 1.1.**

As a practical matter, experiments are better at proving ideas wrong than right. For example, is it a truth that all crows are black? Upon seeing millions of black crows, we may become very confident that all crows are black. The moment we see our first white albino crow, however, this once reasonable idea has been proven false. Of course, learning that our ideas are false is useful information. It can prompt us to double-check our thinking. We can then use our experience and creativity to come up with a more encompassing, alternative explanation. This new explanation may not be perfect, but we can be confident that it is closer to the truth than our previous explanation. The more experiments we conduct, and the more times we refine our explanations, the closer we get to understanding the actual workings of nature.

FOR YOUR INFORMATION

The success of science has much to do with an attitude common to scientists. This attitude is one of inquiry and honest experimentation guided by a confidence that all-natural phenomena can be explained.

The Wheel of Scientific Inquiry

Performing experiments is just one of many activities that scientists use to reach their goal of better understanding nature. As shown in **Figure 1.2**, one of the first activities tends to be the asking of a broad question, such as "Where did the Moon come from?", "Can we efficiently create hydrogen from water using direct solar energy?", or "When did humans first arrive in North America?" All other activities are guided by this broad question. These activities will likely include learning about what is already known, making new observations, narrowing the focus of the research to something manageable, asking specific questions that can be answered by experiment, documenting expectations, performing experiments, confirming the results of experiments, reflecting about what the results might mean, and—perhaps most important—communicating with others.

The order in which these activities are performed is largely up to the scientist. No cookbooks. No algorithms of logic. Just equipment, a blank field journal, some self-discipline, a healthy dose of creative curiosity, and a desire to learn about nature for what it is—not for what we might wish it to be. This is the scientific spirit.

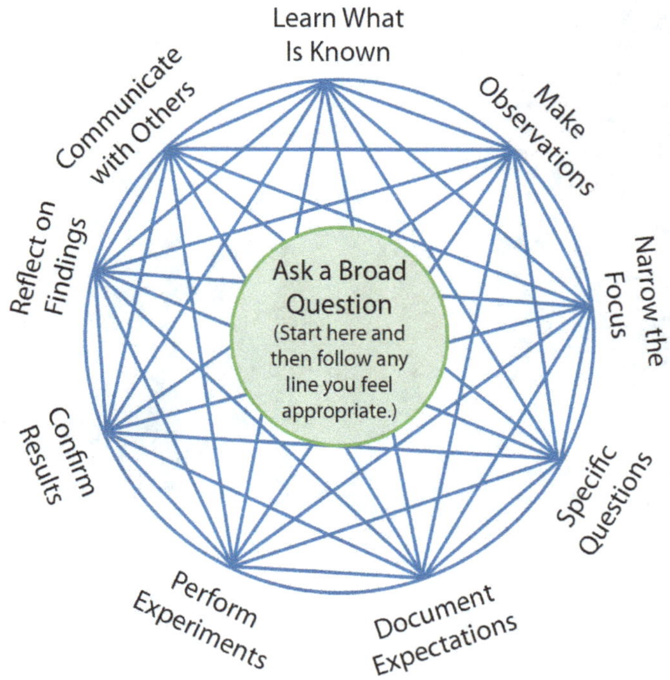

Λ Figure 1.2

This diagram illustrates essential activities conducted by scientists. Commonly, the first activity is the asking of a broad question that defines the scope of the research. It is usually based upon the scientist's particular interests. The scientist can then move among all the various activities in unique paths and repeat activities as often as necessary.

1.2 An Investigation of Sea Butterflies

Let's consider an example of a recent scientific research project that shows how the scientific method can be put to work. Along the way, we'll get a taste of how chemistry and biology are integrated in the physical world. The Antarctic research team headed by James McClintock, Professor of Biology at the University of Alabama at Birmingham, and Bill Baker, Professor of Chemistry at the University of South Florida, was studying the toxic chemicals Antarctic marine organisms secrete to defend themselves against predators (**Figure 1.3**). McClintock and Baker observed an unusual relationship between two animal species, a sea butterfly and an amphipod. This relatioinship led to: a question, a scientific hypothesis, a prediction, tests concerning the chemicals involved, and finally a conclusion. The research generally proceeded according to the steps of the classic scientific method.

1. Observe. The sea butterfly Clione Antarctica is a brightly colored, shell-less snail with winglike extensions used in swimming (**Figure 1.4a**). The amphipod Hyperiella dilatata resembles a small shrimp. McClintock and Baker observed a large percentage of amphipods carrying sea butterflies on their backs, with the sea butterflies held tightly by the legs of the amphipods (**Figure 1.4b**). Any amphipod that lost its sea butterfly would quickly seek another—the amphipods were actively abducting the sea butterflies!

Figure 1.3 >

The Chemical Ecology of Antarctic Marine Organisms Research Project was initiated by James McClintock, shown here (fifth from left) with his team of colleagues and research assistants. He was later joined by Bill Baker (second from right). Baker is shown in the inset dressing for a dive into the icy Antarctic water. Like many other science projects, this one was interdisciplinary, involving the efforts of scientists from a wide variety of backgrounds.

2. Question. McClintock and Baker noted that amphipods carrying butterflies were slowed considerably, making the amphipods more vulnerable to predators and less adept at catching prey. Why then did the amphipods abduct the sea butterflies?

3. Hypothesize. Given their experience with the chemical defense systems of various sea organisms, the research team hypothesized that amphipods carry sea butterflies to produce a chemical that deters a predator of the amphipod.

4. Predict. Based on their hypothesis, they predicted (a) that they would be able to isolate this chemical and (b) that an amphipod predator would be deterred by it.

5. Test predictions. To test their hypothesis and predictions, the researchers captured several predator fish species and conducted the test shown in Figure 1.5. The fish were presented with solitary sea butterflies, which they took into their mouths but promptly spat back out. The fish readily ate uncoupled amphipods but spit out any amphipod coupled with a sea butterfly. These are the results expected if the sea butterfly was secreting some sort of chemical deterrent. The same results would be obtained, however, if a predator fish simply didn't like the feel of a sea butterfly in its mouth. The results of this simple test were therefore ambiguous. A conclusion could not yet be drawn.

All scientific tests need to minimize the number of possible conclusions. Often this is done by running an experimental test along with a control. Ideally, the experimental test and the control differ by only one variable. Any differences in results can then be attributed to how the experimental test differed from the control.

(a)

(b)

< Figure 1.4

(a) The graceful Antarctic sea butterfly is a species of snail that does not have a shell. (b) The shrimplike amphipod attaches a sea butterfly to its back even though doing so limits the amphipod's mobility.

To confirm that the deterrent was chemical and not physical, the researchers made one set of food pellets containing both fish meal and sea butterfly extract (the experimental pellets). For their control test, they made a physically identical set containing only fish meal (the control pellets). As shown in **Figure 1.6**, the predator fish readily ate the control pellets but not the experimental ones. These results strongly supported the chemical hypothesis.

Further processing of the sea butterfly extract yielded five major chemical compounds, only one of which deterred the predator fish from eating the pellets. Chemical analysis of this compound revealed it to be the previously unknown molecule shown in **Figure 1.7**, which they named pteroenone.

6. Draw a conclusion. In addition to running control tests, scientists confirm experimental results by repeated testing. In this case, the Antarctic researchers made many food pellets, both experimental and control, so that each test could be repeated many times. Only after obtaining consistent results in repeated tests can a scientist draw a conclusion. McClintock and Baker were thus able to conclude that amphipods abduct sea butterflies in order to use the sea butterflies' secretion of pteroenone as a defense against predator fish.

Yet, this conclusion would still be regarded with skepticism in the scientific community. Why? There is a great potential for unseen error in any experiment. A laboratory may have faulty equipment that leads to consistently wrong results, for example. Because of the potential for unseen error from any particular research group, experimental results must be reproducible to be considered valid. This means that other scientists must be able to reproduce the same experimental findings in separate experiments. Thus you can see that it is a long road from bright idea to accepted scientific finding! The plodding, painstaking nature of this process is beneficial, though—it is the reason that scientific knowledge is highly trustworthy.

As frequently happens in science, McClintock and Baker's results led to new questions. What are the properties of pteroenone? Does this substance have applications—for example, can it be used as a pest repellent? Could it be useful for treating human disease? In fact, a majority of the medicinals we use were originally discovered in natural sources. This illustrates that there is an important reason for preserving marine habitats, tropical rainforests, and the other diverse natural environments on Earth—they are storehouses of countless yet-to-be-discovered substances.

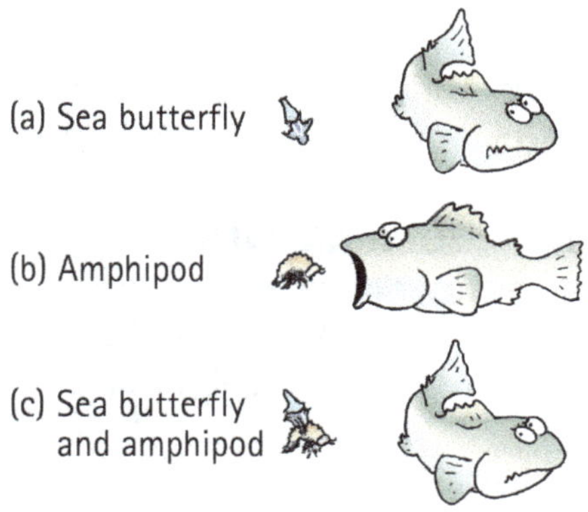

∧ Figure 1.5

In McClintock and Baker's initial experiment, a predatory fish (a) rejected the sea butterfly, (b) ate the free-swimming amphipod, and (c) rejected the amphipod coupled with a sea butterfly.

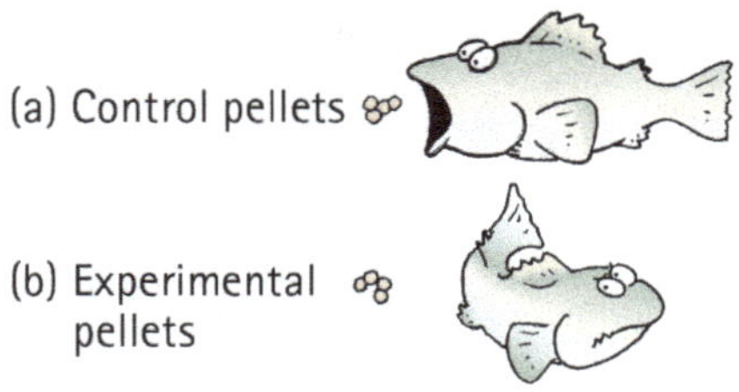

∧ Figure 1.6

The predator fish (a) ate the control pellets but (b) rejected the experimental pellets, which contained sea butterfly extract.

READING CHECK

Why must experimental findings be reproducible to be considered valid?

∧ Figure 1.7

Pteroenone is a molecule produced by sea butterflies as a chemical deterrent against predators. Its name is derived from ptero-, which means "winged" (for the sea butterfly), and -enone, which describes information about the chemical structure. The black spheres represent carbon atoms, the white spheres hydrogen atoms, and the red spheres oxygen atoms.

CONCEPT **CHECK**

1. What variable did the experimental fish pellets contain that was not found in the control pellets?
2. If the fish had eaten the experimental pellets, what conclusion could the scientists have drawn?

CHECK YOUR ANSWER

1. Sea butterfly extract.
2. The scientists would have had to conclude that the predator fish were not deterred by the sea butterfly secretions and thus that the amphipods did not capture the sea butterflies for this reason.

1.3 Technology Is Applied Science

FOR YOUR INFORMATION

Science is a way of knowing. Technology is a way of doing.

Science is concerned with gathering knowledge about the natural world. When we apply this knowledge for practical purposes, we have what we call **technology**. From scientific discoveries, an engineer can design new technologies. New technologies, in turn, can be of assistance to scientists in conducting their research. The computer, for example, is technology scientists use on a daily basis. So, technology arises from science, but technology also supports science—the two are different from each other but very much related.

Technology is a double-edged sword that can be both helpful and harmful. We have the technology, for example, to extract fossil fuels from the ground and then to burn the fossil fuels for the production of energy. Energy production from fossil fuels has benefited our society in countless ways. On the flip side, the burning of fossil fuels endangers the environment. It is tempting to blame technology itself for problems such as pollution, resource depletion, and even overpopulation. These problems, however, are not the fault of technology any more than a shotgun wound is the fault of the shotgun. It is humans who use the technology, and humans who must decide how to use it responsibly.

Remarkably, we already possess the technology to solve many environmental problems. In this 21st century, we are seeing a switch from fossil fuels to more sustainable energy sources, such as photovoltaics, hydroelectric, wind, solar, thermal electric generation, and biomass conversion. Whereas the paper on which the hard copy of a book is printed came from trees, paper can also come from fast-growing weeds, and less of even these materials are needed with the rise of online learning. In some parts of the world, progress is being made to stem the rapid growth of our human population, which is an issue that aggravates almost every problem faced by humans today. We live on a finite planet, and Earth's population carrying capacity is being acknowledged. The greatest obstacle to solving today's problems lies more with social inertia than with a lack of technology. Technology is our tool. What we do with this tool is up to us. The promise of technology is a cleaner and healthier world. Wise applications of technology can lead to a better world.

READING CHECK

What is the promise of technology?

Risk Assessment

The numerous benefits of technology are paired with risks. When the benefits of a technological innovation are seen to outweigh its risks, the technology is accepted and applied. X rays, for example, continue to be used for medical diagnosis despite their potential for causing cancer. But when risks are perceived to outweigh the benefits, the technology tends to be used very sparingly or not at all. Risk can vary for different groups. Aspirin is useful for adults, but for young children it can cause a potentially lethal condition known as Reye's syndrome. Dumping raw sewage into the local river may pose little risk for a town located upstream, but for towns downstream the untreated sewage is a health hazard. Similarly, storing radioactive wastes underground may pose little risk for us today, but for future generations the risks of such storage are greater if there is leakage into groundwater. Technologies involve different risks and benefits for different people, raising questions that are often hotly debated. Which medications should be sold over the counter to the general public, and how should they be labeled? Should food be irradiated in order to minimize food poisoning, which kills more than 3000 Americans each year? The risks to all members of society need to be considered when public policies are decided.

Some people seem to have difficulty accepting the impossibility of zero risk. You cannot go to the beach without risking skin cancer, no matter how much sunscreen you apply. You cannot avoid radioactivity, for it occurs naturally in the air we breathe and the foods we eat. Science, however, can help to determine relative risks. As the tools of science improve, the accuracy of risk assessment improves. Acceptance of risk, on the other hand, is a societal issue. Zero risk is not possible, and a society that accepts no risks receives no benefits.

fyi FOR YOUR INFORMATION

Pharmaceuticals provide measurable benefits. A recently noted risk, however, is that many of these pharmaceuticals are now ending up in our drinking water, lakes, and streams. Although their concentrations are quite low, they are detectable, and these materials are being shown to have a measurable effect on species living in the water. Conclusion: old medicines should be thrown in the trash and never flushed down the drain. Better yet, bring your old prescription drugs to your local pharmacist who will have the means to dispose of these chemicals properly.

CONCEPT **CHECK**

Does technology come from science or does science come from technology?

CHECK YOUR ANSWER

Both! The practical application of knowledge gained through science is technology. In this sense, the science comes first. The tools of technology, however, can be used by scientists to further our understandings of nature. A case in point would be the telescope, which is a tool that permits us to learn about the stars. Science and technology are different, but they complement each other.

1.4 The Natural World

Through science we have already learned much about nature. That said, as any scientist would tell you, there is still much more we have yet to learn. Science is a work in progress. The understandings we have one year will not necessarily be the understandings we have the next year.

For example, it is common to think of a fact as something unchanging and absolute. But in science, a **fact** is something agreed upon by competent observers as being true. Interestingly, what humans accept to be factual changes over time as we learn new ideas. It was once an accepted fact that the universe is unchanging and permanent. Today, we recognize the fact that the universe is expanding and evolving.

READING CHECK

When does a hypothesis become a scientific hypothesis?

A hypothesis is a suggested explanation for an observable phenomenon. A hypothesis becomes a **scientific hypothesis** when, and only when, it can be tested through experiments. The more tests that the scientific hypothesis passes, the greater the confidence we have that the hypothesis is true. However, if the hypothesis fails even one test, then the hypothesis is taken to be false. A new, more encompassing hypothesis is needed.

It was once believed that mass is lost as wood burns, because, clearly, ashes weigh less than the wood that burned. The 18th-century French chemist Antoine Lavoisier (1743–1794), however, was skeptical (**Figure 1.8**). To him it appeared that burning wood lost mass because it was losing gases to the atmosphere. He went on to hypothesize that during any chemical change, such as burning, mass transforms from one substance to another, but it is always conserved. This means that the total mass before the reaction is equal to the total mass after the reaction. To test this hypothesis he conducted experiments during which burning took place in a sealed chamber. He found that the chamber and its contents weighed the same before and after the burning. The old hypothesis didn't fit this observation. Lavoisier's new hypothesis was a better alternative.

Figure 1.8

Antoine Lavoisier, shown here with his wife, Marie-Anne, who assisted him in many of his experiments, was a concerned citizen as well as a first-rate scientist. He established free schools, advocated the use of fire hydrants, and designed street lamps to make travel through urban neighborhoods safer at night.

When a hypothesis has been tested and supported by experimental data over and over again and has not been contradicted, it may become formally stated as a **scientific law**, or *principle*. Lavoisier's conservation of mass hypothesis was repeatedly confirmed over many years, so it became known as the *law of mass conservation*, which we discuss in more detail in Chapter 9. But remember, the goal of science is to describe the rules of nature as accurately as possible. Ideally, a scientific law matches perfectly with the rules of nature. In practice, however, our so-called "laws" are merely our best approximations. If a scientific law is eventually found to be inaccurate through reproducible and verifiable evidence, then the law—in order to be closer to reality—must be modified or discarded.

In the early 20th century, for example, it was found that mass actually does change ever so slightly during a chemical reaction. The law of mass conservation, therefore, is not perfectly accurate. The change in mass during a chemical reaction, however, is exceedingly small and not easily measured. The law of mass conservation, therefore, still holds practical value, which is why it is still widely embraced.

Scientists use the word *theory* in a way that differs from its usage in everyday speech. In everyday speech, a theory is no different from a hypothesis—an explanation for an observable phenomenon. A **scientific theory**, on the other hand, is a well-tested explanation that unifies a broad range of observations within the natural world. Physicists, for instance, speak of the *theory of relativity* and use it to explain how we are held to Earth by gravity and how a strong gravitational field causes time to slow down. Biologists speak of the *theory of natural selection* and use it to explain both the unity and the diversity of life. Chemists speak of the *theory of the atom* and use it to explain how mass is seemingly conserved in a chemical reaction and how one material can transform into another.

FOR YOUR INFORMATION

Critical thinking is the ability to evaluate claims in a rational manner. Creative thinking is the ability to make connections to come up with new ideas.

Theories are a foundation of science, but, like facts, hypotheses, and laws, they are not fixed. Rather, they evolve as they go through stages of redefinition and refinement so as to mirror nature as accurately as possible. Since it was first proposed 200 years ago, for example, the theory of the atom has been repeatedly refined as new evidence about atomic behavior has been gathered. Those who know little about science may argue that scientific theories have little value because they are always being modified. Those who

CONCEPT **CHECK**

Which statement is a scientific hypothesis?

1. The Moon is made of Swiss cheese.
2. Human consciousness arises from an essence that is undetectable.

CHECK YOUR ANSWER

Both statements attempt to explain observed phenomena, so both are hypotheses. Only statement (1) is testable, however, and therefore only statement (1) is a scientific hypothesis.

understand science, however, see it differently. Science is self-correcting, and its theories grow stronger as they are modified. A summary of what we mean by a scientific fact, hypothesis, law, and theory is provided in **Table 1.1**.

The domain of science is restricted to the observable natural world. While scientific methods can be used to debunk various claims, science has no way of verifying testimonies involving the supernatural. The term supernatural literally means "above nature." Science works within nature, not above it. Likewise, science is unable to answer such philosophical questions as "What is the purpose of life?" or such religious questions as "What is the nature of the human spirit?" Though these questions are valid and have great importance to us, they rely on subjective personal experience and do not lead to testable hypotheses.

∧ Figure 1.9

The tree Ayano hugs is made primarily from carbon dioxide and water, the very same chemicals Ayano releases through her breath. In return, the tree releases oxygen, which Ayano uses to sustain her life. We are one with our environment down to the level of atoms and molecules.

Why Should We Learn Science?

Just as you can't enjoy a ball game, computer game, or party game until you know its rules, so it is with nature. Because science helps us learn the rules of nature, it also helps us appreciate nature. You may see beauty in a tree, but you'll see more beauty in that tree when you realize that it was created from substances found not in the ground but primarily in the air—specifically, the carbon dioxide and water put into the air by respiring organisms such as yourself (**Figure 1.9**). Learning science builds new perspectives and is not unlike climbing a mountain. Each step builds on the previous step, while the view grows evermore astounding (**Figure 1.10**).

There are also many practical reasons for which we should become familiar with science. Consider what the world was like before the advent of scientific thinking, around the time of Galileo in the late 1500s. Since that time, science has revealed much about the workings of the natural universe.

TABLE 1.1 A Summary of Scientific Terms

Scientific Fact—An agreed upon truth about the natural world. Example: Wood burns.
Scientific Hypothesis—A testable explanation for an observable phenomenon. Example: Wood transforms mostly to gaseous materials as it burns.
Scientific Law—An experimentally confirmed description of the natural world. Example: The mass of the products of a reaction equals the mass of the reactants.
Scientific Theory—A well-tested explanation that unifies many observations. Example: Matter is made of tiny particles called atoms that are not destroyedduring a chemical transformation.

With this knowledge arose society-changing technologies. For transportation, we've gone from horses to trains to cars to airplanes to spaceships. With the development of agriculture and advances in medicine, the human population has grown from about 500 million to over 8 billion. The materials we've developed provide for everything from skyscrapers to computers to the medicines that protect our health and prolong our lives. For better or for worse, science has had and will continue to have a huge impact on society.

We are now awakening to the fact that the resources of our planet are limited. Should we pay no attention and consume and pollute as we wish? Should we abandon the understandings of nature we've gained through science and merely hope for the best? Or should we embrace these understandings and move toward living on this planet in a sustainable fashion? Are the decisions we now and will soon face better met with ignorance or with knowledge?

Λ Figure 1.10

"Wow, Great Uncle Paul! Are we like unhatched chicks, ready to poke through our shells to a new environment and new understanding of our place in the universe?"

1.5 Chemistry is Integral to Our Lives

READING CHECK

What is the definition of chemistry?

When you wonder what the land, air, or ocean is made of, you are thinking about chemistry. When you wonder how a rain puddle dries up, how a car utilizes the energy of gasoline, how photovoltaics work, or how your body acquires energy from the food you eat, you are again thinking about chemistry. By definition, **chemistry** is the study of matter and the transformations it can undergo. **Matter** is anything that has mass and occupies space. It is the stuff that makes up all material things—anything you can touch, taste, smell, see, or hear is matter. The scope of chemistry, therefore, is very broad.

Chemistry is often described as a central science, because it touches all the other sciences (**Figure 1.11**). It springs from the principles of physics, and it serves as the foundation for the most complex science of all—biology. Indeed, many of the great advances in the life sciences today, such as genetic engineering, are applications of some very exotic chemistry. Chemistry is also the foundation for earth science. It is also an important component of space science. Just as we learned about the origin of the Moon from the chemical analysis of moon rocks in the early 1970s, we are now learning about the history of Mars and other planets from the chemical information gathered by space probes.

Progress in science is made as scientists conduct research. Research is any activity whose purpose is the discovery of new knowledge. Many scientists focus on **basic research**, which leads us to a greater understanding of how the natural world operates. Basic research in chemistry tells us how atoms combine to form molecules or how the structures of molecules can be determined. **Applied research** focuses on the development of useful applications of the knowledge laid down by basic research. The majority of chemists choose applied research as their major focus. Applied research in chemistry has provided us with medicine, food, water, shelter, and so many of the material goods that characterize modern life.

∧ Figure 1.11

Chemistry is a foundation for many other disciplines. (a) Biochemists analyzing DNA profiles. (b) Meteorologist releasing a weather balloon to study the chemistry of the upper atmosphere. (c) Technicians conducting DNA research. (d) Paleontologists preserving uncovered ancient fossils. (e) Astronomer in front of observatory used to study the chemical nature of stars.

Over the course of the past century, we became very good at manipulating atoms and molecules to create materials to match our needs. At the same time, however, mistakes were made when it came to caring for the environment. Waste products were dumped into rivers, buried in the ground, or vented into the air without regard for possible long-term consequences. Many people believed that Earth was so large that its resources were virtually unlimited and that it could absorb wastes without being significantly harmed.

Most nations now recognize this as a dangerous attitude. As a result, government agencies, industries, and concerned citizens are involved in extensive efforts to take care of the environment. For example, members of the American Chemistry Council, who, as a group, produce 90 percent of the chemicals manufactured in the United States, have adopted a program called Responsible Care. Through this program, members of this organization have pledged to manufacture their products without causing environmental damage. The Responsible Care program emblem is shown in **Figure 1.12**. Through the wise use of chemistry, waste products can be minimized, recycled, engineered into useful products, or rendered environmentally safe. This is an area

∧ Figure 1.12

The Responsible Care symbol of the American Chemistry Council.

of research known as *Green Chemistry*, which we explore in detail in the Contextual Chemistry essay at the end of the next chapter.

Chemistry has influenced our lives in many important ways, and it will continue to do so in the future (**Figure 1.13**). For this reason, it is in everyone's interest to become familiar with the basic concepts of chemistry.

Λ Figure 1.13

Most of the material items in any modern house are shaped by some human-devised chemical process.

CONCEPT **CHECK**

Chemists have learned how to produce aspirin using petroleum as a starting material. Is this an example of basic or applied research?

CHECK YOUR ANSWER

This is an example of applied research, because the primary goal was to develop a useful commodity. However, the ability to produce aspirin from petroleum depended on an understanding of atoms and molecules developed from many years of basic research.

1.6 Measuring with Units

Science starts with observations. When possible, it is helpful to quantify observations by taking measurements. By quantifying observations, we are able to make objective comparisons, share accurate information with others, or look for trends that might reveal some inner workings of nature.

Scientists measure *physical quantities*. Some examples of physical quantities you will be learning about and using in this program are length, time, mass, weight, volume, energy, temperature, heat, and density. Any measurement of a physical quantity must always include a number followed by a unit that tells us not only what was measured but also the scale of the measurement. It would be meaningless, for example, to say that your dog weighs 40, because without a specific unit, no one would know what that meant: 40 ounces, 40 pounds, 40 kilograms? A dog that weighed 40 kg would be more than 35 times heavier than one that weighed 40 oz. Units such as ounces, pounds, and kilograms, or feet, yards, and kilometers are all units that allow us to make meaningful comparisons when we measure physical quantities, and they must be included to complete the description.

READING CHECK

Any measurement of a physical quantity must always include two things. What are they?

There are two major unit systems used in the world today. One is the United States Customary System (USCS, formerly called the British System of Units), used in the United States, primarily for nonscientific purposes.* The other is the Système International (SI), which is used in most other nations. This system is also known as the International System of Units or as the metric system. The orderliness of this system makes it useful for scientific work, and it is used by scientists all over the world, including those in the United States. And the International System is beginning to be used for nonscientific work in the United States, as **Figure 1.14** shows. This book uses the SI units given in **Table 1.2**. On occasion, USCS units are also used

TABLE 1.2 Metric Units for Physical Quantities and their USCS Equivalents

PHYSICAL QUANTITY	METRIC UNIT	ABBREVIATION	USCS EQUIVALENT
length	kilometer	km	1 km = 0.621 miles (mi)
	meter	m	1 m = 3.285 feet (ft)
	centimeter	cm	1 cm = 0.3937 inches (in.)
			1 in = 2.54 cm
	millimeter	mm	none commonly used
time	second	s	second also used in USCS
mass	kilogram	kg	1 kg = 2.205 pounds (lb)
	gram	g	1 g = 0.03528 ounces (oz)
			1 oz = 28.345 g
	milligram	mg	none commonly used
volume	liter	L	1 L = 1.057 quarts (qt)
	milliliter	mL	1 mL = 0.0339 fl oz
	cubic centimeter	cm^3	1 cm^3 = 0.0339 fl oz
energy	kilojoule	kJ	1 kJ = 0.239 kilocalories (kcal)
	joule	J	1 J = 0.239 calories (cal)
			1 cal = 4.184 J
temperature	degree Celsius	°C	(°C * 1.8) + 32 = degrees Fahrenheit, °F
	kelvin	K	°C + 273 = K

* Two other countries that continue to use the USCS are Liberia and Myanmar.

Λ Figure 1.14

The metric system has made some headway in the United States where various commercial goods, such as Evan's favorite soda, are commonly sold in metric quantities.

to help you make comparisons. One major advantage of the metric system is that it uses a decimal system, which means all units are related to the next smaller or larger units by a factor of 10. Some of the more commonly used prefixes, along with their decimal equivalents, are shown in **Table 1.3**. From this table, you can see that 1 kilometer is equal to 1000 meters, where the prefix kilo- indicates 1000. Likewise, 1 millimeter is equal to 0.001 meter, where the prefix milli- indicates 1/1000. You need not memorize this table, but you will find it a useful reference when you come across these prefixes in your course of study.

TABLE 1.3 Metric Prefixes

PREFIX	SYMBOL	DECIMAL EQUIVALENT	EXPONENTIAL FORM	EXAMPLE
tera-	T	1,000,000,000,000.	10^{12}	1 **tera**meter (Tm) 1 trillion meters
giga-	G	1,000,000,000.	10^{9}	1 **giga**meter (Gm) 1 billion meters
mega-	M	1,000,000.	10^{6}	1 **mega**meter (Mm) 1 million meters
kilo-	k	1000.	10^{3}	1 **kilo**meter (km) 1 thousand meters
hecto-	h	100.	10^{2}	1 **hecto**meter (hm) 1 hundred meters
deka-	da	10.	10^{1}	1 **deka**meter (dam) ten meters
no prefix	–	1.	10^{0}	1 meter (m) 1 meter
deci-	d	0.1	10^{-1}	1 **deci**meter (dm) 1 tenth of a meter
centi-	c	0.01	10^{-2}	1 **centi**meter (cm) 1 hundredth of a meter
milli-	m	0.001	10^{-3}	1 **milli**meter (mm) 1 thousandth of a meter
micro-	µ	0.000 001	10^{-6}	1 **micro**meter (µm) 1 millionth of a meter
nano-	n	0.000 000 001	10^{-9}	1 **nano**meter (nm) 1 billionth of a meter
pico-	p	0.000 000 000 001	10^{-12}	1 **pico**meter (pm) 1 trillionth of a meter

1.7 Scientific Notation

In science, we often encounter very large and very small numbers. Written in standard decimal notation, these numbers can be cumbersome. There are, for example, about 33,460,000,000,000,000,000,000 water molecules in a thimbleful of water, each having a mass of about 0.00000000000000000000002991 gram. To represent such numbers, scientists use a mathematical shorthand called **scientific notation**. Written in this notation, the number of molecules in a thimbleful of water is 3.346×10^{22} and the mass of a single molecule is 2.991×10^{-23} gram.

To understand how this shorthand notation works, consider the large number 50,000,000. Mathematically, this number is equal to 5 multiplied by $10 \times 10 \times 10 \times 10 \times 10 \times 10 \times 10$. (Check this out on your calculator.) We can abbreviate this chain of numbers by writing all the 10s in an exponential form, which gives us the scientific notation 5×10^7. (Note that 10^7 is the same as $10 \times 10 \times 10 \times 10 \times 10 \times 10 \times 10$.) **Table 1.4** shows the exponential form of some other large and small numbers.

Likewise, the small number 0.0005 is mathematically equal to 5 divided by $10 \times 10 \times 10 \times 10$, which is $5 \div 10^4$. Because dividing by a number is equivalent to multiplying by the reciprocal of that number, $5 \div 10^4$ can be written in the form 5×10^{-4}. So, in scientific notation 0.0005 becomes 5×10^{-4}. (Note the negative exponent.)

All scientific notation is written in the general form:

$$C \times 10^n$$

where C, called the coefficient, is a number less than 10 but equal to or greater than 1 (in other words, between 1 and 9.999 ...) and n is the exponent. A positive exponent indicates a number greater than 1, and a negative exponent indicates a number between 0 and 1 (not a number less than 0). Numbers less than 0 are indicated by putting a negative sign before the coefficient (not in the exponent). Some common examples are shown in **Table 1.5**.

TABLE 1.4 Large and small numbers. Note: All of these numbers are greater than zero.

1,000,000	$= 10 \times 10 \times 10 \times 10 \times 10 \times 10 = 10^6$
100,000	$= 10 \times 10 \times 10 \times 10 \times 10 = 10^5$
10,000	$= 10 \times 10 \times 10 \times 10 = 10^4$
1000	$= 10 \times 10 \times 10 = 10^3$
100	$= 10 \times 10 = 10^2$
10	$= 10 = 10^1$
1	$= 1 = 10^0$
0.1	$= 1/10 = 10^{-1}$
0.01	$= 1/(10 \times 10) = 10^{-2}$
0.001	$= 1/(10 \times 10 \times 10) = 10^{-3}$
0.0001	$= 1/(10 \times 10 \times 10 \times 10) = 10^{-4}$
0.00001	$= 1/(10 \times 10 \times 10 \times 10 \times 10) = 10^{-5}$
0.000001	$= 1/(10 \times 10 \times 10 \times 10 \times 10 \times 10) = 10^{-6}$

To change a decimal number that is greater than +1 or less than -1 to scientific notation, you shift the decimal point to the left until you arrive just before the first digit. For example, to convert the decimal number 45,000 to scientific notation, move the decimal point four places to the left:

$$45{,}000 = 4.5 \times 10^4$$

For decimal numbers between +1 and -1, you move the decimal point to the right until you arrive just before the last digit. This number is the coefficient part of the notation. The exponent part is simply the number of places you moved the decimal point. The number 0.00045, for example, is converted to scientific notation by moving the decimal point four places to the right:

$$0.00045 = 4.5 \times 10^{-4}$$

Note that because you moved the decimal point to the right in this case, you must put a minus sign on the exponent.

TABLE 1.5 Numbers in both decimal and scientific notation

	DECIMAL NOTATION	SCIENTIFIC NOTATION
Large positive number (greater than 1)	6,000,000,000	6×10^9
Small positive number (between 0 and 1)	0.0006	6×10^{-4}
Large negative number (less than −1)	−6,000,000,000	-6×10^9
Small negative number (between −1 and 0)	−0.0006	-6×10^{-4}

TABLE 1.6 Some examples of physical data using scientific notation

Number of molecules in thimbleful of water	3.346×10^{22}
Mass of water molecule	2.991×10^{-23} *gram*
Average radius of hydrogen atom	5×10^{-11} *meter*
Proton mass	1.6726×10^{-27} *kilogram*
Neutron mass	1.6749×10^{-27} *kilogram*
Electron mass	9.1094×10^{-31} *kilogram*
Electron charge	1.602×10^{-19} *coulomb*
Avogadro's number	6.022×10^{23} *particles*
Atomic mass unit	1.661×10^{-24} *gram*

CONCEPT **CHECK**

Express the following exponentials as decimal numbers:

a. 1×10^{-7}

b. 1×10^{8}

c. 8.8×10^{5}

Express the following decimal numbers in scientific notation:

d. 740,000

e. -0.00354

f. 15

CHECK YOUR ANSWER

a. 0.0000001 b. 100,000,000 c. 880,000 d. 7.4×10^{5} e. -3.54×10^{-3} f. 1.5×10^{1}

1.8 Significant Figures

Three kinds of numbers are used in science—those that are counted, those that are defined, and those that are measured. The exact value of a counted or defined number can be stated with absolute certainty. For example, you can count the number of chairs in your classroom or the number of fingers on your hand. Your final count is 100% on the mark. (Of course, this assumes you counted correctly, which is easy when the numbers are small but not so easy when the numbers run into the millions, as occurs during an election.)

Defined numbers are about exact relationships and are defined as being true. The defined number of centimeters in a meter, the defined number of seconds in an hour, and the defined number of sides on a square are examples. Thus, defined numbers are not subject to error (unless you forget a definition).

Every measured number, however, no matter how carefully measured, has some degree of uncertainty. This uncertainty (or margin of error) in a measurement can be illustrated by the two meter sticks shown in **Figure 1.15**. Both sticks are being used to measure the length of a table. Assuming that the zero end of each meter stick has been carefully and accurately positioned at the left end of the table, how long is the table?

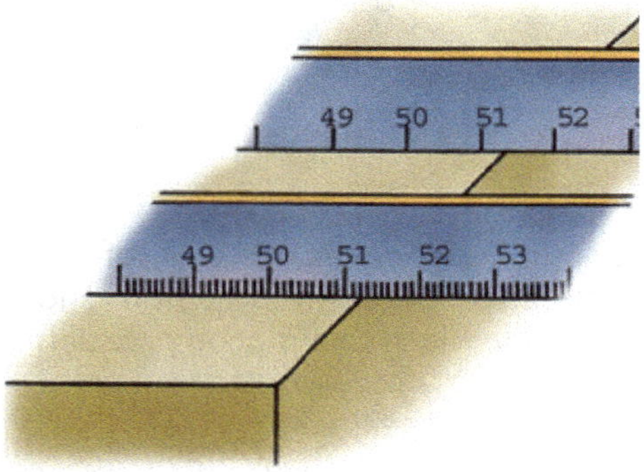

∧ Figure 1.15

Which meter stick offers more precision?

The upper meter stick has a scale marked off in centimeter intervals. Using this scale, you can say with certainty that the length is between 51 and 52 centimeters. You can say further that it is closer to 51 centimeters than to 52 centimeters; you can even estimate it to be 51.2 centimeters.

The scale on the lower meter stick has more subdivisions—and therefore greater precision—because it is marked off in millimeters. With this scale, you can say that the length is definitely between 51.2 and 51.3 centimeters, and you can estimate it to be 51.25 centimeters. With greater precision you are able to reasonably read off more numbers.

With both rulers there are digits that are certain and one digit (the last one) that is estimated. Note also that the uncertainty in the reading from the lower meter stick is less than the uncertainty in the reading from the upper meter stick. The lower meterstick can give a reading to the hundredths place,

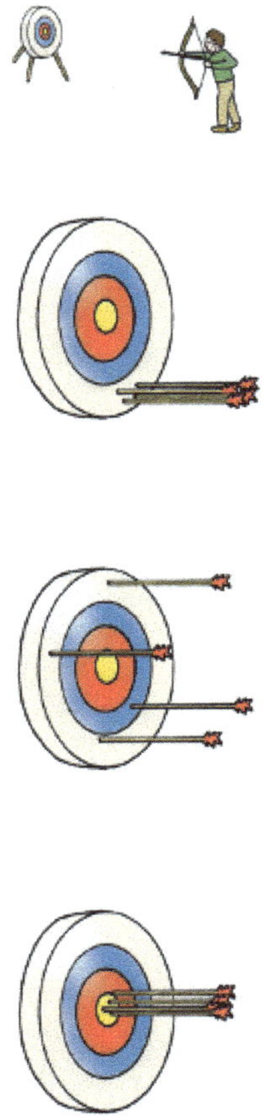

Figure 1.16

Consider archery as a model for the difference between precision and accuracy. Which shows good precision but poor accuracy? Which shows both? Which shows neither?

but the upper one can give a reading only to the tenths place. The lower one is more precise than the top one. So, in any measured number, the digits tell us the magnitude of the measurement and the location of the decimal point tells us the precision of the measurement.

Precision means close agreement in a group of measured numbers; accuracy means a measured value that is very close to the true value of what is being measured. If you measure the same thing several times and get numbers that are close to one another but are far from the true value (perhaps because your measuring device is not working properly), then your measurements are precise but not accurate. An example would be measuring your weight repeatedly on a broken scale.

Significant figures are the digits in any measured value that are known with certainty plus one final digit that is estimated and hence uncertain. These are the digits that reflect the precision of the instrument used to generate the number. They are the digits that have experimental meaning. The measurement 51.2 centimeters made with the upper meter stick in Figure 1.15, for example, has three significant figures, and the measurement 51.25 centimeters made with the lower meter stick has four significant figures. The rightmost digit is always an estimated digit, and only one estimated digit is ever recorded for a measurement. It would be incorrect to report 51.253 centimeters as the length measured with the lower meter stick. This five-significant-figure value has two estimated digits (the final 5 and 3) and is inappropriate because it indicates a precision greater than the meter stick can obtain.

CONCEPT **CHECK**

How many significant figures in a. 43,384 b. 43,084 c. 0.004308 d. 43,084.0 e. 43,000 f. 4.30×10^4

CHECK YOUR ANSWER

a. 5 b. 5 c. 4 d. 6 e. 2 f. 3

In addition to the rules cited on the next page, there is another full set of rules to be followed for significant figures when two or more measured numbers are subtracted, added, divided, or multiplied. Briefly, you can never gain significant figures. You can only lose them. Thus, your final answer should contain no more precision (significant figures) than with what you started. This is explored further within the *Conceptual Chemistry* video on significant figures.

Here are some standard rules for writing and using significant figures.

RULE 1

In numbers that do not contain zeros, all the digits are significant:

4.1327 five significant figures
5.14 three significant figures
369 three significant figures

RULE 2

All zeros between significant digits are significant:

8.052 four significant figures
7059 four significant figures
306 three significant figures

RULE 3

Zeros to the left of the first nonzero digit serve only to fix the position of the decimal point and are not significant:

0.0068 two significant figures
0.0427 three significant figures
0.0003506 four significant figures

RULE 4

In a number that contains digits to the right of the decimal point, zeros to the right of the last nonzero digit are significant:

53.0 three significant figures
53.00 four significant figures
0.00200 three significant figures
0.70050 five significant figures

RULE 5

In a number that has no decimal point and that ends in one or more zeros, the zeros that end the number are not significant. (To scientists, the decimal point is very important. Its presence or absence has great implications in regard to the precision of a measurement.)

3600 two significant figures
290 two significant figures
5,000,000 one significant figure
10 one significant figure
6050 three significant figures

RULE 6

When a number is expressed in scientific notation, all digits in the coefficient are taken to be significant:

4.6×10^{-5} two significant figures
4.60×10^{-5} three significant figures
4.600×10^{-5} four significant figures
2×10^{-5} one significant figure
3.0×10^{-5} two significant figures
4.00×10^{-5} three significant figures

Calculation Corner: Unit Conversion

Often in chemistry, and especially in a laboratory setting, it is necessary to convert from one unit to another. This process is called "unit conversion" or sometimes as "unit analysis" or "dimensional analysis".

To convert from one unit to another, you need only multiply the given quantity by a conversion factor. A conversion factor is a ratio where the numerator and denominator represent the equivalent quantity expressed in different units. Because any amount divided by the same amount is equal to one, all conversion factors are effectively equal to one. For example, the following two conversion factors are both derived from the relationship 100 centimeters = 1 meter. Note that the numerator and denominator represent the same distance, but expressed in different units.

$$\frac{100 \text{ centimeters}}{1 \text{ meter}} = 1 \qquad \frac{1 \text{ meter}}{100 \text{ centimeters}} = 1$$

Because all conversion factors are effectively equal to 1, multiplying a quantity by a conversion factor does not change the value of the quantity. What does change are the units. Suppose you measured an item to be 60 centimeters in length. You can convert this measurement to meters by multiplying it by the conversion factor that allows you to cancel centimeters.

$$\underbrace{(60 \text{ centimeters})}_{\substack{\text{quantity} \\ \text{in centimeters}}} \underbrace{\frac{(1 \text{ meter})}{(100 \text{ centimeters})}}_{\substack{\text{conversion} \\ \text{factor}}} = \underbrace{0.6 \text{ meter}}_{\substack{\text{quantity} \\ \text{in meters}}}$$

You can create two conversion factors for every equality. For example, from **Table 1.2** we see that 1 kilometer equals 0.621 miles. To create these two conversion factors, show a unit as the numerator in one conversion factor but as a denominator in the other:

1 km / 0.621 mi　　　　0.621 mi / 1 km

Multiply the quantity provided to you by the conversion factor of choice, which is the one that shows the original unit in the denominator. This way, the original unit will be canceled, leaving you with the desired new unit. The following example starts with 26.2 miles. To convert this to kilometers, we multiply by appropriate conversion factor, which has miles in the denominator. This allows for the units of miles to cancel. The result is the same quantity expressed in units of kilometers. This calculation tells us that a distance of 26.2 miles is the same as 42.2 kilometers.

(26.2 miles) (1 km/0.621 miles). =. 42.2 km

Always be careful to write down your units. They are your ultimate guide, telling you what numbers go where and whether you are setting up the equation properly. Remember, you should set up a conversion factor so that the desired unit is always in the numerator and the unit to be cancelled is always in the denominator.

Conversion factors can be combined to allow multiple conversions in a single equation.

EXAMPLE

How many pounds are there in 48 ounces? Set up a string of conversion factors using data from Tables 1.2 and 1.3.

ANSWER

Start by writing down the given quantity, which in this case is 48 ounces. Multiply by a series of conversion factors to transform this quantity into the desired unit—from ounces to grams, from grams to kilograms, and then from kilograms to pounds:

$$(48\text{ oz})\left(\frac{28.345\text{ g}}{1\text{ oz}}\right)\left(\frac{1\text{ kg}}{1000\text{ g}}\right)\left(\frac{2.205\text{ lbs}}{1\text{ kg}}\right) = 3\text{ lbs}$$

Of course, the above conversion could be done in a single step knowing that 16 ounces equals 1 pound. A string of conversion factors in a single equation, however, becomes quite useful when the direct conversion factor is unknown.

YOUR TURN

Multiply each physical quantity by the appropriate conversion factor to find its numerical value in the new unit indicated. You will need paper, a pencil, a calculator, and Tables 1.2 and 1.3.

a. Show that 7320 grams is 7.32 kilograms
b. Show that 235 kilograms is 518 pounds
c. Show that 4585 milliliters is 4.846 quarts
d. Show that 100 calories is 0.1 kilocalories
e. Show that 100 calories is 400 joules

Perhaps you are wondering about how many digits to include after you calculate a conversion. Were you perplexed, for example, that the answer to (e) is 400 J and not 418.4 J? Key is that after multiplying or dividing you should never end up with more significant figures than you started with. For example, the number 100 has only one significant figure. Multiplying any number by 100 will give a product that will also have but one significant figure.

Concept Review

Chapter 1

Summary of Terms

Denialism Being doubtful over substantiated claims.

Fact A phenomenon about which competent observers can agree.

Law A general hypothesis or statement about the relationship of natural quantities that has been tested over and over again and has not been contradicted. Also known as a principle.

Hypothesis An educated guess; a reasonable explanation that is not fully accepted as factual until tested over and over again by experiment.

Pseudoscience Fake science that has no tests for its validity.

Science The collective findings of humans about nature, and a process of gathering and organizing knowledge about nature based upon experimental evidence.

Scientific Method An orderly method for gaining, organizing, and applying new knowledge.

Scientific Notation A number written in the general form: $C \times 10^n$ where C, called the coefficient, is a number less than 10 but equal to or greater than 1.

Significant Figure Digits in any measured value that are known with certainty plus one final digit that is estimated and hence uncertain.

Skepticism Being doubtful over unsubstantiated claims.

Technology Method and means of solving practical problems by applying the findings of science.

Theory A synthesis of a large body of information that encompasses well-tested hypotheses about certain aspects of the natural world.

Review Questions

1.1 Science Is a Way of Understanding the Natural World

1. What would have to be done to refute Aristotle's hypothesis that heavier objects fall faster?

2. Are experiments better at proving ideas right or proving them wrong?

3. According to Figure 1.3, what is usually the first step to conducting scientific research?

1.2 An Investigation of Sea Butterflies

4. Is the sea butterfly more closely related to a snail or an insect?

5. According to the results, why do the amphipods abduct the sea butterflies?

1.3 Technology Is Applied Science

6. What is technology?

7. What issue aggravates almost every problem faced by humans today?

8. Are medical X rays used because they carry zero risks?

1.4 We Are Still Learning About the Natural World

9. Is it possible for a fact to change?

10. What do we call a suggested explanation for an observable phenomenon?

11. Which works to unify a broad range of observations: a scientific hypothesis or a scientific theory?

1.5 Chemistry Is Integral to Our Lives

12. Why is chemistry often called the central science?

13. What is the difference between basic research and applied research?

14. What pledge has been made my members of the American Chemistry Council through the Responsible Care program?

1.6 Measuring with Units

15. What are the two major systems of measurement used in the world today?

16. Why are prefixes used in the metric system?

17. A milligram is equal to how many grams?

18. What is true about the numerator and denominator of a conversion factor?

1.7 Scientific Notation

19. When is scientific notation used

20. Is 5×10^{-31} a very large or very small number? Is it greater or less than zero?

1.8 Significant Figures

21. How is precision different from accuracy

22. Is it possible for an uncertain digit to be significant?

Quantitative Questions

23. We can define a "risk/benefit" ratio as the amount of risk taken divided by the amount of potential benefit:

    risk benefit ratio = risk/benefit

    If the risks are equal to the benefits, then the risk/benefit ratio equals one. If an activity offered you a risk/benefit ratio of 100, would you take it? How about if the risk/benefit ratio was 0.01? Is the risk benefit ratio of buying a lottery ticket large or small?

24. Using conversion factors, calculate the age of some one who is exactly 25 years old in units of months. How many days is this? (Assume 1 year = 365 days)

25. Using conversion factors, show that there are about 63 billion seconds in 2000 years. (Assume 1 year = 365 days)

26. A mass of 18.0 grams of water contains 6.02×10^{23} water molecules. Write out two conversion factors from this equality.

27. Using your conversion factors from the previous question, show that 36.0 grams of water contains 1.20×10^{24} molecules of water.

28. Using your conversion factors from question 26, show that 27.0 grams of water contains 9.03×10^{23} molecules of water.

Solutions (Odd-Numbered)

1. An experiment.

3. Scientific research usually begins with the asking of a very broad question.

5. The amphipods abduct the sea butterflies as a means of chemical defense against predators.

7. The rapid growth of our human population.

9. Yes, when competent observers agree that it should.

11. A scientific theory.

13. Basic research leads to a greater understanding of how the natural world works. Applied research focuses on developing applications of knowledge gained through basic research.

15. The two major unit systems are the United States Customary System and the Système International.

17. A milligram is equal to one thousandth of a gram.

19. Scientific notation is used to express vary large or very small numbers conveniently.

21. Accuracy is how close measured values are to the truth. Precision is a function of the reproducibility of the measurements. The greater the precision, the closer the measured values will be to each other regardless of whether they are accurate.

23. A large risk/benefit ratio, such as 100, can be indicated as follows:

RISK/benefit = 100

This is a risky activity that you might choose to avoid. A small risk benefit ratio, such as 0.01, can be indicated as follows:

risk/BENEFIT = 0.01

This activity offers much benefit for only little risk and so may be worthwhile. Regarding the purchase of a lottery ticket, the chances of losing a small amount of money (the risk) far exceeds the chances of gaining a large amount of money (the benefit). The risk/benefit ratio, therefore, is quite large. Furthermore, with continued purchases, the small amount of money lost each time can really add up. Lottery tickets are a clear example of where our powers of wishful thinking win over our powers of analytical thinking. It's true that wishful thinking can be fun. But it's also true that wishful thinking can be damaging. We should be mindful as we seek a proper balance between risks and benefits.

25. (2000 years)(365 days/1 year)(24 hours/1 day)(60 min/1 hour)(60 sec/1 min) = 63,072,000,000 seconds

27. (36.0 grams of water)(6.02×1023 molecules of water/18.0 grams of water) = 1.20×10^{24} molecules of water

Solutions to Calculation Corner

Unit Conversion

a. 7.32 kg

b. 518 lb

c. 4.846 qt

d. 0.1 kcal

e. 400 J

Chapter 1: Contextual Chemistry Essay

Science Denialism *Did we really land on the Moon?*

Common knowledge is that over half a century ago, astronauts landed and walked upon the Moon. This was the culmination of NASA's programs of the 1960s in the context of the "space race" between competing countries—the United States and the then-Soviet Union. The Soviets were the first to launch an orbiting satellite and the first to send a person into space. The Americans, not wanting to be out-done at the height of the Cold War, set the lofty goal of being the first nation to land a human on the Moon. It was a daunting mission performed in an age before advanced computers. Mistakes were made and, tragically but unsurprisingly, astronauts lost their lives. But success was achieved when Neil Armstrong first stepped foot on the Moon on July 20, 1969. It was undeniably one of the greatest achievements in human history. Reaching the Moon was an illustration of what can be accomplished when people work together with reason, creativity, a sense of purpose, and, importantly, an understanding of the rules of nature.

However, not everyone believed or believes that a human walked on the Moon. Thirty years after the landings, a Gallup poll indicated about 6% of Americans believed the lunar missions were staged in a studio, with the help of prominent film makers. This belief has grown with the rise of social media where voices of no authority can receive much attention and even credibility. Young people today were born many years after the lunar missions. Because they weren't there to see the dramatic events unfold, they are perhaps more susceptible to conspiracy theories about the lunar missions.

So, did we really land astronauts on the Moon? And more generally, how can truth versus falsehood be determined in today's world, where just about anyone can say just about anything through a broad array of media channels that do little fact-checking? These are big questions to ponder. However, today as in ancient times, **logic supported by verifiable evidence** provides a reliable path to the truth. Think of this recipe for truth-finding in terms of your own life, as well as on the world stage. Let's practice by applying logic and evidence to assess whether the lunar landing actually occurred. Consider the following:

1. With today's space telescopes we can see the lunar landing sites, as shown in **Figure B**.

2. The Apollo astronauts brought back some 842 pounds of Moon rocks that have been studied (and authenticated) by scientists around the world.

3. The Soviets tracked the progress of our spacecrafts as they traveled to the Moon and back. Given the nature of the space race, it would be most unlikely the Soviets would have been willing to go along pretending that the Americans successfully beat them to the Moon.

< Figure A

Accompanying Armstrong was astronaut Buzz Aldrin, who honors the American flag. The flag is held up against the Moon's gravity by a stick across the top. The waves in the flag resulted not from wind (there is no wind on the Moon) but from how they had to twist the pole into the lunar soil so that it would remain standing.

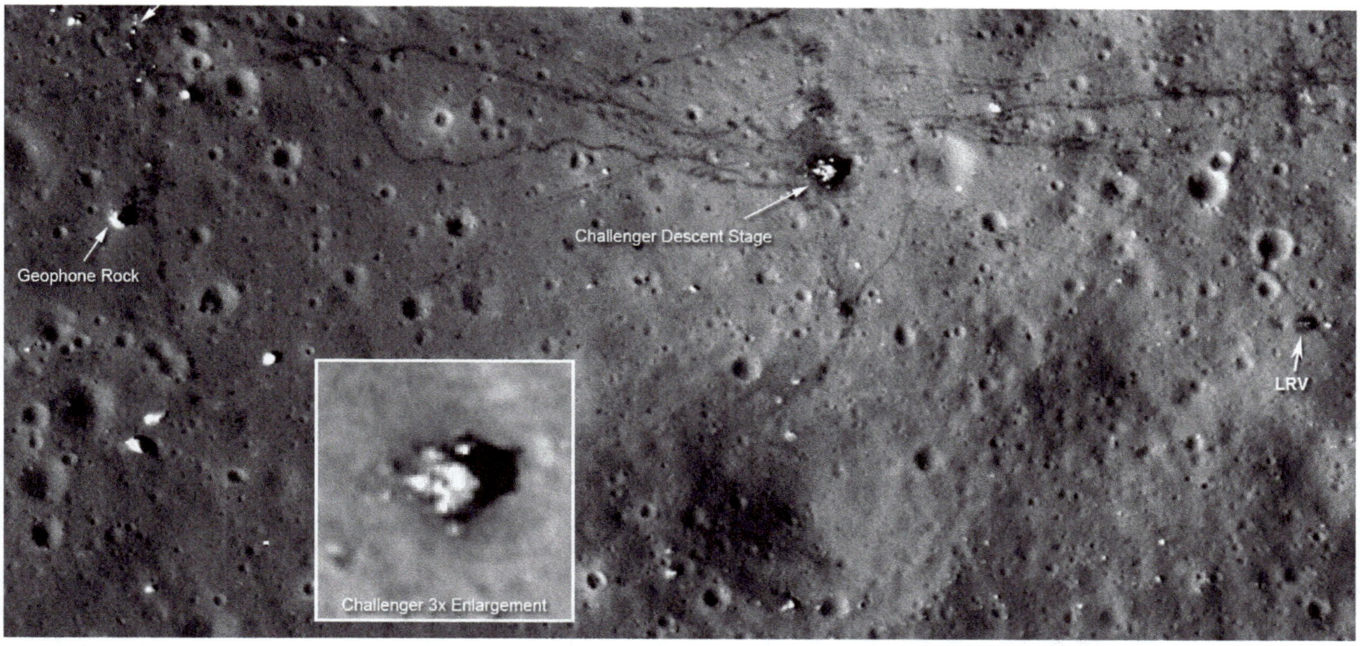

Λ Figure B

The Apollo 17 landing site recently photographed by the Lunar Reconnaissance Orbiter. Lunar rover and astronaut foot trails are evident

4. The astronauts left reflectors on the surface of the Moon. Scientists today routinely reflect laser light off these reflectors to aid measurement of the Moon's distance and for other experiments.

5. If NASA were faking it, why did they continue going back to the Moon six more times, from Apollo 11 in 1969 to Apollo 17 in 1972? With each mission, they would be increasing the chances of error and their forgery revealed.

6. In the late 1960s, there were no modern computers and certainly no CGI to create fantastic computer-generated scenes as in today's movies.

Science is a mode of inquiry and a body of knowledge built upon rational thinking and experimental evidence. Does reasoning and all the evidence support the claim that Neil Armstrong, Buzz Aldrin, and others walked on the Moon? How can you defend your opinion? If you use logic and evidence, then you are thinking in a scientific manner.

Some people might reject science because they don't understand the details. The remedy for this would be more education, which puts the person in a position to accept or reject an argument based on the merits. But you need not be an expert. Even basic knowledge can go a long way to helping you understand the big picture. Furthermore, basic knowledge helps to cement your trust in those who are bonafide experts. From this foundation, there's value in any disagreement you might have with a particular conclusion. Why? Because your disagreement is made from understanding, not ignorance.

There can be an emotional motivation behind denying scientific information. This often occurs when scientific findings carry implications for public policy that are distressing or distasteful. It's analogous to what might happen with your dentist. If your dentist tells you that you need a root canal, you may not want to believe her evidence and reasoning. You are emotionally motivated to deny the report. So, you go for a second opinion, and perhaps a third—which is quite reasonable. You are gathering data to evaluate. But at a certain point, if all of the credible dentists you visit are in consensus about your diagnosis, you will be wise to set your emotions and short-term preferences aside, and try to understand the science of what is happening to your tooth.

Think of a less wise approach: You interview 10,000 dentists until you find one who declares no root canal is needed. You then heed the advice that no root canal is needed. Further, you praise and elevate that dentist's credentials because finally you found someone who knows how to listen to your underlying needs. You convince yourself there are always two sides to a story. You think of yourself as a crafty renegade for bucking a system that's clearly corrupt and self-interested. What you would be doing here is known as "cherry-picking" the evidence. Cherry picking evidence means that you only consider evidence supporting your case, no matter how outlying that evidence is. Cherry-picking evidence is a common way to deny the validity of scientific claims.

As we know, science sometimes reveals facts that are not welcome. More serious than a rotten tooth, is the case of climate change. Is climate change a real threat? If so, how do we respond? How much easier it would be if anthropogenic climate change were not an actual threat to the biosphere? We could keep using fossil fuels, keep cutting down forests to build roads and raise livestock, and otherwise maintain our modern, resource-consuming lifestyles. It can be quite tempting to ignore the science outright without even understanding it—to deny it—because it is threatening and emotionally disturbing.

Denying the lunar landing makes for a juicy conspiracy theory—one that some might find entertaining, shocking, or fun to think about—like a scary movie. Hopefully, the scientific reasoning and evidence presented above regarding the Apollo missions helps to put this conspiracy theory to rest. Just as we must leave the theater after a scary movie in order to function in the real world, we ultimately need to leave unfounded conspiracy theories behind and devote ourselves to understanding the real world—for our impact upon it is immense. The physical world, knowable to a large extent through science, offers real rewards, joys, sustenance, and plenty of hazards too. The human population is booming like never before. This means we are taxing the environment that sustains us like never before.

Can we ultimately use what we learn from science to address our modern challenges and to accomplish great things? If we've been to the Moon and back, then the answer is a resounding yes.

CONCEPT CHECK

Why are long term problems, like over-population, easier to deny than short term problems, like a sewage spill?

CHECK YOUR ANSWER

It's common for people today to think short term. I'm hungry. I want to buy that car. Wow, that sewage stinks. These problems are focused, relatively easy to address, and solving them provides immediate gratification. A problem that develops over a longer time scale, such as too many people in too small a space—overpopulation—is more difficult to grasp, especially if you live in a sparsely populated area. Addressing long-term problems requires a focus on the future and commitment to a sustainable way of life. This is a greater challenge.

Skepticism: Unsubstantiated claims hold no value.

Denialism: Substantiated claims hold no value.

Λ Figure C

Planet Earth as seen from the Apollo 11 mission while in low orbit around the Moon.

Chapter 2

Particles of Matter

THE MAIN IDEA

Matter is made of particles called atoms

Λ Eta Carinae is a supergiant binary star system within our galaxy some 8000 light years away. Explosive turmoil is already apparent within this system, which is likely to supernova within the next million years. The atoms created within such exploding stars are ejected into the universe where they eventually recombine because of gravity to form new star systems. Amazingly, the atoms here on Earth are the remnants of stellar explosions from long ago. We are literally made of stardust.

Hydrogen atoms are the lightest of all atoms, and they make up more than 90 percent of the atoms in the known universe. Their origin goes back to the birth of the universe. Heavier atoms are produced in stars, which are massive collections of hydrogen pulled together by gravitational forces. The great pressures deep in a star's interior cause hydrogen atoms to fuse together, forming heavier atoms. With the exception of hydrogen, nearly all the atoms that occur naturally on the Earth— including those in your body—are the products of stars. A tiny fraction of these atoms came from our own star, the Sun, but most are from stars that ran their course long before our solar system came into being. You are literally made of stardust.

Atoms are so small that we can't see them directly. How then do we know they exist? How do atoms account for the mass of an object or its temperature? How can we use this idea of tiny particles called atoms to explain the nature of solids, liquids, and gases? We explore these sorts of questions in this chapter.

2.1 The Submicroscopic World is Super Small

From afar, a sand dune appears to be made of a smooth, continuous material. Up close, however, the dune reveals itself to be made of tiny grains of sand. In a similar fashion, everything around us—no matter how smooth it may appear— is made of very small fundamental units you know as atoms. Atoms are so small, however, that a single grain of sand contains on the order of 125 million trillion of them, which is 125,000,000,000,000,000,000. In scientific notation, this huge number is written as 1.25×10^{20}. Interestingly, the number of atoms in a single grain of sand is about a quarter of a million times greater than the number of grains of sand shown in the dune of **Figure 2.1**.

As small as atoms are, there is much we have learned about them. We know, for example, that there are more than 100 different types of atoms, and we have arranged them in the widely recognized periodic table. Some atoms link together to form larger but still incredibly small units of matter called *molecules*. As shown in **Figure 2.1**, for example, two hydrogen atoms and one oxygen atom link together to form a single molecule of water, commonly recognized as H_2O. Water molecules are so small that an 8-ounce glass of water contains about a trillion trillion of them, which in scientific notation is 1×10^{24}.

READING CHECK

Are atoms made of molecules, or are molecules made of atoms?

Our world can be studied at different levels of magnification, as illustrated in **Figure 2.2**. At the macroscopic level, matter is large enough to be seen, measured, or handled. A handful of sand and a glass of water are macroscopic samples of matter. At the microscopic level, physical structure is so fine that it can be seen only with a microscope. A biological cell is microscopic, as is the detail on a dragonfly's wing. Below the microscopic level is the submicroscopic—the realm of atoms and molecules, and an important focus of chemistry.

Figure 2.1

There are far more molecules within a glass of water than grains of sand within this towering sand dune.

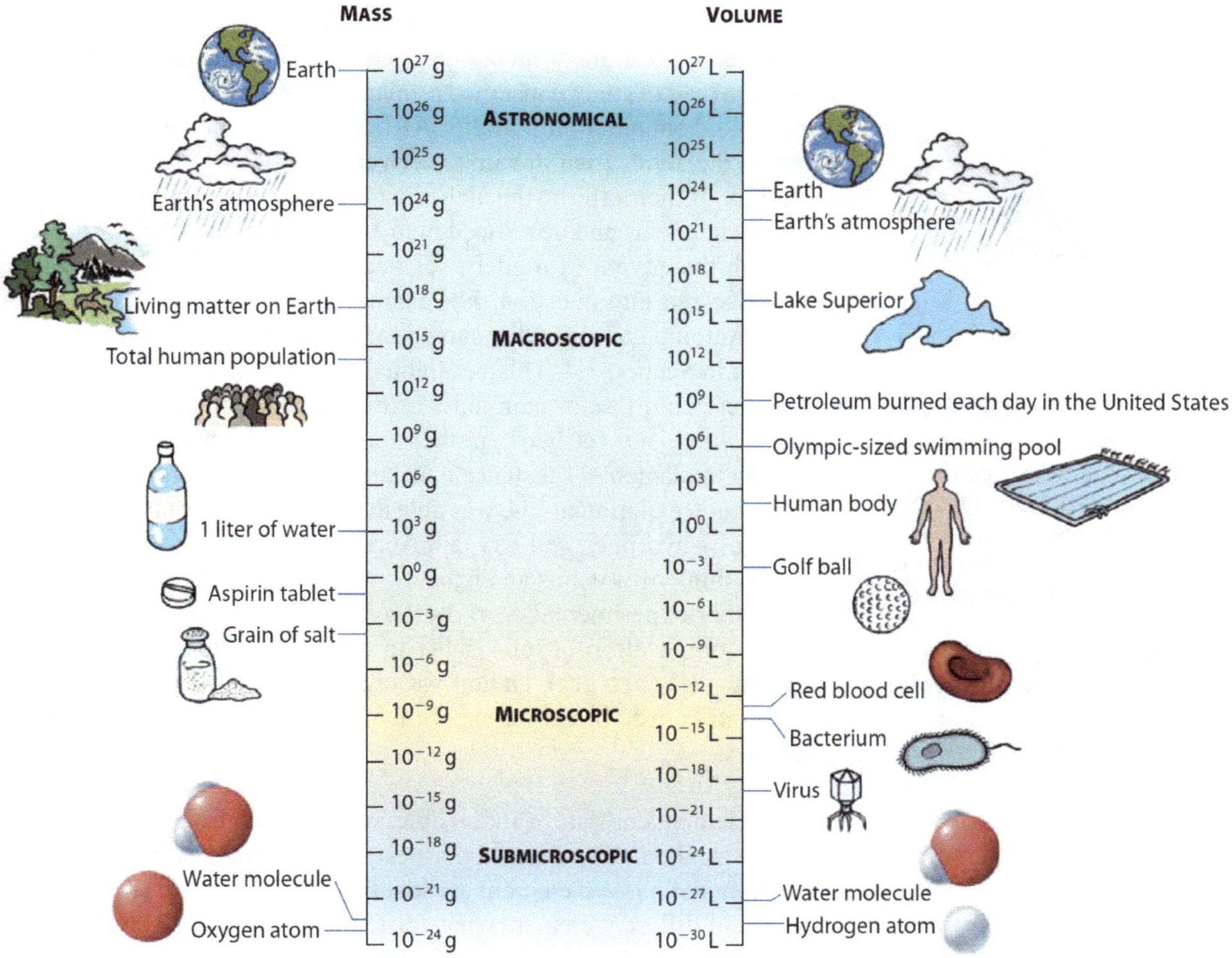

Λ Figure 2.2
Ranges of masses (in grams) and volumes (in liters) from atoms to planet Earth. Note that each successive entry differs by 1000 times.

2.2 Discovering the Atom

In the 4th century BCE, the influential Greek philosopher Aristotle described the composition and behavior of matter in terms of the four qualities shown in **Figure 2.3**: hot, cold, moist, and dry. Aristotle's model was a remarkable achievement for its day, and people using it in Aristotle's time found it made sense. When pottery is made, for example, wet clay changes to ceramic because the heat of the fire drives out the moist quality of the wet clay and replaces it with the dry quality of the ceramic.

Aristotle's views on the nature of matter made so much sense that less obvious views were difficult to accept. One alternative view was the forerunner of our present-day model: matter is composed of a finite number of incredibly small but discrete units we call atoms. This model was advanced by several Greek philosophers, including Democritus (460–370 BCE), who coined the term atom from the Greek phrase *a tomos*, which means "not cut" or "that which is indivisible" (**Figure 2.4**). So compelling was Aristotle's reputation, however, that the atomic model would not reappear for 2000 years.

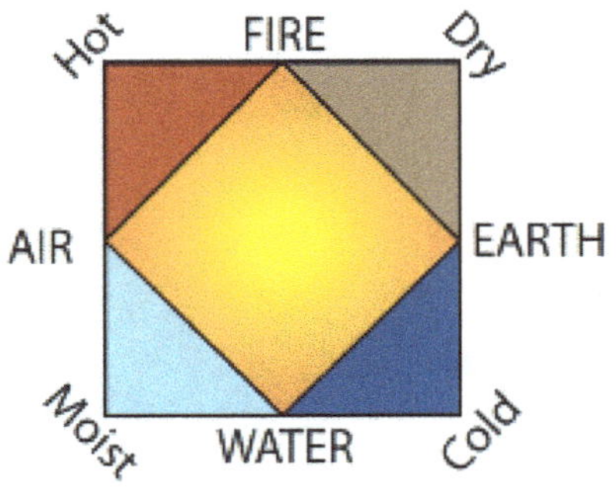

∧ Figure 2.3

Aristotle thought that all materials were made of various proportions of four fundamental qualities: hot, dry, cold, and moist. Various combinations of these qualities gave rise to the four basic elements: hot and dry gave fire, moist and cold gave water, hot and moist gave air, and dry and cold gave earth. He supposed that a hard substance like rock contained mostly the dry quality, for example, and a soft substance like clay contained more of the moist quality.

According to Aristotle, it was theoretically possible to transform any substance to another substance simply by altering the relative proportions of the four basic qualities. This meant that, under the proper conditions, a metal like lead could be transformed to gold. This concept laid the foundation of alchemy, a field of study concerned primarily with finding potions that would produce gold or confer immortality. Alchemists from the time of Aristotle to as late as the 1600s tried in vain to convert various metals to gold. Despite the futility of their efforts, the alchemists learned much about the behavior of many chemicals and developed many useful laboratory techniques.

With the advent of modern science, Aristotle's views on the nature of matter came into question. For example, in the late 1700s the French chemist Antoine Lavoisier discovered the law of mass conservation, as was discussed in Section 1.4. This verifiable law ran counter to Aristotle's idea that matter could lose or gain mass as its hot, dry, cold, or moist qualities changed. Lavoisier further hypothesized that an element is any material made of a fundamental substance that cannot be broken down into anything else. Through experiments, he was able to transform water into two different substances—hydrogen and oxygen. According to Lavoisier, Aristotle was wrong to think of water as an element.

Further experimental work by Lavoisier and others led the English chemist John Dalton (1766–1844) to reintroduce the atomic ideas of Democritus (**Figure 2.5**). Dalton wrote a series of postulates—claims he assumed to be true based on experimental evidence—some of which are as follows:

1. Each element consists of indivisible, minute particles called atoms.
2. Atoms can be neither created nor destroyed in chemical reactions.
3. All atoms of a given element are identical.
4. Atoms of different elements have different masses.

READING CHECK

How did Lavoisier define an element?

It didn't matter that these tiny atoms were too small to be seen. What did matter was that Dalton's atomic model worked to explain much of what was then known about chemical reactions. Where alchemists using Aristotle's model failed, chemists using Dalton's model succeeded—not in making gold but in being able to understand and control the outcome of numerous chemical reactions. Imagine their amazement at finally being able to control the creation of new materials. . . as though finally being given the wizard's cookbook, where the wizard in this case was nature herself.

Figure 2.4 >

In his atomic model, Democritus imagined that atoms of iron were shaped like coils—making iron rigid, strong, and malleable—and that atoms of fire were sharp, lightweight, and yellow.

In 1869, a Russian chemistry professor, Dmitri Mendeleev (1834–1907), produced a chart summarizing the properties of known elements for his students (**Figure 2.6**). Mendeleev's chart was unique in that it resembled a calendar. Elements were listed in horizontal rows in order of increasing mass. The first row contained the lightest elements, the second row contained the next heaviest elements, and so forth. Aligning rows of elements above and below each other (like days of a calendar) revealed that elements within the same vertical column had similar properties, such as chemical reactivity. In order to achieve this pattern, however, he had to shift some elements left or right occasionally. This left gaps—blank spaces that could not be filled by any known element (**Figure 2.7**). Instead of looking on these gaps as defects, Mendeleev boldly predicted the existence of elements that had not yet been discovered. His predictions about the properties of some of those missing elements led to their discovery.

That Mendeleev was able to predict the properties of new elements helped convince many scientists of the accuracy of Dalton's atomic hypothesis, upon which Mendeleev's periodic table was based. This in turn helped promote Dalton's proposed atomic nature of matter from a hypothesis to a more widely accepted theory. Mendeleev's chart ultimately led to our modern periodic table, which we discuss more fully in Chapter 3.

Since the time of Lavoisier, Dalton, and Mendeleev, our understanding of atoms has grown substantially. Although we have not discovered the alchemist's dream of immortality, we have learned how to design medicines that cure numerous diseases. From crude oil we can make fuels, plastics, clothing, and more. From the thin air we can produce fertilizer. Virtually every aspect of modern society has been and will continue to be affected by our ability to manipulate atoms to meet our needs. Of all the discoveries made by humans, our discovery of the atom is arguably one of our greatest and most profound. Are atoms for real? Today we have the technology to capture images of individual atoms, as shown in **Figure 2.8**.

Just as important is the fact that atoms and molecules can be used to explain common observations. For example, heat transforms moist clay into ceramic by driving off water molecules. In ice, water molecules are stuck together in a fixed orientation. Warmth melts the ice by helping the water molecules break away from each other. Consider placing a fragrance, such as cinnamon oil, within an inflated rubber balloon. The balloon is sealed. How then does the outer surface of the balloon smell like cinnamon? We can explain how this occurs by assuming the fragrance consists of tiny molecules that can pass through the micropores of the inflated rubber balloon. Similarly, we can use the idea of molecules to explain how moisture can collect on a tabletop before disappearing, as shown in **Figure 2.9**, or how a dark-colored powdered drink mix dissolves in water with no stirring, as shown in **Figure 2.10**. The explanatory powers of the atomic model are great. This, along with our hi-tech evidence, leads us to trust that matter is made of these super-small particles we call atoms.

∧ Figure 2.5

John Dalton was born into a very poor family. Although his formal schooling ended at age 11, he continued to learn on his own and even began teaching others when he was only 12. His primary research interest was weather, which led him to conduct many experiments with gases. Soon after publishing his conclusions on the atomic nature of matter, his reputation as a first-rate scientist increased rapidly. In 1810, he was elected into Britain's premiere scientific organization, the Royal Society.

Figure 2.6 >

Dmitri Mendeleev was a devoted and highly effective teacher. Students adored him and would fill lecture halls to hear him speak about chemistry. Much of his work on the periodic table occurred in his spare time following his lectures. Mendeleev taught not only in the university classrooms but anywhere he traveled. During his journeys by train, he would travel third class with peasants to share his findings about agriculture.

> **fyi FOR YOUR INFORMATION**
>
> To help finance his scientific projects, Lavoisier took part-time employment as a tax collector, in which position he introduced reforms to help ease the tax burden on peasants. But because of this employment, he was beheaded in 1794 during the French Revolution. After hearing appeals to spare Lavoisier's life, the judge determined that "the Republic needs neither scientists nor chemists; the course of justice cannot be delayed." About 18 months after Lavoisier's execution, the French government sent a formal apology to his widow, Marie-Anne.

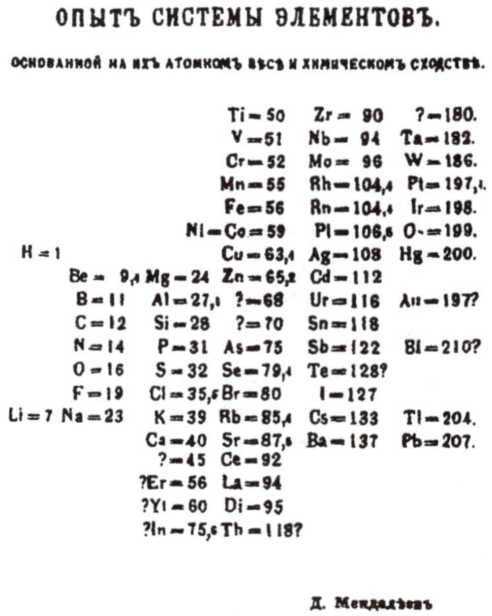

ОПЫТЪ СИСТЕМЫ ЭЛЕМЕНТОВЪ.

ОСНОВАННОЙ НА ИХЪ АТОМНОМЪ ВѢСѢ И ХИМИЧЕСКОМЪ СХОДСТВѢ.

			Ti=50	Zr=90	?=180.
			V=51	Nb=94	Ta=182.
			Cr=52	Mo=96	W=186.
			Mn=55	Rh=104,4	Pt=197,4.
			Fe=56	Rn=104,4	Ir=198.
			Ni=Co=59	Pl=106,6	Os=199.
H=1			Cu=63,4	Ag=108	Hg=200.
	Be=9,4	Mg=24	Zn=65,2	Cd=112	
	B=11	Al=27,4	?=68	Ur=116	Au=197?
	C=12	Si=28	?=70	Sn=118	
	N=14	P=31	As=75	Sb=122	Bi=210?
	O=16	S=32	Se=79,4	Te=128?	
	F=19	Cl=35,5	Br=80	I=127	
Li=7	Na=23	K=39	Rb=85,4	Cs=133	Tl=204.
		Ca=40	Sr=87,6	Ba=137	Pb=207.
		?=45	Ce=92		
		?Er=56	La=94		
		?Yt=60	Di=95		
		?In=75,6	Th=118?		

Д. Менделѣевъ

Figure 2.7

An early draft of Mendeleev's periodic table.

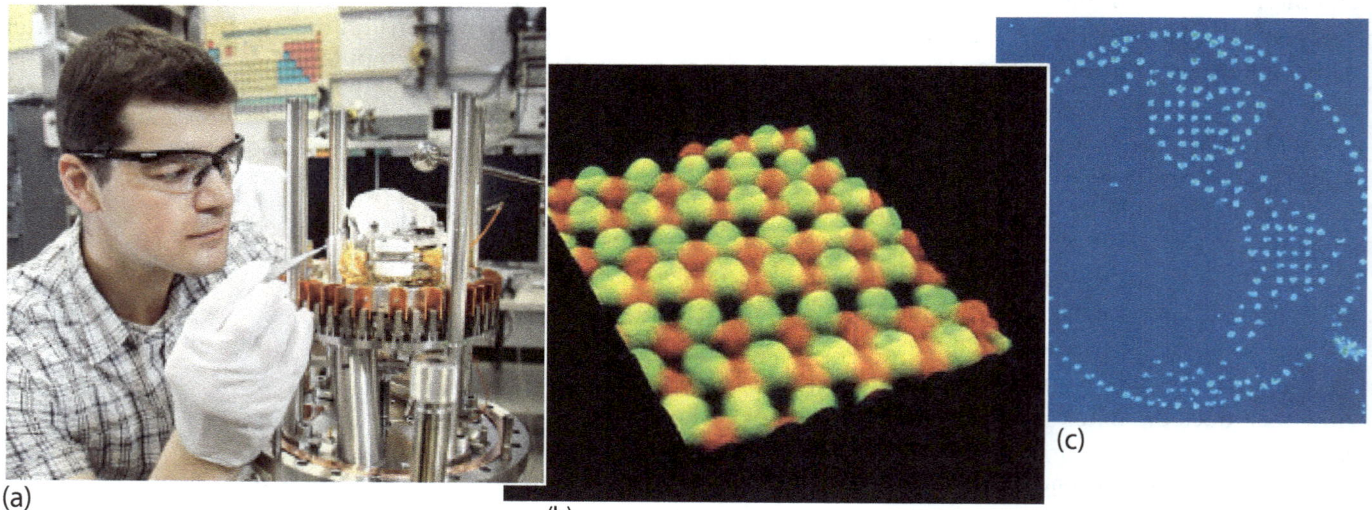

(a) (b) (c)

Figure 2.8

(a) Scanning probe microscopes are relatively simple devices used to create submicroscopic imagery. (b) An image of gallium and arsenic atoms. (c) Each dot in the world's tiniest map consists of a few thousand gold atoms, each dot moved into its proper place by a scanning probe microscope.

> **fyi FOR YOUR INFORMATION**
>
> Some atoms are larger than others, but they are all exceedingly small. Gold atoms, for example, are so small that about 4,000,000,000,000 (4 trillion) of them could fit within the period at the end of this sentence.

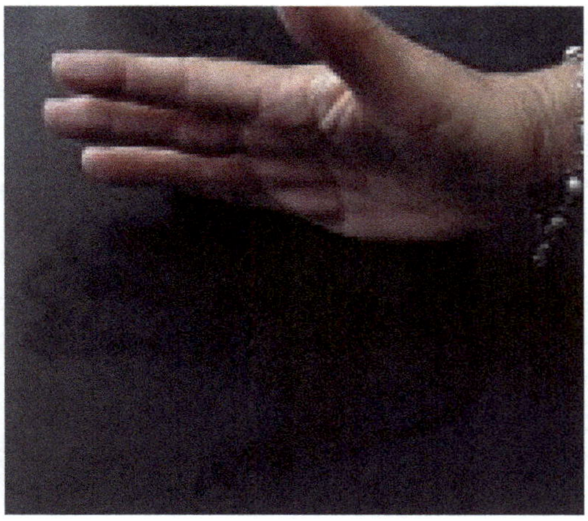

< **Figure 2.9**

Place your palms down on a cool, dark, and reflective table such as a slate lab benchtop. Water molecules exiting from your skin collect onto this surface. Lift your hands to see this moisture, which quickly disappears as the water molecules evaporate into the air.

fyi FOR YOUR INFORMATION

Most atoms are ancient. They have existed through imponderable ages, recycling through the universe in innumerable forms, both nonliving and living. In this sense, you don't "own" the atoms that make up your body–you are simply their present caretaker. There will be many caretakers to follow.

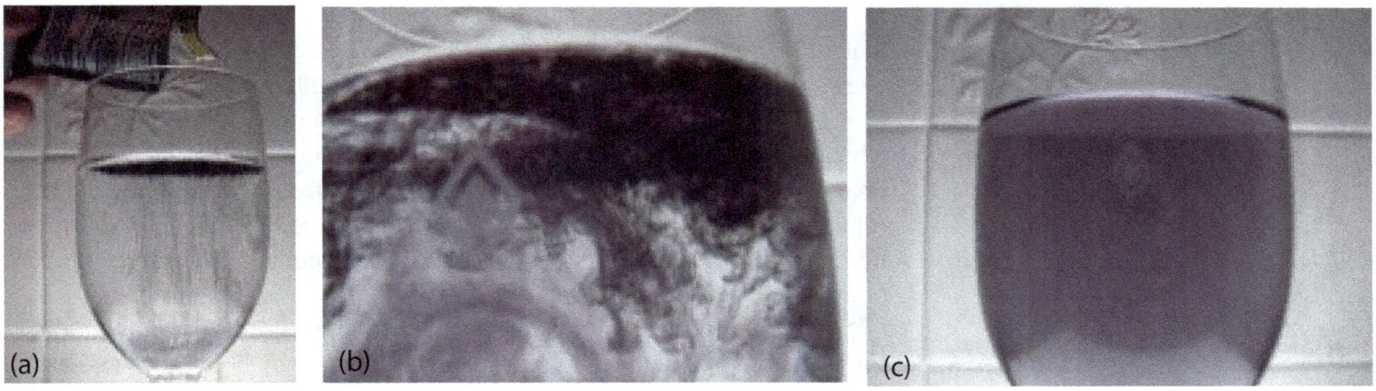

Λ **Figure 2.10**

(a) Kool-Aid crystals settle to the bottom of a container of water. (b) Without stirring, the crystals begin to dissolve as they are bombarded by water molecules in the liquid phase. (c) The bustling movement of the water molecules eventually causes the Kool-Aid to be uniformly mixed with the water.

CONCEPT **CHECK**

Lavoisier hypothesized that an element was a material made of a fundamental substance that cannot be broken down into anything else. According to Dalton, this fundamental substance was made of

a. water. b. fire. c. atoms. d. molecules

CHECK YOUR ANSWER

The answer is (c). Dalton reintroduced the concept of atoms put forth by Democritus some 2000 years earlier. Unlike Democritus, however, Dalton assumed that the atoms of different elements differed from each other by their mass.

2.3 Mass and Volume

READING CHECK
What is mass?

To describe a material object, we can quantify any number of properties, but perhaps the most fundamental of these is mass. **Mass is the quantitative measure of how much matter a material object contains.** The greater the mass of an object, the greater the amount of matter in it. A gold bar that is twice as massive as another gold bar, for example, contains twice as many gold atoms.

Mass is also a measure of an object's *inertia*, which is the resistance the object has to any change in its motion. A cement truck, for example, has a lot of mass (inertia), which is why it requires a powerful engine to get it moving and powerful brakes to cause it to come to a stop.

The standard unit of mass is the kilogram. A replica of the cylinder used to determine exactly what mass "1 kilogram" describes is shown in **Figure 2.11**. An average-sized adult human male has a mass of about 70 kilograms (154 pounds). For smaller quantities, we use the gram. Table 1.3 in the previous chapter tells us that the prefix kilo- means "1000," so we see that 1000 grams is equivalent to 1 kilogram (1000 grams = 1 kilogram). For even smaller quantities, the milligram is used (1000 milligrams = 1 gram).

∧ Figure 2.11

The standard kilogram is defined as the mass of a platinum-iridium cylinder kept at the International Bureau of Weights and Measures in Sèvres, France. The cylinder is removed from its very safe location only once a year for comparison with duplicates, such as the one shown here, which is housed at the National Institute of Standards and Technology in Washington, D.C.

Since mass is simply a measure of the amount of matter in a sample, which is a function of how many atoms the sample contains, the mass of an object remains the same no matter where it is located. A 1-kilogram gold bar, for example, has the same mass whether it is on the Earth, on the Moon, or floating "weightless" in space. This is because it contains the same number of atoms in each location.

Weight is more complicated. By definition, **weight** is the gravitational force exerted on an object by the most massive nearby body, such as Earth. The weight of an object, therefore, depends entirely upon its location, as is shown in **Figure 2.12**. On the Moon, a gold bar weighs less than it does on Earth. This is because the Moon is much less massive than Earth; hence, the gravitational force exerted by the Moon on the bar is much less. On Jupiter, the gold bar would weigh more than it does on Earth, because of the greater gravitational force exerted on the bar by this very massive planet.

The amount of space a material object occupies is its **volume**. The SI unit of volume is the liter, which is only slightly larger than the USCS unit

∧ Figure 2.12

(a) A 1-kilogram gold bar resting on the Earth weighs 2.2 pounds. (b) On the Moon, this same gold bar would weigh 0.37 pound. (c) Deep in space, far from any planet, the gold bar would weigh 0 pounds, though it would still have a mass of 1 kilogram.

< Figure 2.13

The volume of an object, no matter what its shape, can be measured by its displacement of water. When this object is immersed in the water, the rise in the water level equals the volume of the object, which in this example measures about 7 mL.

of volume, the quart. A liter is the volume of space marked off by a cube measuring 10 centimeters by 10 centimeters by 10 centimeters, which is 1000 cubic centimeters. A smaller unit of volume is the milliliter, which is one-thousandth of a liter, or 1 cubic centimeter.

A convenient way to measure the volume of an irregular object is shown in **Figure 2.13**. The volume of water displaced is equal to the volume of the object.

fyi FOR YOUR INFORMATION

If you counted by ones at a rate of one number each second, counting to a million would take you 11.6 days (assuming no time off for sleeping). How about a billion? Is there a dramatic difference? Get this: counting to a billion would take you 31.8 years. Counting to a trillion, which is a thousand billion, would take you 31,800 years! Now you know why there are far more millionaires than billionaires, and no trillionaires yet.

CONCEPT **CHECK**

Is there gravity on the Moon?

CHECK YOUR ANSWER Yes, absolutely! The Moon exerts a downward gravitational pull on any body near its surface, as evidenced by the fact that astronauts were able to land and walk on the Moon. This NASA photograph shows an astronaut jumping. Without gravity, this jump would have been his last, because he would never have come back down. With the Moon's weaker gravity he is able to jump relatively high.

2.4 Density: Mass to Volume

READING CHECK
Density is a measure of what?

The relationship between an object's mass and the amount of space it occupies is the object's density. **Density** is a measure of compactness, of how tightly mass is squeezed into a given volume. A block of lead has much more mass than does a same-sized block of aluminum. A reason for this is because each lead atom has more mass. The lead, having more mass per volume, is therefore more dense. We can think of density as the "lightness" or "heaviness" of objects of the same size, as **Figure 2.14** shows.

Density is the amount of mass contained in a sample divided by the volume of the sample:

$$\text{Density} = \text{Mass} / \text{Volume}$$

An object having a mass of 1 gram and a volume of 1 milliliter, for example, has a density of

$$\text{Density} = 1 \text{ g} / 1 \text{ mL} = 1 \text{ g/mL}$$

An object having a mass of 2 grams and a volume of 1 milliliter is denser; its density is

$$\text{Density} = 2 \text{ g} / 1 \text{ mL} = 2 \text{ g/mL}$$

Figure 2.14

The amount of mass in a block of lead far exceeds the amount of mass in a block of aluminum of the same size. Hence, the lead block weighs much more and is more difficult to lift.

Other units of mass and volume besides grams and milliliters may be used in calculating density. The densities of gases, for example, because they are so low, are often given in grams per liter. In all cases, however, the units are a unit of mass divided by a unit of volume.

The densities of some substances are given in **Table 2.1**. Which would be more difficult to pick up: a liter of water or a liter of mercury? Note how water has a density of 1.000 g/mL at 4°C. Why do we specify the temperature? As discussed in Section 2.6, the volume of a material changes with temperature. A change in temperature, therefore, can also result in a change in density, even though the mass of the object stays the same.

Gas densities are much more affected by temperature as well as pressure than are the densities of solids and liquids. With an increase in pressure, gas molecules are squeezed closer together. This makes for less

Figure 2.15

The hot air inside these hot air balloons is less dense than the surrounding colder air, which is why the balloons rise.

CONCEPT **CHECK**

Which occupies a greater volume: 1 kilogram of lead or 1 kilogram of aluminum?

CHECK YOUR ANSWER

The aluminum. Think of it this way. Because lead is so dense, you need only a little bit in order to have 1 kilogram. Aluminum, by contrast, is far less dense, so 1 kilogram of aluminum occupies much more volume than the same mass of lead.

volume (but the same mass) and therefore greater density. The density of the air inside a diver's breathing tank, for example, is much greater than the density of air at normal atmospheric pressure. With an increase in temperature, gas molecules move faster and thus have a tendency to push outward, thereby occupying a greater volume. Thus, hot air is less dense than cold air, which is why hot air rises and the balloons in **Figure 2.15** can take passengers for breathtaking rides.

TABLE 2.1 Densities of Some Solids, Liquids, and Gases

SUBSTANCE	DENSITY (G/ML)	DENSITY (G/L)
Solids		
osmium	22.5	22,500
gold	19.3	19,300
lead	11.3	11,300
copper	8.92	8920
iron	7.86	7860
zinc	7.14	7140
aluminum	2.70	2700
ice	0.92	920
Liquids		
mercury	13.6	13,600
seawater	1.03	1030
fresh water at 4°C	1.00	1000
ethyl alcohol	0.81	810
Gases*		
oxygen at 0°C	0.00143	1.43
dry air		
0°C	0.00129	1.29
20°C	0.00121	1.21
helium at 0°C	0.000178	0.178

*All values at sea-level atmospheric pressure.

CONCEPT **CHECK**

1. Which has greater density: 1 gram of water or 10 grams of water?
2. Which has greater density: 1 gram of lead or 10 grams of aluminum?

CHECK YOUR ANSWER

1. The density is the same for any amount of water. Whereas 1 gram of water occupies a volume of 1 milliliter, 10 grams occupies a volume of 10 milliliters. The ratio 1 gram/1 milliliter is the same as the ratio 10 grams/10 milliliters.
2. The lead. Density is mass per volume, and this ratio is greater for any amount of lead than for any amount of aluminum.

2.5 Energy Moves Matter

READING CHECK
How did Aristotle classify motion?

Matter is substance, and **energy** is that which can move a substance. The concept of energy is abstract and therefore not as easy to define as the concepts of mass and volume. One definition of energy is the capacity to do work. If something has energy, it can do work on something else—it can exert a force and move that something else. Accordingly, energy is not something we observe directly. Rather, we only witness its effects.

An object may store energy by virtue of its position. This stored energy is called potential energy because it has the "potential" for doing work. As shown in Figure 2.16, a boulder perched on the edge of a cliff has potential energy due to the force of gravity, just as the poised arrow has potential energy due to the tension of the bow. The **potential energy** of an object increases as the distance over which the force is able to act increases. The higher a boulder is positioned above level ground, the more potential energy it has to do work as it falls downward under the pull of gravity. Similarly, an arrow in a fully drawn bow has more potential energy than does one in a half-drawn bow.

(a) (b)

Figure 2.16

(a) An elevated boulder's potential energy becomes apparent when the boulder is perched in a position where it might easily fall downhill. When it does fall, this potential energy is converted to kinetic energy. (b) Much of the potential energy in Tenny's drawn bow will be converted to the kinetic energy of the arrow upon its release.

Kinetic energy is the energy of motion. Both a falling boulder and a flying arrow have kinetic energy. The faster a body moves, the more kinetic energy it has and therefore the more work it can do. For example, the faster an arrow flies, the more work it can do to a target, as evidenced by its deeper penetration.

Substances possess what is known as *chemical potential energy*, which is the energy that is stored within atoms and molecules. For example, any material that can burn has chemical potential energy. The firecracker in Figure 2.17 , for instance, has chemical potential energy. This energy gets released when the firecracker is ignited. During the explosion, some of the chemical potential energy is transformed to the kinetic energy of flying particles. Much of the chemical potential energy is also transformed to light and heat. We explore the relationship between energy and chemical reactions in Chapter 9.

The SI unit of energy is the joule, which is roughly the amount of energy released from a candle burning for only a moment. In the United States, a common unit of energy is the calorie. One calorie is by definition the amount of energy required to raise the temperature of 1 gram of water by 1 degree Celsius. One calorie is 4.184 times larger than 1 joule. Put differently, 4.184 joules of energy is equivalent to 1 calorie (4.184 joules = 1 calorie). So, a joule is about one-fourth of a calorie.

In the United States, the energy content of food is measured by the Calorie (note the capital C). One Calorie equals 1 kilocalorie, which is 1000 calories (note the small c). The candy bar in **Figure 2.18** offers 150 Calories (150 kilocalories), providing a total of 150,000 calories to the consumer in one serving. How many joules is this? Multiply by the appropriate conversion factor, which would be 4.184 joules/1 calorie.

CONCEPT **CHECK**

How does a flying arrow have potential energy as well as kinetic energy?

CHECK YOUR ANSWER It has gravitational potential energy as long as it remains above the ground.

< Figure 2.17

A firecracker is a mixture of solids that possess chemical potential energy. When a firecracker explodes, the solids react to form gases that fly outward and so possess a great deal of kinetic energy. Light and heat (both of which are forms of energy) are also formed.

< Figure 2.18

The energy content of this candy bar, 150 Calories per serving (150,000 calories), when released through burning, is enough to heat up 150,000 grams (about 330 pounds) of water by 1 degree Celsius.

2.6 Temperature and Heat

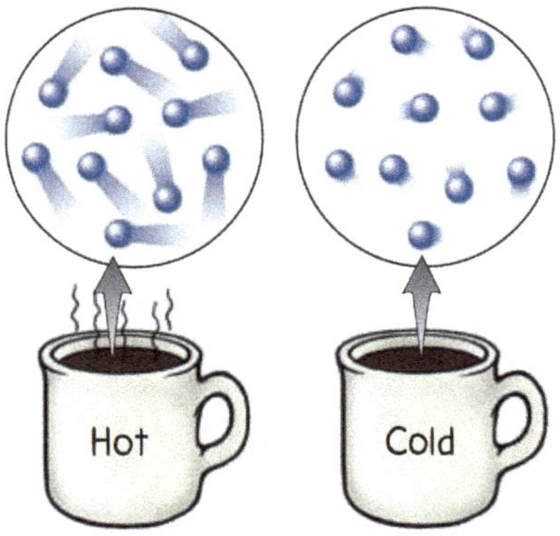

∧ Figure 2.19

The difference between hot coffee and cold coffee is the average speed of the molecules. In the hot coffee, the molecules are moving faster on average than they are in the cold coffee. (The "motion trails" on the molecules of hot coffee indicate their higher speed.)

Atoms and molecules are in constant motion, jiggling to and fro or bouncing from one position to another. By virtue of their motion, these particles possess kinetic energy. Their average kinetic energy is directly related to a property you can sense: how hot something is. Whenever something becomes warmer, the kinetic energy of its submicroscopic particles increases. For example, strike a penny with a hammer and the penny becomes warm because the hammer's blow causes its atoms to jostle faster, increasing their kinetic energy. (The hammer becomes warmer for the same reason.) Put a flame to a liquid and the liquid becomes warmer because the energy of the flame causes the particles of the liquid to move faster, increasing their kinetic energy. For example, the molecules in the hot coffee in **Figure 2.19** are moving faster on average than those in the cold coffee.

Temperature tells us how warm or cold an object is relative to some standard. We express temperature by a number that corresponds to the degree of hotness on some chosen scale. Just touching an object certainly isn't a good way of measuring its temperature, as **Figure 2.20** illustrates. To measure temperature, therefore, we take advantage of the fact that nearly all materials expand when their temperature is raised and contract when lowered. With increasing temperature, the particles move faster and are on average farther apart—the material expands. With decreasing temperature, the particles move more slowly and are on average closer together—the material contracts. An old-style **thermometer** exploits this characteristic of matter, measuring temperature by means of the expansion and contraction of a liquid, usually mercury or colored alcohol.

The most common thermometer scale in the world is the Celsius scale, named in honor of the Swedish astronomer Anders Celsius (1701–1744). He first suggested the scale of 100 degrees between the freezing point and boiling point of fresh water. In a Celsius thermometer, the number 0 is assigned to the temperature at which pure water freezes and the number 100 is assigned to the temperature at which it boils (at standard atmospheric pressure), with 100 equal divisions called degrees between these two points.

< Figure 2.20

Can we trust our sense of hot and cold? Will both fingers feel the same temperature when they are put in the warm water? Try this yourself, and you will see why we use a thermometer for an objective measurement.

CONCEPT **CHECK**

You may have noticed telephone wires sagging on a hot day. This happens because the wires are longer in hot weather than in cold. What is happening on the atomic level to cause such changes in wire length?

CHECK YOUR ANSWER

On a hot day, the atoms in the wire are moving faster, and as a result the wire expands. On a cold day, those same atoms are moving more slowly, which causes the wire to contract.

In the United States, we use the Fahrenheit thermometer, named after its originator, the German scientist G. D. Fahrenheit (1686–1736). Fahrenheit chose to assign 0 to the temperature of a mixture of water, ice, and a salt and to assign 100 to the body temperature of a human. Because these reference points are not dependable, the Fahrenheit scale has since been modified such that the freezing point of pure water is designated 32°F and the boiling point of pure water is designated 212°F. On this recalibrated scale, the average human body temperature taken orally is around 98.2°F.

A temperature scale favored by scientists is the **Kelvin scale**, named after the British physicist Lord Kelvin (1824–1907). This scale is calibrated not in terms of the freezing and boiling points of water but rather in terms of the motion of atoms and molecules. On the Kelvin scale, zero is the temperature at which there is no atomic or molecular motion. This is a theoretical limit called **absolute zero**, which is the temperature at which the particles of a substance have absolutely no kinetic energy to give up. Absolute zero corresponds to –459.7°F on the Fahrenheit scale and to –273.15°C on the Celsius scale. On the Kelvin scale, this temperature is simply 0 K, which is read "zero kelvin" or "zero K." Marks on the Kelvin scale are the same distance apart as those on the Celsius scale, so the temperature of freezing water is 273 kelvin. (Note that the word degree is not used with the Kelvin scale. To say "273 degrees kelvin" is incorrect. To say "273 kelvin" is correct.) The three scales are compared in **Figure 2.21**.

READING CHECK
What does temperature measure?

It is important to understand that temperature is a measure of the *average* amount of energy in a substance, not the total amount of energy, as **Figure 2.22** shows. The total energy in a swimming pool full of boiling water is much more than the total energy in a cupful of boiling water even though both are at the same temperature. Your utility bill after heating the swimming pool water to 100°C would show this. Whereas the total amount of energy in the pool is much more than in the cup, the average molecular motion is the same in both water samples. The water molecules in the swimming pool are moving on average just as fast as the water molecules in the cup. The only difference is that the swimming pool contains more water molecules and hence a greater total amount of energy.

Heat is energy that flows from a higher-temperature object to a lower temperature object. If you touch a hot stove, heat enters your hand because the stove is at a higher temperature than your hand. When you touch a piece of ice, energy passes out of your hand and into the ice because the ice is at a lower temperature than your hand. From a human perspective, if you are absorbing heat, you experience warmth; if you are losing heat, you experience cooling. The next time you touch the hot forehead of a sick, feverish friend, ask him or her whether your hand feels hot or cold. Whereas temperature is absolute, hot and cold are relative.

In general, the greater the temperature difference between two bodies in

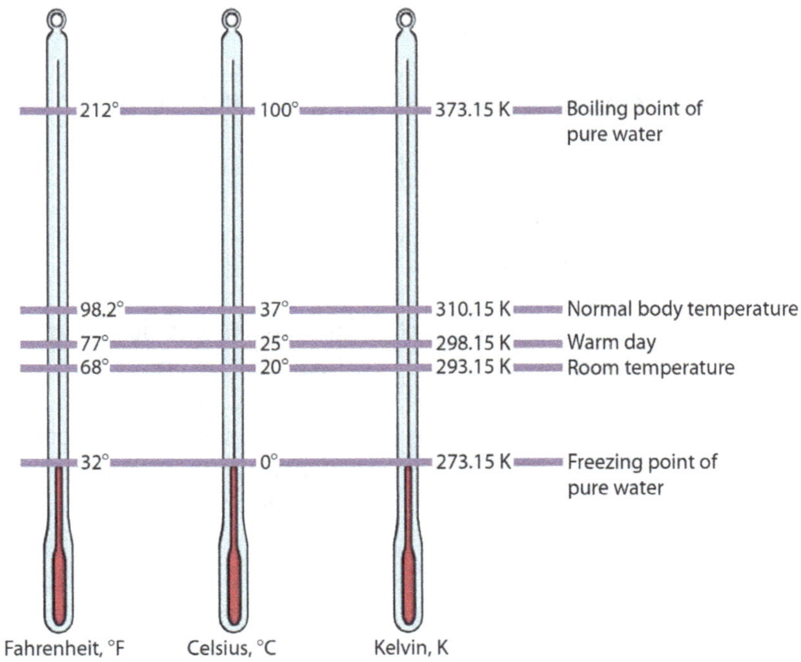

Figure 2.21
Some familiar temperatures measured on the Fahrenheit, Celsius, and Kelvin scales.

contact with each other, the greater the rate of heat flow. This is why a hot stove can cause much more damage to your skin than a warm stove. Because heat is a form of energy, its unit is the *joule*.

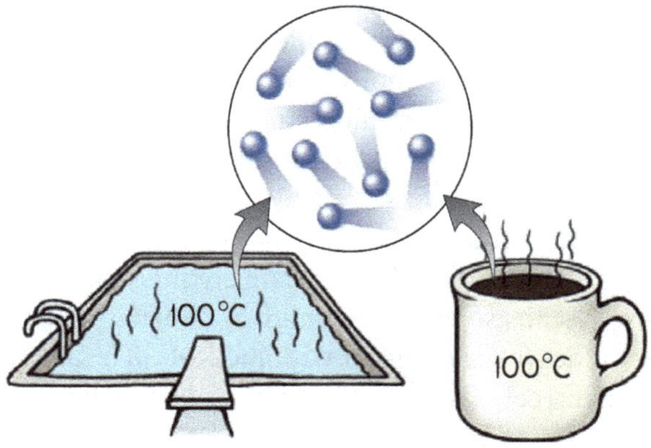

Figure 2.22 >

Bodies of water at the same temperature have the same average molecular kinetic energies. The total energy, however, depends upon how much water you have. For example, a swimming pool of water has much more total energy than a cupful of water, even when at the same temperature. Consider what your electric bill would look like after heating a swimming pool full of water to 100°C.

CONCEPT **CHECK**

When you enter a swimming pool, the water may feel quite cold. After a while, though, your body "gets used to it," and the water no longer feels so cold. Use the concept of heat to explain what is going on.

CHECK YOUR ANSWER

Heat flows because of a temperature difference. When you enter the water, your skin temperature is much higher than the water temperature. The result is a significant flow of heat from your body to the water, which you experience as cold. Once you have been in the water awhile, your skin temperature is much closer to the water temperature (due to the cooling effects of the water and your body's ability to conserve heat), so the flow of heat from your body is less. With less heat fl owing from your body, the water no longer feels so cold.

2.7 Phases of Matter

One of the most evident ways we can describe matter is by its physical form, which may be one of three *phases* (also sometimes described as physical states): solid, liquid, or gas. A solid material, such as a rock, occupies a constant amount of space and does not readily deform upon applying pressure. In other words, a solid has both a definite volume and a definite shape. A liquid also occupies a constant amount of space (it has a definite volume), but its form changes readily (it has an indefinite shape). A liter of milk, for example, may take the shape of its carton or the shape of a puddle, but its volume is the same in both cases. A gas is diffuse, having neither a definite volume nor a definite shape. Any sample of gas assumes both the shape and the volume of the container it occupies. A given amount of air, for example, may assume the volume and shape of a toy balloon or be compressed into the volume and shape of a bicycle tire. Released from its container, a gas diffuses into the atmosphere, which is a collection of various gases held to our planet by the force of gravity.

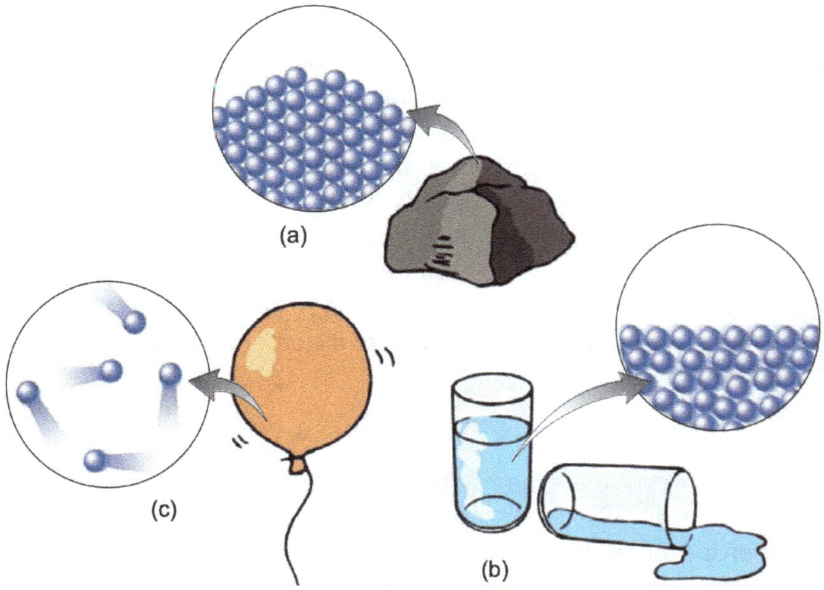

< Figure 2.23

The familiar bulk properties of a solid, liquid, and gas. (a) The particles of the solid phase vibrate about fixed positions. (b) The particles of the liquid phase slip past one another. (c) The fast-moving particles of the gaseous phase are separated by large average distances.

READING CHECK

On a submicroscopic level, how are solids, liquids, and gases distinguished?

On the submicroscopic level, the solid, liquid, and gaseous phases are distinguished by the extent of interaction between the submicroscopic particles (the atoms or molecules). This is illustrated in **Figure 2.23**. In solid matter, the attractions between particles are strong enough to hold all the particles together in some fixed 3-dimensional arrangement. The particles are able to vibrate about fixed positions, but they cannot move past one another. Adding heat causes these vibrations to increase until, at a certain temperature, the vibrations are rapid enough to disrupt the fixed arrangement. The particles can then slip past one another and tumble around much like a bunch of marbles in a bag. This kind of motion is representative of the liquid phase of matter, and it is the mobility of the particles that gives rise to the liquid's fluid character—its ability to flow and take on the shape of its container.

Further heating causes the particles in the liquid to move so fast that any attractions they have for one another are unable to hold them together. They then separate from one another, forming a gas. Moving at an average speed of 500 meters per second (1100 miles per hour), the particles of a gas are widely separated from one another. Matter in the gaseous phase therefore occupies much more volume than it does in the solid or liquid phase, as **Figure 2.24** shows. Applying pressure to a gas squeezes the gas particles closer together, which reduces their volume. Enough air for an

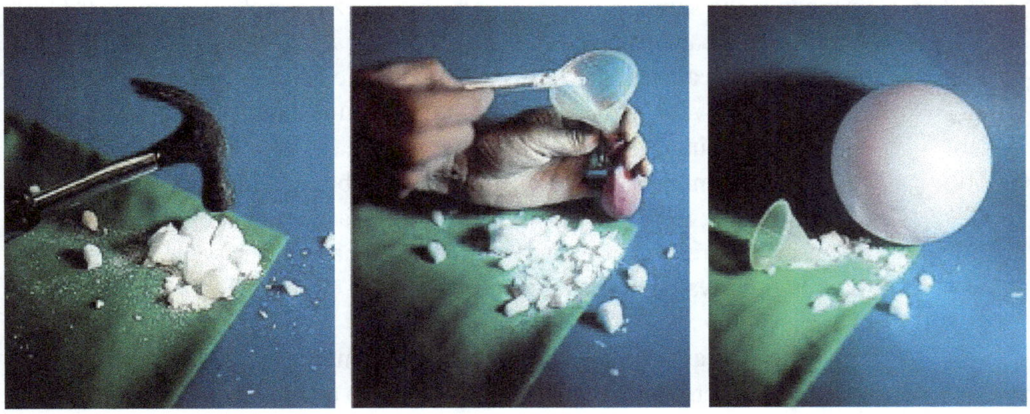

< Figure 2.24

The gaseous phase of any material occupies significantly more volume than either its solid or liquid phase. (a) Solid carbon dioxide (dry ice) is broken up into powder form. (b) The powder is funneled into a balloon. (c) The balloon expands as the contained carbon dioxide becomes a gas as the powder warms up.

CHEMICAL CONNECTIONS

How is your sense of smell connected to temperature?

underwater diver to breathe for many minutes, for example, can be squeezed (compressed) into a tank small enough to be carried on the diver's back.

Although gas particles move at high speeds, the speed at which they can travel from one side of a room to the other is relatively slow. This is because the gas particles are continually hitting one another, and the path they end up taking is circuitous. At home, you get a sense of how long it takes for gas particles to migrate each time someone opens the oven door after baking, as Figure 2.25 shows. A shot of aromatic gas particles escapes the oven, but there is a notable delay before the aroma reaches the nose of someone sitting in the next room.

CONCEPT **CHECK**

Why are gases so much easier to compress into smaller volumes than solids and liquids?

CHECK YOUR ANSWER Because there is a lot of space between gas particles. The particles of a solid or liquid, on the other hand, are already close to one another, meaning there is little room left for a further decrease in volume.

∧ Figure 2.25

In traveling from point A to point B, the typical gas particle travels an indirect path because of numerous collisions with other gas particles—about 8 billion collisions every second! The changes in direction shown here represent only a few of these collisions. Although the particle travels at very high speeds, it takes a relatively long time to cross between two distant points because of these numerous collisions.

Terms Used to Describe Phase Changes

Figure 2.26 illustrates that you must either add heat to a substance or remove heat from it if you want to change its phase. The process by which a solid transforms into a liquid is called **melting**. To visualize what happens when heat begins to melt a solid, imagine that you are holding hands with a group of people and each of you starts jumping around randomly. The more violently you jump, the more difficult it is to hold onto one another. If everyone jumps violently enough, keeping hold is impossible. Something like this happens to the particles of a solid when it is heated. As heat is added to the solid, the particles vibrate more and more violently. If enough heat is added, the attractive forces between the particles are no longer able to hold them together. The solid melts.

A liquid can be changed to a solid by the removal of heat. This process is called **freezing**. As heat is withdrawn from the liquid, particle motion diminishes until the particles are moving slowly enough on average for the attractive forces between them to take firm hold. The only motion the particles are capable of then is vibration about fixed positions, which means the liquid has solidified, or frozen. As we explore further in Chapter 8, freezing is merely the reverse of melting. The temperature at which a substance freezes or melts is the same. For pure water, this freezing/melting point is 0°C.

A liquid can be heated so that it becomes a gas—a process called **evaporation**. As heat is added, the particles of the liquid acquire more kinetic energy and move faster. Particles at the liquid surface eventually gain enough energy to fly out of the liquid. In other words, they enter the gaseous phase. As more and more particles absorb the heat being added, they too acquire enough energy to escape from the liquid surface and become gas particles. Because a gas results from evaporation, this phase is also sometimes referred to as *vapor*. Water in the gaseous phase, for example, may be referred to as *water vapor*.

The rate at which a liquid evaporates increases with temperature. A puddle of water, for example, evaporates from a hot pavement more quickly than it does from your cool kitchen floor. When the temperature is hot enough, evaporation occurs beneath the surface of the liquid. As a result,

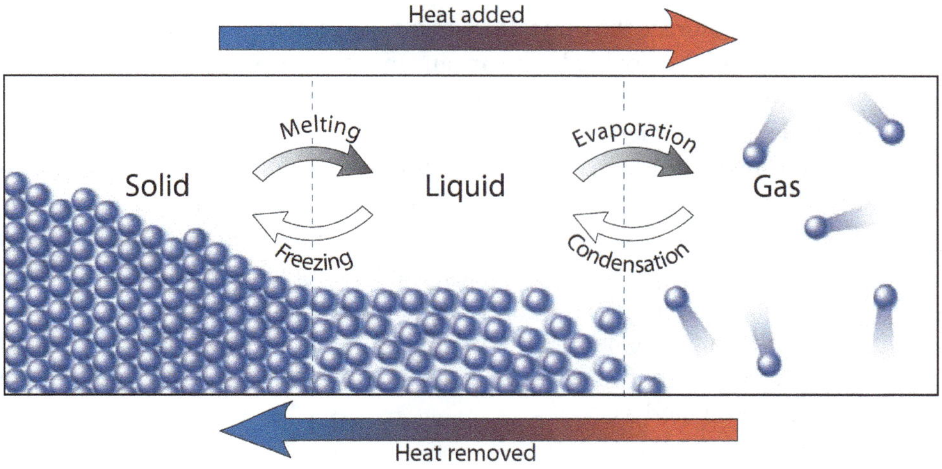

< **Figure 2.26**

Melting and evaporation involve the addition of heat; condensation and freezing involve the removal of heat.

bubbles form and are buoyed up to the surface. We say that the liquid is **boiling**. A substance is often characterized by its boiling point, which is the temperature at which it boils. At sea level, the boiling point of fresh water is 100°C.

The transformation from gas to liquid—the reverse of evaporation—is called **condensation**. This process can occur when the temperature of a gas decreases. The water vapor held in the warm daylight air, for example, may condense to form a wet dew in the cool of the night.

Sublimation and deposition are two less commonly mentioned phase changes. **Sublimation** is the transformation of a solid directly to a gas. Snow readily sublimes from mountain tops on sunny days. Dry ice, as shown in Figure 2.24 , also readily sublimes. **Deposition** is the transformation of a gas directly to a solid. An example of deposition is the formation of frost. The various phase transformations are summarized in **Figure 2.27**.

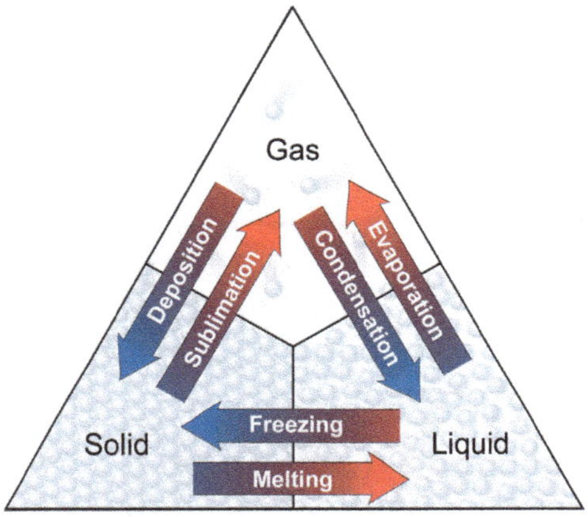

∧ **Figure 2.27**

A summary of terms used to describe phase changes.

2.8 Gas Laws

Scientists of the 17th, 18th, and 19th centuries investigated the relationships among the pressure, volume, and temperature of gaseous materials. Their observations are summed up by a set of gas laws named in their honor. These laws help us to understand the behavior of gases, including the air we breathe.

Boyle's Law: Pressure and Volume

Think of the molecules of air inside the inflated tire of an automobile. Inside the tire, the molecules behave like zillions of tiny ping pong balls, perpetually moving helter-skelter and banging against the inner walls. Their impacts on the inner surface of the tire produce a jittery force that appears to our coarse senses as a steady outward push. Averaging this pushing force over a unit of area gives the pressure of the enclosed air.

Suppose the temperature is kept constant and twice as many molecules are pumped into the same volume, as shown in **Figure 2.28**. Then the air density—the number of molecules per given volume—is doubled. If the molecules move with the same average kinetic energy, or, equivalently, if

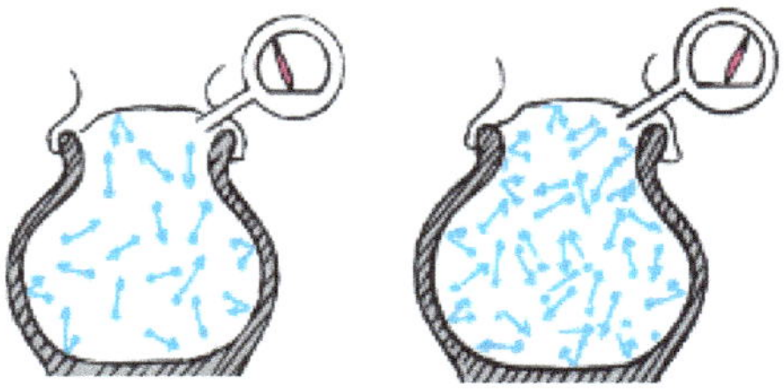

∧ **Figure 2.28**

When the density of gas in the tire is doubled, the pressure is doubled.

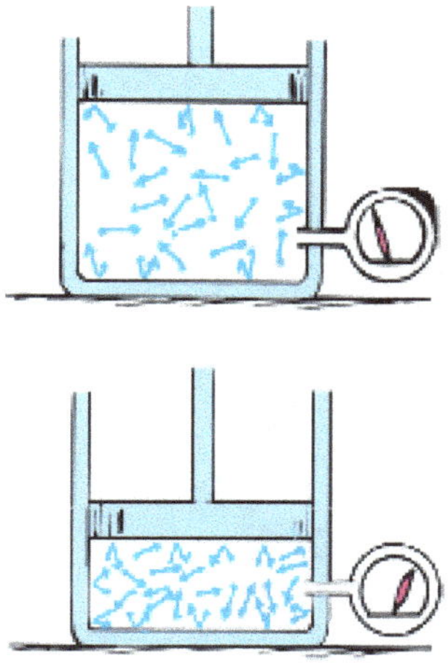

∧ **Figure 2.29**

When the volume of gas is decreased, the density, and therefore the pressure, are increased.

their temperature is the same, then, to a close approximation, the number of collisions will be doubled. This means that the pressure is doubled.

We can also double the air density by compressing air to half its volume. Consider the cylinder with the movable piston in **Figure 2.29**. If the piston is pushed downward so that the volume is half the original volume, the density of molecules will double and the pressure will correspondingly double. In general, a lessening of the volume means a correspondingly greater pressure (because the gas density is greater), while a larger volume means a correspondingly smaller pressure (because the gas density is less).

Notice in the example involving the piston that volume and pressure are inversely proportional—as one gets bigger, the other gets smaller. This can be represented as

$$V \propto \frac{1}{\mathrm{P}}$$

This relationship is known as **Boyle's Law**, after the 17th-century scientist Robert Boyle (1627–1691) who first described it. Boyle discovered that as volume decreases, pressure increases, and as volume increases, pressure decreases. Similarly, an increase in pressure results in a smaller volume, while a decrease in pressure results in a larger volume. Boyle's Law, however, holds true only assuming that the temperature and number of gas particles remain the same.

Charles's Law: Volume and Temperature

The 18th-century French scientist and daring balloon aviator Jacques Charles (1746–1823) discovered the direct relationship between the volume of a gas and its temperature at constant pressure. Charles showed that the volume of a gas increases as its temperature increases. Likewise, the volume of a gas decreases as its temperature decreases, as is shown in **Figure 2.30**.

Remarkably, when the volume and temperature of various gases at constant pressure are plotted on a graph, it appears that at a rather cold temperature, –273.15°C, the volume of any gas dwindles down to zero, as shown in **Figure 2.31**.

Of course, the volume of a gas never reaches zero. Somewhere along the way the gas transforms into a liquid. Nonetheless, in 1848, William Thomson, also known as Lord Kelvin (1824–1907), recognized that this convergent temperature of –273.15°C would be a convenient zero point for an alternative temperature scale by which the absolute motion of particles could be measured. The number 0 is assigned to this lowest possible temperature—absolute zero. Accordingly, this scale became known as the Kelvin temperature scale, as was introduced in Section 2.6.

According to **Charles's Law**, the volume and absolute temperature (measured in kelvins) of a gas are directly proportional—as one gets bigger, so does the other. This can be represented as

$$V \propto T$$

CONCEPT CHECK

1. You bring a long snorkel with you to the bottom of a deep swimming pool. The snorkel provides a direct path between your lungs and the air above, but you still can't breathe through it. Why?

2. A scuba diver swimming underwater breathes compressed air, which counteracts the water pressure, allowing her to breathe. She then holds her breath while returning to the surface. What happens to the volume of air in her lungs?

CHECK YOUR ANSWER

1. The weight of the water above you squeezes you from all directions. This water is very heavy, and your lungs are not strong enough to expand against all this pressure. That's why all snorkels are short.

2. As she rises to the surface, the water pressure decreases, which allows the compressed air in her lungs to expand in accordance with Boyle's Law. Rising 10 m to the surface results in half the pressure, which results in a doubling of the volume. This is enough to burst the lungs. A first lesson in scuba diving is not to hold your breath while ascending—to do so can be fatal.

∧ Figure 2.30

Evan inflates two balloons to the same size and heats one over some boiling water while the other cools down in the freezer. A quick re-comparison of the two balloons shows that while the heated balloon expanded, the cooled balloon became smaller. What do you suppose happens when an air-filled balloon is dipped into −196°C liquid nitrogen and then pulled back out?

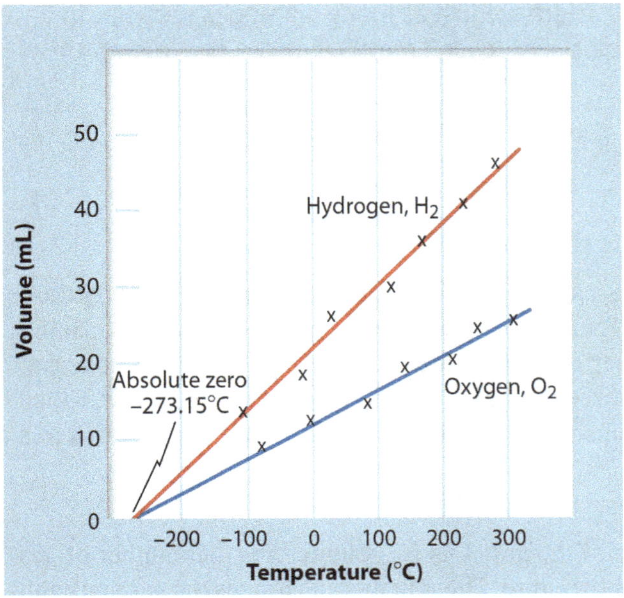

< Figure 2.31

A plot of experimental data showing volume versus the temperature (at constant pressure) for hydrogen, H_2, and oxygen, O_2, in their gaseous phases. Note how the gases converge to zero volume at the same temperature, −273.15°C.

CONCEPT **CHECK**

A perfectly elastic balloon holding helium is warmed. What happens to its volume?

CHECK YOUR ANSWER According to Charles's Law, as a gas is warmed, its volume expands.

FOR YOUR INFORMATION

In 1783, Jacques Charles was the first to fly in a balloon filled with hydrogen gas, which he created by reacting scrap iron with sulfuric acid. (Two weeks earlier, the Mongolfier brothers had been responsible for the first balloon flight using hot air.) One of Charles's first flights over the French countryside lasted about 45 minutes and took him 15 miles to a small village, where the people were so terrified they tore the balloon to shreds.

Avogadro's Law: Volume and Number of Particles

The 19th-century Italian scientist Amedeo Avogadro hypothesized that the volume of a gas is a function of the number of gas particles it contains. In other words, as the number of gas particles increases, so does the volume, assuming a constant pressure and temperature. This relationship is known as **Avogadro's Law**. An easy way to demonstrate this law is to grasp the opening of an empty plastic bag and blow into it. The more air molecules you blow into the bag, the bigger its volume. This can be represented as

$$V \propto n$$

CONCEPT **CHECK**

Assuming the same pressure and temperature, which has a greater number of gas particles: 5 liters of gaseous water, H_2O, or 5 liters of gaseous oxygen, O_2?

CHECK YOUR ANSWER

According to Avogadro's Law, the same volume means the same number of gas particles. Each particle here refers to a single molecule of either water, H_2O, or oxygen, O_2. There are just as many water gas particles in 5 liters as there are oxygen gas particles. The nature of the gas particles doesn't matter. But can you answer this question: which has a greater number of atoms?

READING CHECK

What four interrelated quantities can be used to describe the properties of a gas?

The Ideal Gas Law and Kinetic Molecular Theory

The properties of a gas can be described by four interrelated quantities: Boyle's, Charles's, and Avogadro's gas laws each describe how one quantity varies relative to another as long as the remaining two are held constant. Mathematically, these three laws can be combined into a single law, called the ideal gas law, which shows the relationship of all these quantities in a single equation:

$$PV = nRT$$

where P is the pressure of the gas, V is its volume, n is the number of gas particles, and T is the temperature. The R in this equation is the gas constant, which is the same no matter what the identity of the gas. Its value depends only upon the chosen units of pressure, volume, and temperature.*

The ideal gas law gets its name from the fact that it accurately describes only an ideal gas. Such a gas is one in which individual gas particles are considered to occupy no volume in their container. Furthermore, the particles of an ideal gas experience no attractions to or repulsions from one another. Each particle of a real gas, however, does have volume, albeit very, very small. Also, these particles do interact with each other. Real gases, therefore, do not follow any of the gas laws perfectly. At normal pressures, however, the contribution of particle size to total volume is insignificant and gas laws are good predictors of gas behavior. Furthermore, when the temperature of the gas is well above its boiling point, the gas particles are moving so fast that they rebound off one another without sticking. The interactions, therefore, are negligible, so the particles closely resemble those of an ideal gas.

CONCEPT **CHECK**

Which do gas laws describe more accurately: a gas at high pressure and low temperature or a gas at low pressure and high temperature?

CHECK YOUR ANSWER

Gas laws work best for gases at low pressures and high temperatures. The air we breathe is a good example. At atmospheric pressure, the distances between air molecules are much greater than the sizes of the air molecules. Also, air is well above its components' boiling points (N_2 boils at –196°C and O_2 at –183°C). This behavior permitted the gas law discoveries of Boyle, Charles, Avogadro, and others.

From the gas laws a model has been developed that gives us insight as to why gases behave as they do. This model, known as the **kinetic molecular theory**, can be summarized as a set of five postulates:

1. A gas consists of tiny particles, either atoms or molecules or both.

2. Gas particles are in constant random motion, colliding with one another and with the walls of their container.

3. The impacts of gas particles on the walls of the container produce a jittery force that appears as a steady push against the inner surface. This pushing force provides the pressure of the enclosed gas.

4. Deviations from gas laws arise primarily because of the interactions occurring among gas particles and because gas particles are not infinitely small.

5. The average kinetic energy (energy due to motion) of the gas particles is directly proportional to the temperature of the gas.

We can use these postulates to rationalize the various gas laws. For Boyle's Law, pressure decreases with increasing volume because collisions of the gas particles with the walls of the container occur less frequently, and those impacts are spread out over a greater area. For Charles's Law, volume increases with increasing temperatures because faster-moving particles push against the inner sides of the container more forcefully, which causes

fyi FOR YOUR INFORMATION

With your mouth wide open, blow air from your lungs onto the palm of your hand. Now repeat the same procedure with your lips puckered so that your breath is compressed within your mouth and expands upon exiting. You'll discover that as air expands, it cools. In more technical terms, as the pressure decreases, so does the temperature. In regard to the weather, as warm air rises, it expands into less dense higher-altitude air and thus cools. This cooling causes atmospheric water vapor to condense into tiny suspended droplets, which is how clouds are formed.

*A commonly used value for the gas constant is 0.082057 L atm/K mol. Regarding the units, L is volume in liters, atm is pressure in atmospheres, K is temperature in kelvin, and 1 mole (mol) equals 6.02×10^{23} particles. The concept of the mole is introduced in Chapter 7.

Figure 2.32

This gap in the roadway (called an expansion joint); allows a bridge to expand and contract. (Can you tell whether this photo was taken on a warm or cold day?)

an expansion. Similarly, for Avogadro's Law, volume increases with an increasing number of particles because each particle produces a force on the inner sides of the container, and when there are a greater number of particles, the net force on the inner surface is also greater.

The kinetic molecular theory was developed from the study of gases, but this theory also nicely sums up what we know about liquids and solids. As discussed earlier in this chapter, liquids and solids are also made of tiny particles, such as atoms or molecules. These particles are in constant random motion. This motion is confined, however, because the particles are held together by electrical attractions. We'll be discussing the nature of these electrical attractions in later chapters. In the liquid phase, the particles can tumble over one another like marbles in a bag, which gives the liquid its fluid character. In the solid phase, the particles are held together so strongly that they can only vibrate about fixed positions. Kinetic molecular theory also tells us that the higher the temperature, the faster the movement of the particles, as was discussed in Section 2.6. This explains why liquids and solids, like gases, also tend to expand when heated and contract when cooled, as illustrated in Figure 2.32.

CONCEPT **CHECK**

Assuming a constant volume and number of particles, as the temperature of a gas increases, what happens to the pressure? Why?

CHECK YOUR ANSWER

As the temperature increases in a gas, so does the average kinetic energy (speed) of its particles. Faster-moving particles collide more frequently with the inner surface of a container and strike it harder. More frequent and more forceful impacts mean greater pressure. So, as the temperature of a gas increases, so does its pressure, provided the volume and number of gas particles remain constant. Understanding this, can you explain why the pressure inside an automobile tire is greatest just after driving?

Dalton's Law of Partial Pressures

The ideal gas law assumes the distance between gas particles is vastly larger than the size of each particle. It also assumes that the gas particles have no "stickiness". In other words, the particles bounce in a perfectly elastic manner upon colliding with each other as well as the walls of the container. Experiments show that the ideal gas law works well to describe real gases at pressures of 1 atmosphere or lower and also at temperatures of about 273 K (0°C, 32°F) or higher. At higher pressures and lower temperatures, the size and stickiness of real atoms and molecules begin to show their effects.

But when a gas is behaving under ideal gas law conditions, it is interesting to note that the chemical identity of each gas particles makes no difference. This was first discovered by the English chemist John Dalton in the early 1800s. Dalton's discovery is called the law of partial pressures, which can be summarized as follows: For a mixture of gases in a container, the total pressure exerted is the sum of the partial pressures of the gases present. In equation form:

$$\text{Pressure}_{\text{total}} = \text{Pressure}_{\text{A}} + \text{Pressure}_{\text{B}} + \text{Pressure}_{\text{C}}$$

$$P_{\text{total}} = P_{\text{A}} + P_{\text{B}} + P_{\text{C}}$$

The air we breathe is a mixture of many gases, most notably nitrogen, N_2, oxygen, O_2, water vapor, H_2O, and argon, Ar. At sea level on a humid day, the percent of these gases might be 74%, 20%, 5%, and 1%, respectively.

If the atmospheric pressure on that day is 1 atmosphere (1 atm), then what might be the pressure exerted by each gaseous component? The answer is easy. Just multiply the total pressure (1 atm) by the percentages, as shown by the partial pressures in the margin.

We can summaries like this:

$$P_{\text{atm}} = P_{\text{Nitrogen}} + P_{\text{Oxygen}} + P_{\text{Water}} + P_{\text{Argon}}$$

$$1.00 \text{ atm} = 0.74 \text{ atm} + 0.20 \text{ atm} + 0.05 \text{ atm} + 0.01 \text{ atm}$$

Percent Composition

Nitrogen, N_2 = 74%
Oxygen, O_2 = 20%
Water, H_2O = 5%
Argon, Ar = 1%

Partial Pressures

Nitrogen, N_2 = 0.74 atm
Oxygen, O_2 = 0.20 atm
Water, H_2O = 0.05 atm
Argon, Ar = 0.01 atm

Calculation Corner: Manipulating Algebraic Equations

With a little algebraic manipulation, it is easy to change the equation for density around so that it can be solved for either mass or volume. Here are the results:

Density = Mass / Volume   Mass = Density × Volume   Volume = Mass / Density

A good way to remember these relationships is to use the following diagram. Use your finger to cover the quantity you want to know, and that quantity's relationship to the other quantities is revealed. For example, covering the M shows that mass is equal to density times volume, D × V.

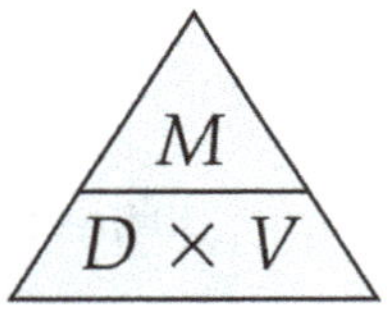

EXAMPLE 1
A pre-1982 penny has a density of 8.92 grams per milliliter and a volume of 0.392 milliliters. What is its mass?

ANSWER 1
M = D × V = (8.92 g/mL)(0.392 mL) = 3.50 g

EXAMPLE 2
A post-1982 penny has a density of 7.40 grams per milliliter and a mass of 2.90 grams. What is its volume?

ANSWER 2
V = M/D = (7.40 g/mL)/(2.90 g) = 0.392 mL

YOUR TURN

1. Show that the average density of a loaf of bread having a mass of 500 grams and a volume of 1000 milliliters is 0.5 g/mL.

2. The loaf of bread in the previous question loses all its moisture after being toasted. Its volume remains at 1000 mL but its new density is 400 g/mL. Show that its mass has been reduced to 400 grams.

3. A sack of groceries accidentally set on a 500-gram loaf of white bread increases the average density of the loaf to 5 g/mL. Show that its new volume is 100 mL.

Calculation Corner: SCUBA Diving and Hot Air Balloons

We can express Boyle's Law and Charles's Law mathematically as follows:

$$P_1V_1 = P_2V_2 \qquad \frac{V_1}{T_1} = \frac{V_2}{T_2}$$

Boyle's Law   Charles's Law

Here P_1, V_1, and T_1 represent an original pressure, volume, and temperature, respectively, while P_2, V_2, and T_2 represent a new pressure, volume, and temperature, respectively. Each of the preceding equations contains four variables. For such equations, if three of the variables are known we can calculate the fourth (assuming all temperatures are expressed in kelvins).

EXAMPLE

What would be the new volume of a 1.00 liter balloon if it were brought from sea level, where the air pressure is 1.00 atmosphere, to an altitude of 2500 meters, where air pressure is about 0.743 atmospheres? Assume there is no change in temperature. (As we explore in Chapter 16, the "atmosphere" is a common unit of pressure and is equal to the average atmospheric pressure at sea level.)

P_1 = 1.00 atm
P_2 = 0 .743 atm
V_1 = 1.00 liter
V_2 = ?

Use algebra to rearrange the equation for Boyle's Law to solve for V_2:

$V_2 = (P_1)(V_1) / (P_2)$

= (1.00 atm)(1.00 liter) / (0 .743 atm)

= 1.35 liters

YOUR TURN

1. A scuba diver swimming underwater in the ocean breathes compressed air at a pressure of 2 atmospheres. If she holds her breath while returning to the surface, show that the volume of her lungs will double.

2. A 5.00-liter rubber balloon is submerged 5.00 meters under ocean water, where its new volume is measured to be 3.38 liters. Show that the pressure at this depth is 1.48 atmospheres.

3. A perfectly elastic 419-liter balloon is heated from 25°C (298 K) to 50°C (323 K). Show that the volume expands to 454 liters.

4. A hot air balloon 401,000 liters in volume is warmed from 298 K to 398 K. As the air inside the balloon expands, it is unable to stretch the fabric, which is not very elastic. Instead, the expanded air escapes out of a hole placed at the top of the balloon. Show that 135,000 liters of air escapes.

5. Air has a density of 1.18 g/L. Show that the hot air balloon in the previous question is now lighter by 159 kilograms, which helps the balloon to rise.

Concept Review

Chapter 2

Summary of Terms

Absolute Zero The lowest possible temperature, which is the temperature at which the atoms of a substance have no kinetic energy: 0 K = -273.15°C = -459.7°F.

Alchemy A medieval endeavor concerned with turning other metals to gold.

Atoms Extremely small fundamental units of matter.

Avogadro's Law A gas law that describes the direct relationship between the volume of a gas and the number of gas particles it contains at constant pressure and temperature. The greater the number of particles, the greater the volume.

Boiling Evaporation in which bubbles form beneath the liquid surface.

Boyle's Law A gas law that describes the indirect relationship between the pressure of a gas sample and its volume at constant temperature. The smaller the volume, the greater the pressure.

Charles' Law A gas law that describes the direct relationship between the volume of a gas sample and its temperature at constant pressure. The greater the temperature, the greater the volume.

Condensation The transformation of a gas to a liquid.

Density The amount of mass contained in a sample divided by the volume of the sample.

Deposition The process of a material transforming from a gas directly to a sold without passing through the liquid phase.

Energy The capacity to do work.

Evaporation The transformation of a liquid to a gas.

Freezing The transformation of a liquid to a solid.

Gas Matter that has neither a definite volume nor a definite shape, always filling any space available to it.

Heat The energy that flows from one object to another because of a temperature difference between the two.

Ideal Gas Law A gas law that summarizes the pressure, volume, temperature, and number of particles of a gas within a single equation often expressed as $PV = nRT$, where P is presure, V is volume, n is number of molecules, R is the gas constant and T is temperature given in kelvins.

Kinetic energy Energy due to motion.

Kinetic Molecular Theory A theory that explains the properties of solids, liquids, and gases by proposing that they consist of rapidly moving tiny particles, either atoms or molecules or both.

Liquid Matter that has a definite volume but no definite shape, assuming the shape of its container.

Mass The quantitative measure of how much matter an object contains.

Melting The transformation of a solid to a liquid.

Molecule An extremely small fundamental structure built of atoms.

Potential energy Stored energy.

Solid Matter that has a definite volume and a definite shape

Sublimation The process of a material transforming from a solid directly to a gas without passing through the liquid phase.

Submicroscopic The realm of atoms and molecules, where objects are smaller than can be detected by optical microscopes.

Temperature How warm or cold an object is relative to some standard. Also, a measure of the average kinetic energy per molecule of a substance, measured in degrees Celsius, degrees Fahrenheit, or kelvins.

Thermometer An instrument used to measure temperature.

Volume The amount of space an object occupies.

Weight The gravitational force of attraction between two bodies (where one body is usually the Earth).

Review Questions

2.1 The Submicroscopic World Is Super-Small

1. It would take you 31,800 years to count to a trillion. About how many times would you have to do this to have counted all atoms there are in a single grain of sand?

2. Is a biological cell macroscopic, microscopic, or submicroscopic

2.2 Discovering the Atom

3. The term "atom" was derived from what Greek phrase?

4. What 18th century chemist discovered the law of mass conservation?

5. What did Mendeleev predict based upon his newly created periodic table?

2.3 Mass Is How Much and Volume Is How Spacious

6. What is inertia, and how is it related to mass?

7. Which can change from one location to another: mass or weight?

8. What is the difference between an object's mass and its volume?

2.4 Density Is the Ratio of Mass to Volume

9. The units of density are a ratio of what two quantities?

10. What happens to the volume of a loaf of bread that is squeezed? The mass? The density?

2.5 Energy Is the Mover of Matter

11. What do we call the energy an object has because of its position?

12. What do we call the energy an object has because of its motion?

13. Which represents more energy: a joule or a calorie?

2.6 Temperature and Heat

14. In which is the average speed of the molecules less: in cold coffee or in hot coffee?

15. Which temperature scale has its zero point as the point of zero atomic and molecular motion?

16. Is it natural for heat to travel from a cold object to a warmer object?

2.7 The Phase of a Material Depends on the Motion of Its Particles

17. How does the arrangement of particles in a gas differ from the arrangements in liquids and solids?

18. Which requires the removal of thermal energy: melting or freezing?

19. What is it called when evaporation takes place beneath the surface of a liquid?

2.8 Gas Laws Describe the Behavior of Gases

20. What happens to the pressure inside a tire as more air molecules are pumped into the tire?

21. What happens to the volume of a gas as its temperature is increased? (Assume constant pressure and number of particles.)

22. At what temperature do gases theoretically cease to occupy any volume?

23. What happens to the volume of a gas as more gas particles are added to it? (Assume constant pressure and temperature.)

24. Why do real gases not obey the ideal gas law perfectly?

Quantitative Questions

25. What is the mass in kilograms of a 130-pound human standing on Earth?

26. Gravity on the Moon is only one-sixth as strong as gravity on the Earth. What is the mass of a 10-kilogram object on the moon, and what is its mass on the Earth?

27. Some one wants to sell you a piece of gold and says it is nearly pure. Before buying the piece, you measure its mass to be 52.3 grams and find that it displaces 4.16 mL of water. Calculate its density and consult Table 2.1 to assess its purity.

28. What volume of water would a 52.3 gram sample of pure gold displace?

29. How many joules are there in a candy bar containing 230,000 calories?

30. How many milliliters of dirt are there in a hole that has a volume of 5 liters? How many milliliters of air?

31. You measure the pressure of the four tires of your car each to be 35.0 pounds per square inch (psi). You then roll your car forward so that each tire is upon a sheet of paper. You outline the surface area of contact between each tire and the paper, which you later measure to be 32.0 square inches. What is the weight of your car?

32. A perfectly elastic balloon holding 1.0 liters of helium at 298K is warmed to 348K. What is the new volume of the helium-filled balloon?

33. A SCUBA tank is filled with 79.7% oxygen and 20.3% helium to a total pressure of 11.3 atm.What are the partial pressures of the oxygen and the helium?

Solutions (Odd-Numbered

1. It would take you 31,800 years to count to a trillion. Do this 125 million times and you would have counted to about the number of atoms there are in a single grain of sand.

3. The term atom was derived from the Greek phrase a tomos, which means "not cut" or "that which is indivisible."

5. Mendeleev predicted the existence of elements that had not yet been discovered.

7. Weight can change from one location to the next because it is dependent on gravity.

9. Density is the ratio between the mass of a substance and its volume. Note, that as the mass of the substance increases, so does its volume. The ratio of the mass to volume, which is its density, remains the same.

11. The energy due to position is potential energy.aa

13. A calorie is 4.184 times greater than a joule.

15. The Kelvin scale places zero at the point of zero atomic and molecular motion.

17. The particles in a gas have so much energy that they overcome their attractions to each other and expand to fill all of the space available. In a liquid the particles tumble loosely around one another. In a solid, the particles are fixed in a three-dimensional arrangement.

19. When evaporation occurs beneath the surface of a liquid it is called boiling.

21. The volume increases as the temperature increases.

23. The volume of a gas increases as more particles are added to it.

25. Multiply by the conversion factor to arrive at the answer:

(130 lb) × (1 kg/2.205 lbs) = 59 kg

27. Divide the mass by volume to arrive at the density:

density = mass/volume =
52.3 grams/4.16 mL = 12.6 g/mL

From Table 2.1, we see that this is substantially less than the accepted density of pure gold, which is 19.3 g/mL. This evidence indicates that the piece they were trying to sell you was far from pure.

29. Multiply by the conversion factor to arrive at the answer:

230,000 calories × 4.184 joule/1 calorie =
960,000 joules

31. Multiply the pressure of each tire by its surface area of contact to find the weight placed downward on each tire. Since there are four tires, multiply by four to arrive at the total weight of your car.

Weight upon each tire

(35.0 lbs/sq inch)(32.0 sq inches) = 1120 lbs

Weight upon all four tires (weight of car)

1120 pounds × 4 = 4480 pounds

33. P_{Oxygen} = (11.3 atm)(0.797) = 9.01 atm

P_{Helium} = (11.3 atm)(0.203) = 2.29 atm

Solutions to Chapter 2 Calculation Corners

Manipulating Algebraic Equations

1. 0.5 grams per milliliter
2. 400 grams
3. 100 milliliters

SCUBA Diving and Hot Air Balloons

1. The new volume of the scuba diver will be two times that of the original volume:

$$\begin{aligned} V_2 &= P_1 V_1/P_2 \\ &= (2.00\ \text{atm})(V_1)/(1\ \text{atm}) \\ &= 2.00\ V_1 \end{aligned}$$

2. Plug the following values into Boyles law and solve to the new pressure, P_2:

$$\begin{aligned} P_1V_1 &= P_2V_2 \\ (1\ \text{atm})(5.00\ \text{L}) &= (P_2)(3.38\ \text{L}) \\ P_2 &= (1.00\ \text{atm})(5.00\ \text{L})/3.38\ \text{L}) \\ P_2 &= 1.48\ \text{atm} \end{aligned}$$

3. Plug the following values in Charles's Law and solve for the new volume, V_2. Be sure to use the absolute temperatures given in kelvin:

$$\frac{V1}{T1} = \frac{V2}{T2}$$

$$\begin{aligned} (419\ \text{L})/(298\ \text{K}) &= V_2/(323\ \text{K}) \\ V_2 &= (419\ \text{L})(323\ \text{K})/(298\ \text{K}) \\ V_2 &= 454\ \text{L} \end{aligned}$$

4. Use Charles's Law to show that the new volume at the higher temperature would be 536,000 liters. The answer needs to have three significant figures, which is why 535,564 liters rounds up to 536,000 liters. Take the difference to find the volume of gas that escapes from the balloon.

$$\frac{V1}{T1} = \frac{V2}{T2}$$

$$\begin{aligned} (401{,}000\ \text{L})/(298\ \text{K}) &= V_2/(398\ \text{K}) \\ V_2 &= (401{,}000\ \text{L})(398\ \text{K})/(298\ \text{K}) \\ V_2 &= 536{,}000\ \text{L} \end{aligned}$$

Volume of air that escapes = $V_2 - V_1 = 135{,}000$ liters

5. The density of air is 1.18 grams per liter. In other words, 1.18 grams equals 1 liter. As per the Calculation Corner of Chapter 1, you should be able to recognize this as a conversion factor. Put this together with the conversion factor of 1000 grams equals 1 kilogram and we can convert from liters to grams to kilograms in a single equation:

(135,000 L of air)(1.18 grams of air/L of air)(1 kilogram/1000 grams) = 159 kg

Chapter 2: Contextual Chemistry Essay

Green Chemistry

Over the past couple decades there has been a growing effort among industries, governmental agencies, and universities to develop technologies that allow materials to be manufactured with reduced or no negative environmental impacts. This is an area of research known as *green chemistry*. To ensure implementation, green chemistry technologies are also designed to be cost competitive and profitable.

Green chemistry applied to the manufacture of materials follows a set of high standards, as outlined here:

1. The raw materials used in the manufacturing process should be sustainably renewable (such as from agriculture) rather than depleting (such as from fossil fuels).

2. Waste products from the manufacturing process should be minimized. Any waste products should be either recycled or rendered environmentally safe.

3. The manufacturing process should be energy efficient. For example, chemical reactions that run at easy-to-attain ambient temperatures and pressures are most desirable. Also, the fewer steps there are to a manufacturing process, the better. Energy can also be saved by using catalysts and by monitoring the manufacturing process carefully to ensure optimal operating conditions.

4. The manufacturing process should minimize hazards. This can be accomplished by choosing chemical reactions that involve fewer toxic chemicals and are not explosive.

5. The desired final products of the manufacturing process should have little or no toxicity and be recyclable. Furthermore, the materials of these products should degrade to innocuous substances after use and not accumulate in the environment.

To illustrate the power and potential of green chemistry, it is useful to look into a few of the growing number of green chemistry projects. To learn about many other projects, be sure to visit either www.epa.gov/greenchemistry or www.acs.org/greenchemistry.

Renewable Toner

Renewable Toner

Laser printers and copiers consume about 400 million pounds of toner each year in the United States. Most of this toner is made from nonrenewable petroleum. This toner is also not easily de-inked from the paper, which means expensive and harsh chemicals are needed for recycling. As a green chemistry alternative, the Battelle Institute (www.Battelle.org) oversaw the development of a high-quality toner made from soybeans. Much less energy is required to create this soy-based toner, which is made from a renewable resource. Furthermore, soy-based toner is easily de-inked from paper, thereby simplifying the recycling process.

Benign Pesticides

As described in Chapter 15, we rely on pesticides to protect our crops. Many modern pesticides, however, are harmful to a broad range of species, including beneficial insects, such as those that pollinate. A number of these pesticides also persist in the environment, where they are taken up into the food chain. To counter these disadvantages, scientists at Dow AgroSciences (www.corteva.com) have been able to produce pesticides that are specific to the nervous system of certain detrimental insects. As a result, toxicity to other species, including mammals, is many times lower than that of traditional pesticides.

Furthermore, the starting materials for the synthesis of these pesticides are created from fermentation broths. In other words, the starting materials are grown much like a yogurt culture. Also, due in part to their natural origin, these pesticides decompose well, so that they do not persist in the environment.

Affordable Hydrogen Peroxide

Almost everyone is familiar with disinfecting solutions of hydrogen peroxide, H_2O_2, available from the local drugstore. What most people don't know is that the manufacture of this simple molecule on a large scale is costly, dangerous, and involves numerous steps. This is unfortunate, because H_2O_2 is an effective and environmentally friendly oxidizing agent that would be used extensively by industries if not for its high cost (see Chapter 11). Instead, companies turn to other oxidizing agents, such as chlorine. Although cheaper, chlorine tends to produce highly toxic by-products, such as dioxins. Using nanotechnology, chemists at Headwaters Technology (www.ecomaterial.com) have designed catalysts that allow the efficient and safe production of H_2O_2 in a single step directly from hydrogen, H_2, and oxygen, O_2 (see Section 3.8). Cheaper and readily available H_2O_2 is good news for the profit margins of the chemical industry as well as for the environment, where H_2O_2 simply decomposes to water (H_2O) and oxygen.

Versatile Ionic Liquids

Salts, such as sodium chloride, NaCl, are commonly solids at room temperature. These salts can be transformed into a liquid solution by adding water. There are some salts, however, that are already liquids at room temperature—no water needed! These ionic liquids have attracted much attention over the past couple of decades for their many unusual properties and potential applications, especially when it comes to green chemistry.

Many ionic liquids are organic compounds, which means it is easy to modify their properties by tweaking their chemical structures (see Chapter 12). An ionic liquid made of the compound shown below, for example, has been found to have great potency against the polymer coats, called biofilms, that colonies of bacteria build around themselves for protection. Biofilms are a major problem in hospitals. They also foul the hulls of ships and pipes in industrial machinery. In these situations, biofilms are treated with harsh chemicals. The ionic liquids offer a potentially more effective and environmentally friendly alternative

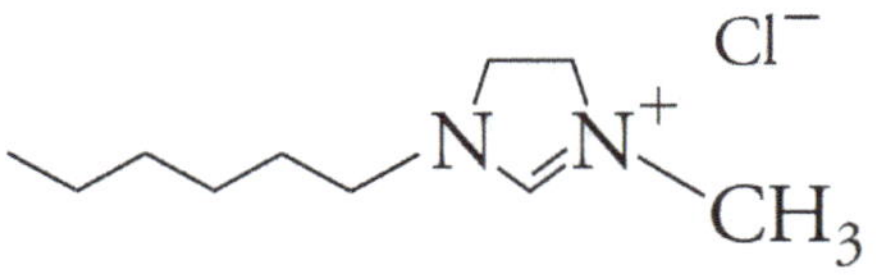

1-hexyl-3-methylimidazolium chloride

Very exciting is the fact that certain ionic liquids are able to dissolve cellulose, which is, by far, the most abundant and renewable material and energy resource on this planet. As discussed in Chapter 13, cellulose is the structural biopolymer of plants. Strands of cellulose bind very tightly to each other, which is good for the wood we use to build our houses or the cotton in our clothes. When the cellulose is dissolved in ionic liquids, however, this tight binding relaxes, allowing the separation of individual strands. Separated strands of cellulose can then be easily broken down into sugars that can be fermented into ethanol. (Ethanol from such sources is known as cellulosic ethanol.) They can also serve as a starting material for nonpetroleum-based polymers. Both these applications would help move us away from our dependence upon nonrenewable and greenhouse gas-emitting petroleum.

Switchgrass is an efficient source of cellulose in that it grows year-round in many environments and never needs replanting.

CONCEPT **CHECK**

How can we be sure that green chemistry technology will get implemented by industries?

CHECK YOUR ANSWER

Green chemistry technology is good for the pocketbooks of the industries that implement this technology as well as for the environment.

Think and Discuss

1. Specify which of the principles of green chemistry are illustrated by the example of affordable hydrogen peroxide.

2. To what extent should green chemistry technologies be exported to developing nations? Why?

3. As a consumer, what can you do to support green chemistry?

4. There is now much research being dedicated to the transformation of ever-abundant and renewable cellulose into biofuels and bioplastics. Why wasn't this research started in earnest during the 1960s, which was a decade of much environmental awakening?

5. About 40 percent of all our solid waste sent to landfills consists of cellulose-containing products, such as paper, cardboard, and packaging. Might our landfills one day be considered a valuable source of cellulosic ethanol?

Author Responses to Think and Discuss

1. Using hydrogen peroxide over chlorine minimizes the formation of toxic wastes, such as dioxins. Hydrogen peroxide itself decomposes to harmless water. Furthermore, this new method of producing hydrogen peroxide is less costly, less dangerous, and energy efficient requiring fewer steps than traditional means.

2. Green chemistry technologies should be exported to developing nations to the fullest extent, especially in light of the fact that the developing nations have the greatest rates of population growth. Petroleum is energy rich and has fueled the great advancements within developed nations. One can easily imagine developing nations demanding the rights to use petroleum for the same sorts of advancements. Their being denied such rights, say, through environmental regulations, would likely lead to much political tension. In this light, green chemistry technologies are more than a nice development—they are essential to the future well-being of human civilization.

3. Consumers owe it to themselves and society at large to become knowledgeable about the environmental impacts of the products they buy. Now that you are familiar with soy-based toners, perhaps in the future you will be more likely to choose it over petroleum-based toner if you are given the option. Your taking this chemistry course is a step in a positive direction.

4. Hindsight is 20/20. It's easy to look back a half-century and offer advice as to what we should have done. During the 1960s, however, most people viewed fossil fuels as ever-abundant—there was plenty petroleum available to last at least for their own lifetimes. A more immediate issue they faced was the pollution the burning of these fossil fuels was causing. This led to important legislation, such as the Clean Air Act of 1970.

5. We are unlikely to start digging up buried cellulose when it is ever abundant at the surface. An important implication to this question, however, is that future cellulose-based waste, rather than going to landfills, might find its way to cellulosic ethanol processing plants.

Chapter 3

Elements of Chemistry

THE MAIN IDEA

Elements combine to form compounds, which blend together to form mixtures

Λ Gold and diamond are elements in that they each consist of only one type of atom.

Chapter 1 introduced 12 key terms, Chapter 2 provided 30, and in this chapter you'll find another 26. Why all these new terms? In the laboratory, chemists perform experiments, make many observations, and then draw conclusions. Over time, the result is a growing body of new knowledge that inevitably exceeds the capacity of everyday language. New terms are needed as we attempt to describe the nature of matter beyond its casual appearance.

Instead of just memorizing the formal definitions of terms, you will serve yourself better by making sure that you understand the underlying concepts. Practice articulating and paraphrasing these concepts aloud to yourself or to a friend without looking at the pages of this book. When you are able to express these concepts in your own words, you will have insight that goes beyond memorization allowing you to do well in this course and beyond.

3.1 Matter Has Physical and Chemical Properties

Properties that describe the look or feel of a substance, such as color, hardness, density, texture, and phase, are called **physical properties**. Every substance has its own set of characteristic physical properties that we can use to identify that substance (**Figure 3.1**).

The physical properties of a substance can change when conditions change, but that does not mean a different substance is created. Cooling liquid water to below 0°C causes the water to change to

< **Figure 3.1**

Gold, diamond, and water can be identified by their physical properties.

∧ Figure 3.2

Two physical changes. (a) Liquid water and ice might appear to be different substances, but at the submicroscopic level, both consist of water molecules. (b) At 25°C, the atoms in a sample of mercury are a certain distance apart, yielding a density of 13.5 g per mL. At warmer tempeatures, the atoms are farther apart, meaning the mercury becomes less dense as it expands. Old style thermometers calibrate this expansion allowing for a measurement of temperature. The physical property we call density has changed with temperature, but the identity of the substance remains unchanged: mercury is mercury.

ice, but the substance is still water, no matter what the phase. The only difference is the orientation of the H_2O molecules and how rapidly they are moving. In the liquid phase, the water molecules tumble around one another, whereas in the ice phase, they vibrate about fixed positions. Water freezing is an example of what chemists call a **physical change**. During a physical change, a substance changes its phase or some other physical property but not its chemical identity, as shown in **Figure 3.2.**

Chemical properties are those that characterize the ability of a substance to react with other substances or to transform from one substance to another. **Figure 3.3** shows three examples. The methane of natural gas reacts with oxygen to produce carbon dioxide and water, along with appreciable heat energy. Baking soda reacts with vinegar to produce carbon dioxide and water while absorbing a small amount of heat energy. Copper reacts with carbon dioxide and water to form a greenish-blue solid known as patina. Copper statues exposed to the carbon dioxide and water in the air become coated with patina. The patina is not copper, it is not carbon dioxide, and it is not water. It is a new substance formed by the reaction of these chemicals with one another.

READING CHECK

What happens to the atoms within a molecule undergoing a chemical reaction?

During a chemical change, there is a change in the way the atoms are *chemically bonded* to one another. A **chemical bond** is the force of attraction between two atoms that holds them together in a compound. A methane molecule, for example, is made of a single carbon atom bonded to four hydrogen atoms, and an oxygen molecule is made of two oxygen

Methane

Reacts with oxygen to form carbon dioxide and water, giving off lots of heat during the reaction.

Baking soda

Reacts with vinegar to form carbon dioxide and water, absorbing heat during the reaction.

Copper

Reacts with carbon dioxide and water to form the greenish-blue substance called patina.

∧ Figure 3.3

The chemical properties of substances determine the ways in which they can change into new substances. Natural gas and baking soda, for example, can both undergo chemical reactions in which they are transformed into carbon dioxide and water. Similarly, copper can be transformed into patina.

atoms bonded to each other. **Figure 3.4** shows the chemical change in which the atoms in a methane molecule and those in two oxygen molecules first pull apart and then form new bonds with different partners, resulting in the formation of molecules of carbon dioxide and water.

Any change in a substance that involves a rearrangement of the way atoms are bonded is called a **chemical change**. Thus, the transformation of methane and oxygen to carbon dioxide and water is a chemical change, as are the other two transformations shown in Figure 3.3.

The chemical change shown in **Figure 3.5** occurs when an electric current passes through water. The energy of the current is sufficient to pull bonded atoms away from each other. Loose atoms then form new bonds with different atoms, which results in the formation of new molecules. Thus, water molecules are changed to molecules of hydrogen and oxygen, two substances that are very different from water. The hydrogen and oxygen are both gases at room temperature, and they can be seen as bubbles rising to the surface.

In the language of chemistry, materials undergoing a chemical change are said to be *reacting*. Methane reacts with oxygen to form carbon dioxide and water. Water reacts when exposed to electricity to form hydrogen gas and oxygen gas. Thus, the

CHEMICAL CONNECTIONS

Chemically speaking, how is your fingernail connected to the air?

CONCEPT **CHECK**

The melting of gold is a physical change. Why?

CHECK YOUR ANSWER

During a physical change, a substance changes only one or more of its physical properties; its chemical identity does not change. Because melted gold is still gold but in a different form, its melting is only a physical change.

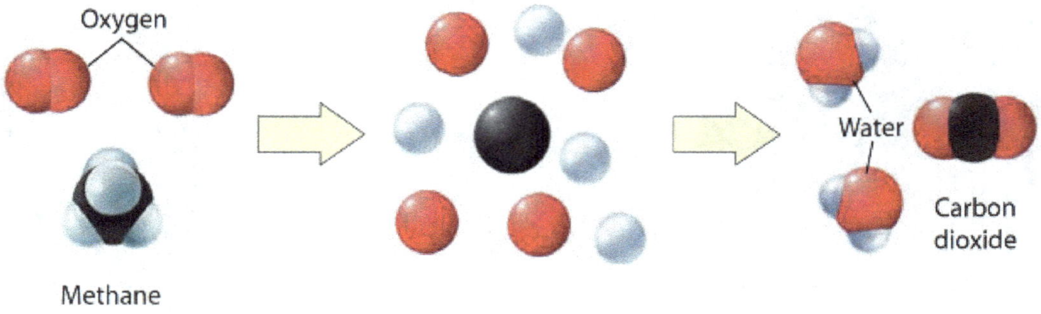

∧ **Figure 3.4**

The chemical change in which molecules of methane and oxygen transform to molecules of carbon dioxide and water, as atoms break old bonds and form new ones. In this sort of illustration, note that each sphere represents an atom, while a set of joined spheres represents a molecule.

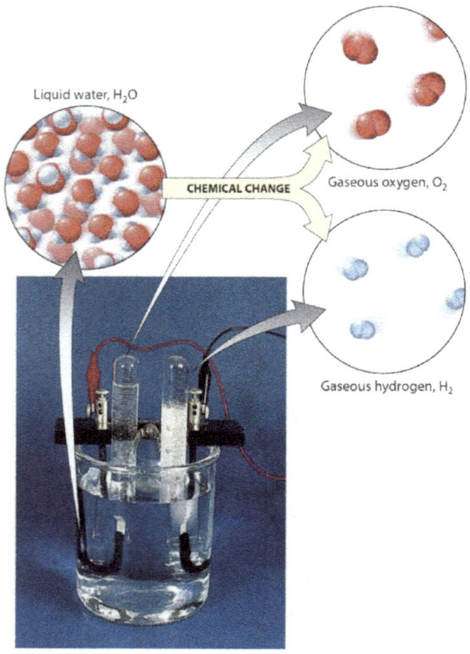

∧ **Figure 3.5**

Water can be transformed to hydrogen gas and oxygen gas by applying the energy of an electric current. This is a chemical change, because new materials (the two gases) are formed as the atoms originally found in the water molecules are rearranged. From the photograph, note that the volume of hydrogen generated is twice that of oxygen. This tells us that each water molecule, H_2O, is made of twice as many hydrogen atoms as oxygen.

term *chemical change* means the same thing as *chemical reaction*. During a **chemical reaction**, new materials are formed by a change in the way atoms are bonded together. Again, for emphasis, *during a chemical reaction, new materials are formed by a change in the way atoms are bonded together*. We will explore chemical bonds and their role in chemical reactions when we get to Chapters 6 and 9.

Distinguishing Physical and Chemical Changes

After a physical change, the molecules are the same as the ones you started with. After a chemical change, the original molecules no longer exist and new ones are in their place. In both cases there is a change in physical appearance. Frozen water and melted water, for example, look very different. Likewise, iron and rust look very different (**Figure 3.6**). So how can you quickly determine whether an observed change is physical or chemical? After all, we can't see the individual molecules.

Two powerful guidelines can help you distinguish between physical and chemical changes. First, in a physical change, a change in appearance is the result of a new set of conditions imposed on the same material. Restoring the original conditions restores the original appearance: frozen water melts upon warming. Second, in a chemical change, a change in appearance is the result of the formation of a new material that has its own unique set of physical properties. The more evidence you have suggesting that a different material has been formed, the greater the likelihood that the change is a chemical change. Iron is a moldable metal that can be used to build cars. Rust is a reddish solid that readily falls apart. This suggests that the rusting of iron is a chemical change.

Figure 3.7 shows potassium chromate, a material whose color depends on its temperature. At room temperature, potassium chromate is a bright canary yellow. At higher temperatures (above 660°C), it is a deep reddish orange. Upon cooling, the canary color returns, suggesting that the change is physical. With a chemical change, reverting to the original conditions does not restore the original appearance. Ammonium dichromate, shown in **Figure 3.8**, is an orange material that, when heated, explodes into ammonia, water vapor, and green chromium (III) oxide. When the test tube is returned to the original temperature, there is no trace of orange ammonium dichromate. In its place are new substances having completely different physical properties.

∧ **Figure 3.6**

The transformation of water to ice and the transformation of iron to rust both involve changes in physical appearance. The formation of ice is a physical change, whereas the formation of rust is a chemical change. But how do we know?

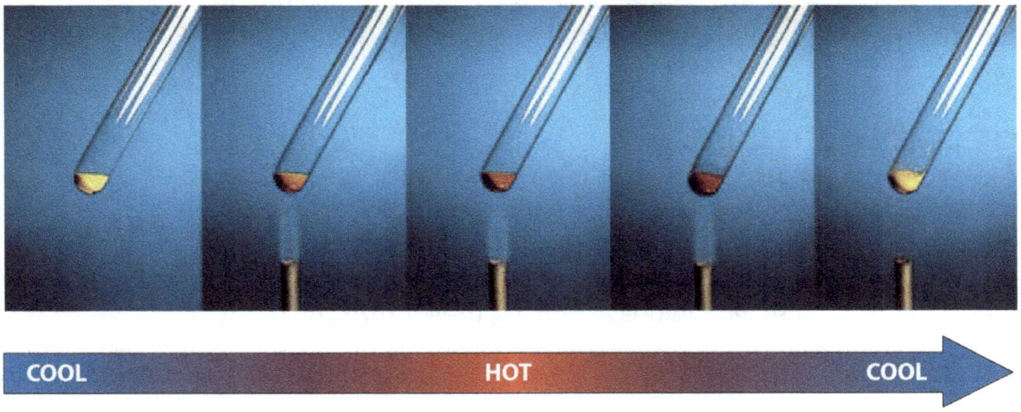

Figure 3.7 >

When heated, potassium chromate changes from yellow to orange. After cooling, the original yellow color is restored.

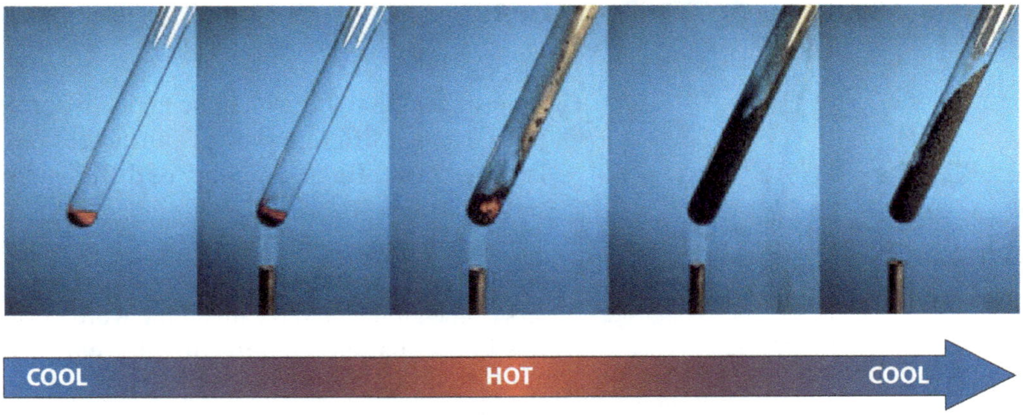

Figure 3.8 >

When heated, orange ammonium dichromate undergoes a chemical change to ammonia, water vapor, and chromium (III) oxide. A return to the original temperature does not restore the orange color, because the ammonium dichromate is no longer there.

CONCEPT CHECK

Each sphere in the following diagrams represents an atom. Joined spheres represent molecules. One set of diagrams shows a physical change, and the other shows a chemical change. Which is which?

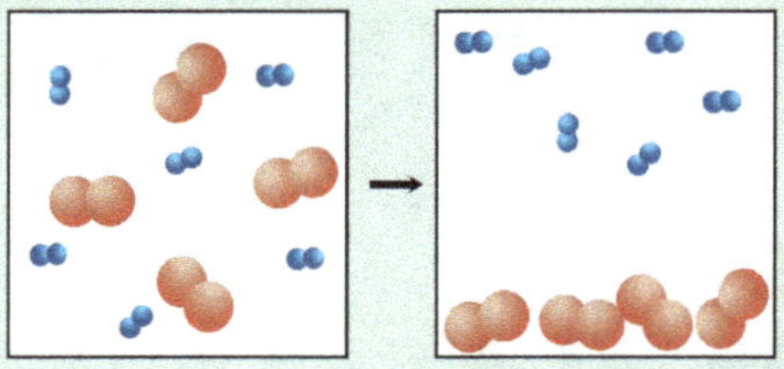

A

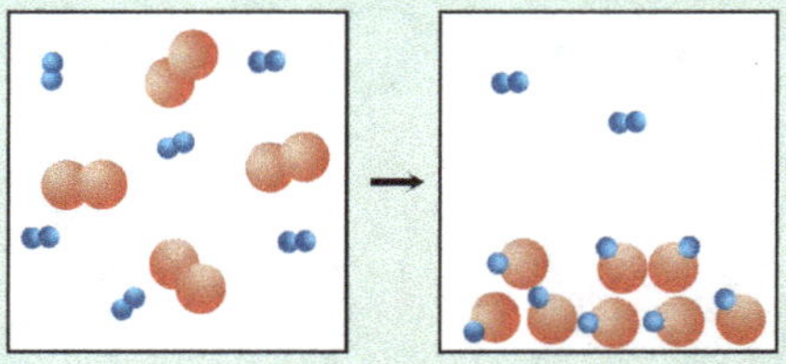

B

CHECK YOUR ANSWER

In set A, the molecules before and after the change are the same. They differ only in their positions. Set A, therefore, represents only a physical change. In set B, new molecules, consisting of bonded red and blue spheres, appear after the change. These molecules represent a new material, so set B represents a chemical change.

CONCEPT CHECK

Evan has grown an inch in height over the past year. Is this change best described as a physical or a chemical change?

CHECK YOUR ANSWER

Are new materials being formed as Evan grows? Absolutely—created out of the food he eats. His body is very different from, say, the spaghetti he ate yesterday. Yet, through some very advanced chemistry, his body is able to absorb the atoms of that spaghetti and rearrange them into new materials. Biological growth, therefore, is the result of chemical changes.

3.2 Elements Are Made of Atoms

READING CHECK

How are there so many types of materials yet so few elements of the periodic table?

It may seem that there must be many different kinds of atoms to account for the many different type of substances—from wood to steel to chocolate ice cream. But the number of different kinds of atoms is surprisingly small. The great variety of substances results from the many ways a few kinds of atoms can be combined. Just as the three colors red, green, and blue can be combined to form any color on a computer screen or the 26 letters of the alphabet make up all the words in a dictionary, only a relatively few kinds of atoms combine in different ways to produce all substances. To date, we know of 118 different kinds of atoms. Of these, about 90 are found in nature. The remaining atoms have been created in the laboratory.

Any material consisting of only one type of atom is classified as an **element**. A few examples are shown in Figure 3.9. Pure gold, for example, is an element—it contains only gold atoms. Similarly, one of the gases in air is nitrogen, an element. Nitrogen gas is an element because it contains only nitrogen atoms. Likewise, the graphite in your pencil is an element—carbon. Graphite is made up solely of carbon atoms. All of the elements are organized in a chart called the **periodic table**, shown in **Figure 3.10.**

As you can see from the periodic table, each element is designated by its **atomic symbol**, which comes from the letters of a name associated with that element. For example, the atomic symbol for carbon is C, and that for chlorine is Cl. In many cases, the atomic symbol is derived from the element's Latin name. Gold has the atomic symbol Au after its Latin name, *aurum*. Lead has the atomic symbol Pb after its Latin name, *plumbum* (**Figure 3.11**). Elements having symbols derived from Latin names are usually those discovered earliest. Note that only the first letter of an atomic symbol is capitalized. The symbol for the element cobalt, for instance, is Co, while CO is a combination of two elements: carbon, C, and oxygen, O.

The terms *element* and *atom* are often used in a similar context. You might hear, for example, that gold is an element made of gold atoms. Generally, *element* is used in reference to an entire macroscopic or microscopic sample, and *atom* is used when speaking of the submicroscopic particles in the sample. The important distinction is that elements are made of atoms and not the other way around.

The number of atoms that arrange themselves in a unit of an element is shown by the **elemental formula**. For elements in which two or more atoms are bonded into molecules, the elemental formula is the chemical symbol followed by a subscript indicating the number of atoms in each molecule. For example, elemental nitrogen, as shown in Figure 3.9, consists of molecules containing two nitrogen atoms per molecule. Thus, N_2 is the elemental formula given for atmospheric nitrogen. Similarly, atmospheric oxygen has the elemental formula O_2, while the elemental formula for sulfur is S_8. For elements in which the basic units are individual atoms (not molecules), the elemental formula is simply the chemical symbol. This is the case for most elements. To name two examples, Au is the elemental formula for gold, and Li is the elemental formula for lithium.

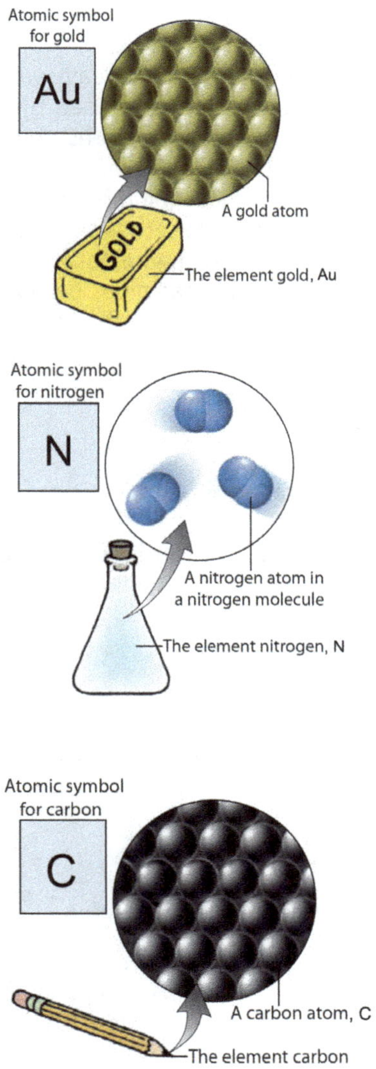

∧ Figure 3.9

Any element consists of only one kind of atom. Gold consists of only gold atoms, a flask of gaseous nitrogen consists of only nitrogen atoms, and the carbon of a graphite pencil consists of only carbon atoms.

1 H																	2 He
3 Li	4 Be											5 B	6 C	7 N	8 O	9 F	10 Ne
11 Na	12 Mg											13 Al	14 Si	15 P	16 S	17 Cl	18 Ar
19 K	20 Ca	21 Sc	22 Ti	23 V	24 Cr	25 Mn	26 Fe	27 Co	28 Ni	29 Cu	30 Zn	31 Ga	32 Ge	33 As	34 Se	35 Br	36 Kr
37 Rb	38 Sr	39 Y	40 Zr	41 Nb	42 Mo	43 Tc	44 Ru	45 Rh	46 Pd	47 Ag	48 Cd	49 In	50 Sn	51 Sb	52 Te	53 I	54 Xe
55 Cs	56 Ba	57 La	72 Hf	73 Ta	74 W	75 Re	76 Os	77 Ir	78 Pt	79 Au	80 Hg	81 Tl	82 Pb	83 Bi	84 Po	85 At	86 Rn
87 Fr	88 Ra	89 Ac	104 Rf	105 Db	106 Sg	107 Bh	108 Hs	109 Mt	110 Ds	111 Rg	112 Cn	113 Nh	114 Fl	115 Mc	116 Lv	117 Ts	118 Og
				58 Ce	59 Pr	60 Nd	61 Pm	62 Sm	63 Eu	64 Gd	65 Tb	66 Dy	67 Ho	68 Er	69 Tm	70 Yb	71 Lu
				90 Th	91 Pa	92 U	93 Np	94 Pu	95 Am	96 Cm	97 Bk	98 Cf	99 Es	100 Fm	101 Md	102 No	103 Lr

∧ Figure 3.10

The periodic table lists the known elements. Interestingly, only about 90 occur in nature.

FOR YOUR INFORMATION

Carbon is the only element that can form bonds with itself indefinitely. Sulfur's practical limit is S_8, and nitrogen's limit is around N_{12}. The elemental formula for a 1-carat diamond, however, is about $C_{10,000,000,000,000,000,000,000}$.

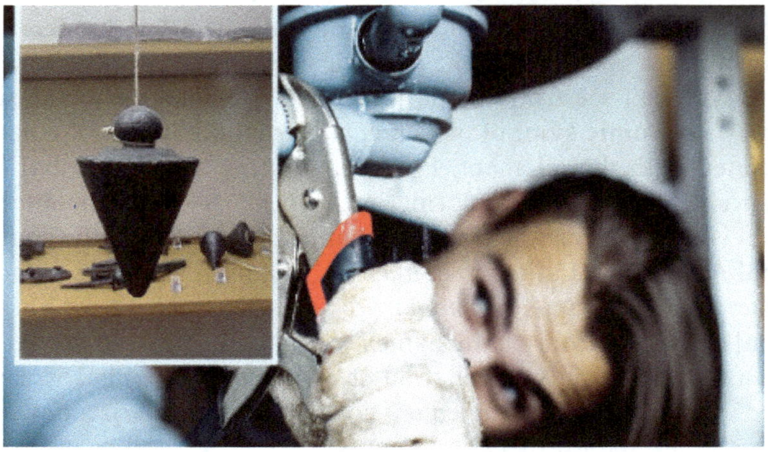

Figure 3.11 >

A plumb bob (inset) is a heavy weight attached to a string and used by carpenters and surveyors to establish a straight vertical line. The name "plumb" comes from the lead (plumbum, Pb) that is still sometimes used as the weight. Plumbers got their name because they once worked with lead pipes. Because of lead's toxicity, copper or PVC pipes are now used.

3.3 The Periodic Table Helps Us to Understand the Elements

READING CHECK
Name two physical properties of metals.

The periodic table is so much more than a mere listing of known elements. Most notably, the elements are organized in the table based upon their physical and chemical properties. One of the most apparent examples is the way the elements are grouped as metals, nonmetals, and metalloids.

As shown in **Figure 3.12**, most of the known elements are **metals**, which are characterized as being shiny, opaque, and good conductors of electricity and heat. Metals are malleable, which means they can be hammered into different shapes or bent without breaking. They are also ductile, which means they can be drawn into wires. All but a few metals are solid at ambient temperature. The exceptions include mercury, Hg; gallium, Ga; cesium, Cs; and francium, Fr, which are all liquids at 30°C (86°F). Another interesting exception is hydrogen, H, which takes on the properties of a liquid metal only at very high pressures (**Figure 3.13**). Under normal conditions, hydrogen behaves as a nonmetallic gas. The nonmetallic elements, with the exception of hydrogen, are on the right-hand side of the periodic table. Most **nonmetals** are poor conductors of electricity and heat. Solid nonmetals tend to be neither malleable nor ductile. Rather, they are brittle and shatter when hammered. At 30°C (86°F), some nonmetals are solid (carbon, C), others are liquid (bromine, Br), and still others are gaseous (helium, He).

Six elements are classified as **metalloids**: boron, B; silicon, Si; germanium, Ge; arsenic, As; antimony, Sb; and tellurium, Te. Situated between the metals and the nonmetals in the periodic table, the metalloids

CONCEPT **CHECK**

The oxygen we breathe, O_2, is converted to ozone, O_3, in the presence of an electric spark. Is this a physical or chemical change?

CHECK YOUR ANSWER

When atoms regroup, the result is an entirely new substance, and that is what happens here. The oxygen we breathe, O_2, is odorless and life-giving. Ozone, O_3, can be toxic and has a pungent smell commonly associated with electric motors. The conversion of O_2 to O_3 is therefore a chemical change. However, both O_2 and O_3 are elemental forms of oxygen.

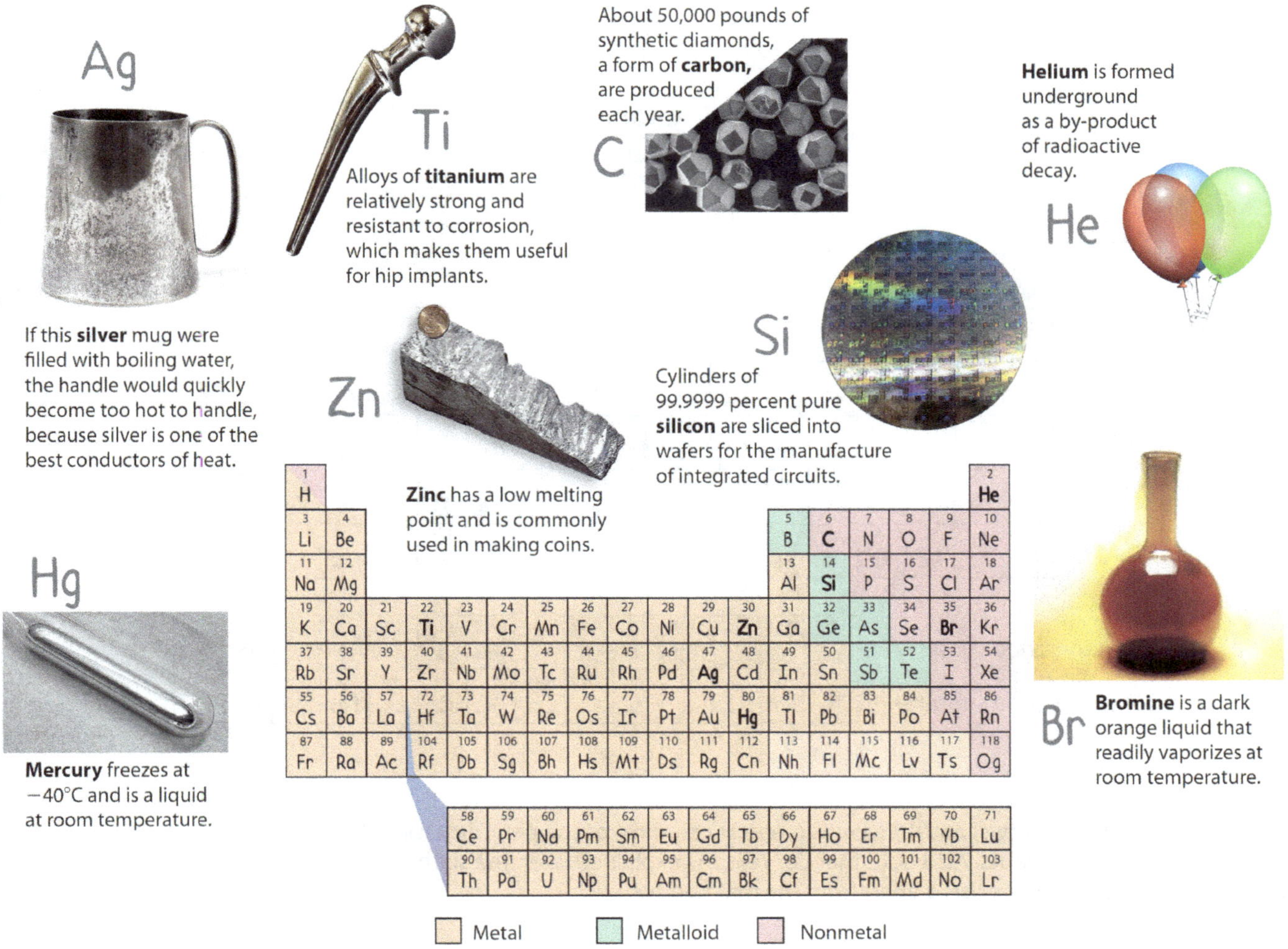

Λ Figure 13.12

The periodic table color-coded to show metals, nonmetals, and metalloids.

have both metallic and nonmetallic characteristics. For example, these elements are weak conductors of electricity, which makes them useful as semiconductors in the integrated circuits of computers. Note from the periodic table how germanium (number 32) is closer to more metals than nonmetals. Because of this positioning, we can deduce that germanium has more metallic properties than silicon (number 14) and is a slightly better conductor of electricity. So, we find that integrated circuits fabricated with germanium operate faster than those fabricated with silicon. Because silicon is much more abundant and less expensive to obtain, however, silicon computer chips remain the industry standard.

Λ Figure 13.13

Hydrogen exists as a liquid metal deep beneath the surfaces of Jupiter (shown here) and Saturn, where pressures are exceedingly high.

Periods and Groups

Two other important ways in which the elements are organized in the periodic table are by horizontal rows and vertical columns. Each horizontal row is called a **period**, and each vertical column is called a **group** (or a *family*). As shown in **Figure 3.14**, there are 7 periods and 18 groups.

Across any period, the properties of elements gradually change. This gradual change is called a **periodic trend**. For example, as is shown in **Figure 3.15**, the size of atoms generally decreases from left to right across any period. Note that the trend repeats from one horizontal row to the next. This phenomenon of repeating trends is called *periodicity*, a

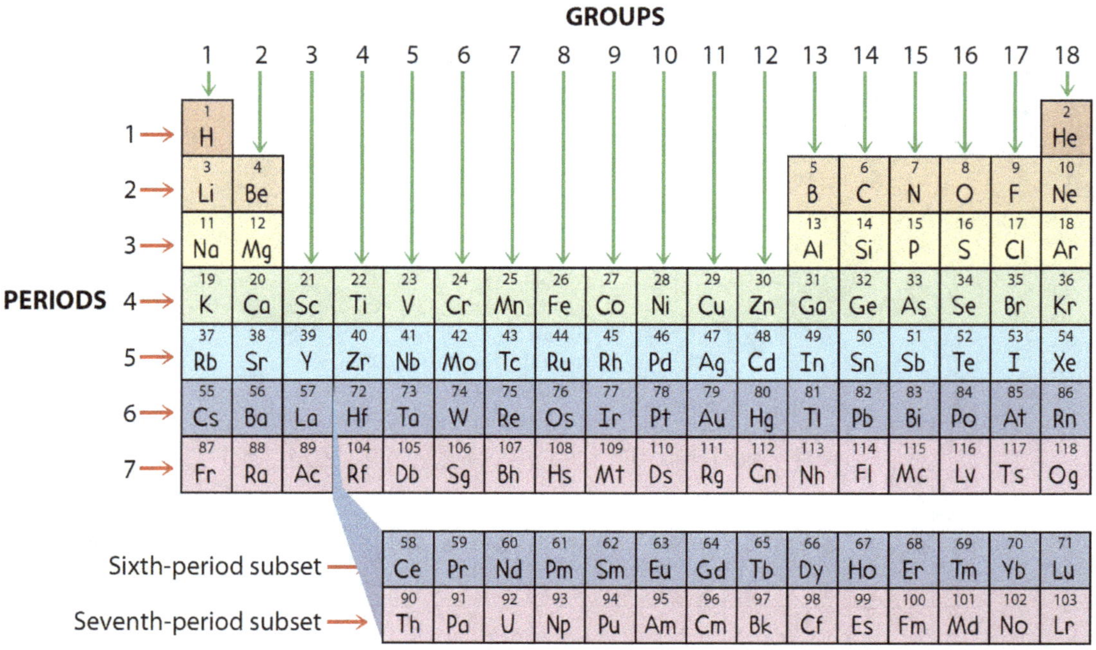

∧ Figure 3.14

The 7 periods (horizontal rows) and 18 groups (vertical columns) of the periodic table. Note that not all periods contain the same number of elements. Also note that, for reasons explained later, the sixth and seventh periods each include a subset of elements, which are listed apart from the main body.

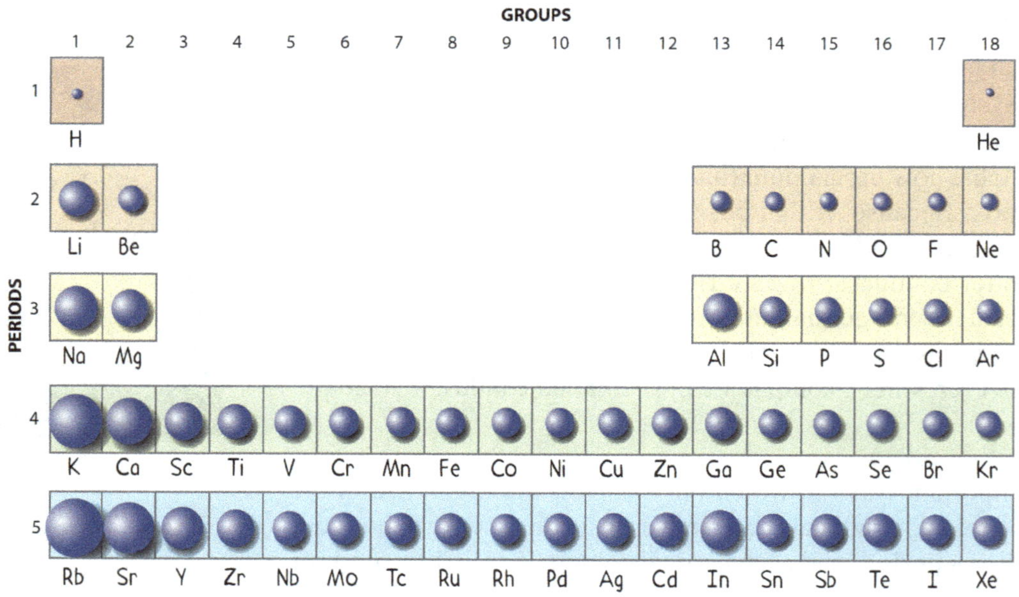

∧ Figure 3.15

The size of atoms gradually decreases in moving from left to right across any period. Atomic size is a periodic (repeating) property.

term used to indicate that the trends recur in cycles. Each horizontal row is called a *period* because it corresponds to one full cycle of a trend. As we explore further in Chapter 4, there are many other properties of elements that change gradually in moving across a row of the periodic table.

Down any group (vertical column), the properties of elements tend to be remarkably similar, which is why these elements are said to be "grouped" or "in a family." As **Figure 3.16** shows, several groups have traditional names that describe the properties of their elements. Early in human history, people discovered that ashes mixed with water produce a slippery solution useful for removing grease. By the Middle Ages, such mixtures were described as being alkaline, a term derived from the Arabic word for ashes, al-qali. Alkaline mixtures found many uses, particularly in the preparation of soaps (**Figure 3.17**). We now know that alkaline ashes contain compounds of group 1 elements, most notably potassium carbonate, also known as potash. Because of this history, group 1 elements, which are metals, are called the *alkali metals*.

Elements of group 2 also form alkaline solutions when mixed with water. Furthermore, medieval alchemists noted that certain minerals (which we now know are made up of group 2 elements) do not melt or change when put in fire. These fire-resistant substances were known to the alchemists as "earth." As a holdover from these ancient times, group 2 elements are known as the *alkaline-earth metals*.

Over toward the right side of the periodic table, elements of group 16 are known as the *chalcogens* ("ore-forming" in Greek), because the top two elements of this group, oxygen and sulfur, are so commonly found in ores. Elements of group 17 are known as the *halogens* ("salt-forming" in Greek), because of their tendency to form various salts. Group 18 elements are all unreactive gases that tend not to combine with other elements. For this reason, they are called the inert *noble gases*, presumably because the nobility of earlier times were above interacting with common folk.

The elements of groups 3 through 12 are all metals that do not form alkaline solutions with water. These metals tend to be harder than the alkali metals and less reactive with water; hence, they are used for structural purposes. Collectively, they are known as the *transition metals*, a name

FOR YOUR INFORMATION

Most nonmetallic materials do not conduct electricity. An important exception is a material called graphene, which is a one atom thick sheet of carbon atoms. The graphite in your pencil is made of stacks of graphene. The nanotubes are graphene sheets rolled into tubes. All these materials conduct electricity rather well, which leads to many useful applications. Sheets of graphene, for example, are used to make ultra-thin and flexible OLED display screens.

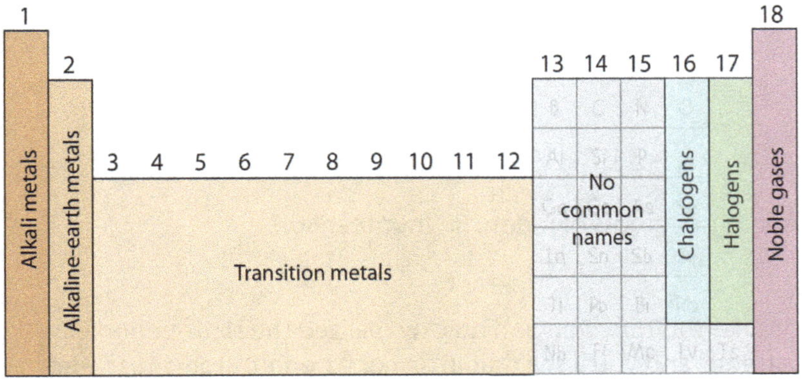

Figure 3.16

The common names for various groups of elements.

Figure 3.17

Ashes and water make a slippery alkaline solution once widely used to clean grime and oil from skin—a precursor to modern soap.

that denotes their central position in the periodic table. The transition metals include some of the most familiar and important elements—iron, Fe; copper, Cu; nickel, Ni; chromium, Cr; silver, Ag; and gold, Au. They also include many lesser-known elements that are nonetheless important in modern technology. People with hip implants appreciate the transition metals titanium, Ti; molybdenum, Mo; and manganese, Mn, because these noncorrosive metals are used in implant devices.

READING CHECK
List several common transition metals

Within the sixth period is a subset of 14 metallic elements (numbers 58 to 71) that are quite unlike any of the other transition metals. A similar subset (numbers 90 to 103) is found within the seventh period. These two subsets are the *inner transition metals*. Inserting the inner transition metals into the main body of the periodic table, as in **Figure 3.18**, produces a long and cumbersome table. So that the table can fit nicely on a standard paper size, these elements are commonly placed below the main body of the table, as shown in **Figure 3.19.**

FOR YOUR INFORMATION
The carat is the common unit used to describe the mass of a gem. A 1.0 carat diamond, for example, has a mass of 0.20 grams. The karat is the common unit used to describe the purity of a precious metal, such as gold. A 24-karat gold ring is as pure as can be. A gold ring that is 50 percent pure is 12-karat gold.

The sixth-period inner transition metals are called the *lanthanides*, because they fall after lanthanum, La. Because of their similar physical and chemical properties, they tend to occur mixed together in the same locations in the Earth. Also because of their similarities, lanthanides are unusually difficult to purify. Recently, the commercial use of lanthanides has increased. Several lanthanide elements, for example, are used in the fabrication of the light-emitting diodes (LEDs) of computer monitors and television screens. Neodymium is useful for making powerful magnets.

The lanthanides were first discovered within rare ores containing the elements scandium and yttrium. Collectively, these elements have come to be known as the "rare earth metals". These metals, however, are actually relatively abundant in the Earth's crust—about as abundant as copper. The problem is that they are distributed throughout Earth's crust rather evenly, which is to say, mineral ores containing easy to isolate concentrated samples are not common. Of great commercial use, about 300,000 tons of these rare Earth metals are produced every year.

The seventh-period inner transition metals are called the *actinides*, because they fall after actinium, Ac. They, too, all have similar properties and hence are not easily purified. The nuclear power industry faces this obstacle because it requires purified samples of two of the most publicized actinides: uranium, U, and plutonium, Pu. Actinides heavier than uranium are not commonly found in nature but are synthesized in the laboratory.

CONCEPT CHECK
Which are larger: atoms of cesium, Cs (number 55), or atoms of radon, Rn (number 86)?

CHECK YOUR ANSWER
Perhaps you tried looking to Figure 3.15 to answer this question and quickly realized the sixth-period elements are not shown. Well, relax. Look at the trends and you'll see that in any one period, atoms to the left tend to be larger than those to the right. Accordingly, cesium is positioned at the far left of period 6, and you can reasonably predict that its atoms are larger than those of radon, which is positioned at the far right of period 6. The periodic table is a road map to understanding the elements.

FOR YOUR INFORMATION

The air inside a traditional incandescent light bulb is a mixture of nitrogen and argon. As the tungsten filament is heated, minute particles of tungsten evaporate—much like steam leaving boiling water. Over time, these particles are deposited on the inner surface of the bulb, causing the bulb to blacken. Losing its tungsten, the filament eventually breaks and the bulb has "burned out."

CONCEPT CHECK

The elements copper, Cu; silver, Ag; and gold, Au, are three of the few metals that can be found naturally in their elemental state. These three metals have found great use as coins and jewelry for a number of reasons, including their resistance to corrosion and their remarkable colors. How is the fact that these metals have similar properties reflected in the periodic table?

CHECK YOUR ANSWER

Copper (number 29), silver (number 47), and gold (number 79) are all in the same group in the periodic table (group 11), which suggests they should have similar—though not identical—properties.

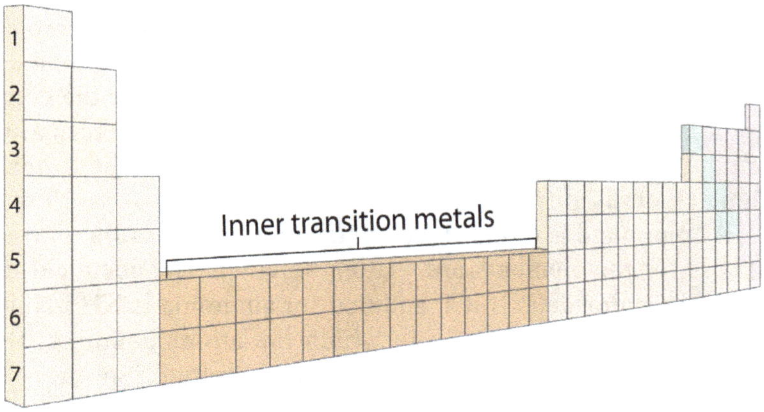

< Figure 3.18

Inserting the inner transition metals between atomic groups 3 and 4 results in a periodic table that is not easy to fit on a standard sheet of paper.

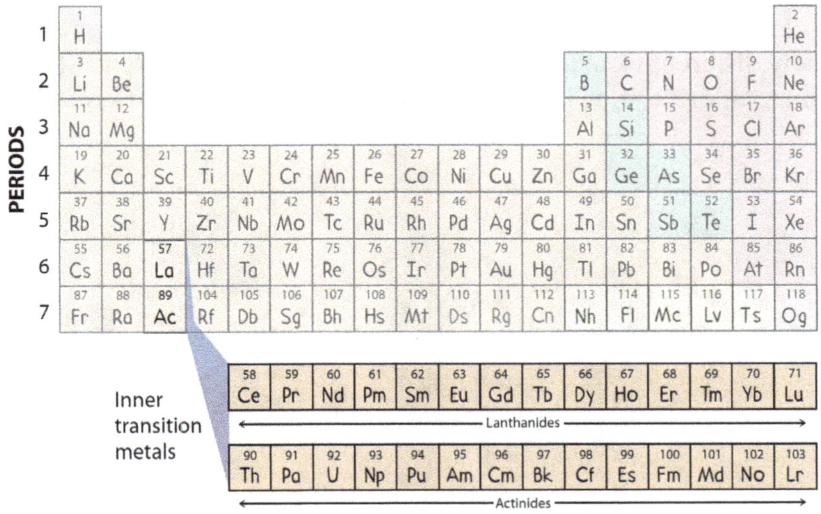

< Figure 3.19

The typical display of the inner transition metals. The count of elements in the sixth period goes from lanthanum (La, 57) to cerium (Ce, 58), on through to lutetium (Lu, 71), and then back to hafnium (Hf, 72). A similar jump is made in the seventh period.

FOR YOUR INFORMATION

The Hanford nuclear facility in central Washington state, from 1943 to 1986, produced 72 tons of plutonium, nearly two-thirds the nation's supply. Creating this much plutonium generated an estimated 450 billion gallons of radioactive and hazardous liquids, which were discharged into the local environment. Today, some 53 million gallons of high-level radioactive (see Chapter 5) and chemical wastes are stored in 177 underground tanks.

Hanford Nuclear Facility, 1945

3.4 Elements Can Combine to Form Compounds

READING CHECK

Why is the formation of a chemical compound an example of a chemical change?

When atoms of *different* elements bond to one another, they make a compound. Sodium atoms and chlorine atoms, for example, bond to make the compound sodium chloride, commonly known as table salt. Nitrogen atoms and hydrogen atoms join to make the compound ammonia, which is a common household cleaner. The formation of a compound is a chemical change, because it involves the formation of a fundamentally different material.

A compound is represented by its **chemical formula**, in which the symbols for the elements are written together. The chemical formula for sodium chloride is NaCl, and that for ammonia is NH_3. Numerical subscripts indicate the ratio in which the atoms combine. By convention, the subscript 1 is understood and omitted. So, the chemical formula NaCl tells us that in the compound sodium chloride there is one sodium atom for every chlorine atom. The chemical formula NH_3 tells us that in the compound ammonia there is one nitrogen atom for every three hydrogen atoms, as **Figure 3.20** shows.

Compounds have physical and chemical properties that are completely different from the properties of their constituent elements. The sodium chloride, NaCl, shown in **Figure 3.21** is very different from

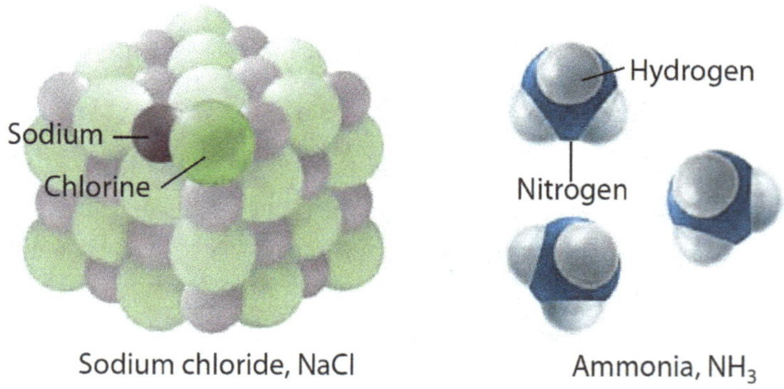

Figure 3.20 >

The compounds sodium chloride and ammonia are represented by their chemical formulas, NaCl and NH3. A chemical formula shows the ratio of atoms that constitute the compound.

Figure 3.21

Sodium metal and chlorine gas react together to form sodium chloride. Although the compound sodium chloride is made of sodium and chlorine, the physical and chemical properties of sodium chloride are very different from the physical and chemical properties of either sodium metal or chlorine gas.

elemental sodium and elemental chlorine. Elemental sodium, Na, consists of nothing but sodium atoms, which form a soft, silvery metal that can be cut easily with a knife. Its melting point is 97.5°C, and it reacts violently with water. Elemental chlorine, Cl_2, consists of chlorine molecules. This material, a yellow-green gas at room temperature, is very toxic, and it was used as a chemical warfare agent during World War I. Its boiling point is −34°C. The compound sodium chloride, NaCl, is a transparent, brittle, colorless crystal with a melting point of 800°C. Sodium chloride does not react chemically with water the way sodium does. It is not toxic like chlorine—in fact, sodium chloride is an essential nutrient for all living organisms. Sodium chloride is not sodium, nor is it chlorine; it is uniquely sodium chloride, a tasty chemical when sprinkled lightly over popcorn.

CONCEPT CHECK

Hydrogen sulfide, H_2S, is an offensively smelly compound. Rotten eggs get their characteristically unpleasant smell from the hydrogen sulfide they release. Can you conclude from this information that elemental sulfur, S_8, is just as smelly?

CHECK YOUR ANSWER

No, you cannot. In fact, the odor of elemental sulfur is negligible compared with that of hydrogen sulfide. Compounds are truly different from the elements from which they are formed. Hydrogen sulfide, H_2S, is as different from elemental sulfur, S_8, as water, H_2O, is from elemental oxygen, O_2.

3.5 There Is a System for Naming Compounds

A system for naming the countless number of possible compounds has been developed by the International Union for Pure and Applied Chemistry (IUPAC). This system is designed so that a compound's name reflects the elements it contains and how those elements are joined. This system can be used to figure out the chemical identity of a compound from its systematic name.

To accommodate the immense number of different compounds, this system is rather complex. However, there is no need for you to learn all its rules. Instead, learning some guidelines will prove most helpful. These guidelines alone will not enable you to name every compound. However, they will acquaint you with how the system works for most of the compounds presented in any introductory chemistry course. In Chapter 12 we'll be looking at how carbon-based compounds are named. For now, we'll focus on some of the simpler known compounds.

READING CHECK

How can you tell whether an atom within a compound takes on a positive or negative charge?

As we will discuss in Chapter 6, atoms are held together in a chemical compound by their electrical charges. Whether an atom within a compound takes on a positive or negative charge can be predicted by its place in the periodic table. The atom of an element located closer to the left side of the periodic table tends to take on a positive charge, while the atom of an element placed closer to the right side of the periodic table takes on a negative charge. This applies to the naming of compounds in that the more positively charged atom is, by convention, listed first. That's why we have "sodium chloride," rather than "chloride sodium" as spelled out in guideline 1—note that sodium, Na, is on the left-hand side of the periodic table and chlorine, Cl, is on the right-hand side.

GUIDELINE 1: *Read Left to Right*

The name of the element farther to the left in the periodic table is followed by the name of the element farther to the right, with the suffix "-ide" added to the name of the latter:

NaCl	Sodium chloride
MgO	Magnesium oxide
Li_2O	Lithium oxide
HCl	Hydrogen chloride
CaF_2	Calcium fluoride
Sr_3P_2	Strontium phosphide

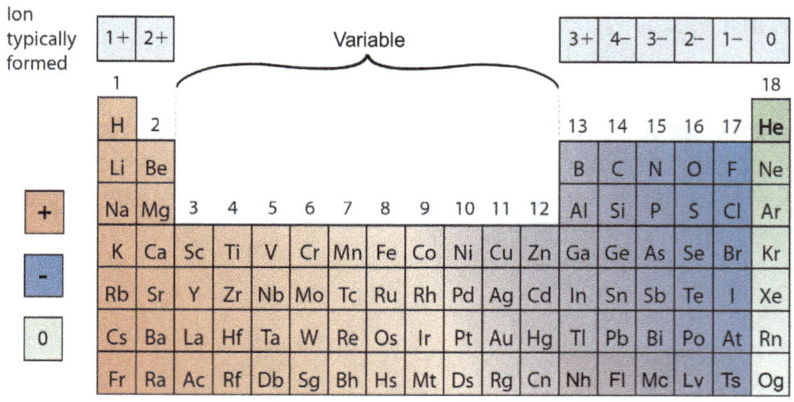

Figure 3.22 >

The type of charge an atom forms is given by its placement in the periodic table.

GUIDELINE 2: *Use the Periodic Table*

Looking at the subscripts given in the formula and knowing the periodic table, you can often deduce the charge on each of the atoms in a compound. Recall from the previous section that no subscript implies a subscript of one. So, sodium chloride can be represented as Na_1Cl_1. We see that sodium and chlorine are bound in a 1:1 ratio. As shown in **Figure 3.22**, elements in the first group of the periodic table tend to form a 1+ charge and elements in group 17 tend to form a 1- charge. So, for sodium chloride, the 1+ charge on the sodium is balanced by the 1- charge on the chlorine.

Consider lithium oxide, Li_2O. Lithium, being in the first group, forms a 1+ charge. Why then are two lithiums needed when combined with oxygen? The answer is that the oxygen forms a 2- charge (See Figure 3.22), which means two lithiums are needed to make the charges balanced.

$$Li^{1+} \quad O^{2-} \quad Li^{1+} \longrightarrow Li_2O$$

So, there's a pattern. Take the subscript of one element and you'll often find that this is the charge of the other element, as follows:

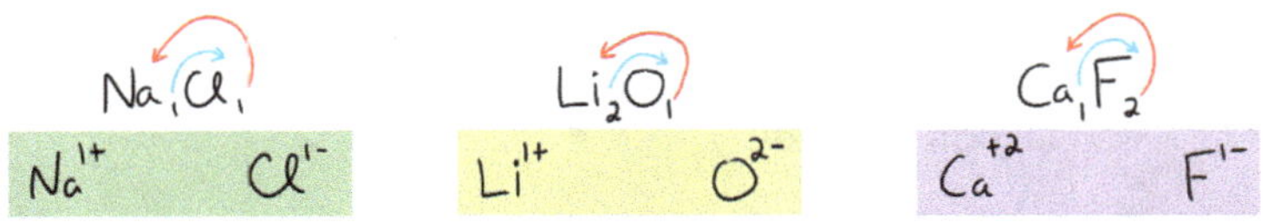

But you need to be careful. Magnesium, for example, is in the second group of the periodic table and so it forms a 2+ charge. Magnesium combines with chlorine to form magnesium chloride, $MgCl_2$. But what about when magnesium binds with oxygen, which forms the 2- charge? The result is magnesium oxide, Mg_2O_2. By convention, the subscripts are reduced to the lowest common denominator, so magnesium oxide is generally presented as MgO, which is short-hand for Mg_1O_1. This tells us accurately that the magnesium and oxygen bind in a 1:1 ratio. What's hidden, however, is that they do so by each carrying a double charge.

$$Mg^{2+} + O^{2-} \longrightarrow Mg_2O_2 \longrightarrow MgO$$

The way to figure this out is to use the periodic table as a tool, as shown in Figure 3.22. Because magnesium, Mg, is in group 2, this tells you it forms the 2+ ion.

So, do the elements within group 3 form the 3+ ion? Yes and no. The elements of the transition metals form a variety of different ions. Recall the transition metals are those found in groups 3 through 12. Iron, for example, can form the Fe^{2+} and the Fe^{3+} ions depending upon the conditions. This leads us to the next guideline.

GUIDELINE 3: *Transition Metals*

If the compound contains a transition metal, then roman numerals are used to indicate the charge on the transition metal. The compound iron (II) chloride has the formula $FeCl_2$. Here the roman numerals explicitly tell us the charge on the transition metal.

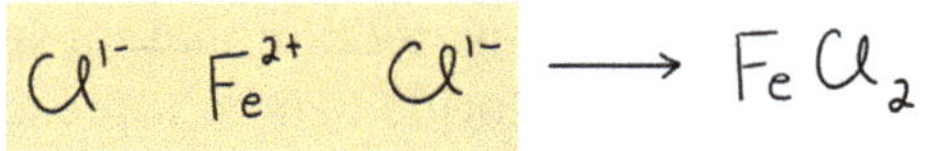

Similarly, the compound iron (III) chloride has the formula $FeCl_3$. Notice that with $FeCl_2$, two chloride ions, Cl^{1-}, are needed to balance the one Fe^{2+} and that with $FeCl_3$, three chloride ions are needed.

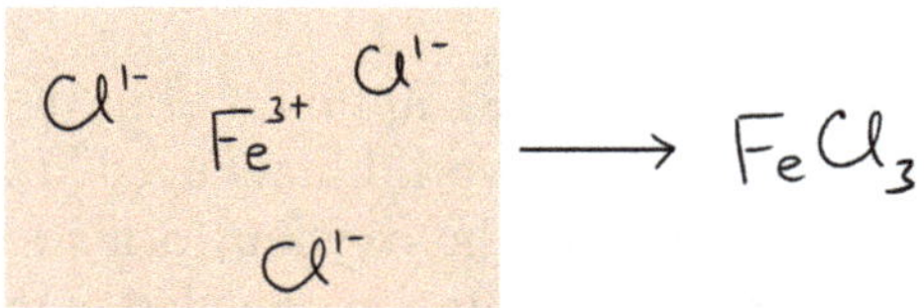

GUIDELINE 4: *Solving Ambiguity*

When two or more compounds have different numbers of the same elements, prefixes are added to remove the ambiguity. This occurs primarily with compounds of nonmetals. The first four prefixes are mono- (one), di- (two), tri- (three), and tetra- (four). The prefix mono-, however, is commonly omitted from the beginning of the first word of the name:

Carbon and oxygen
CO Carbon monoxide
CO_2 Carbon dioxide

Nitrogen and oxygen
NO_2 Nitrogen dioxide
N_2O_4 Dinitrogen tetroxide

Sulfur and oxygen
SO_2 Sulfur dioxide
SO_3 Sulfur trioxide

CONCEPT CHECK

Name the following compounds: $BeCl_2$; K_2O; CaS, $CaBr_2$, CuCl, $SnCl_4$, $PbCl_2$, Fe_2O_3

CHECK YOUR ANSWER

Beryllium chloride, potassium oxide, calcium sulfide; calcium bromide; copper (I) chloride; tin (IV) chloride; lead (II) chloride; iron (III) oxide.

GUIDELINE 5: *Polyatomic Ions*
Atoms can clump together to form a molecular unit that acts as a single electrically charged group, called a *polyatomic ion*. For example, a carbon atom can join with three oxygen atoms to form what is known as a carbonate ion, CO_3^{2-}. Some commonly encountered polyatomic ions are shown in **Table 3.1**. Note that most of them are negatively charged.

We'll be exploring the nature of these polyatomic ions in greater detail in Chapters 6 and 10. For now it suffices to recognize that positively charged polyatomic ions are listed first within the name (without the word ion). An example is ammonium chloride, NH_4Cl. Negatively charged polyatomic ions are placed at the end of the name. An example is lithium nitrate, $LiNO_3$.

A polyatomic ion may appear more than once within a compound. This is indicated by placing the polyatomic ion within parentheses. A subscript just outside the parentheses indicates the number of times the polyatomic ion appears. To keep it simple, the prefixes of mono-, di-, tri-, and tetra- are commonly not included for polyatomic ions.

K_2CO_3 Potassium carbonate
$AuPO_4$ Gold phosphate
$Mg(CN)_2$ Magnesium cyanide
$Al_2(SO_4)_3$ Aluminum sulfate

TABLE 3.1 Common Polyatomic Ions

NAME	FORMULA
Acetate ion	$CH_3CO_2^-$
Ammonium ion	NH_4^+
Bicarbonate ion	HCO_3^-
Carbonate ion	CO_3^{2-}
Cyanide ion	CN^-
Hydroxide ion	OH^-
Nitrate ion	NO_3^-
Phosphate ion	PO_4^{3-}
Sulfate ion	SO_4^{2-}

GUIDELINE 6: *Common Names*
Many compounds are not usually referred to by their systematic names. Instead, they are assigned common names that are more convenient or have been used traditionally for many years. Some common names are water for H_2O, ammonia for NH_3, and methane for CH_4.

CONCEPT **CHECK**
What is the systematic name for $Ca(CH_3CO_2)_2$? How many oxygen atoms does it have?

CHECK YOUR ANSWER
The systematic name for this compound, which consists of calcium and the polyatomic acetate ion, is calcium acetate. Each acetate ion has two oxygen atoms. This compound has two acetate ions, which means it has a total of four oxygen atoms.

3.6 Most Materials Are Mixtures

A **mixture** is a combination of two or more substances in which each substance retains its properties. Most materials we encounter are mixtures: mixtures of elements, mixtures of compounds, or mixtures of elements and compounds. Stainless steel, for example, is a mixture of the elements iron, chromium, nickel, and carbon. Seltzer water is a mixture of the liquid compound water and the gaseous compound carbon dioxide. Our atmosphere, as **Figure 3.23** illustrates, is a mixture of the elements nitrogen, oxygen, and argon plus small amounts of such compounds as carbon dioxide and water vapor.

Tap water is a mixture containing mostly water but also many other compounds. Depending on your location, your water may contain compounds of calcium, magnesium, fluorine, iron, and potassium;

CHEMICAL CONNECTIONS

Chemically speaking, how is an icy winter road connected to the ocean?

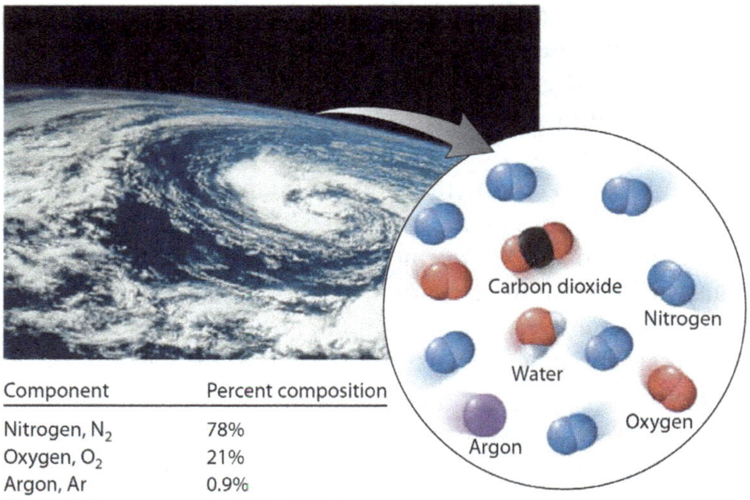

Figure 3.23 >

The Earth's atmosphere is a mixture of gaseous elements and compounds. Some of them are shown here.

Component	Percent composition
Nitrogen, N_2	78%
Oxygen, O_2	21%
Argon, Ar	0.9%
Water, H_2O	0–4% (variable)
Carbon dioxide, CO_2	0.034% (variable)

∧ Figure 3.24

Most of the oxygen, O_2, in the air bubbles produced by an aquarium aerator escapes into the atmosphere. Some of the oxygen, however, mixes with the water. It is this oxygen the fish depend on to survive. Without this dissolved oxygen, which they extract with their gills, the fish would promptly drown. So, fish don't "breathe" water. They breathe the oxygen, O_2, dissolved in the water.

chlorine disinfectants; trace amounts of compounds of lead, mercury, and cadmium; organic compounds; and dissolved gases like oxygen, nitrogen, and carbon dioxide (**Figure 3.24**).

While it is surely important to minimize any toxic components in your drinking water, it is unnecessary, undesirable, and impossible to remove all other substances from it. Some of the dissolved solids and gases give water its characteristic taste, and many of them promote human health: fluoride compounds protect teeth, chlorine destroys harmful bacteria, and as much as 10 percent of our daily requirement for iron, potassium, calcium, and magnesium is obtained from drinking water (**Figures 3.25**).

There is a difference between the way substances—either elements or compounds—combine to form mixtures and the way elements combine to form compounds. Each substance in a mixture retains its chemical identity. The sugar molecules in the teaspoon of sugar in **Figure 3.26**, for example, are identical to the sugar molecules already in the tea. The only difference is that the sugar molecules in the tea are mixed with other substances, mostly water. The formation of a mixture, therefore, is a physical change.

As was discussed in Section 3.4, when elements join to form compounds, there is a change in chemical identity. Sodium chloride is not a mixture of sodium and chlorine atoms. Instead, sodium chloride is a compound, which means it is entirely different from the elements used to make it. The formation of a compound is therefore a chemical change.

Figure 3.25 >

Tap water provides us with H_2O as well as a large number of other compounds, many of which are flavorful and help us grow, as shown by Graham across ages. Bottoms up!

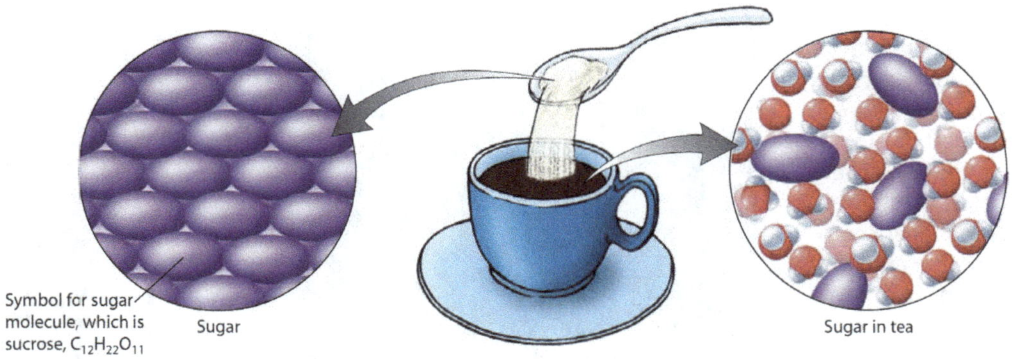

∧ Figure 3.26

Table sugar is a compound consisting of only sucrose molecules. Once these molecules are mixed into hot tea, they become interspersed among the water and tea molecules and form a sugar–tea–water mixture. No new compounds are formed, so this is an example of a physical change.

Mixtures Can Be Separated by Physical Means

The components of mixtures can be separated from one another by taking advantage of differences in the components' physical properties. A mixture of solids and liquids, for example, can be separated using filter paper through which the liquids pass but the solids do not. This is how coffee is often made: the caffeine and flavor molecules after being extracted by the hot water pass through the filter and into the coffee pot while the solid coffee grounds remain behind. This method of separating a solid–liquid mixture is called *filtration* and is a common technique used by chemists.

Mixtures can also be separated by taking advantage of a difference in boiling or melting points. Seawater is a mixture of water and a variety of compounds, mostly sodium chloride. Whereas pure water boils at 100°C, sodium chloride doesn't even melt until 800°C. One way to separate pure water out of the mixture we call seawater, therefore, is to heat the seawater to about 100°C. At this temperature, the liquid water readily transforms to water vapor but the sodium chloride stays behind, dissolved in the remaining water. As the water vapor rises, it can be channeled into a cooler container, where it condenses to a liquid without the dissolved solids. This process of collecting a vaporized substance, called *distillation*, is illustrated in **Figure 3.27**. After all the water has been distilled from seawater, what remains consists of dry solids. These solids, also a mixture of compounds, contain a variety of valuable materials, including sodium chloride, potassium bromide, and a small amount of gold! A commercial application of this concept is shown in **Figure 3.28**.

READING CHECK

How do we separate the components of a mixture?

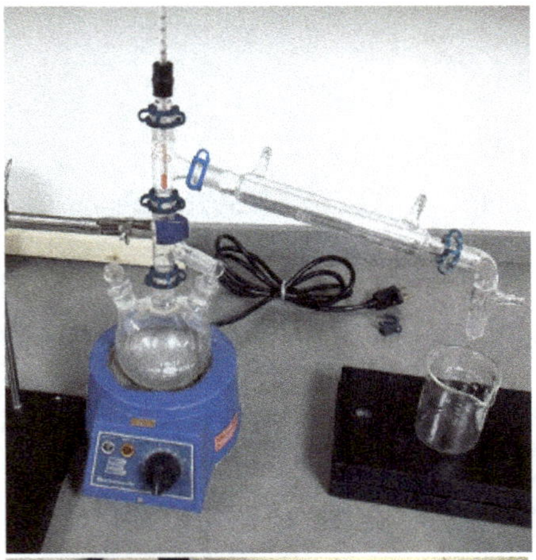

Figure 3.27 >

(a) The mixture is boiled in the flask on the left. The rising water vapor is channeled into a downward-slanting tube kept cool by cold water flowing across its outer surface. The water vapor inside the cool tube condenses and collects in the flask on the right. (b) A whiskey still functions on the same principle. A mixture containing alcohol is heated to the point where the alcohol, some flavoring molecules, and some water are vaporized. These vapors travel through copper tubes, where they then condense to a liquid high in alcohol content.

Figure 3.28 >

At the southern end of San Francisco Bay, there are areas where the seawater has been partitioned off by earthen dikes. These are evaporation ponds, where the water is allowed to evaporate, leaving behind the solids that were dissolved in the seawater, mostly salt. These solids are further refined for commercial sale. The remarkable colors of the ponds are due to organic pigments made by salt-loving bacteria.

CONCEPT CHECK

So far, you have learned about three kinds of matter: elements, compounds, and mixtures. Which of the following boxes contains only an element? Which contains only a compound? Which contains a mixture?

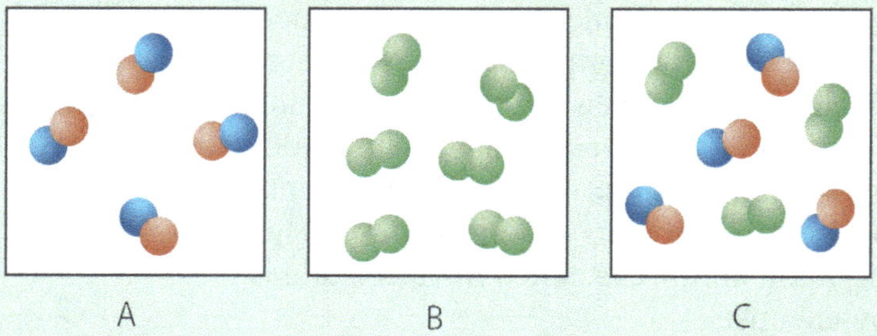

CHECK YOUR ANSWER

The molecules in box A each contain two different types of atoms and so are representative of a compound. The molecules in box B each consist of the same atoms and so are representative of an element. Box C is a mixture of the compound and the element. Note how the molecules of the compound and those of the element remain intact in the mixture. That is, upon the formation of the mixture, there is no exchange of atoms between the components.

3.7 Matter Can Be Classified as Pure or Impure

From a chemist's point of view, if a material is **pure**, it consists of only a single element or a single compound. In pure gold, for example, there is nothing but the element gold. In pure table salt, there is nothing but the compound sodium chloride. If a material is **impure**, it is a mixture and contains two or more elements or compounds. These concepts are mapped out in the classification scheme shown in Figure 3.29.

READING CHECK
How does a chemist define a "pure" material?

Because atoms and molecules are so small, there is a countless number of them in even a tiny sample. If just one atom or molecule out of many were different, then this sample could not be classified as 100 percent pure. Samples can be "purified," however, by various methods, such as distillation. When we say pure, it is understood to be a relative term. Comparing the purity of two samples, the purer one contains fewer impurities. A sample of water that is 99.9 percent pure has a greater proportion of impurities than does a sample of water that is 99.9999 percent pure. The 99.9999 percent pure water would be much more expensive, because this high degree of purity is rather difficult to attain.

Sometimes naturally occurring mixtures are labeled as being pure, as in "pure orange juice." Such a statement means that nothing artificial has been added. According to a chemist's definition, however, orange juice is anything but pure, as it contains a wide variety of materials, including water, pulp, flavorings, vitamins, and sugars. Also, outside the language of chemistry, sometimes a mixture can be identified as pure. A cook, for example, might ask for pure baking powder. To a chemist, however, this doesn't make sense because baking powder is a mixture of baking soda and sodium aluminum sulfate, plus many other chemicals.

Mixtures may be heterogeneous or homogeneous. In a **heterogeneous mixture**, the different components can be seen as individual substances, such as pulp in orange juice, sand in water, or oil globules dispersed in vinegar. The different components are visible. **Homogeneous mixtures** have the same composition throughout as

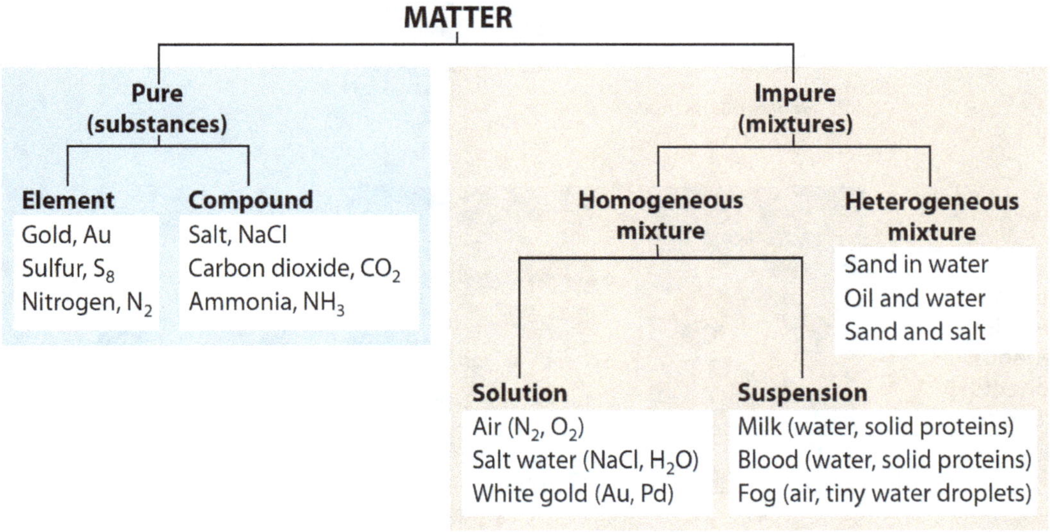

< **Figure 3.29**
The chemical classification of matter.

FOR YOUR INFORMATION

White gold is a mixture of gold with smaller amounts of white metals, such as silver, palladium, or rhodium, that increase its hardness. Unlike gold, the precious metal platinum is used in jewelry in almost its pure form, about 95 percent. Platinum is a very white metal and is also very dense. A platinum ring will feel heavier than a typical yellow or white gold ring, but it will also be much more expensive.

judged by the unaided eye. Any one region of the mixture has the same ratio of substances as does any other region. The reason for this is because the different components are mixed at a very fine level such that the components cannot be seen as individual identifiable entities. The distinction is shown in **Figure 3.30**.

A homogeneous mixture may be either a solution or a suspension. In a **solution**, all components are in the same phase. The atmosphere we breathe is a gaseous solution consisting of the gaseous elements nitrogen and oxygen as well as minor amounts of other gaseous materials. Salt water is a liquid solution, because both the water and the dissolved sodium chloride are found in a single liquid phase. An example of a solid solution is white gold. We will discuss solutions in more detail in Chapter 7.

A **suspension** forms when the particles of a substance are finely mixed but not dissolved. The components of a suspension can be of different phases, such as solid particles suspended within a liquid or liquid droplets suspended within a gas. In a suspension, the mixing can be so thorough that the different phases are not readily distinguished. Milk is a suspension because it is a homogeneous mixture of proteins and fats finely dispersed in water. Blood is a suspension composed of finely dispersed blood cells in water. Another example of a suspension is clouds, which are homogeneous mixtures of tiny water droplets suspended in air. Light shining through a suspension, as is shown in **Figure 3.31**, results in a visible cone as the light is reflected by the suspended components.

The easiest way to distinguish a suspension from a solution in the laboratory is to spin a sample in a centrifuge. This device, spinning at thousands of revolutions per minute, separates the components of suspensions but not those of solutions, as **Figure 3.32** shows.

Figure 3.30 >

(a) In heterogeneous mixtures, the different components can be seen with the naked eye. (b) In homogeneous mixtures, the different components are mixed at a much finer level and so are not readily distinguished.

∧ **Figure 3.31**

The path of light becomes visible when it passes through a suspension.

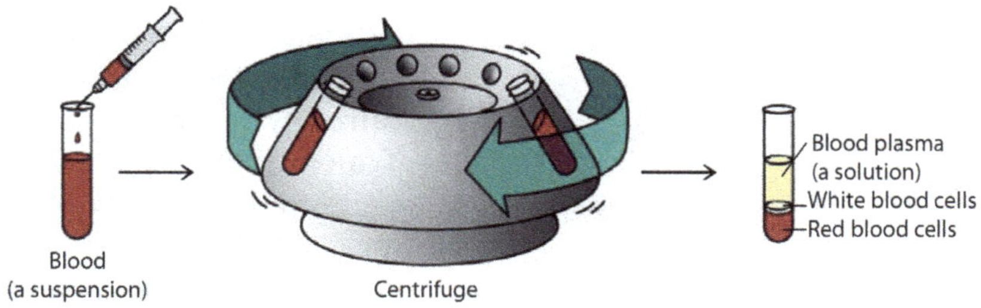

∧ **Figure 3.32**

Blood, because it is a suspension, can be centrifuged into its components, which include the blood plasma (a yellowish solution) and white and red blood cells. The components of the plasma cannot be separated from one another because a centrifuge has no effect on solutions. Notice that blood plasma, white blood cells, and red blood cells can be isolated from blood. None of these components of blood, however, are in themselves pure materials.

CONCEPT **CHECK**

Impure water can be purified by

a. removing the impure water molecules.
b. removing everything that is not water.
c. breaking down the water to its simplest components.
d. adding some disinfectant, such as chlorine.

CHECK YOUR ANSWER

The answer is b: impure water can be purified by removing everything that isn't water. H_2O is a compound made of the elements hydrogen and oxygen in a 2-to-1 ratio. Every H_2O molecule is the same as every other, and there's no such thing as an impure H_2O molecule. Just about anything, including you, beach balls, rubber ducks, dust particles, and bacteria, can be found in water. When something other than water is found in water, we say that the water is impure. It is important to see that the impurities are in the water and not part of the water, which means that it is possible to remove them by a variety of physical means, such as filtration or distillation.

3.8 The Advent of Nanotechnology

The age of microtechnology was ushered in some 65 years ago with the invention of the solid state transistor, a device that serves as a gateway for electronic signals. Engineers were quick to grasp the idea of integrating many transistors together to create logic boards that could perform calculations and run programs. The more transistors they could squeeze into a circuit, the more powerful the logic board. The race thus began to squeeze more and more transistors together into tinier and tinier circuits. The scales achieved were in the realm of the micron (10^{-6} meters)—thus the term microtechnology. At the time of the transistor's invention, few people realized the impact microtechnology would have on society—from personal computers to smart phones to the internet.

Today, we are at the beginning of a similar revolution. Technological advances have recently brought us past the realm of microns to the realm of the nanometer (10^{-9} meters), which is the realm of individual atoms and molecules—a realm where we have reached the basic building blocks of matter. Technology that works on this scale is called *nanotechnology*. No one knows exactly what impact nanotechnology will have on society, but people are quickly coming to realize its vast potential, which is likely to be much greater than that of microtechnology.

READING CHECK
How large are the objects that are the focus of nanotechnology?

Nanotechnology generally concerns the manipulations of objects from 1 to 100 nanometers in scale. For perspective, a DNA molecule is about 2.0 nm wide (though about a meter long!), while a water molecule is only about 0.2 nm. Like microtechnology, nanotechnology is interdisciplinary, requiring the cooperative efforts of chemists, engineers, physicists, molecular biologists, and many others. Interestingly, there are already many products on the market that contain components developed through nanotechnology. These include sunscreens, mirrors that don't fog, dental bonding agents, automotive catalytic converters, stain-free clothing, water filtration systems, the heads of computer hard drives, and much more. Nanotechnology, however, is still in its infancy, and it will likely be decades before its potential is fully realized (**Figure 3.33**). Consider, for example, that personal computers didn't blossom until the 1990's, some 40 years after the first solid state transistor.

FOR YOUR INFORMATION

Before he died in 2005, Rick Smalley, co-discoverer of the buckyball, advocated that carbon nanotubes, if developed into wires, could be an ideal material for efficiently transporting electricity over vast distances. If such an infrastructure were in place, the wind energy of the Great Plains of the United States would be sufficient to supply the electrical needs of the entire country.

There are two main approaches to building nanoscale materials and devices: top-down and bottom-up. The top-down approach is an extension of microtechnology techniques to smaller and smaller scales. A nano-sized circuit board, for example, might be carved out from a larger block of material. The bottom-up approach involves building nanosized objects atom by atom. A very important tool for either of these approaches is the **scanning probe microscope**, which detects and characterizes the surface atoms of materials by way of an ultrathin probe tip, as shown in **Figure 3.34**. The tip is mechanically dragged over the surface. Interactions between the tip and the surface atoms cause movements in the probe that are detected by way of a laser beam and translated by a computer into a topographical image. Scanning probe microscopes can also be used to move individual atoms into desired positions.

Nanotechnology will enable the continued miniaturization of integrated circuits needed for ever smaller and more powerful computers. But a computer need not rely on an integrated circuit of nanowires for processing power. A wholly new approach involves designing logic boards in which molecules (not electric circuits) read, process, and write information. One molecule that has proved most promising for such molecular computation is DNA, the same molecule that holds our genetic code. An advantage that molecular computing has over conventional computing is that it can run a massive number of calculations in parallel (at the same time). Because of such fundamental differences, molecular computing may one day outshine even the fastest of integrated circuits. Molecular computing, in turn, may then soon be eclipsed by other novel approaches, such as quantum or photon computing, also made possible by nanotechnology. Consider how this might impact the continued development of artificial intelligence.

The ultimate expert on nanotechnology is nature. Living organisms, for example, are complex systems of interacting biomolecules all functioning on the scale of nanometers. In this sense, the living organism is nature's nanomachine. We need look no further than our own

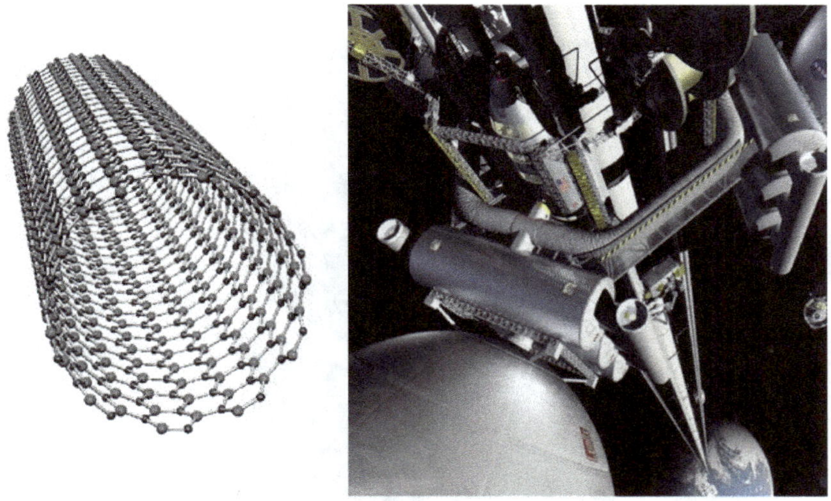

< **Figure 3.33**

Carbon nanotubes can be nested within each other to provide the strongest fiber known—a thread 1 millimeter in diameter can support a weight of about 13,000 pounds. A network of such strong fibers could be used to build the once science-fictional space elevator.

bodies to find evidence of the feasibility and power of nanotechnology. With nature as our teacher, we have much to learn. Such knowledge will be particularly applicable to medicine. By becoming nanotechnology experts ourselves, we would be well equipped to understand exact causes of nearly any disease or disorder (aging included) and empowered to develop innovative cures.

What are the limits of nanotechnology? As a society, how will we deal with the impending changes nanotechnology may bring? Consider the possibilities. Wall paint that can change color or be used to display video. Smart dust that the military could use to seek out and destroy an enemy. Solar cells that capture sunlight so efficiently that they render fossil fuels obsolete. AI models with so much processing power that we begin to wonder whether they experience consciousness. Nanobots that roam our circulatory systems destroying cancerous tumors or arterial plaque. Nanomachines that can "photocopy" 3-dimensional objects, including living organisms. Medicines that more than double the average human life span. Stay tuned for a new revolution in human capabilities.

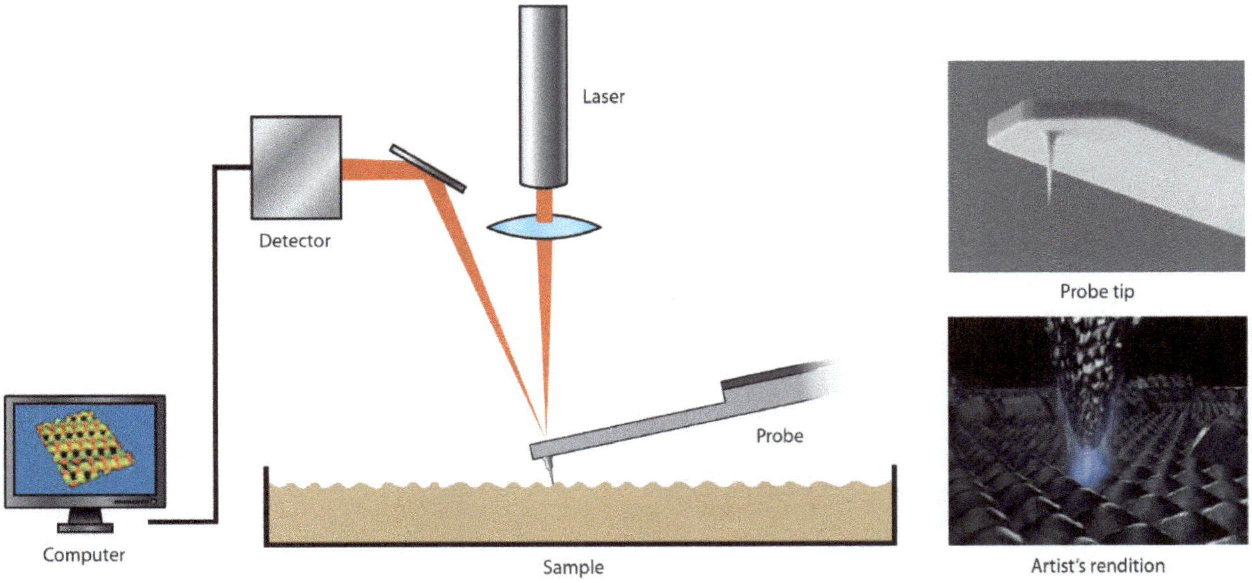

∧ **Figure 3.34**

A schematic of a scanning probe microscope that detects and characterizes the surface atoms of a material by way of an ultrathin probe tip attached to a miniature cantilever.

FOR YOUR INFORMATION

An interesting discovery of nanoscience is that the properties of a material at the level of its atoms can be different from its properties in bulk quantities. A bar of gold, for example, is gold in color. A thin sheet of gold atoms, by contrast, is dark red. There is much research currently being directed toward the discovery of the unique nanoproperties of materials. Many novel applications of these nanoproperties are sure to follow.

Figure 3.35

In the 1950s, physicist Richard Feynman foresaw the development of nanotechnology in his talk "There's Plenty of Room at the Bottom" where he noted the great potential of being able to manipulate matter by its individual atoms.

CONCEPT **CHECK**

How believable would our present technology be to someone living 200 years ago? How believable might the technology of 200 years in the future be to us right now?

CHECK YOUR ANSWER

Hindsight is 20/20. It's always easy to look back over time and see the progression of events that led to our present state. Much more difficult is it to think forward and project possible scenarios. Perhaps the future technology of 200 years from now will be just as unbelievable to us as our present technology is unbelievable to someone of 200 years ago. This assumes we survive as a civilization for that long.

Calculation Corner: How Pure Is Pure?

A 100-gram sample of water contains about 3×10^{24} molecules. If this sample were ideally pure, every one of those molecules would be water. Atoms and molecules, however, are so amazingly small, and hence numerous, that the formation of a truly pure sample of macroscopic quantity is virtually impossible.

For example, consider a 100-gram sample of water that is 99.9999 percent pure. What this means is that the sample contains 99.9999 grams of water, which is still nearly 3×10^{24} water molecules. Pretty good, right? However, if the remaining 0.0001 grams were made of dissolved lead, Pb, then this would correspond to about 3×10^{17} (300,000 trillion) atoms of Pb, which is quite small compared to the number of water molecules but is still an amazingly large number.

Any material that is seemingly pure will inevitably contain impurities. Sometimes these impurities are of particular interest. For example, minor impurities in a solution might be toxic and their presence might need to be monitored. How much is present is frequently measured in units of milligrams per liter (mg/L), micrograms per liter (mg/L), or nanograms per liter (ng/L) of solution.

One liter of water contains one million milligrams of water. Because 1 L of water equals 1 million mg of water, the units of mg/L can also be expressed as 1 mg per 1 million mg. Another way of saying this is one part per million, or simply 1 ppm. The units of mg/L and ppm, therefore, are equivalent. Similarly, "micrograms per liter" is often expressed as parts per billion, ppb, and "nanograms per liter" is often expressed as parts per trillion, ppt. (As Table 1.3 shows, 1 milligram equals 1000 micrograms, while 1 microgram equals 1000 nanograms.)

EXAMPLE

There are about 35 grams of salts in every liter of ocean water. Express this concentration in units of ppm.

ANSWER

Convert grams of salt into milligrams of salt:

(35 g salts)(1000 mg/1g) = 35,000 mg

There are about 35,000 mg of salts in a liter of ocean water, which equals 35,000 ppm.

YOUR TURN

1. Typical levels of fluoride found in fluoridated public drinking water are about 1.0 ppm. If you were to drink a liter of this water, show that you have ingested 1 milligram of fluoride.

2. Aquatic organisms require a dissolved oxygen concentration of about 6 ppm. At this concentration, show that there are 0.006 grams of oxygen present in each liter of water.

3. Chloroform, $CHCl_3$, is a common contaminant of chlorinated drinking water. A usual concentration in municipal tap water may be around 25 ppb. At this concentration, show that there are 0.025 milligrams of chloroform are present in each liter of water.

4. Assume that the population of the United States is 330 million. Show that each U.S. citizen has a concentration of 3.0 ppb. Assume a world human population of 8 billion. Show that each human has a concentration of 125 ppt. If you can imagine how small in number you are compared to national or world populations, then you have a sense of how dilute a dissolved substance is when its concentration is measured by the ppb or ppt.

Concept Review

Chapter 3

Summary of Terms

Atomic symbol An abbreviation for an element or atom.

Chemical bond The force of attraction between two atoms that holds them together.

Chemical change The formation of new substance(s) by rearranging the atoms of the original material(s).

Chemical formula A notation that indicates the composition of a compound, consisting of the atomic symbols for the different elements of the compound and numerical subscripts indicating the ratio in which the atoms combine.

Chemical property A type of property that characterizes the ability of a substance to change into a different substance under specific conditions.

Chemical reaction A term synonymous with chemical change.

Compound A material in which atoms of different elements are bonded to one another.

Element A material consisting of only one type of atom.

Elemental formula A notation that uses the atomic symbol and (sometimes) a numerical subscript to denote how many atoms are bonded in one unit of an element.

Group A vertical column in the periodic table, also known as a family of elements.

Heterogeneous mixture A mixture in which the different components can be seen as individual substances.

Homogeneous mixture A mixture in which the components are so finely mixed that anyone region of the mixture has the same ratio of substances as any other region.

Impure The state of a material that is a mixture of more than one element or compound.

Metal An element that is shiny, opaque, and able to conduct electricity and heat.

Metalloid An element that exhibits some properties of metals and some properties of nonmetals. Six elements recognized as metalloids include boron, B; silicon, Si; germanium, Ge; arsenic, As; antimony, Sb; and tellurium, Te.

Mixture A combination of two or more substances in which each substance retains its properties.

Nonmetal An element located toward the upper right of the periodic table, with the exception of hydrogen, that is neither a metal nor a metalloid.

Period A horizontal row in the periodic table.

Periodic table A chart in which all known elements are organized by physical and chemical properties.

Periodic trend The gradual change of any property in the elements across a period of the periodic table.

Physical change A change in which a substance changes its physical properties without changing its chemical identity.

Physical property Any physical attribute of a substance, such as color, density, or hardness.

Pure The state of a material that consists solely of a single element or compound.

Scanning Probe Microscope A tool of nanotechnology that detects and characterizes the surface atoms of materials by way of an ultrathin probe tip, which is detected by laser light as it is mechanically dragged over the surface.

Solution A homogeneous mixture in which all components are dissolved in the same phase.

Suspension A homogeneous mixture in which the various components are finely mixed, but not dissolved.

Review Questions

3.1 Matter Has Physical and Chemical Properties

1. What happens to the chemical identity of a substance during a physical change?
2. What changes during a chemical reaction?
3. Why is it sometimes difficult to decide whether an observed change is physical or chemical?
4. Why is melting ice an example of a physical change?

3.2 Elements Are Made of Atoms

5. How many types of atoms can you expect to find in a pure sample of any element?
6. Distinguish between an atom and an element.
7. What information is given by the elemental formula?

3.3 The Periodic Table Helps Us to Understand the Elements

8. How many periods are there in the periodic table? How many groups?
9. Do properties change or remain the same for elements across any period of the periodic table?
10. Why are the lanthanides and actinides placed beneath the main body of the periodic table?
11. What groups of the periodic table comprise the transition metals?

3.4 Elements Can Combine to Form Compounds

12. What is the difference between an element and a compound?
13. What does the chemical formula of a substance tell us about that substance?
14. How are the properties of a compound related to the properties of the elements used to make that compound?

3.5 There Is A System for Naming Compounds

15. What is the chemical formula for the compound titanium dioxide?
16. What is the name of the compound with the formula $NaNO_3$?
17. How many carbon atoms are there in magnesium cyanide, $Mg(CN)_2$?

3.6 Most Materials Are Mixtures

18. What defines a material as being a mixture?
19. How can the components of a mixture be separated from one another?
20. How does distillation separate the components of a mixture?

3.7 Matter Can Be Classified as Pure or Impure

21. Why is it not practical to have a macroscopic sample that is 100 percent pure?
22. How is a solution different from a suspension?
23. How can a solution be distinguished from a suspension?

3.8 The Advent of Nanotechnology

24. How soon will nanotechnology give rise to commercial products?
25. What are the two main approaches to building nanoscale materials and devices?
26. Who is the ultimate expert at nanotechnology?

Quantitative Questions

27. The Colorado River water in Colorado has a salinity of about 50 ppm. By the time this water passes into Mexico its salinity has increased to about 1000 ppm. How many milligrams of salts have been added to each liter of water?
28. Rainwater is naturally acidic containing about 48 micrograms of acidity per liter. Express this concentration in units of ppm, ppb, and ppt.
29. Dioxins are highly toxic compounds that form upon the burning of certain plastics, especially PVC. Dioxins bioaccumulate, which means that animals higher in the food chain tend to have greater concentrations within their bodies. Most of our exposure to dioxins comes from the food we eat rather than the air we breathe. How many milligrams of dioxins are there in a liter of milk containing 0.16 ppt?
30. Drinking water is routinely disinfected by adding chlorine to a concentration of about 2 ppm. How many milligrams is this per liter of water?

31. Sea water has a salt concentration of about 3.5%. Express this in units of ppm.

32. As a follow up to the previous question, express the concentration of salt in seawater in units of ppb.

Solutions (Odd-Numbered)

1. Nothing. During a physical change, the chemical identity of a substance remains the same.

3. Determining whether a change is physical or chemical can be difficult because both involve changes in appearance.

5. An element consists of only one type of atom.

7. The elemental formula tells us how the atoms within an element are grouped. Within nitrogen, N_2, for example, nitrogen atoms are paired.

9. Across any period, the properties of elements gradually change.

11. The transition metals are those from group 3 – 10.

13. The chemical formula tells us the ratio in which atoms come together to form one unit a particular substance.

15. The formula for titanium dioxide is TiO_2.

17. There are two carbon atoms within magnesium cyanide.

19. Filtration or distillation can separate mixtures. These methods take advantage of differences in the components' physical properties.

21. Atoms and molecules are very small so if one atom or molecule out of a trillion is different then the sample is no longer pure.

23. A centrifuge can be used to determine if a mixture is a solution or a suspension because it will separate the components of a suspension.

25. The top-down approach in which nanostructures are carved out of larger materials and the bottom-up approach in which nanostructures are pieced together atom by atom.

27. The difference between these two concentrations is 950 ppm or 950 milligrams per liter. The amount of salts added to the river water is 950 milligrams per liter.

29. A concentration of 0.16 ppt is 0.16 nanograms per liter. Convert nanograms to milligrams using these equalities: 1 gram = 10^3 mg, 1 gram = 10^9 ng.

(0.16 ng)(1 gram/10^9 ng)(10^3 mg/1 gram) = 1.6×10^{-7} mg = 0.00000016 mg

31. The "per cent" is "per 100". So the concentration of salt in seawater is about 3.5 parts per hundred, or 3.5 pph. There are 10,000 hundreds in a million so the conversion factors will look like this:

(3.5 parts/hundred)(10,000 hundreds/1 million) = 35,000 ppm

Solutions to Chapter 3 Calculation Corner

How Pure Is Pure?

1. One liter of 1.0 ppm fluoridated drinking water contains 1.0 mg of fluoride.

2. A dissolved oxygen concentration of 6 ppm corresponds to 6 mg/L. With one liter, you have 6 mg:

   (6 mg/L)(1 L) = 6 mg

   Use a conversion unit to express 6 mg as 0.006 grams:

   (6 mg)(1 gram/1000 mg) = 0.006 grams

3. A concentration of 25 ppb is 25 micrograms per liter. Convert from micrograms into milligrams as follows:

   (25 micrograms)(1 milligram/1000 micrograms) = 0.025 milligrams

   Thus, there are 0.025 milligrams of chloroform in each liter of this water.

4. Assuming you live in the U.S., which has a population of about 330 million, then your concentration is about 1/0.33 billion = 3.0 ppb. The concentration of you in the whole world is about 1/8 billion = 0.125 ppb, which is about 125 ppt. If you can imagine how few of you there are compared to national or world populations, then you have a sense of how dilute a dissolved substance is when its concentration is measured by the ppb or ppt

Chapter 3: Contextual Chemistry Essay

Extending the Human Life Span

Living isn't easy, especially at the level of the molecules that make us. Our cells and the molecules they contain are constantly exposed to a hostile environment of viruses, bacteria, free radicals, radiation, and random chemical reactions. We live because our bodies are able to repair themselves from perpetual molecular damage. Over time, however, our bodies lose the ability to self-heal. We age. We grow frail and eventually die.

Since the introduction of modern medicine and better health habits, the average life expectancy of humans has increased dramatically—in the United States from about 48 years old in 1900 to about 78 years old in 2020. The maximum attainable human life span of about 120 years, however, appears to have remained fairly constant. Is there truly a limit to how long we can live? Is it possible to change how long we live while still maintaining youthful vigor and resilience? After all, the quality of life matters just as much, if not more, than the quantity of life.

Scientists have long known that a healthier and longer life can be attained by reducing the intake of calories by at least one-third of the number in a normal, healthy diet while maintaining necessary nutrients. Worms fed such a calorie-restricted diet live up to 5 months longer, which for them is a life span increase of about 60 percent. Mice live about 14 months longer (50 percent increase) and dogs also about 14 months longer (10 percent increase). Would calorie restriction also help humans extend their maximum life span? The answer is likely yes, but by how much is questionable. Some scientists are optimistic that it could add 10 to 15 years. Others are more cautious in thinking 2 to 3 years may be more reasonable. Either way, there are potential benefits to be had, and every gain counts.

Scientists are working to unravel the molecular mysteries of why calorie restriction works. This, in turn, should help them to discover compounds that mimic the effects of calorie restriction. Interestingly, one class of compounds that mimics those effects are polyphenols, which are abundant in highly pigmented foods, such as pomegranates, or beverages, such as red wine.

Gerontologists who study the aging process have come to recognize at least six categories of damage sustained by our cells and biomolecules (see Table A). These are likely to be the underlying causes of our becoming frail and more susceptible to death as we grow older. Prevent or reverse these damages, and the result would be a rejuvenated body, which, in turn, would have a greater chance of experiencing an extended life span. The important thing to note here is that we're not talking about keeping older people alive in their frailty. Instead, we're talking about strategies that would permit people to both look and feel better despite having lived for so many years. Humans would remain productive and active for much longer periods of time.

Let's take a brief look at senescent cell therapy, which is just one example of longevity research. In the early 1960s, Professor Leonard Hayflick, one of the founders of the biotech revolution, noted that a human cell is

TABLE A Six Categories of Cellular Damage

PROBLEM	CURRENT AND POTENTIAL REMEDIES*
Cells die and are not replaced.	Improved health habits; growth factors; **gene doping,** stem cells; apoptosis active dephosphorylation inhibitors **(salubrinal); senescent cell** therapy
DNA within the cell nucleus is altered, giving rise to cancer.	Improved health habits; chemotherapy; radiation therapy; surgery; **telomere restoration therapy;** gene therapy; **nanoshells; angiogenesis inhibitors**
DNA within cellular mitochondria is altered, resulting in cell death.	Gene therapy to produce mitochondrial proteins from nuclear DNA
Unwanted cells, such as fat cells, accumulate.	Improved health habits; **calorie restriction diet;** surgery; target cell surface differences
Collagen loses elasticity because of cross-linking.	Sunscreen; advanced glycosylation end-product breaker drugs, such as **ALT-711,** to break crosslinks or inhibit their formation
Unwanted junk, such as atherosclerotic and amyloid plaques, accumulates.	**Beta-sheet breaker peptides;** inhibition of **ABAD beta amyloid complex;** genetically modified white blood cells; **lysosome replacement therapy**

only able to replicate about 60 times. Once this limit is reached the cell may self-destruct or it may be attacked by the immune system. Some of these cells, however, remain in a quasi-dormant state, called *senescence*, where they produce chemicals that are generally not good for the organism. As we grow older, senescent cells in our body accumulate. We are then exposed to greater amounts of the dangerous chemicals these cells produce. This, in turn, gives rise to numerous age-related problems such as the weakening of tissues, arthritis, and cataracts. Recently, scientists found a way to selectively kill senescent cells as they formed in mice. Remarkably, the treated mice remained healthier as they aged compared to mice who received no treatment. If a similar treatment could be applied to humans it would be a huge advance in the treatment of age-related diseases.

Might scientists be interested in finding remedies to the maladies of aging? How about the businesses or organizations that sponsor the scientists? Do you suppose more or less money will be channeled into these efforts as more promising discoveries are made? Will there be a sufficient demand for resulting products that provide for a longer and healthier life? With wrinkle-free skin? Will people of impoverished nations be able to afford this anti-aging technology? Should it be offered freely? Might we have more or fewer remedies available to us by the time you are 40 years older? What if medical science advances faster than you age? Stay tuned. Future advances in biotech are sure to have a profound impact on your quality of living.

Think and Discuss

1. There are millions of people who don't exercise or eat right even though they know such habits will likely extend how long they live in good health. Why?

2. In the past 20 years, the average life expectancy within most nations has risen by a couple of years, but so has the "healthy life expectancy," which is a measure of how long people remain in good health. Are the two necessarily related? How so?

3. How might cures for age-related diseases also be a solution to the problem of overpopulation?

4. How would society be able to support so many people living well into their 100s?

5. The Pentagon today owes its soldiers about a trillion in future retirement benefits. What might happen to this cost if the soldiers actually lived some 40 years longer than expected? How about 100 years? 500 years? At what point should the Pentagon no longer "owe" these benefits to the soldiers? What trends do you foresee in company retirement plans?

6. How are younger people going to be able to find work when so many older people are not retiring? And how might a healthier upper age group affect the wealth distribution between generations?

CONCEPT **CHECK**
How are people like cars?

CHECK YOUR ANSWER
Professor Leonard Hayflick says people are like cars because they age reliably "even though there's nothing in the blueprints that shows a process for doing it." In other words, there is no "death gene," no mechanism that kills us off after a certain time limit; aging is a consequence of living in a corrosive environment. If, however, you want to keep a car like "new," what do you do? Wait until it's ready to fall apart? Or repair as necessary despite the expense of new body parts or skilled labor? Both people and cars need daily maintenance if their life expectancies are to be maximized. If a car can be nurtured to live for many centuries, can people too? Do we have the resources?

Author Responses to Think and Discuss

1. Time is a funny thing. Millions of people living in modern society are so pressed for time, they put off what they perceive as secondary priorities. "I'll get to it as soon as I finish ______." (fill in the blank). The irony is that if they would exercise, eat right and just relax, they would then have longer lives with even more time to "do what they need to do." Like most animals on Earth, we are adept at focusing on the present moment and our present concerns. Unlike most other animals on Earth, we have the ability to think long term anticipating the future. This is a talent we are wise to practice.

2. Average life expectancies have risen not so much because medicine has been able to keep frail people alive longer. Instead, life expectancies have risen primarily because people are living healthier lives. This trend will likely continue. In other words, as the average life expectancies continue to increase, so will the average healthy life expectancies.

3. There is a direct relationship between infant mortality and birth rate—if half of your babies are dying, it's only natural to have more babies. But babies are expensive and they consume a lot of resources. Once more babies start surviving, then the birth rate goes down, which is exactly what happened soon after Louis Pasteur introduced hygiene to the world in the 1870s. Today we still see the relationship. Countries with the greatest infant mortality, such as developing nations, are countries with the greatest birth rates. Cultural inertia in many of these countries, however, has prevented a proportional drop in birth rates. As a consequence, these countries are experiencing unprecedented hardships. So what happens when the elders start living much longer? Might these elders have grown wise enough to understand the problems of overpopulation? Might they retain their old ways thereby reinforcing the cultural inertia to change? In a world where resources for everyone are growing more scarce, the pressure to keep birth rates down will be great. The crux of the issue, however, is whether or not the elderly will be like the babies in that they are expensive and consume a lot of resources.

4. If people are living well into their 100s then they would likely also have the energy to provide some sort of remunerable contribution back to society on up through to their 90s. With their experience and potential wisdom they would have much to offer. An interesting issue we face today at the beginning of the 21st century is what to do, if anything, with our social security system or various retirement plans. What happens when we expect to live to 80 and thus retire at 65, when in fact we live a healthy life up to 120? Social security and retirement funds won't be sufficient. As our expectations change, so will our plans of action. We don't know what to expect!

5. If in the future people live significantly longer than they live today, then government and corporate retirement plans are destined for bankruptcy. To avoid such bankruptcy, the government and corporations will likely move towards privatized retirement accounts where the impending "lack of sufficient funds" falls squarely onto the retiree, who would thus feel pressure to go back to work. Interestingly, the Pentagon predicts that soldier-replacing robots will be a major fighting force in the American military in less than a decade. Their current robot project is the biggest military contract in American history with a price tag of $127 billion.

6. The older folks have a life's worth of work experience that may be of benefit to society. As they remain in the work force longer, they have more opportunities to help in training younger individuals. The older folks are also hopefully in a position to hire younger folks through companies they may own or manage. Mostly, however, new sorts of jobs and careers will need to be created that younger people would be best suited to fill.

Chapter 4

Subatomic Particles

THE MAIN IDEA

Atoms are made of electrons, protons, and neutrons

Λ Each element emits its own characteristic spectral pattern, which can be used to identify the element just as fingerprint can be used to identify a person. The inset image shows the spectral pattern from strontium, which is a common ingredient of fireworks.

You have already learned that matter is made of fundamental units we call atoms. In this chapter we'll be exploring how atoms themselves are made of even more fundamental units called subatomic particles, which include electrons, protons, and neutrons. We'll be looking at this from a historical perspective, beginning with the discovery of the electron in the early 20th century. You will learn that an atom is mostly empty space with nearly all of its mass concentrated in a tiny center called the atomic nucleus.

As electrons whiz around the atomic nucleus, they can absorb or emit energy in the form of light, which travels in distinct little packets called photons. Evidence of this is seen by looking at glowing elements through a glass prism or diffraction grating, as shown in the above photograph. By studying this light, scientists have developed models of the atom. A few of these models are presented in this chapter. Through these models, which continue to be refined even today, we gain a powerful understanding of how atoms behave.

4.1 Physical and Conceptual Models

Atoms are so amazingly small that the number of them in a baseball is roughly equal to the number of ping pong balls that could fit inside a hollow sphere as big as the Earth, as **Figure 4.1** illustrates.

Atoms are so small that we can never see them in the usual sense. This is because light travels in waves, and atoms are smaller than the wavelengths of visible light. As illustrated in **Figure 4.2**, a single object, at any magnification will remain invisible if it is smaller than the wavelength of light used to illuminate it. We could stack microscope on top of microscope and still not see an individual atom.

Although we cannot see atoms directly, we can generate images of them indirectly. As was discussed in Section 3.8, the scanning probe microscope does this by dragging an ultra-thin needle back and forth over the surface of a sample. The result is a computer-generated image of the positions of the atoms on the surface. While such an image is both

Figure 4.1 >
If the Earth were filled with nothing but Ping-Pong balls, the number of balls would be huge and roughly equal to the number of atoms in a baseball. Said differently, if a baseball were the size of the Earth, one of its atoms would be the size of a Ping-Pong ball.

remarkable and useful, it is not a photograph showing the actual appearance of the atoms. Atoms do not have solid surfaces, as implied by these images. Rather, as we'll be discussing in this chapter, atoms are made of mostly empty space. Envision what empty space looks like. Then you will have a more accurate picture of what it's really like down there at the level of atoms and molecules.

A very small or very large visible object can be represented with a **physical model**, which is a model that replicates the object at a more convenient scale. **Figure 4.3a**, for instance, shows a large-scale physical model of a microorganism that a biology student might use to study the microorganism's structure. Because atoms are invisible, however, we cannot use a physical model to represent them. In other words, we cannot simply scale up the atom to a larger size, as we might with a microorganism. So, rather than describing the atom with a physical model, chemists use a **conceptual model**, which is a representation of a system that helps us predict how the system behaves. The more accurate a conceptual model, the more accurately it predicts the behavior of the system.

READING CHECK
What is a conceptual model?

For example, the weather is best described using a conceptual model like the one shown in **Figure 4.3b**. Such a model shows how the various components of the system—humidity, atmospheric pressure, temperature, electric charge, the motion of large masses of air—interact with one another. Other systems that can be described by conceptual models are the economy, population growth, the spread of diseases, and team sports.

Like the weather, the atom is a complex system of interacting ultra-tiny components, notably electrons, protons, and neutrons. The atom, therefore, is best described with a conceptual model. Thus, you should be careful not to interpret any visual representation of an atomic conceptual model as

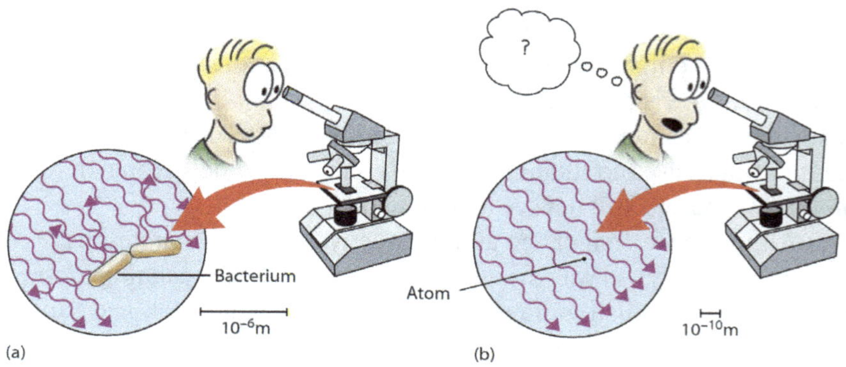

Figure 4.2 >
(a) A bacterium is visible because it is much larger than the wavelengths of visible light. We can see the bacterium through the microscope because the bacterium reflects visible light back toward the eye. (b) An atom is invisible because it is smaller than the wavelengths of visible light, which pass by the atom with no reflection.

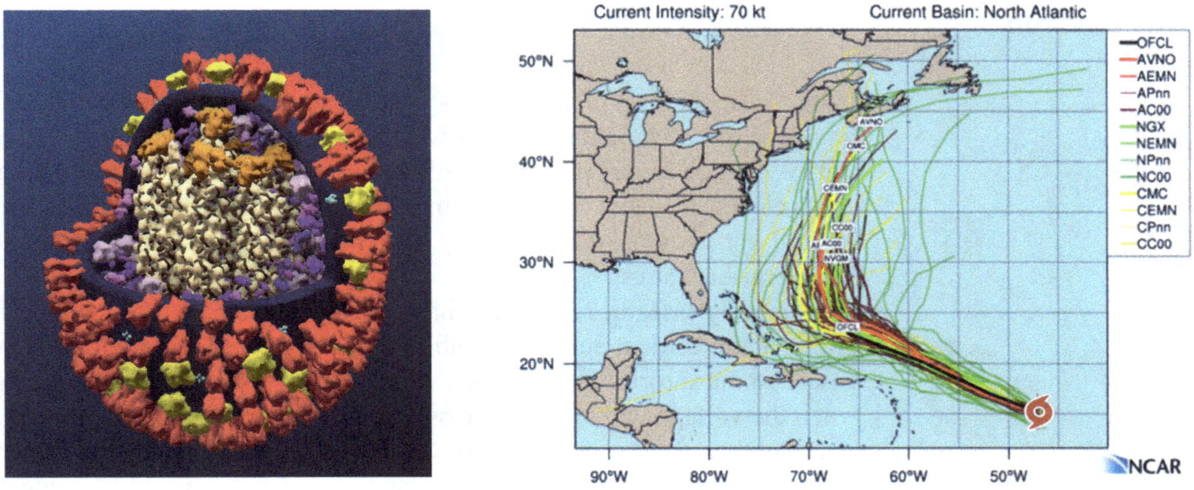

(a) (b)

Λ Figure 4.3

(a) This large-scale model of an influenza virus is a physical model rendered by a computer. To play with this model in 3D, visit: https://3d.nih.gov/entries/3DPX-013373. (b) Weather forecasters rely on conceptual models to predict the behavior of weather systems, such as the potential paths of a hurricane.

a recreation of an actual atom. In Section 4.6, for example, you will be introduced to the planetary model of the atom, wherein electrons are shown orbiting a dense center of protons and neutrons (the atomic nucleus) much as planets orbit the Sun. This planetary model is limited, however, in that it fails to explain many properties of atoms. Therefore, newer and more accurate (and more complicated) conceptual models of the atom have since been introduced. In these models, electrons appear as a cloud, but even these models have their limitations.

In this textbook, our focus is on conceptual atomic models that are easily represented by visual images, including the planetary model and a slightly more sophisticated model in which electrons are grouped in units called shells. Despite their limitations, such images are excellent guides to learning chemistry, especially for the beginning student. We will introduce these models as they were developed historically, beginning with the discovery of the subatomic particles.

fyi FOR YOUR INFORMATION

Atoms are so small that there are more than the number of breaths in Earth's atmosphere. Within a few years, the atoms of your breath are uniformly mixed throughout the atmosphere. What this means is that anyone anywhere on the earth inhaling a breath of air takes in numerous atoms that were once part of you. And, of course, the reverse is true: you inhale atoms that were once part of everyone who has ever lived. We are literally breathing one another.

CONCEPT **CHECK**

A basketball coach describes a playing strategy to her team by way of sketches on a game card. Do the illustrations represent a physical model or a conceptual model?

CHECK YOUR ANSWER

The sketches are a conceptual model the coach uses to describe a system (the players on the court), with the hope of achieving an outcome (winning the game).

4.2 The Electron

A popular myth holds that Franklin discovered the electrical nature of lightning by flying a kite during a lightning storm. Franklin, however, was smart enough to know the extreme danger posed by such a foolish act. As a practical application of his discovery, Franklin invented the lightning rod, which is a sharp point of metal placed on a rooftop and connected to the ground by a long wire. Houses equipped with such rods are protected from the danger of lightning strikes.

READING CHECK

The greater a particle's mass, the greater its resistance to what?

In 1752, Benjamin Franklin (1706–1790) learned from experiments with thunderstorms that lightning is a flow of electrical energy through the atmosphere. This discovery prompted 19th-century scientists to explore whether or not electrical energy could travel through gases other than the atmosphere. To find out, they passed electricity across glass tubes in which they had sealed various gases.

In every case, the result was a brightly glowing ray (**Figure 4.4a**). This meant that electrical energy was able to travel through different types of gases. To the surprise of these early investigators, a ray was also produced when the voltage was applied across a glass tube that had been evacuated and was thus empty of any gas (**Figure 4.4b**). This implied that the ray was not simply the glowing of a contained gas but rather an entity in and of itself. Experiments showed that the ray emerged from the end of the tube that was negatively charged. Because this negatively charged end was called the cathode, the apparatus, shown in **Figure 4.5a**, was named a *cathode ray tube*. Magnetic fields deflected the ray, as did small electrically charged metal plates. When such plates were used, the ray was always deflected toward the positively charged plate and away from the negatively charged plate. Because identical charges repel each other, this meant the cathode ray was negatively charged. The speed of the ray was found to be considerably less than the speed of light. Because of these characteristics, it appeared that the ray behaved more like a beam of particles than a beam of light. In 1897, J. J. Thomson (1856–1940) measured the deflection angles of cathode ray particles in a magnetic field, using a magnet positioned as shown in **Figure 4.5b**. He reasoned that the deflection of the particles depended on their mass and electric charge. The greater a particle's mass (inertia), the greater its resistance to a change in motion, which gives rise to a smaller deflection. The greater a particle's charge, the stronger the magnetic interactions, which gives rise to a larger deflection. The angle of deflection, he concluded, was proportional to the ratio of the particle's charge to its mass:

$$\text{angle of deflection} \propto \frac{\text{charge}}{\text{mass}}$$

Knowing only the angle of deflection, however, Thomson was unable to calculate either the charge or the mass of each particle. In order to calculate the mass, he needed to know the charge, but in order to calculate the charge, he needed to know the mass.

In 1909, the American physicist Robert Millikan (1868–1953) calculated the fundamental unit of electric charge on the basis of the innovative experiment shown in **Figure 4.6**. Millikan sprayed tiny oil droplets into a specially designed chamber. The droplets picked up a negative charge after passing through a hole in a charged plate. Millikan then reversed the charge on the plate so that the droplets would be pulled upward. By adjusting the strength of the plate's charge he could get droplets of various sizes to hover motionless. In such cases the upward electric force exactly balanced the downward force of gravity.

Repeated measurements showed that the electric charge on any droplet was always some multiple of a single very small value, 1.60×10^{-19} coulomb, which Millikan proposed to be the fundamental increment of all electric charge. (The coulomb is a unit of electric charge.) Using this value and the

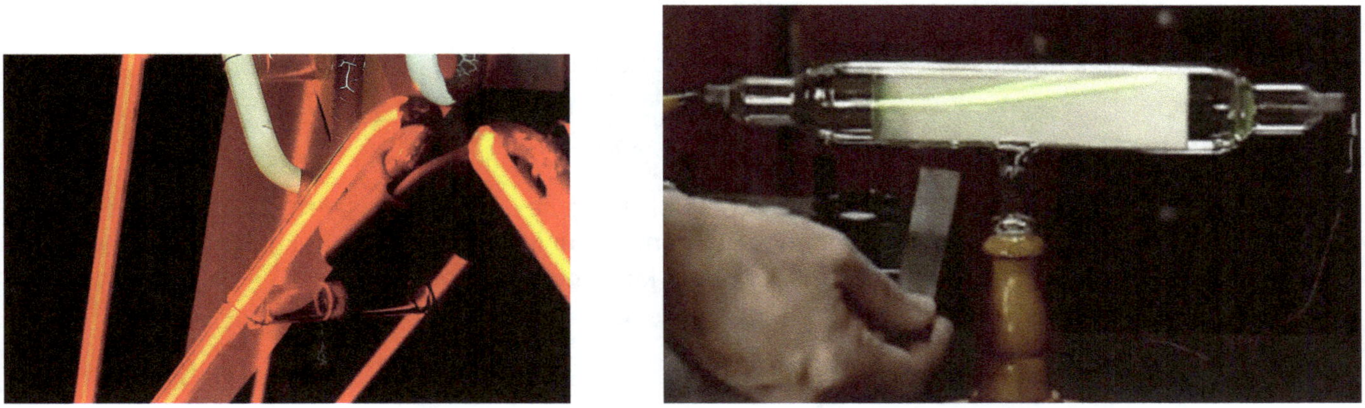

∧ Figure 4.4
(a) Electrical energy passing through a glass tube filled with neon gas generates a bright glowing red. (b) The ray passing through an evacuated glass tube is not usually visible. In the tube shown here, however, the ray is highlighted by a fluorescent backing that glows green as the ray passes over it.

charge-to-mass ratio discovered by Thomson, Millikan calculated the mass of a cathode ray particle to be considerably less than that of the smallest known atom, hydrogen. This was startling because at the time the atom was thought to be the smallest particle of matter. Here scientists had discovered a particle smaller than the smallest atom.

The cathode ray particle is known today as the **electron**, a name that comes from the Greek word for amber (*electrik*), which is a material the early Greeks used to study the effects of static electricity. The electron is a fundamental component of all atoms. All electrons are identical, each having a negative electric charge and an incredibly small mass of 9.1×10^{-31} kilograms. The arrangement of electrons in atoms determine many of a material's properties, including chemical reactivity and such physical attributes as taste, texture, appearance, and color.

The cathode ray—a stream of electrons—has found a great number of applications. Most notably, the original television sets (not the modern LCD screens) were cathode ray tubes with one end widened out into a phosphor-coated screen. Signals from the television station would cause electrically charged plates in the tube to control the direction of the ray in such a way that images were traced onto the screen.

Joseph John Thomson, known to his colleagues as J. J., was one of the first directors of the famous Cavendish Laboratory of Cambridge University in England, where almost all the early discoveries concerning subatomic particles and their behavior were made. Seven of Thomson's students went on to receive Nobel prizes for their scientific work. Thomson himself won a Nobel prize in 1906 for his work with the cathode ray tube.

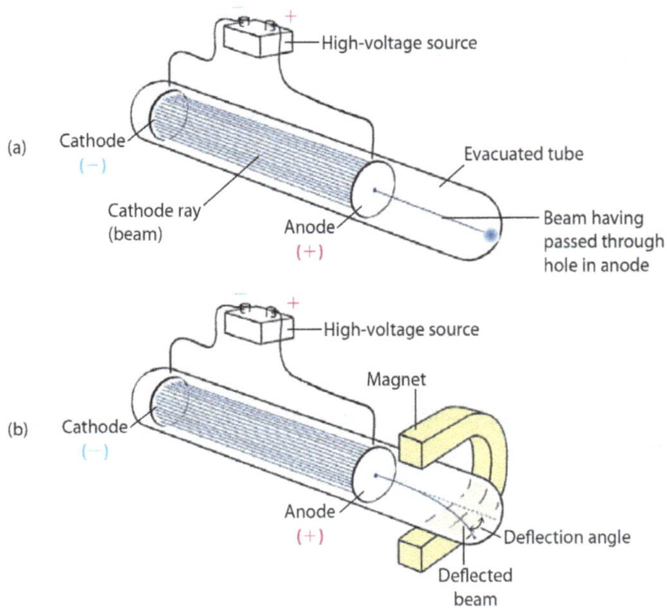

< Figure 4.5
(a) A simple cathode ray tube. The small hole in the positively charged end of the tube, the anode, permits the passage of a narrow beam that strikes the end of the tube, producing a glowing dot as the beam interacts with the glass. (b) The cathode ray is deflected by a magnetic field.

CONCEPT **CHECK**

For which equation is it not possible to calculate one specific value for x?

$4 = x/2$

$3 = x/y$

CHECK YOUR ANSWER In the first equation, it is possible to figure that x = 8 (because 8/2 = 4). In the second equation, one specific value for x cannot be determined unless the value of y is known. Similarly, Thomson could not calculate the electron's mass without knowing its charge.

Figure 4.6 >
Millikan determined the charge of an electron with this oil-drop experiment.

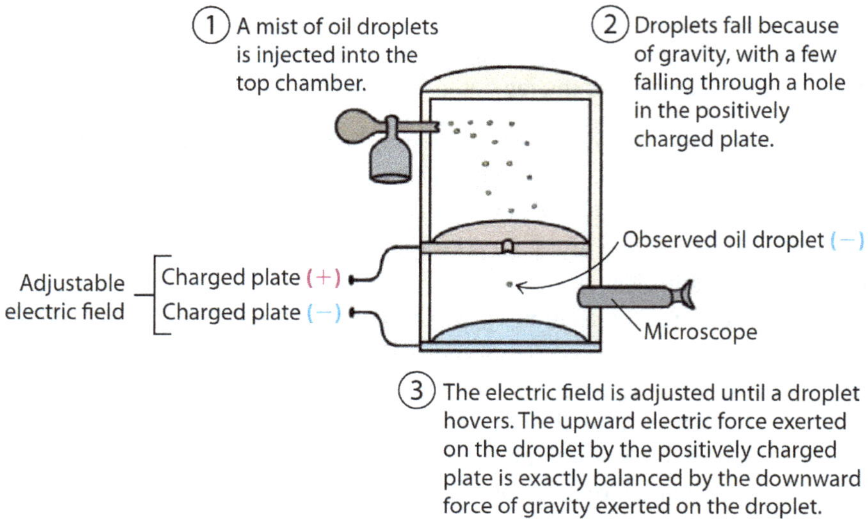

The European science community of the 1800s viewed most American scientists as inventors—clever, but not profound in their thinking or discoveries. This attitude began to change at the turn of the 20th century, principally because of the work of American scientists such as Robert Millikan, who excelled in his experimental designs and conclusions. In addition to research, he also spent much time preparing textbooks so that his students did not have to rely so much on lectures. He won a Nobel prize in 1923 and served as the president of Caltech from 1921 to 1945.

CONCEPT **CHECK**

What do the numbers 45, 30, 60, 75, 105, 35, 80, 55, 90, 20, and 65 have in common?

CHECK YOUR ANSWER

They are all multiples of 5. In a similar fashion, Millikan noted electric charges that were multiples of a very small number, which he calculated to be 1.60×10^{-19} coulomb.

4.3 The Atomic Nucleus

It was reasoned that if atoms contained negatively charged particles, some balancing positively charged matter must also exist. From this, Thomson put forth what he called a *plum-pudding model* of the atom, shown in **Figure 4.7.** Further experimentation, however, soon proved this model to be wrong.

Around 1910, a more accurate picture of the atom came to one of Thomson's former students, the New Zealand physicist Ernest Rutherford (1871–1937). Rutherford oversaw the now-famous gold-foil experiment, which was the first experiment to show that the atom is mostly empty space and that most of its mass is concentrated in a tiny central core called the **atomic nucleus**.

In Rutherford's experiment, shown in **Figure 4.8**, a beam of positively charged particles, called alpha particles, was directed through a very thin sheet of gold foil. Since alpha particles were known to be thousands of times more massive than electrons, it was expected that the alpha-particle stream would not be impeded as it passed through the "atomic pudding" of gold foil. This was indeed observed to be the case—for the most part. Nearly all alpha particles passed through the gold foil. However, some particles were deflected from their straight-line path as they passed through the foil. A few of them were even deflected straight back toward the source! These alpha particles must have hit something relatively massive, but what?

Rutherford reasoned that undeflected particles traveled through regions of the gold foil that were empty space, as **Figure 4.9** shows, and the deflected ones were repelled by extremely dense positively charged centers. Each atom, he concluded, must contain one of these centers, which he named the *atomic nucleus*.

Rutherford guessed that atomic nuclei must be positively charged to balance the negative charge of the electrons in the atom. He also guessed that the electrons were not part of this nucleus but still somewhere in the atom. Today we know that, as **Figure 4.10** illustrates, the electrons do indeed exist outside the nucleus, swirling around it at ultrahigh speeds.

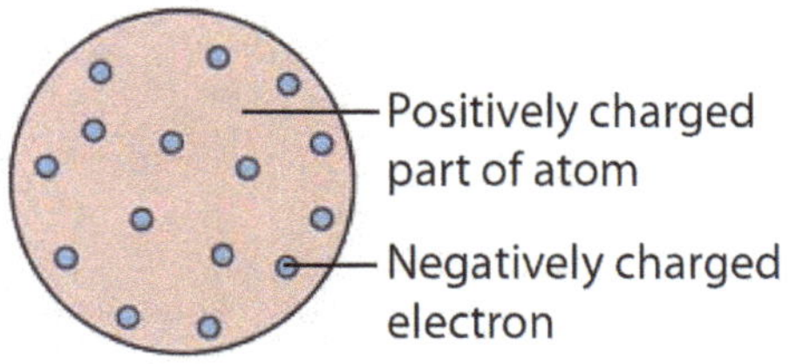

Λ Figure 4.7

Thomson's plum-pudding model of the atom. Thomson proposed that the atom might be made of thousands of tiny, negatively charged particles swarming within a cloud of positive charge, much like plums and raisins in an old-fashioned Christmas plum pudding.

READING CHECK

What did Rutherford reason about the deflected alpha particles?

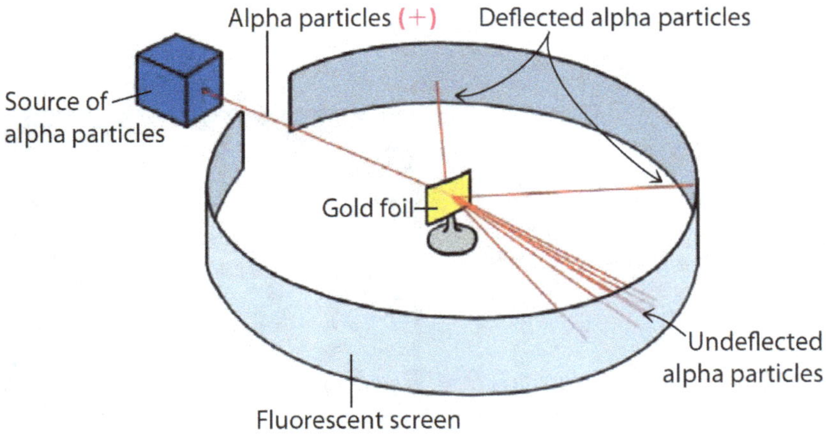

Λ Figure 4.8

Rutherford's gold-foil experiment. A beam of positively charged alpha particles was directed at a piece of gold foil. Most of the particles passed through the foil undeflected, but some were deflected. This result implied that each gold atom was mostly empty space with a concentration of mass at its center—the atomic nucleus.

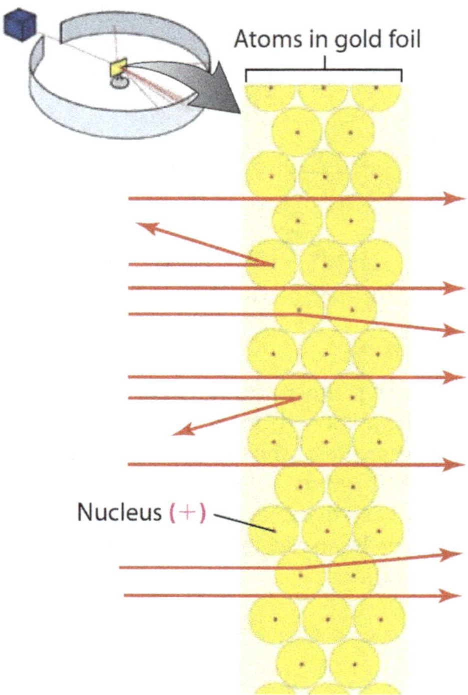

^ Figure 4.9
Rutherford's interpretation of the results from his gold-foil experiment. Most alpha particles passed through the empty space of the gold atoms undeflected, but a few were deflected by atomic nuclei.

What Figure 4.10 does not show is that an atom is mostly empty space, with the diameter of the whole atom being about 10,000 times greater than the diameter of its nucleus. If a nucleus were the size of the period at the end of this sentence, the outer edges of the atom would be located 3.3 meters (11 feet) away. Because electrons are even smaller than the nucleus, and because they are widely separated from each other (as well as from the nucleus), atoms are indeed mostly empty space—just as our solar system is mostly empty space.

So, if atoms are mostly empty space, why don't they simply pass through one another? They don't because electrons repel the electrons of neighboring atoms. Therefore, two atoms can get only so close to each other before they are repelled. This explains why gravity doesn't pull you through the floor as you stand. While the force of gravity pulls you down, the electric force of repulsion between the atoms of the floor and those of your feet pushes you up. As you stand on the floor, these two forces are balanced and you find yourself neither falling nor rising—there's more to standing than most people realize!

Similarly, when the atoms of your hand push against the atoms of a wall, repulsions between electrons in your hand and electrons in the wall prevent your hand from passing through the wall. You sense this repulsion as a pressure that pushes back. Our sense of touch comes from these electrical repulsions. Interestingly, when you touch someone, your atoms and those of the other person do not meet. Instead, atoms from the two of you get close enough so that you sense an electrical repulsion. There is still a tiny, though imperceptible, gap between the two of you (**Figure 4.11**).

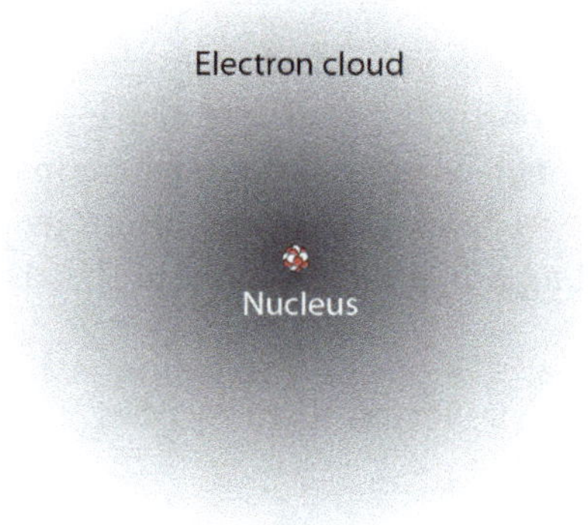

< Figure 4.10
Electrons whiz around the atomic nucleus, forming what can be best described as an ultrathin cloud. If this illustration were drawn to scale, the atomic nucleus would be too small to be seen. An atom is mostly empty space.

< Figure 4.11
Stephanie and Kempton make contact by pushing together against the electric repulsive forces of their atoms, which never actually meet.

When Ernest Rutherford was 24, he placed second in a New Zealand scholarship competition to attend Cambridge University in England, but the scholarship was awarded to Rutherford after the winner decided to stay home and get married. In addition to discovering the atomic nucleus, Rutherford was also first to characterize and name many of the nuclear phenomena discussed in Chapter 5. He won a Nobel prize in 1908 for showing how elements such as uranium can become different elements through the process of radioactive decay. At the time, the idea of one element transforming to another was shocking and met with great skepticism, because it seemed reminiscent of alchemy.

4.4 Protons and Neutrons

The positive charge of any atomic nucleus was found to be equal in magnitude to the combined negative charge of all the electrons in the atom. It was thus reasoned, and then experimentally confirmed, that the nucleus contains positively charged subatomic particles, which we call **protons**. Protons are nearly 2000 times more massive than electrons. The number of protons in a nucleus is equal to the number of electrons whirling about it, so the charges are balanced.

Scientists have agreed to identify elements by **atomic number**, which is the number of protons in the nucleus of each atom of a given element. The modern periodic table lists the elements in order of increasing atomic number. Hydrogen, with one proton per atom, has atomic number 1; helium, with two protons per atom, has atomic number 2; and so on.

CONCEPT **CHECK**

How many protons are there in an iron atom, Fe (atomic number 26)?

CHECK YOUR ANSWER The atomic number of an atom and its number of protons are the same. Thus, there are 26 protons in an iron atom. Another way to put this is that all atoms that contain 26 protons are, by definition, iron atoms.

If we compare the electric charges and masses of different atoms, we see that an atomic nucleus must be made up of more than just protons. Helium, for example, has twice the electric charge of hydrogen but four times the mass. The added mass is due to another subatomic particle found in the nucleus, the neutron, which was first detected in 1932 by the British physicist James Chadwick (1891–1974).

Neutrons have about the same mass as protons, but they have no electric charge. Any object that has no net electric charge is said to be *electrically neutral*, and that is where the neutron got its name. We discuss the important role that neutrons play in holding the atomic nucleus together in Chapter 5.

Both protons and neutrons are called **nucleons**, a term that denotes their location in the atomic nucleus. Figure 4.12 shows an illustration of a nucleus with many nucleons (protons and neutrons). Table 4.1 summarizes the basic facts about electrons, protons, and neutrons.

For any element, the number of neutrons in the nucleus may vary. For example, most hydrogen atoms (atomic number 1) have no neutrons. A small percentage, however, have one neutron, and a smaller percentage have two neutrons. Similarly, most iron atoms (atomic number 26) have 30 neutrons, but a small percentage have 29 neutrons. Atoms of the same element that contain different numbers of neutrons are **isotopes** of one another.

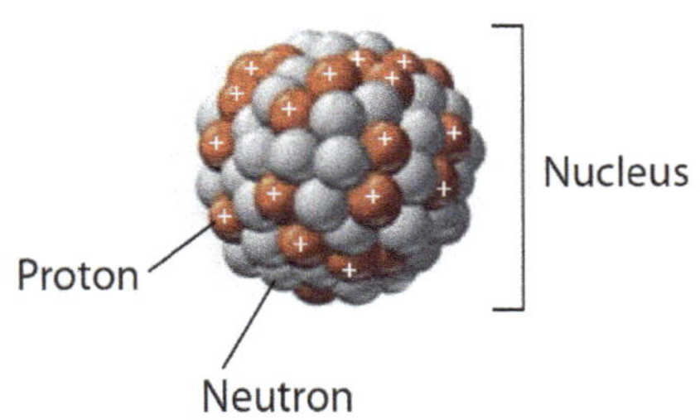

Figure 4.12
An illustration of an atomic nucleus showing its protons and neutrons all clumped together. Warning! This is not what the nucleus really looks like. Interestingly, the atomic nucleus, just like the atom, is also made of mostly empty space, but is filled with a very high concentration of energy, as we discuss in Chapter 5.

TABLE 4.1 Subatomic Particles

	PARTICLE	CHARGE	RELATIVE MASS	ACTUAL MASS (KG)
	Electron	−1	1	9.11×10^{-31}*
Nucleons	Proton	+1	1836	1.673×10^{-27}
	Neutron	0	1841	1.675×10^{-27}

*9.11×10^{-31} kg = 0.000000000000000000000000000000911 kg

READING CHECK

How are isotopes identified?

We identify isotopes by their **mass number**, which is the total number of protons plus neutrons (in other words, the number of nucleons) in the nucleus. As **Figure 4.13** shows, a hydrogen isotope with only one proton is called hydrogen-1, where 1 is the mass number. A hydrogen isotope with one proton and one neutron is therefore hydrogen-2, and a hydrogen isotope with one proton and two neutrons is hydrogen-3. Similarly, an iron isotope with 26 protons and 30 neutrons is called iron-56, and one with only 29 neutrons is iron-55.

An alternative method of indicating isotopes is to write the mass number as a superscript and the atomic number as a subscript to the left of the atomic symbol. For example, an iron isotope with a mass number of 56 and atomic number of 26 is written as

$$\text{Mass number} \searrow {}^{56}_{26}\text{Fe} \text{—Atomic symbol}$$
$$\text{Atomic number} \nearrow$$

The total number of neutrons in an isotope can be calculated by subtracting its atomic number from its mass number:

Mass Number – Atomic Number = Number of Neutrons

For example, uranium-238 has 238 nucleons. The atomic number of uranium is 92, which tells us that 92 of these 238 nucleons are protons. The remaining 146 nucleons must be neutrons:

$$\begin{array}{l} 238 \text{ protons and neutrons} \\ \underline{-92 \text{ protons}} \\ 146 \text{ neutrons} \end{array}$$

Atoms interact with one another electrically. Therefore, the way any atom behaves in the presence of other atoms is determined largely by the charged particles it contains, especially its electrons. Isotopes of an element differ only by mass, not by electric charge. For this reason, isotopes of an element share many characteristics—in fact, as chemicals they cannot be distinguished from one another. For example, a sugar molecule containing carbon atoms with seven neutrons is digested no differently from a sugar molecule containing carbon atoms with six neutrons. Interestingly, about 1 percent of the carbon we consume in food is the carbon-13 isotope, containing seven

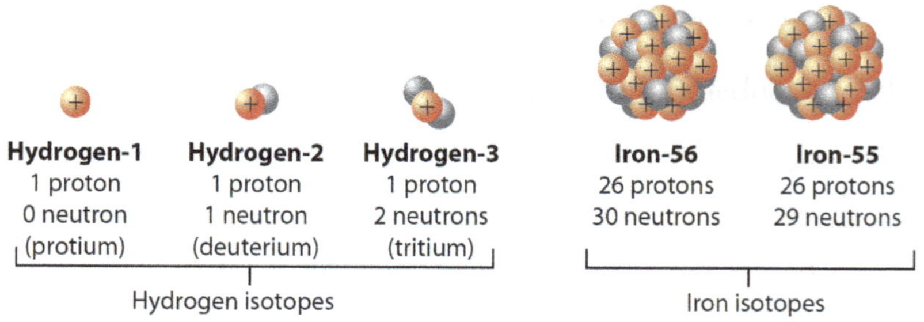

Figure 4.13 >
Isotopes of an element have the same number of protons but different numbers of neutrons and hence different mass numbers. The three hydrogen isotopes have special names: protium for hydrogen-1, deuterium for hydrogen-2, and tritium for hydrogen-3. Of these three isotopes, hydrogen-1 is the most common. For most elements, such as iron, the isotopes have no special names and are indicated merely by mass number

neutrons per nucleus. The remaining 99 percent of the carbon in our diet is the more common carbon-12 isotope, containing six neutrons per nucleus.

The total mass of an atom is called its **atomic mass**. This is the sum of the masses of all the atom's components (electrons and the nucleus). Because electrons are so much less massive than the nucleus, their contribution to atomic mass is negligible.

As we explore further in Section 9.2, a special unit has been developed for atomic masses. This is the *atomic mass unit*, amu; 1 atomic mass unit is equal to 1.661×10^{-24} gram, which is slightly less than the mass of a single proton. As shown in **Figure 4.14**, the atomic masses listed in the periodic table are in atomic mass units. As is explored in the Calculation Corner, the atomic mass of an element as presented in the periodic table is actually the average atomic mass of the various isotopes of that element occurring in nature.

CONCEPT **CHECK**

Distinguish between mass number and atomic mass.

CHECK YOUR ANSWER

Both terms include the word mass, so they are easily confused. Focus your attention on the second word of each term, however, and you'll get it right every time. Mass number is a count of the number of nucleons in an isotope. An atom's mass number requires no units because it is simply a count. Atomic mass is a measure of the total mass of an atom, which is given in atomic mass units. If necessary, atomic mass units can be converted to grams using the relationship 1 atomic mass unit = 1.661×10^{-24} gram.

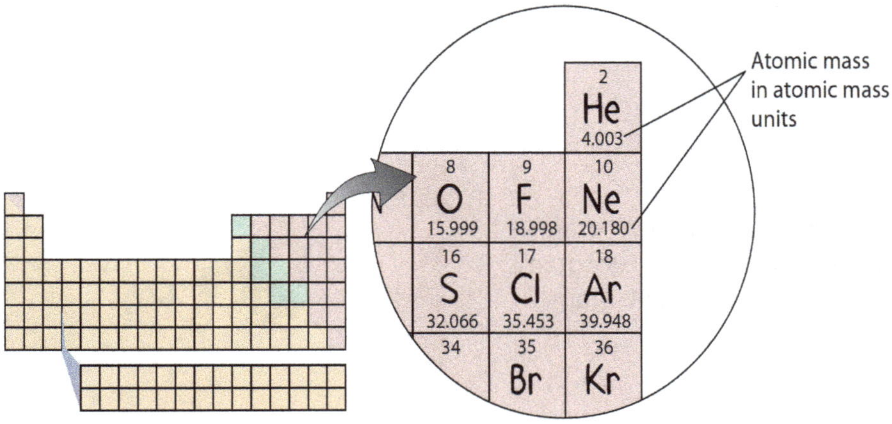

< Figure 4.14
Helium, He, has an atomic mass of 4.003 atomic mass units, and neon, Ne, has an atomic mass of 20.180 atomic mass units.

Calculation Corner: Calculating Atomic Mass

Roughly 99 percent of all carbon atoms, are the isotope carbon-12, and most of the remaining 1 percent are the heavier isotope carbon-13. This small amount of carbon-13 raises the *average* mass of carbon from 12.000 amu to the slightly greater value 12.011 amu. By analogy, consider a quiz where everyone scores a 12—the average is 12. But what if one student scored a 13? The class average would be weighted a little bit higher than 12 because of that one student.

To arrive at the atomic mass presented in the periodic table, you first multiply the mass of each naturally occurring isotope of an element by the fraction of its abundance and then add up all the fractions. This gives you the *weighted average* of all the isotopes.

EXAMPLE
Carbon-12 has a mass of 12.0000 atomic mass units and makes up 98.89 percent of naturally occurring carbon. Carbon-13 has a mass of 13.0034 atomic mass units and makes up 1.11 percent of naturally occurring carbon. Use this information to show that the atomic mass of carbon shown in the periodic table, 12.011 amu, is correct.

ANSWER
Recognize that 98.89 percent and 1.11 percent expressed as fractions are 0.9889 and 0.0111, respectively. Hence:
$(12.0000 \text{ amu})(0.9889) + (13.0034)(0.0111) = 12.011$

YOUR TURN
Chlorine-35 has a mass of 34.97 amu, and chlorine-37 has a mass of 36.95 amu. Determine the atomic mass of chlorine, Cl (atomic number 17), if 75.53 percent of all chlorine atoms are the chlorine-35 isotope and 24.47 percent are the chlorine-37 isotope.

From the periodic table, you can calculate the average mass of an individual atom in units of grams using the atomic mass unit (amu) to grams conversion factor:

EXAMPLE
In units of grams, what is the mass of a single oxygen atom? Assume an atomic mass of 15.999 amu.

ANSWER
$(15.999 \text{ amu})(1.661 \times 10^{-24} \text{ g/1 amu}) = 2.657 \times 10^{-23} \text{ g}$

YOUR TURN
Show that the mass of two hydrogen atoms is 3.348×10^{-24} grams.
Show that the mass of a water molecule, H_2O, is 2.992×10^{-23} grams.
Show that there are 3.342×10^{22} water molecules in 1.000 gram.

4.5 Light Is a Form of Energy

Modern models of the atom were developed by scientists to explain how atoms emit light. To help you understand and appreciate these models, it is good to review the basic concepts regarding the nature of light.

From physics we learn that surrounding every charged particle is an *electric field*, which is a region of space through which the electric force of that charged particle may act. If an electron were to start vibrating, the electric field surrounding that electron would also start to vibrate and at the same frequency. Interestingly, such a vibrating electric field generates a complementary vibrating magnetic field, which in turn reinforces the vibrating electric field. The net result is a series of self-reinforcing electric and magnetic waves that propagate away from the vibrating electron.

By analogy, you know that if you shake the end of a stick back and forth in still water, you create waves that travel outward on the water's surface. If you similarly shake an electrically charged rod in empty space, you create electromagnetic waves in space that also travel outward. Electromagnetic waves, however, are oscillations (vibrations) of electric and magnetic fields, not oscillations of a material medium such as water. These waves are called *electromagnetic radiation*. Most of the electromagnetic radiation we encounter is generated by electrons, which can oscillate at exceedingly high rates because of their small size.

READING CHECK
What are electromagnetic waves?

The distance between two crests of an electromagnetic wave is called the *wavelength* of the wave. **Figure 4.15** labels two wavelengths—one very long, the other very short—on a fictitious wave drawn for illustration only. Electromagnetic waves can also be characterized by their *wave frequency*, a measure of how rapidly they oscillate. The basic unit of wave frequency is the hertz (abbreviated Hz), where 1 hertz equals 1 cycle per second. The higher the frequency of a wave, the shorter its wavelength and the greater its energy.

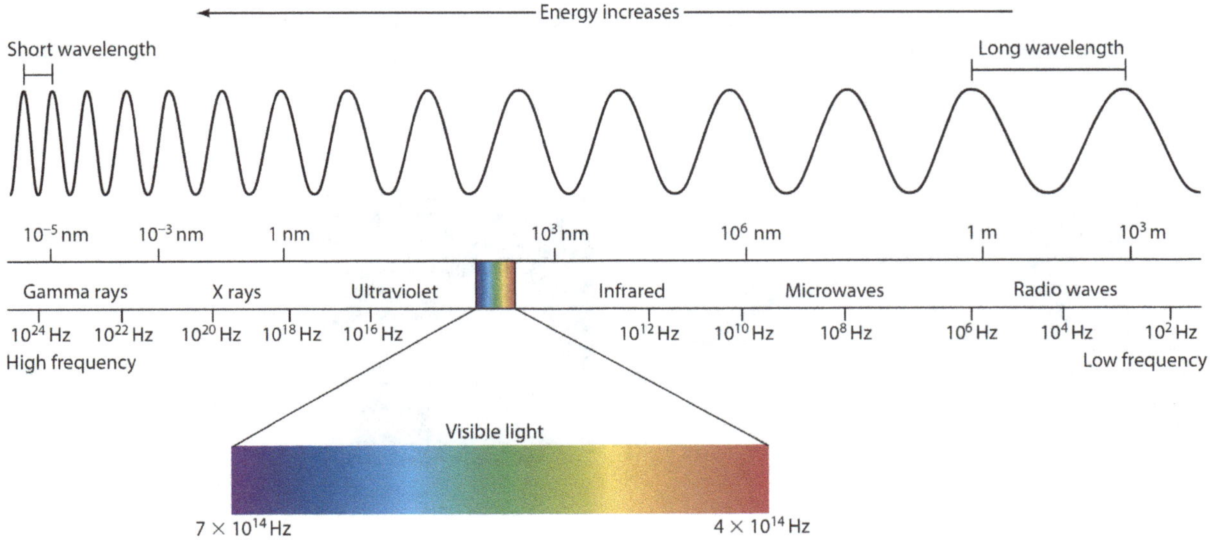

Λ Figure 4.15
The electromagnetic spectrum is a continuous band of wave frequencies extending from high- energy gamma rays, which have short wavelengths and high frequencies, to low-energy radio waves, which have long wavelengths and low frequencies. The descriptive names of these regions are merely a historical classification, for all waves are the same in nature, differing only in wavelength and frequency. Interestingly, the speed at which they travel is also the same. This is the speed of light, which is about 3×10^8 meters per second.

> **FOR YOUR INFORMATION**
>
> Tap the end of a stick on the surface of some still water. Do this repeatedly and you will generate waves that emanate outward. The faster you tap your stick, the closer together the waves are to one another, but the speed at which they travel outward remains the same. Try it and see! Similarly, as electrons oscillate back and forth in an atom, they generate electromagnetic waves that emanate from the atom. All electromagnetic waves travel at the same speed, which is the speed of light.

Figure 4.15 shows a full range of frequencies and wavelengths of electromagnetic radiation in a display known as the **electromagnetic spectrum**. The most energetic region of the electromagnetic spectrum consists of gamma rays. Next is the region of slightly lower energy, where we find X rays, and following that is the electromagnetic radiation we call ultraviolet light. Within a narrow region from about 7×10^{14} (700 trillion) hertz to about 4×10^{14} (400 trillion) hertz are the frequencies of electromagnetic radiation known as visible light. This region includes the rainbow of colors our eyes are able to detect, from violet at 700 trillion hertz to red at 400 trillion hertz. Lower in energy than visible light are infrared waves (detected by our skin as "heat waves"), then microwaves (used to cook foods), and finally radio waves (through which radio signals are sent)—the waves of lowest energy.

We see white light when all frequencies of visible light reach our eye at the same time. By passing white light through a prism or through a diffraction grating, which is a glass plate or plastic sheet with microscopic lines etched into it, the color components of the light can be separated, as shown in **Figure 4.16**. Remember—each color of visible light corresponds to a unique frequency.

A **spectroscope**, shown in **Figure 4.17**, is an instrument used to observe the color components of any light source. As we discuss in the following section, a spectroscope allows us to analyze the frequencies of light emitted by elements as they are made to glow.

> CONCEPT **CHECK**
>
> Can you see radio waves? Can you hear them?
>
> **CHECK YOUR ANSWER**
>
> Your eyes are equipped to see only the narrow range of frequencies of electromagnetic radiation from about 700 trillion to 400 trillion hertz—the range of visible light. Radio waves are one type of electromagnetic radiation, but their frequency is much lower than what your eyes can detect. Thus, you can't see radio waves. Neither can you hear them. You can, however, turn on an electronic gizmo called a radio. The radio translates radio waves into signals that drive a speaker to push air molecules around in the form of sound waves your ears can detect.

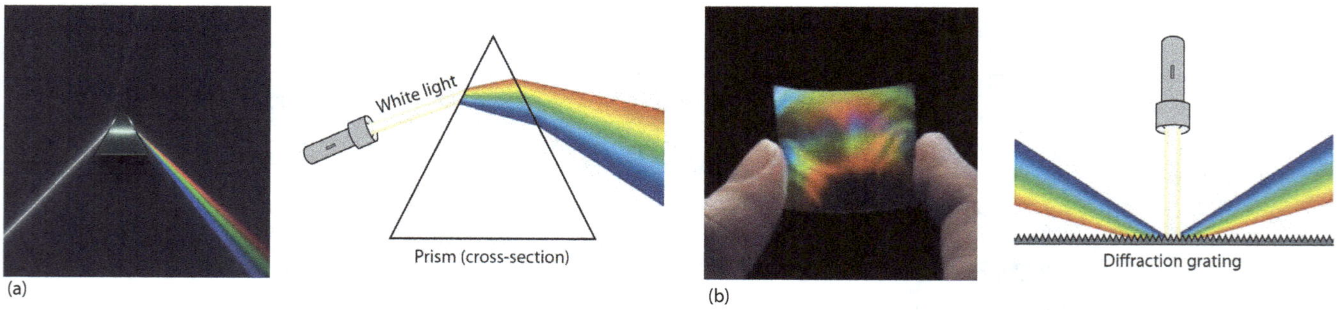

Λ Figure 4.16
White light is separated into its color components by (a) a prism and (b) a diffraction grating.

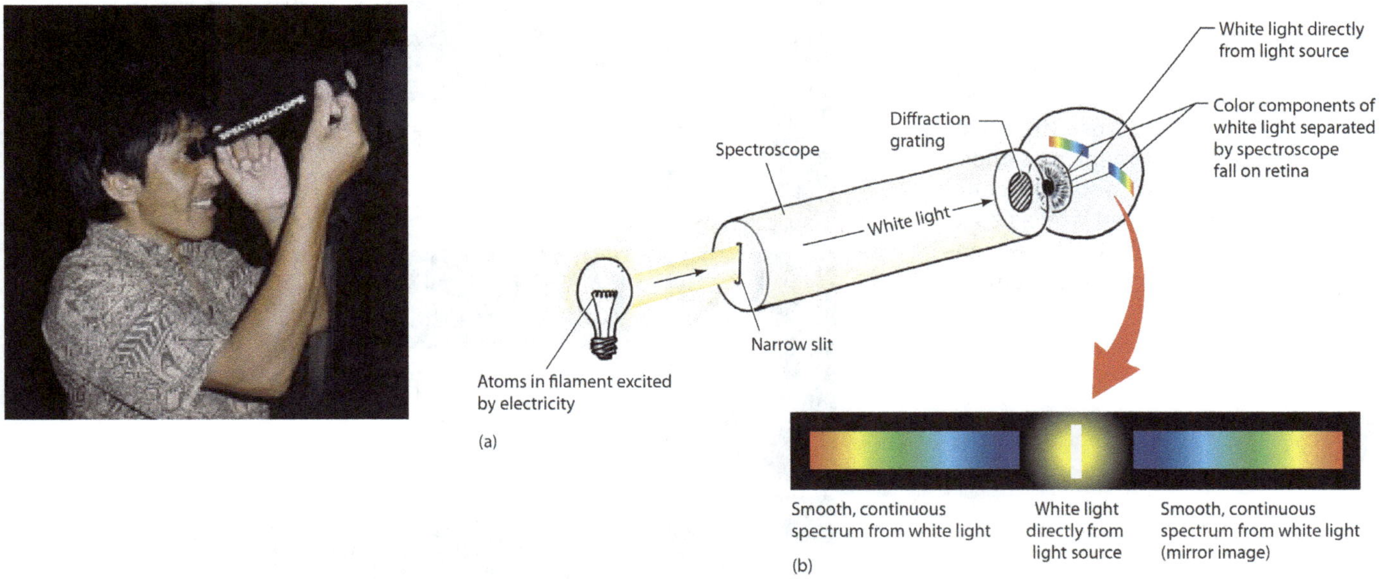

Λ Figure 4.17
(a) In a spectroscope, light emitted by atoms passes through a narrow slit before being separated into particular frequencies by a prism or (as shown here) a diffraction grating. (b) This is what the eye sees when the slit of a diffraction-grating spectroscope is pointed toward a white-light source. Spectra of colors appear to the left and right of the slit.

4.6 Atomic Spectra and the Quantum

Atoms give off light as they are subjected to various forms of energy, such as heat or electricity. The atoms of any given element in the gaseous phase emit only certain frequencies of light, however. As a consequence, each element emits its own distinctive glow when energized. Sodium atoms emit bright yellow light, which makes them useful as the light source in street lamps because our eyes are very sensitive to yellow light. To name just one more example, neon atoms emit a brilliant red-orange light, which makes them useful as the light source in neon signs.

When we view the light from glowing atoms through a spectroscope, we see that the light consists of a number of discrete (separate from one another) frequencies rather than a continuous spectrum like the one shown in Figure 4.17. The pattern of frequencies formed by a given element—some of which are shown in **Figure 4.18**—is referred to as that element's **atomic spectrum**. The atomic spectrum is an element's fingerprint. You can identify the elements in a light source by analyzing the light through a spectroscope and looking for characteristic patterns.

> **FOR YOUR INFORMATION**
>
> A star's age is revealed by its elemental makeup. The first and oldest stars were composed of hydrogen and helium, because those were the only elements that existed at that time. Heavier elements were produced after many of these early stars exploded in supernovae. Later stars incorporated these heavier elements in their formation. In general, the younger a star, the greater the amounts of these heavier elements it contains.

The Quantum Hypothesis

An important step toward our present-day understanding of atoms and their spectra was taken by the German physicist Max Planck (1858–1947). In 1900, Planck hypothesized that light energy is *quantized* in much the same way matter is quantized. The mass of a gold brick, for example, equals some whole-number multiple of the mass of a single gold atom. Similarly, an electric charge is always some whole-number multiple of the charge on a single electron. Mass and electric charge are therefore said to be *quantized*, in that they consist of some number of fundamental units.

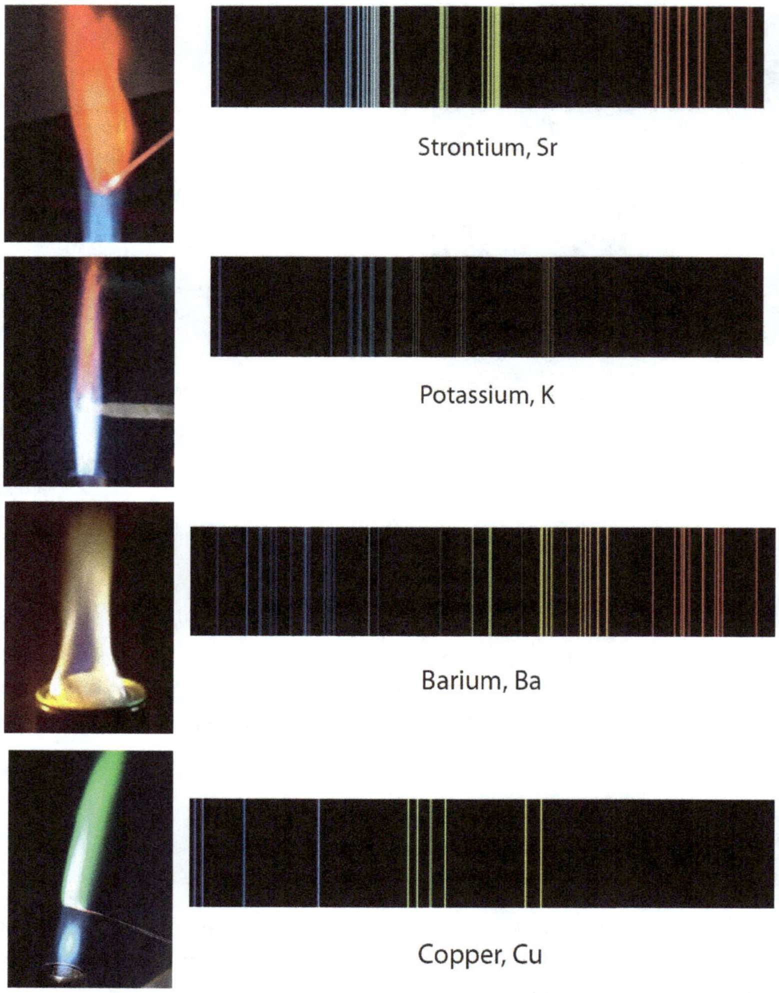

Figure 4.18 >
Elements heated by a flame glow their characteristic color. This is commonly called a flame test and is used to test for the presence of an element in a sample. When viewed through a spectroscope, the color of each element is revealed to consist of a pattern of distinct frequencies known as an atomic spectrum.

Planck identified each discrete parcel of light energy as a **quantum**, represented in **Figure 4.19**. A few years later, Einstein recognized that a quantum of light energy behaves much like a tiny particle of matter. To emphasize its particulate nature, each quantum of light was called a *photon*, a name coined because of its similarity to the word electron.

READING CHECK
What does an electron have when farther from the nucleus?

Using Planck's quantum hypothesis, the Danish scientist Niels Bohr (1885–1962) explained the formation of atomic spectra as follows. First, an electron has more potential energy when farther from the nucleus. This is analogous to the greater potential energy an object has when held higher above the ground. Second, Bohr recognized that when an atom absorbs a photon of light, it is absorbing energy. This energy is acquired by one of the electrons. Because this electron has gained energy, it must move away from the nucleus.

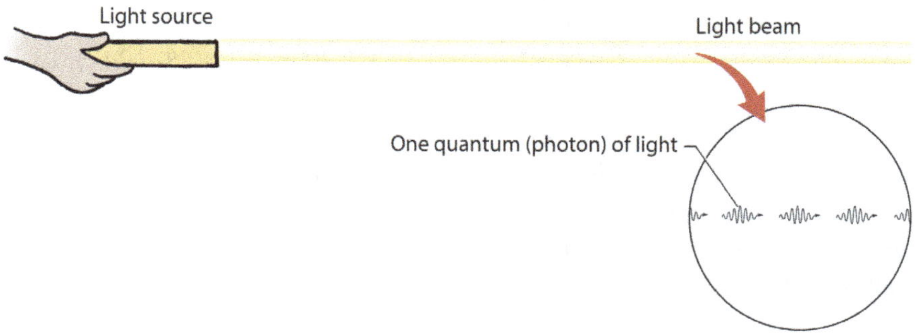

Figure 4.19 >
Light is quantized, which means it consists of a stream of energy packets. Each packet is called a quantum, also known as a photon.

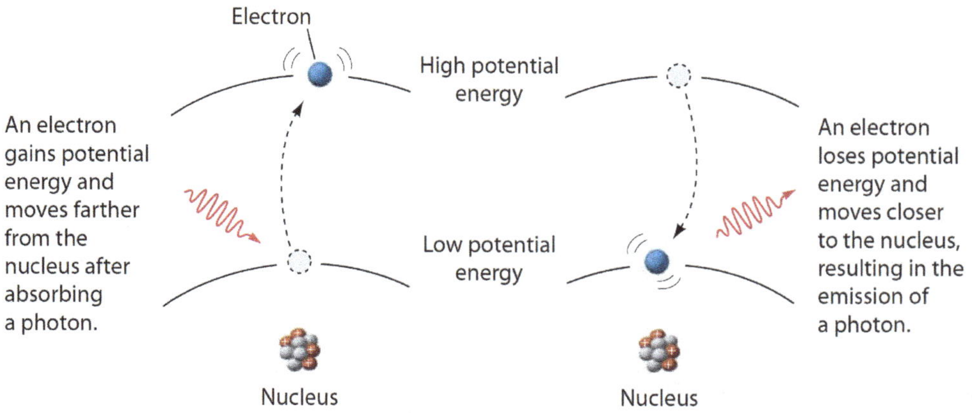

< **Figure 4.20**
An electron is lifted away from the nucleus as the atom it is in absorbs a photon of light and drops closer to the nucleus as the atom releases a photon of light.

Bohr realized that the opposite is also true: when a high-potential-energy electron in an atom loses some of its energy, the electron moves closer to the nucleus and the energy lost from the electron is emitted from the atom as a photon of light. Both absorption and emission are illustrated in **Figure 4.20.**

Bohr reasoned that because light energy is quantized, the energy of an electron in an atom must also be quantized. In other words, an electron cannot have just any amount of potential energy. Rather, within the atom there must be a number of distinct energy levels, analogous to steps on a staircase. Where you are on a staircase is restricted to where the steps are—you cannot stand at a height that is, say, halfway between any two adjacent steps. Similarly, there are only a limited number of permitted energy levels in an atom, and an electron can never have an amount of energy between these permitted energy levels. Bohr gave each energy level a **quantum number** n, where n is always some integer. The lowest energy level has a quantum number $n = 1$. An electron for which $n = 1$ is as close to the nucleus as possible, and an electron for which $n = 2$, $n = 3$, and so forth is farther away, in a step-wise fashion, from the nucleus.

Using these ideas, Bohr developed a conceptual model in which an electron moving around the nucleus is restricted to certain distances from the nucleus, with these distances determined by the amount of energy the electron has. Bohr saw this as similar to the way the planets are held in orbit around the Sun at given distances from the Sun. The allowed energy levels for any atom, therefore, could be graphically represented as orbits around the nucleus, as shown in **Figure 4.21**. Bohr's quantized model of the atom thus became known as the *planetary model.*

∧ Niels Bohr (1885–1962) and Albert Einstein (1879–1955) were good friends and colleagues, but they differed on their views of quantum theory and its philosophical implications. Einstein accepted quantum theory but was one of its strongest critics. His critiques were answered for the most part by Niels Bohr through a series of exchanges, called the Bohr–Einstein debates, spanning the latter parts of their lives.

CONCEPT **CHECK**
What is released as an electron transitions from a higher to a lower energy level?

CHECK YOUR ANSWER A photon of light.

Bohr used his planetary model to explain why atomic spectra contain only a limited number of light frequencies, as shown in **Figure 4.22**. According to the model, photons are emitted by atoms as electrons move from higher-energy outer orbits to lower-energy inner orbits. The energy of an emitted photon is equal to the difference in energy between the two orbits. Because an electron is restricted to discrete orbits, only particular light frequencies are emitted, as atomic spectra show.

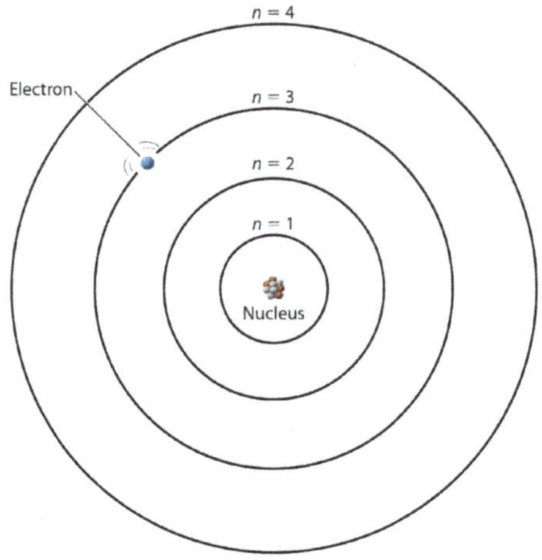

Λ Figure 4.21
Bohr's planetary model of the atom, in which electrons orbit the nucleus much as planets orbit the Sun, is a graphical representation that helps us understand how electrons can possess only certain quantities of energy.

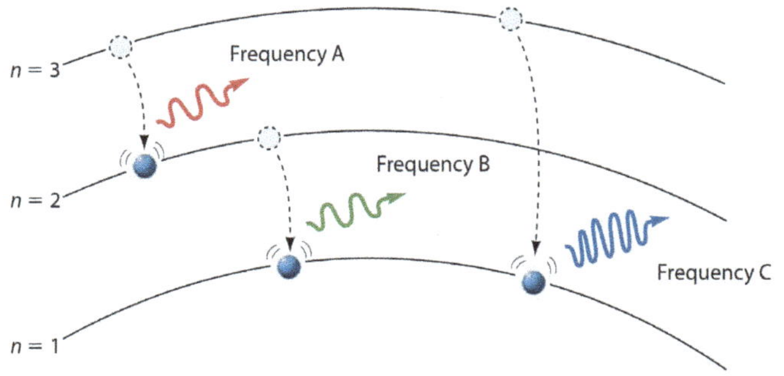

Λ Figure 4.22
The frequency of light emitted (or absorbed) by an atom is proportional to the energy difference between electron orbits. Because the energy differences between orbits are discrete, the frequencies of light emitted (or absorbed) are also discrete. The electron here can emit only three discrete frequencies of light—A, B, and C. The greater the transition, the higher the frequency of the photon emitted.

Interestingly, any transition between two orbits is always instantaneous. In other words, the electron doesn't "jump" from a higher to a lower orbit the way a squirrel jumps from a higher branch in a tree to a lower one. Rather, it takes no time for an electron to move between two orbits. Bohr was serious when he stated that electrons could never exist between permitted energy levels!

Bohr's planetary atomic model proved to be a tremendous success. By utilizing Planck's quantum hypothesis, Bohr's model solved the mystery of atomic spectra. Despite its successes, though, Bohr's model was limited, because it did not explain why energy levels in an atom are quantized. Bohr himself was quick to point out that his model was to be interpreted only as a crude beginning and that the picture of electrons whirling about the nucleus like planets about the Sun was not to be taken literally (a warning to which popularizers of science paid no heed).

In the remaining sections of this chapter you will be introduced to an abbreviated discussion of a more modern understanding of how electrons behave within an atom. Though abbreviated, these discussions are still most unusual. So be sure to study this material with a fresh mind. What you will get for your effort is tremendous insight into how and why atoms behave as they do. Most notably, this includes powerful insight into the formation of chemical bonds as described in Chapter 6.

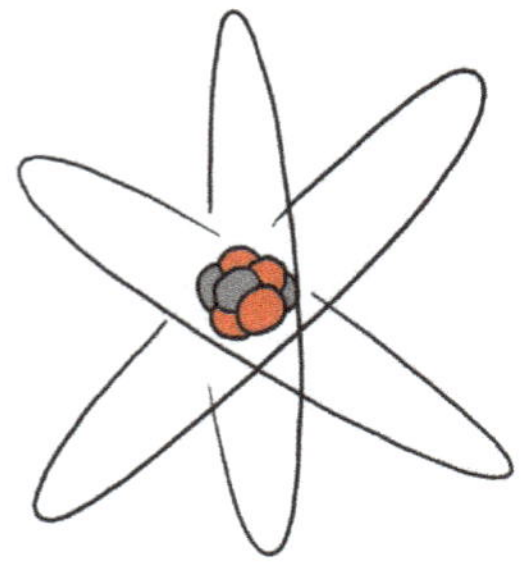

Not to be taken literally

CONCEPT **CHECK**

Is the Bohr model of the atom a physical model or a conceptual model?

CHECK YOUR ANSWER The Bohr model is a conceptual model. It is not a scaled-up version of an atom but instead is a representation that accounts for the atom's behavior.

4.7 Electrons Exhibit Wave Properties

If light has both wave properties and particle properties, why can't a material particle, such as an electron, also have both? This question was posed by the French physicist Louis de Broglie (1892–1987) while he was still a graduate student in 1924. His revolutionary answer was that every moving particle of matter—by virtue of its energy of motion—is endowed with the characteristics of a wave.

We now speak of waves as an essential feature of any bit of matter. An electron, or any particle, can show itself as a wave or as a particle, depending on how we examine it. This is called the wave–particle duality. Just a few years after de Broglie's suggestion, researchers in Great Britain and the United States confirmed the wave nature of electrons by observing diffraction and interference effects when electrons bounced from crystals. A practical application of the wave properties of electrons is the electron microscope, which focuses not visible-light waves but rather electron waves. Electron microscopes are able to show far greater detail than optical microscopes, as **Figure 4.23** shows.

An electron's wave nature can be used to explain why electrons in an atom are restricted to particular energy levels. Permitted energy levels are a natural consequence of the need for electron waves to form stable patterns around the atomic nucleus.

READING CHECK
What does the wave nature of an electron help explain?

As an analogy, consider the wire loop shown in **Figure 4.24a**. This loop is affixed to a mechanical vibrator that can be adjusted to create a series of waves traveling through the wire. If the lengths of the waves are equal to the length of the wire (or some fraction thereof), then these waves form what is called a *standing wave*, as shown in **Figure 4.24b**. Within a standing wave, identically sized waves overlap one another, and they do so in a manner that creates regions of maximum intensity and other regions of minimum intensity. A region of maximum intensity occurs where crests are overlapping with crests and troughs are overlapping with troughs, as shown in **Figure 4.25**. (Note: a "crest" is the high point of a wave while a "trough" is the low point of a wave.) A region of minimum intensity occurs where crests and troughs overlap so as to cancel each other out. At the center of minimum intensity is a point of zero intensity, called a node.

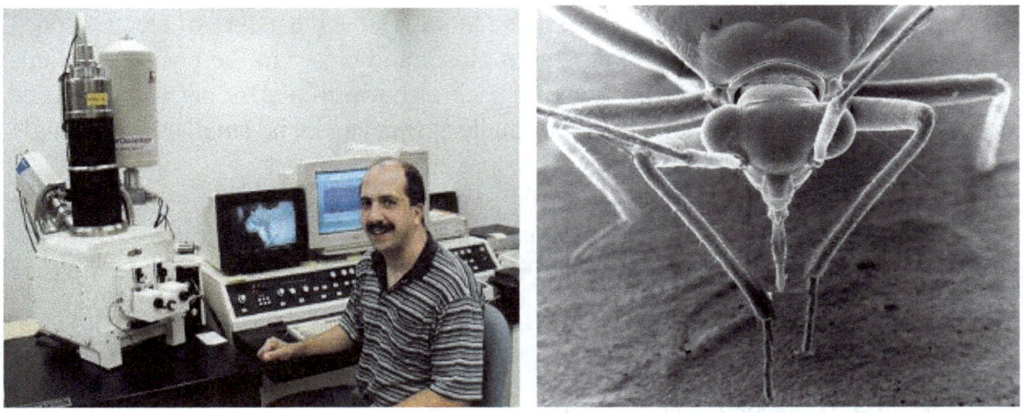

Figure 4.23
An electron microscope makes practical use of the wave nature of electrons. The wavelengths of electron beams are typically thousands of times shorter than the wavelengths of visible light, so the electron microscope is able to distinguish detail not visible with optical microscopes. (b) Detail of an insect's head as seen with an electron microscope at a "low" magnification of 200 times. Note the remarkable resolution.

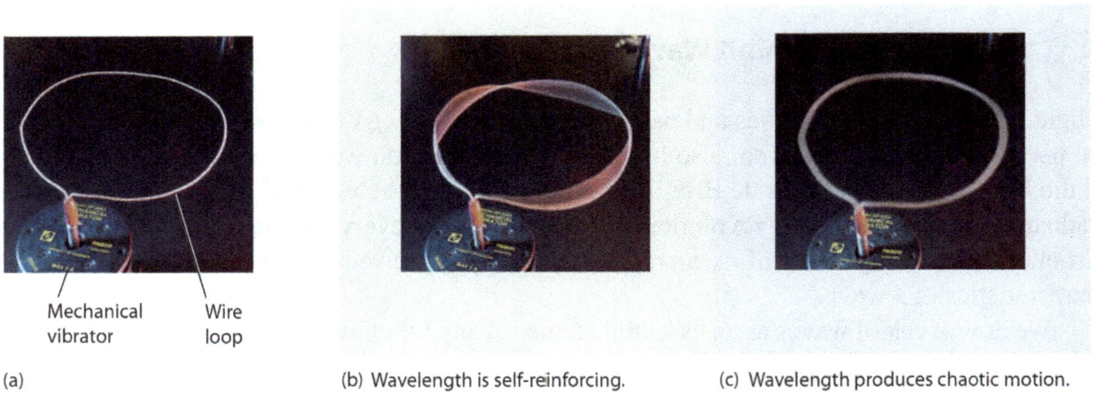

∧ Figure 4.24
A wire loop affixed to the post of a mechanical vibrator at rest. Waves are sent through the wire when the post vibrates. (b) Only waves of certain wavelengths are able to form standing waves. The one shown here is 2/3 the length of the loop. (c) Most other wavelengths result in chaotic motion with no specific regions of maximum and minimum intensities.

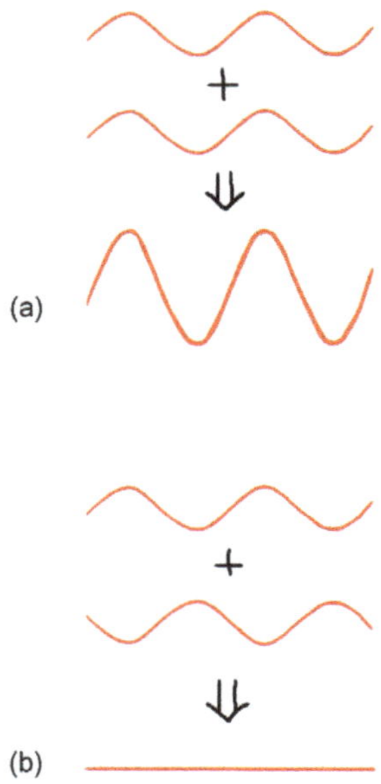

∧ Figure 4.25
(a) Crests overlapping crests and troughs overlapping troughs results in a wave of maximum intensity. (b) Crests overlapping troughs results in a wave of zero intensity.

For a given size of the wire loop, only certain wavelengths will give rise to a stable standing wave. Waves of any other wavelength are unable to align properly and the result is chaotic motion, as suggested in **Figure 4.24c**. These same ideas apply to electron waves within an atom. We find that only certain electron wavelengths form standing waves. Because of their stability, these particular electron wavelengths are favored.

But why does an electron have wavelike characteristics? Recall that according to de Broglie, the wavelength of an electron is a consequence of its energy. Significantly, if only certain wavelengths are favored, then it follows that only certain amounts of energy are favored. In other words, the energy of an electron confined to an atom is *quantized.* Consider what happens when whistling down a long tube. The whistler soon discovers that only certain frequencies can be generated. Interestingly, when a sound wave is confined to a tube, the consequence is a quantization of its frequencies. This is an important principle behind any wind instrument. Likewise, when an electron is confined to an atom, the consequence is a quantization of the electron's energy.

Each standing wave of an electron corresponds to one of the permitted energy levels seen in Figures 4.21 and 4.22. Only the frequencies of light that match the difference between any two of these permitted energy levels can be absorbed or emitted by an atom. The wave nature of electrons also explains why they do not spiral closer and closer to the positive nucleus that attracts them. By viewing each electron as a standing wave, we see that the circumference of the smallest orbit can be no smaller than a single wavelength.

In summary, the properties of electrons—and all atoms made of those electrons—arise from their quantum nature. In this sense, the bizarre quantum nature of the super-small is what makes our natural world possible.

CONCEPT **CHECK**
What must an electron be doing in order to have wave properties?

CHECK YOUR ANSWER
According to de Broglie, particles of matter behave like waves by virtue of their motion, which is a form of energy. An electron must therefore be moving in order to have wave properties.

4.8 Orbitals and Energy-Level Diagrams

Electron waves are 3-dimensional, which makes them difficult to visualize, but scientists have come up with two ways of visualizing them: as *probability clouds* and as *atomic orbitals*.

Look carefully at the standing wave of Figure 4.24b. Note that there are regions where the wire vibrates intensely and others where it doesn't appear to vibrate at all. In a similar fashion, electron waves in an atom vibrate more intensely in some regions than in others. In 1926, the Austrian–German scientist Erwin Schrodinger (1887–1961) formulated a remarkable equation from which the intensities of electron waves in an atom could be calculated. It was soon recognized that the wave intensity at any given location determined the probability of finding the electron at that location. In other words, the electron was most likely to be found where its wave intensity was greatest and least likely to be found where its wave intensity was smallest.

If we could plot the positions of an electron of a given energy over time as a series of tiny dots, the resulting pattern would resemble what is called a *probability cloud*. **Figure 4.26a** shows a probability cloud for hydrogen's electron. The denser a region of the cloud, the greater the probability of finding the electron in that region. The densest regions correspond to where the electron's wave intensity is greatest. A probability cloud is therefore a close approximation of the actual shape of an electron's 3-dimensional wave. Interestingly, this cloud has no distinct surface. Rather, there is some probability that the electron could be very far away from the nucleus, but that probability quickly decreases with increasing distance.

An **atomic orbital**, like a probability cloud, specifies a volume of space where the electron is most likely to be found. By convention, atomic orbitals are drawn to delineate the volume inside which the electron is located 90 percent of the time. This gives the atomic orbital an apparent border, as shown in **Figure 4.26b**. This border is arbitrary, because the electron may exist on either side of it. Almost all of the time, though, the electron remains within the border. So, probability clouds and atomic orbitals are essentially the same thing. They differ only in that atomic orbitals specify an outer limit, which makes them easier to depict graphically.

As **Table 4.2** shows, atomic orbitals come in a variety of shapes, some quite exquisite. We categorize these orbitals using the letters *s*, *p*, *d*, and *f*. The simplest is the spherical *s* orbital. The *p* orbital consists of two lobes and resembles an hourglass at the center of which there is a node. There are three kinds of *p* orbitals, which differ from one another only by their

fyi FOR YOUR INFORMATION

The German physicist Werner Heisenberg (1901–1976) found that we can't precisely know the position of an electron if we know its momentum. And vice versa, if we know the momentum of an electron, we can't precisely know its position. Heisenberg spelled out this relationship in the following equation where momentum is p, position is x, and h is Planck's constant:

$$(\Delta p)(\Delta x) \geq h/2\pi$$

Heisenberg's equation, called the *uncertainty principle*, was as helpful as Schrodinger's in allowing physicists to map out probability clouds, and, hence, a more developed atomic model.

READING CHECK

What does a probability cloud closely approximate?

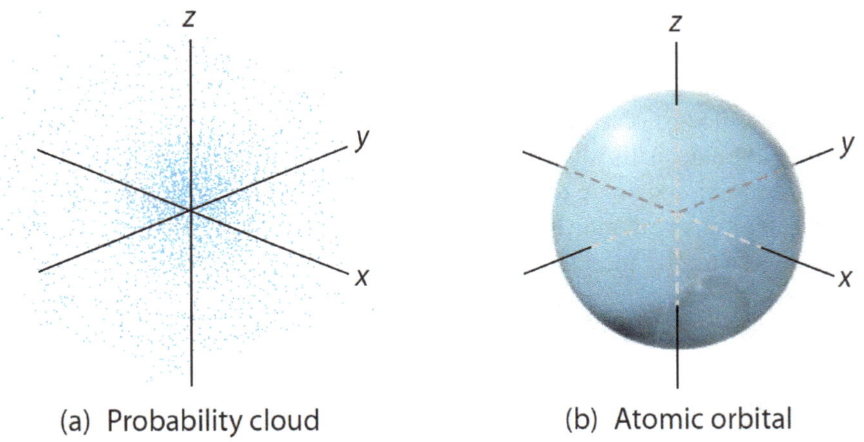

< Figure 4.26
(a) The probability cloud for hydrogen's electron in an x, y, z coordinate system. The nucleus is located at the origin, which is where all three lines intersect. The more concentrated the dots, the greater the chance of finding the electron at that location. (b) The atomic orbital for hydrogen's electron. The electron is somewhere inside this spherical volume 90 percent of the time.

TABLE 4.2 Four Categories of Atomic Orbitals: *s, p, d, f*

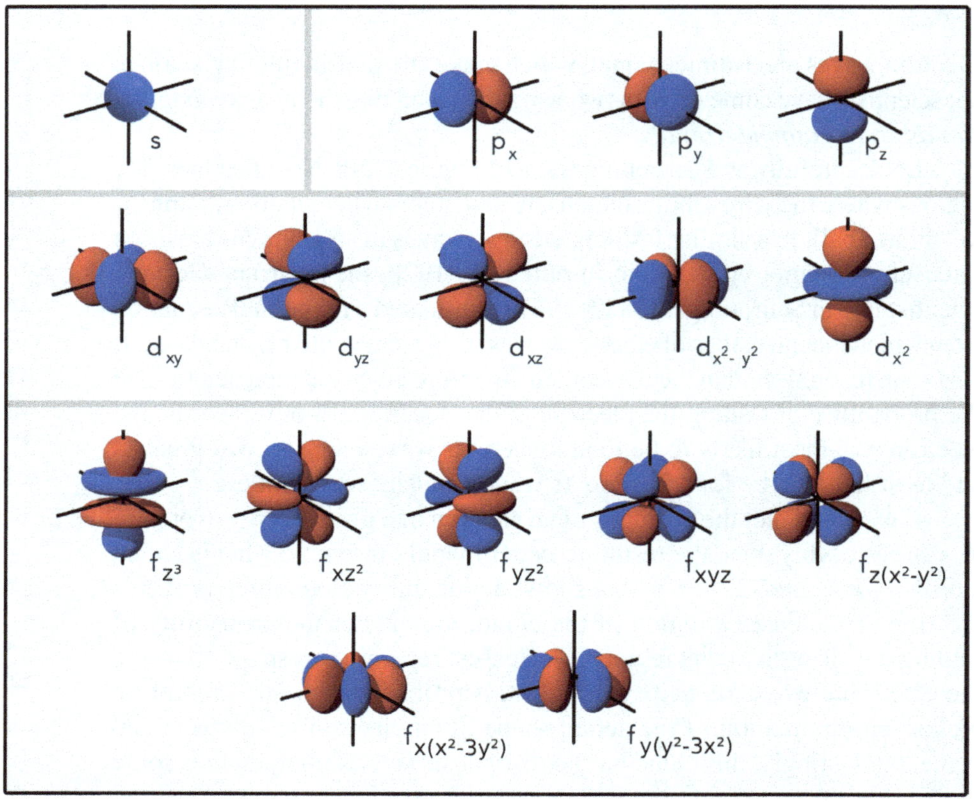

orientation in 3-dimensional space. The more complex *d* orbitals have five possible shapes, and the *f* orbitals have seven. Remember, these are just the potential shapes of electron waves in an atom.

In addition to a variety of shapes, atomic orbitals also come in a variety of sizes. In general, energetic electrons are able to spend more of their time farther away from the attracting nucleus, which means they are distributed over a greater volume of space. The higher the energy of the electron, therefore, the larger its atomic orbital. Because electron energies are quantized, however, the possible sizes of the atomic orbitals are also quantized. The size of an orbital is indicated by Bohr's quantum number $n = 1, 2, 3, 4$, and so on. The first two *s* orbitals are shown in **Figure 4.27**. The smallest *s* orbital is the 1*s* (pronounced "one" "ess"), for which the quantum number is 1. The next largest *s* orbital is the 2*s* , and so forth.

Each Orbital Can Hold up to Two Electrons

In the early 20th century, quantum physicists learned that electrons have a property called *spin*. Spin can have two states, analogous to a ball spinning either clockwise or counterclockwise, as shown in **Figure 4.28**. Because an electron has an electric charge, the spinning generates a tiny magnetic field. Two electrons spinning in opposite directions have oppositely aligned magnetic fields. This allows those two electrons—but no more than those two electrons—to come together as a pair within the same atomic orbital.

For atoms with multiple electrons, this idea of two electrons per orbital is of central importance. Consider the helium atom, which has two electrons. At their lowest energy states, both of these electrons will reside within the same 1*s* orbital because this orbital has a capacity of two electrons. What about lithium, which has three electrons? Two of its electrons will fill the *1s* orbital. Lithium's third electron, however, must enter the higher energy 2*s* orbital.

fyi FOR YOUR INFORMATION

Electron waves. Probability clouds. Atomic orbitals. Electron configurations. Electron spin. And more! This is certainly complicated information. But what it all boils down to is the study of how electrons are organized within the atom. Hang in there. Read this material slowly and for comprehension. Consider that one of the best ways to learn new ideas is to practice explaining them to others.

Energy-Level Diagrams Describe How Orbitals Are Occupied

The increasing energy levels of the various atomic orbital are depicted in an **energy-level diagram**, as shown in **Figure 4.29** (next page). Note that each orbital is represented schematically as a box and that each electron is represented by an arrow. This arrow points either up or down depending upon the spin of the electron. If two electrons are in the same orbital (box), their spins (arrows) must point in opposite directions. It is the natural tendency of electrons to occupy the lowest energy orbitals first. These are the orbitals that bring the electron closest to the atomic nucleus, which attracts electrons because of its positive charge. So, the electrons of an atom fill the orbitals of an energy-level diagram from bottom to top. Figure 4.29 shows how the orbitals of a rubidium atom are filled.

The arrangement of an atom's electrons within orbitals is called its **electron configuration**. Figure 4.29 shows the electron configuration for rubidium. An abbreviated way of presenting an electron configuration is to write the quantum number and letter of occupied orbitals and then use a superscript to indicate the number of electrons in each of those orbitals. For each of the group 1 elements, the notation is as follows:

Hydrogen, H	$1s^1$
Lithium, Li	$1s^2 2s^1$
Sodium, Na	$1s^2 2s^2 2p^6 3s^1$
Potassium, K	$1s^2 2s^2 2p^6 3s^2 3p^6 4s^1$
Rubidium, Rb	$1s^2 2s^2 2p^6 3s^2 3p^6 4s^2 3d^{10} 4p^6 5s^1$
Cesium, Cs	$1s^2 2s^2 2p^6 3s^2 3p^6 4s^2 3d^{10} 4p^6 5s^2 4d^{10} 5p^6 6s^1$
Francium, Fr	$1s^2 2s^2 2p^6 3s^2 3p^2 4s^2 3d^{10} 4p^6 5s^2 4d^{10} 5p^6 6s^2 4f^{14} 5d^{10} 6p^6 7s^1$

Look carefully at rubidium's electron configuration as shown previously and also in Figure 4.29 and see how they correspond. Note also that all the superscripts for an atom must add up to the total number of electrons in the atom—1 for hydrogen, 3 for lithium, 11 for sodium, and so forth. Also note that the orbitals are listed in order of their energy levels as seen on the energy level diagram. This explains why the 4s orbital is listed before the 3d orbitals even though it has a higher quantum number. Lastly, to simplify the task of writing out an electron configuration, the inner electrons can be represented using the symbol for a noble gas element in brackets. For example, the configuration for aluminum, Al, $1s^2 2s^2 2p^6 3s^2 2p^1$, can be written as [Ne] $3s^2 2p^1$, where [Ne] is shorthand for $1s^2 2s^2 2p^6$.

An atom's outermost electrons are the ones that interact most strongly with the external environment. Because of this, these outermost electrons play a key role in determining the properties of the atom—both chemical and physical. Elements that have similar electron configurations in their outermost orbitals, therefore, have similar properties. For example, in the alkaline metals of group 1, shown previously, the outermost occupied orbital (shown in blue) is an *s* orbital containing a single electron. In general, elements in the same group of the periodic table have similar electron configurations, which explains why elements in the same group have similar properties—a concept first presented in Section 3.3.

There are many interesting details to be said about energy-level diagrams. For example, why are there multiple 2*p* orbitals and no 1*p* orbitals? Why does the 4*s* orbital have a lower energy than the 3*d* orbital when it has a higher quantum number? And why does the 2*s* orbital have a slightly lower energy than any of the 2*p* orbitals when their quantum numbers are the same? Also, in what order do electrons fill the three 2*p* orbitals when

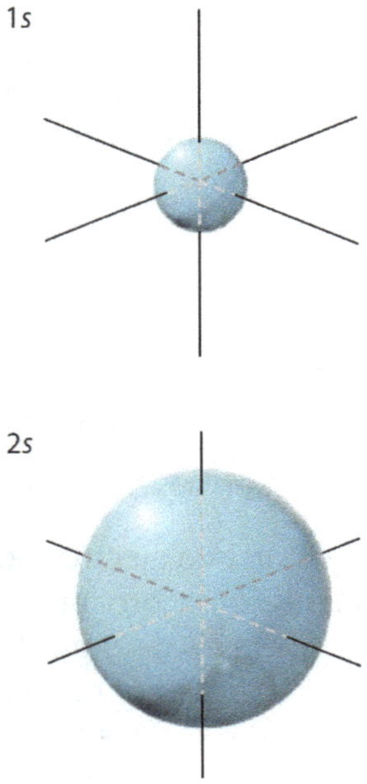

Figure 4.27
The 2*s* orbital is larger than the 1*s* orbital because the electrons in the 2*s* orbital have greater energy.

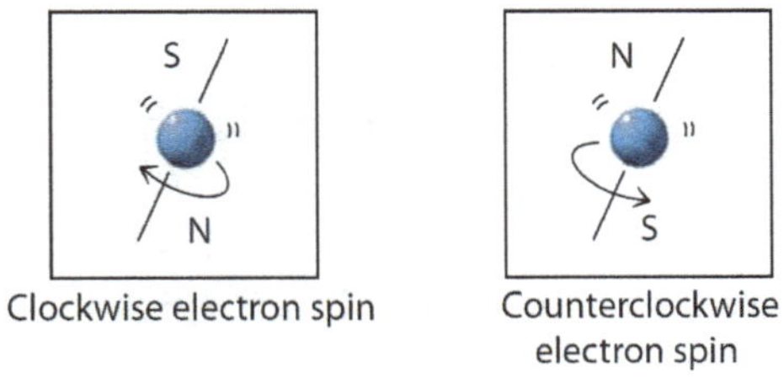

Figure 4.28
Two electrons can pair together when they have opposite spins.

Figure 4.29 >

This energy-level diagram shows the relative energy levels of atomic orbitals in a multielectron atom (in this case rubidium, Rb, atomic number 37).

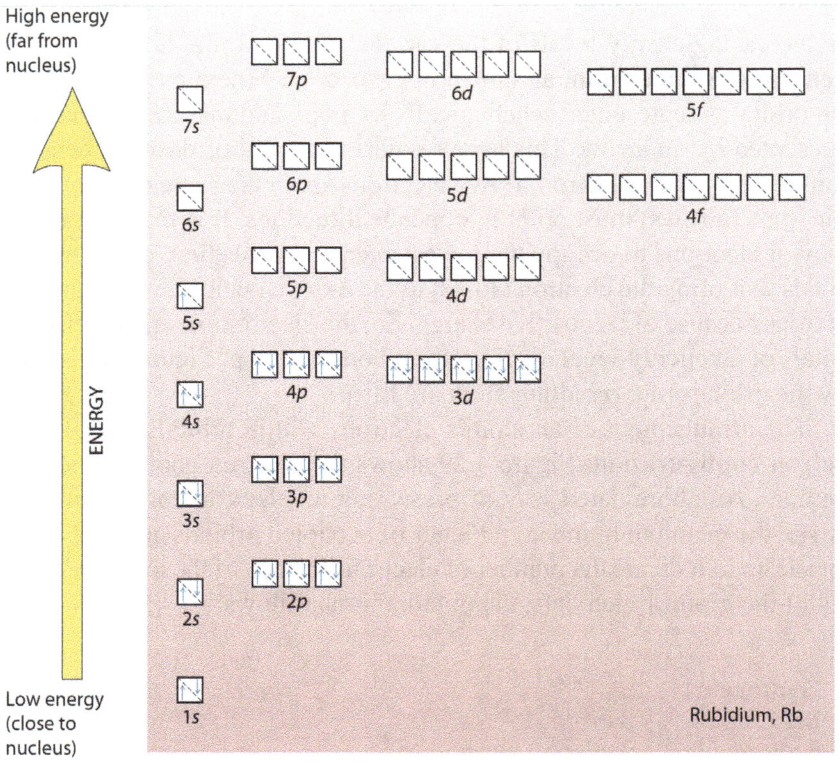

KEY POINTS

- An orbital is a region of space within which an electron may reside.
- No more than two electrons can fit into one atomic orbital.
- The energy-level diagram tells us which orbitals accommodate what level of electron energy.
- How electrons are distributed through numerous orbitals is given by the atom's electron configuration.

CONCEPT **CHECK**

What is the electron configuration for bromine, Br (atomic number 35)?

CHECK YOUR ANSWER The electron configuration for bromine is $1s^2 2s^2 2p^6 3s^2 3p^6 4s^2 3d^{10} 4p^5$. This can also be represented as [Ar] $4s^2 3d^{10} 4p^5$.

fyi FOR YOUR INFORMATION

In more advanced textbooks and also online you may see orbitals listed in order of quantum numbers so that the 3d is written before the 4s. This is a common convention but conceptually inconsistent with energy-level diagrams.

each of these orbitals has the same energy level? Probing into these sorts of details under the guidance of your instructor will provide you with keen insight into the workings of chemistry. But there is also value in getting to the main point as soon as possible. For this reason, we now move onward to a discussion of how the energy-level diagram is intimately connected to that all-important tool of chemistry: the periodic table.

Simplifying the Energy-Level Diagram

The energy-level diagram can be organized so that orbitals of similar energy are grouped together. As shown in **Figure 4.30**, no orbital has an energy level similar to that of the 1*s* orbital, so this orbital is grouped by itself. The energy level of the 2*s* orbital, however, is close to that of the three 2*p* orbitals, so these four orbitals are grouped together. Likewise, the 3*s* and three 3*p* orbitals are at a similar energy level as are the 4*s*, five 3*d*, and three 4*p* orbitals, and so on. The result is a set of seven distinct horizontal rows of orbitals.

Interestingly, the seven rows in Figure 4.30 correspond to the seven periods in the periodic table, with the bottom row corresponding to the first period, the next row up corresponding to the second period, and so on. Furthermore, the maximum number of electrons each row can hold is equal to the number of elements in the corresponding period. The bottom row in Figure 4.30 can hold a maximum of two electrons, so there are only two elements, hydrogen and helium, in the first period of the periodic table. The second and third rows up from the bottom each have a capacity for eight electrons, and so eight elements are found in both the second and third periods. Continue analyzing Figure 4.30 in this way, and you will find 18 elements in the fourth and fifth periods, and 32 elements in the sixth and seventh periods.

Recall from Section 4.7 that the higher the energy level of an orbital, the farther its electrons are from the nucleus. Electrons in the same row of orbitals in Figure 4.30, therefore, are roughly the same distance from the nucleus. Graphically, this can be represented by converging all the orbitals in a given row into a single 3-dimensional hollow shell, as shown in **Figure 4.31**. Each shell is a graphic representation of a collection of orbitals of comparable energy in a multielectron atom. To distinguish this shell from one used in more advanced chemistry courses, we call it a **noble gas shell**. We say "noble gas" because, when filled, such a shell represents the electron configuration of a noble gas element.

The seven rows of orbitals in Figure 4.29 can thus be represented either by a series of seven concentric shells or by a series of seven cross-sectional circles of these shells, as shown in Figure 4.31. The number of electrons each shell can hold is equal to the number of orbitals it contains multiplied by 2 (because there can be two electrons per orbital).

FOR YOUR INFORMATION

What do poets and scientists have in common? They both use metaphors to help us understand abstract concepts and relationships. The shell model of the atom is a metaphor that helps us visualize an invisible reality. Conceptual models are essentially equivalent to the metaphorical language used in poetry.

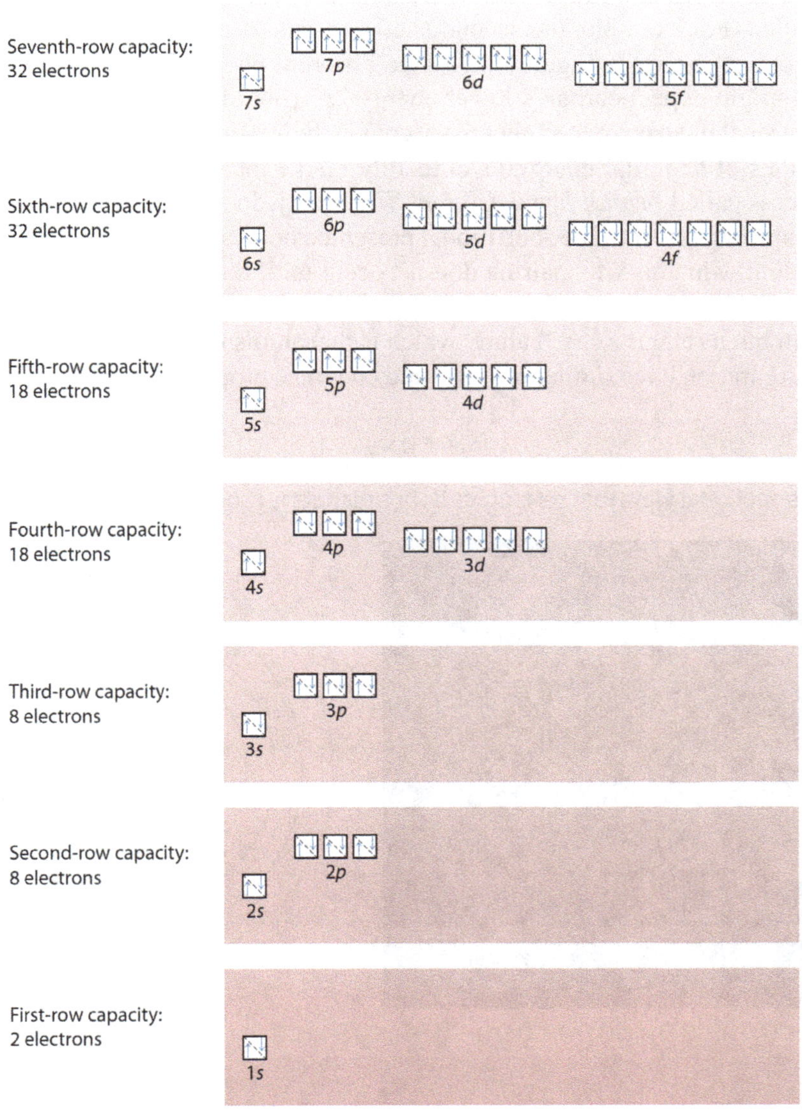

Figure 4.30
Orbitals of comparable energy levels can be grouped together to give rise to a set of seven rows of orbitals.

Figure 4.31 >
The second row of orbitals, which consists of the 2*s* and three 2*p* orbitals, can be represented either as a single smooth, spherical shell or as a cross section of such a shell. (Note that this shell model ignores the slight difference in energy levels of closely matched orbitals.)

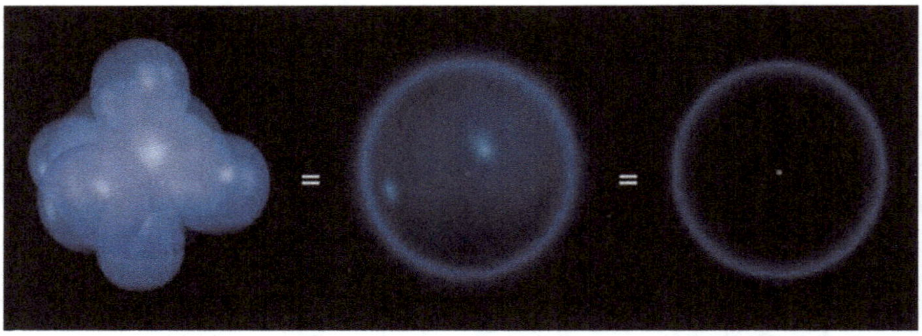

Second row of orbitals (all 2*s* and 2*p* orbitals combined)

Second row of orbitals (highly simplified perspective)

Second row of orbitals (cross section of highly simplified perspective)

You add electrons to a shell diagram just as in an energy-level diagram—electrons first fill the shells closest to the nucleus. **Figure 4.32** shows how this works for the first three periods. As with the energy-level diagram, there is one shell for each period, and the number of elements in a period is equal to the maximum number of electrons that the shell representing that period can hold.

Notice how electrons in the outermost shell begin to pair only after that shell is half filled. Carbon, for example, has four outer-shell electrons, none of which are paired. This differs from how an energy-level diagram is filled. For example, the second shell consists of the 2*s* and 2*p* orbitals, which, as shown in Figure 4.29, have different energy levels. Therefore, you might expect carbon's lower energy 2*s* orbital to fill with two paired electrons. In an advanced chemistry course, however, you would learn that orbitals of a similar energy level readily merge into each other through a process called *orbital hybridization*. When they do so, their energy levels become equivalent. The shell model presented here takes hybridization into account, which is why pairing doesn't occur until the shell is half filled.

READING CHECK
When do electrons in the outermost shell of an atom begin to pair with each other?

This noble gas shell model is a simplification of the energy-level diagram but it retains a key feature, which is that atoms with a similar electron configuration have similar physical and chemical properties. Look carefully

V Figure 4.32
(a) A cutaway view of the seven shells, with the number of electrons each shell can hold indicated. (b) A 2-dimensional, cross-sectional view of the shells. (c) An easy-to-draw cross-sectional view that resembles Bohr's planetary model.

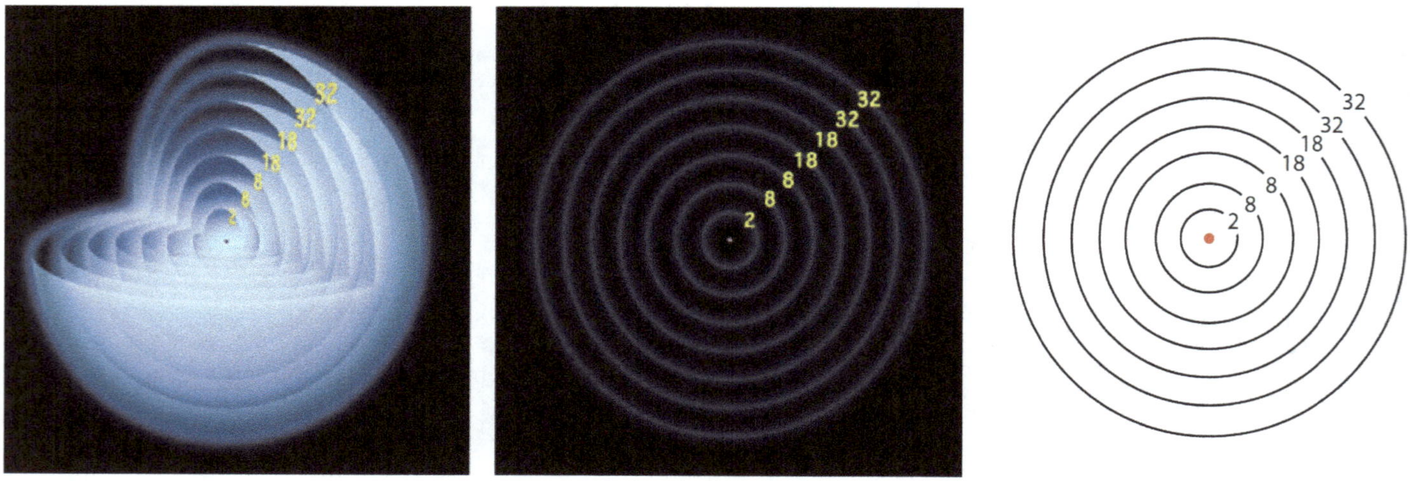

(a) (b) (c)

at **Figure 4.33**. Can you see that the outer-shell electrons of atoms above and below one another in the periodic table—that is, within the same group—are similarly organized? For example, atoms of the first group—which includes hydrogen, lithium, and sodium—each have a single outer-shell electron. The atoms of the second group, including beryllium and magnesium, each have two outer-shell electrons. Similarly, atoms of the last group—including helium, neon, and argon—each have their outermost shells filled to capacity with electrons, two for helium and eight for both neon and argon. In general, the outer-shell electrons of atoms in the same group of the periodic table are similarly organized, which is why these atoms have similar properties.

Remember that the shell model is not to be interpreted as an actual representation of the atom's physical structure—electrons, for example, are not really confined to the surface of a shell. Rather, this model serves as a tool to help us understand and predict how atoms behave. In Chapter 6 we will be using an even simpler version of this model, called the *electron-dot structure*, to show how atoms join together to form molecules, which are tightly held groups of atoms. For now let's use this model to explain the remarkable organization of the periodic table.

FOR YOUR INFORMATION

Congratulations for making it to the end of this challenging chapter section. Remember, the main idea is how electrons are organized within an atom. This, in turn, is the key to understanding much in chemistry.

CONCEPT CHECK

Why are there only two elements in the first period of the periodic table?

CHECK YOUR ANSWER

The number of elements in each period corresponds to the number of electrons each shell can hold. The first shell has a capacity of only two electrons, which is why the first period has only two elements.

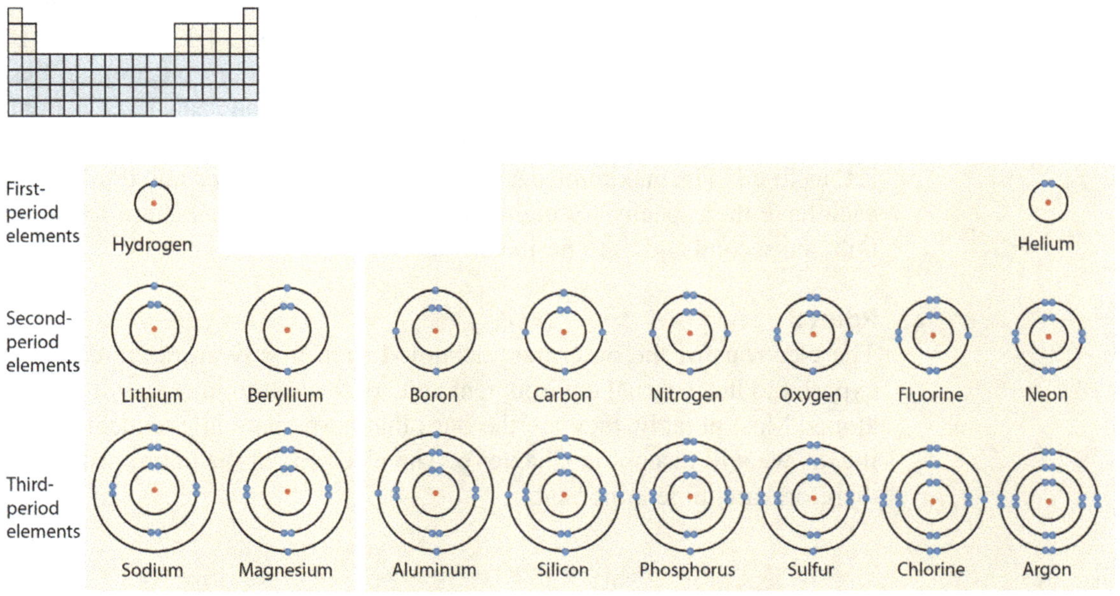

Figure 4.33

The first three periods of the periodic table according to the noble gas shell model. Elements in the same period (horizontal rows) have electrons in the same shells. Elements in the same period differ from one another by the number of electrons in the outermost shell. Elements in the same group (vertical column) have the same number of electrons in the outermost shell. These outer-shell electrons determine the character of the atom, which is why elements in the same group have similar properties.

4.9 The Shell Model and Periodic Table

Two-time Nobel laureate Linus Pauling (1901–1994) was an early proponent of teaching beginning chemistry students a shell model, from which the organization of the periodic table could be described. In 1954 Pauling won the Nobel Prize in Chemistry for his research into the nature of the chemical bond. In 1962 he was awarded the Nobel Peace Prize for his campaign against the testing of nuclear bombs, which was introducing massive amounts of radioactivity into the environment.

Chemistry is the study of how atoms combine to form new materials. But how exactly does this happen? In order to answer this question, we first need to understand a bit about the properties of the atoms themselves. For this, we turn to the shell model, made popular by the noted chemistry and two-time Nobel laureate Linus Pauling (1901-1994). This model, described in detail in the previous section, is similar to Bohr's planetary model in that it shows electrons restricted to particular distances from the nucleus. The shell model, however, is a bit more sophisticated because it incorporates the wave nature of electrons.

First, let's take some time to review the three main points of the previous section:

Point 1
According to the shell model, electrons behave as though they are arranged in a series of concentric shells. A shell is a region of space around the atomic nucleus within which electrons may reside. An important aspect of this model is there are at least seven shells and each shell can hold only a limited number of electrons. As was shown in Figure 4.32, the innermost shell can hold 2 electrons, the second and third shells 8 each, the fourth and fifth shells 18 each, and the sixth and seventh shells 32 each.

Point 2
A series of seven such concentric shells accounts for the seven periods of the periodic table. Furthermore, the number of elements in each period is equal to the shell's capacity for electrons. The first shell, for example, has the capacity for only two electrons. That's why we find only two elements, hydrogen and helium, in the first period, as was shown in Figure 4.33. Hydrogen is the element whose atoms have only one electron. This one electron resides within the first shell, which is the shell closest to the nucleus. Each helium atom has two electrons, both of which are also within the first shell, which is thus filled to its maximum capacity. Similarly, the second and third shells each have the capacity for eight electrons, so eight elements are found in both the second and third periods.

Point 3
The electrons of the outermost occupied shell in any atom are directly exposed to the external environment and are the first to interact with other atoms. Most notably, they are the ones that participate in chemical bonding, as we will discuss in Chapter 6. The electrons in the outermost shell, therefore, are quite important.

CONCEPT **CHECK**
Do atoms really consist of shells that look like those depicted in Figure 4.32?

CHECK YOUR ANSWER
No. The shell model is not a depiction of the "appearance of an atom." Rather, it is a conceptual model that allows us to account for observed behavior. An atom, therefore, does not actually contain a series of concentric shells; it merely behaves as though it does.

The quality of a song depends on the arrangement of musical notes. In a similar fashion, the properties of an element depend on the arrangements of electrons in its atoms, especially the outer-shell electrons. Look carefully at Figure 4.33 and you'll see that the outer-shell electrons of atoms above and below one another on the periodic table (within the same group) are similarly organized. This explains why elements of the same group have similar properties—a concept first introduced in Section 3.3.

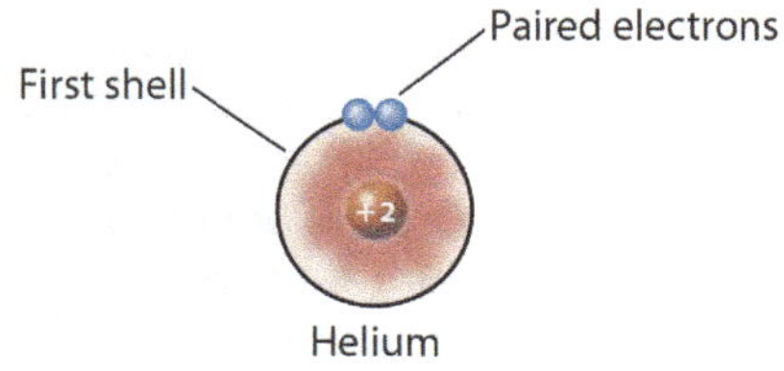

∧ Figure 4.34
The two electrons in a helium atom have equal exposure to the nucleus; hence, they experience the same degree of attraction, represented by the pink shading in the space between the nucleus and the shell boundary.

Effective Nuclear Charge

Recall from Section 3.3 that properties of elements gradually change across the periodic table. The atoms of elements toward the upper right of the periodic table, for example, tend to be smaller than the atoms of elements toward the lower left (see Figure 3.15). Knowing this, you can reasonably predict that selenium atoms (atomic number 34) are smaller than calcium atoms (atomic number 20), which is indeed the case.

Recall that such a gradual change in moving across the periodic table is called a *periodic trend*. Underlying most periodic trends are two important concepts: *inner-shell shielding* and *effective nuclear charge*.

Consider the two electrons within the first shell of a helium atom. As shown in **Figure 4.34**, both of these electrons have an unobstructed "view" of the nucleus. Their attractions for this atomic nucleus are therefore the same. The situation gets more complicated for atoms beyond helium, which have more than one shell occupied by electrons. In these cases, inner-shell electrons weaken the attraction between outer-shell electrons and the nucleus. Imagine, for example, you are that second-shell electron in the lithium atom shown in **Figure 4.35**. Looking toward the nucleus, what do you sense? Not just the attractive force of the positively charged nucleus. You also feel the effect of the two electrons in the first shell—which, because of their negative charge, exert a repulsive force. Thus, the two inner electrons have the effect of weakening your electrical attraction to the nucleus. This is **inner-shell shielding**—inner-shell electrons shield electrons farther out from some of the attractive pull exerted by the positively charged nucleus.

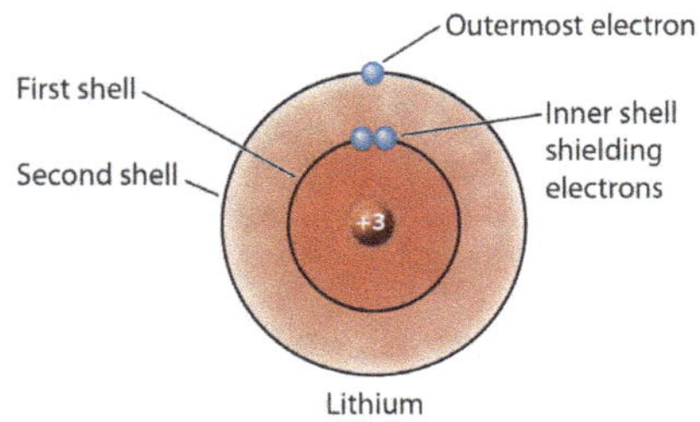

∧ Figure 4.35
Lithium's two first-shell electrons shield the second-shell electron from the nucleus. The nuclear attraction, again represented by pink shading, is less intense in the second shell.

Because inner-shell electrons diminish the attraction outer-shell electrons have for the nucleus, the nuclear charge sensed by outer-shell electrons is always less than the actual charge of the nucleus. This diminished nuclear charge experienced by outer-shell electrons is called the **effective nuclear charge** and is abbreviated Z* (pronounced zee-star), where Z stands for the nuclear charge and the asterisk indicates that this charge appears to be less than it actually is. A rough estimate of the Z* for any electron can be calculated by subtracting the total number of inner-shell electrons from the nuclear charge. For lithium (atomic number 3), the total number of inner-shell electrons, 2, is subtracted from the charge of the nucleus, +3, to give a calculated effective nuclear charge of +1. The second-shell electron of lithium, therefore, senses a nuclear charge of about +1, which is much less than the actual nuclear charge of +3. For most elements, subtracting the total number of inner-shell electrons from the nuclear charge provides a convenient estimate of the effective nuclear charge, as **Figure 4.36** illustrates.

READING CHECK
How is it possible to make a rough estimate of the effective nuclear charge experienced by an electron?

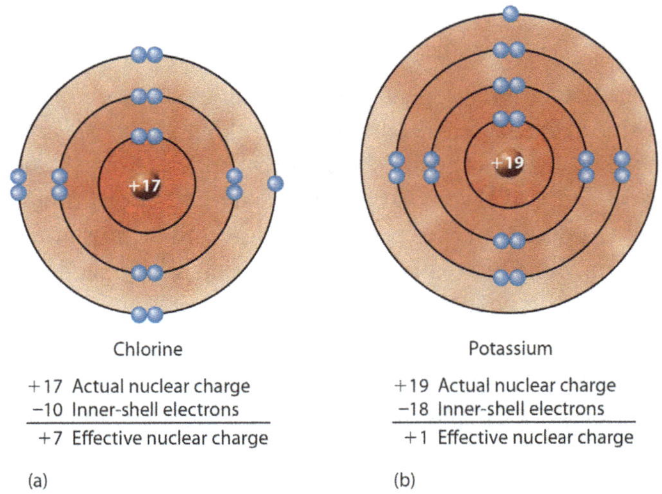

Figure 4.36 >
A chlorine atom has three occupied shells. The ten electrons of the inner two shells shield each of the seven electrons of the third shell from the +17 nucleus. The third-shell electrons therefore experience an effective nuclear charge of about 17 - 10 = + 7. (b) In a potassium atom, the fourth-shell electron experiences an effective nuclear charge of about 19 - 18 = + 1.

Why Atoms toward the Upper Right Are Smaller

From left to right across any row of the periodic table, the atomic diameters get *smaller*. Let's look at this trend from the point of view of effective nuclear charge. Consider lithium's outermost electron, which experiences an effective nuclear charge of about +1. Then look across period 2 to neon, in which each outermost electron experiences an effective nuclear charge of about +8, as **Figure 4.37** shows. Because the outer-shell electrons in neon experience a greater attraction to the nucleus, they are pulled in closer to the nucleus. So, neon, although nearly three times as massive as lithium, has a considerably smaller diameter.

In general, across any period from left to right, atomic diameters become smaller because of an increase in effective nuclear charge. Look back at Figure 4.33 and you will see this trend illustrated for the first three periods. In addition, **Figure 4.38** shows relative atomic diameters obtained from experimental data. Note that there are some exceptions to this trend, especially between groups 12 and 13. These exceptions can be explained by probing further into the shell model than we need to for our purposes. Moving down a group, atomic diameters get larger because of an increasing number of occupied shells. Whereas lithium has a small diameter because it has only two occupied shells, francium (atomic number 87) has a much larger diameter because it has seven occupied shells.

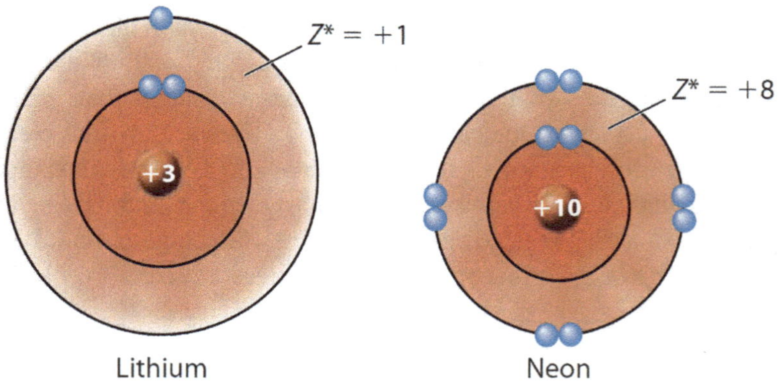

Figure 4.37 >
Lithium's outermost electron experiences an effective nuclear charge of about +1, while those of neon experience an effective nuclear charge of about +8. As a result, the outer-shell electrons in neon are closer to the nucleus and the diameter of the neon atom is smaller than the diameter of the lithium atom.

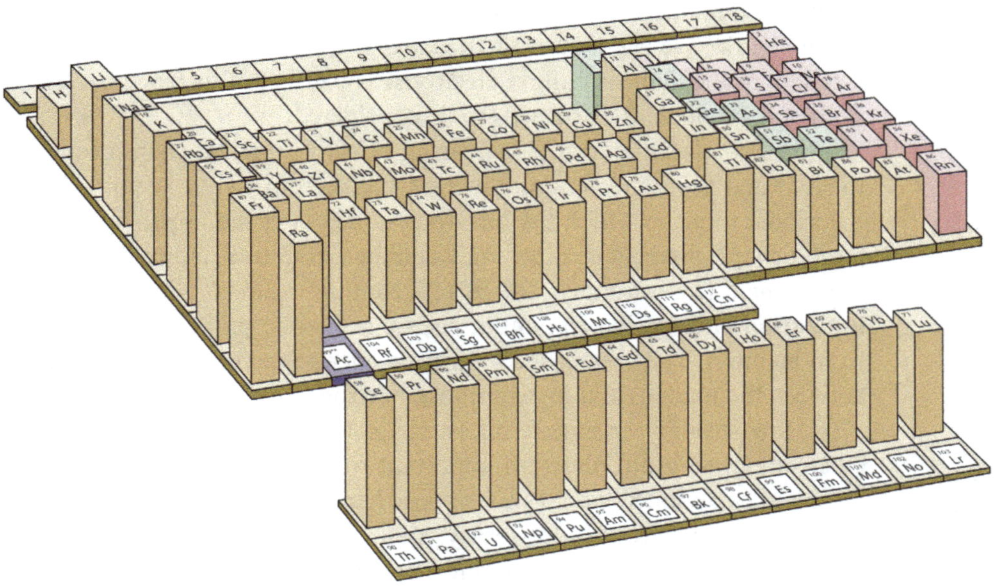

< **Figure 4.38**
This chart shows the relative atomic sizes, indicated by height. Note that atomic size generally decreases in moving to the upper right of the periodic table.

The Smallest Atoms Have the Most Strongly Held Electrons

How strongly electrons are bound to an atom is another property that changes gradually across the periodic table. In general, the trend is that the smaller the atom, the more tightly bound are its electrons.

As discussed earlier, effective nuclear charge increases in moving from left to right across any period. Thus, not only are atoms toward the right smaller, but their electrons are held more strongly. It takes about four times as much energy to remove an outer electron from a neon atom, for example, than to remove the outer electron from a lithium atom.

Moving down any group, the effective nuclear charge—as calculated by subtracting the charge of inner-shell electrons from the charge of the nucleus—stays the same. The effective nuclear charge for all group 1 elements, for example, works out to +1. Because of their greater number of shells, however, elements toward the bottom of a group (vertical column) are larger. The electrons in the outermost shell are therefore *farther* from the nucleus by an appreciable distance. From physics we learn that the electric force weakens rapidly with increasing distance. As **Figure 4.39** illustrates, an outer-shell electron in a larger atom, such as cesium, is not held as tightly as an outer-shell electron in a smaller atom, such as lithium. As a consequence, the energy needed to remove the outer electron from a

CHEMICAL CONNECTIONS

How is the weather connected to an atom?

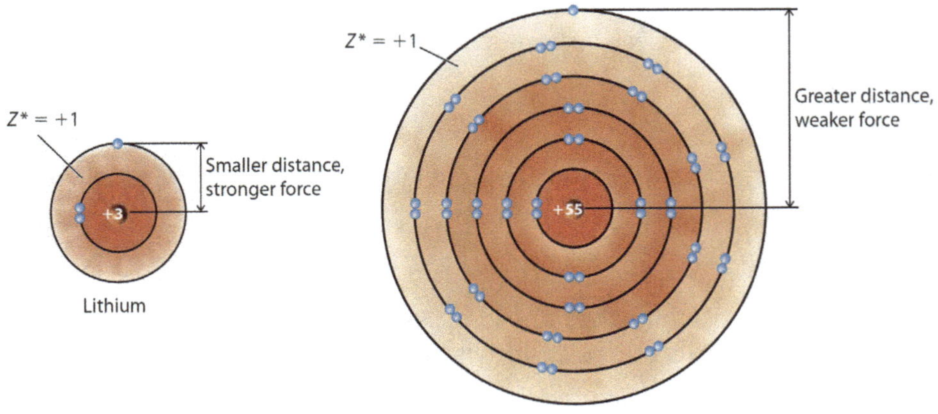

< **Figure 4.39**
In both lithium and cesium, the outermost electron has a calculated effective nuclear charge of +1. The outermost electron in a cesium atom, however, is not held as strongly to the nucleus because of its greater distance from the nucleus.

cesium atom is about half the energy needed to remove the outer electron from a lithium atom. This is true even though they both have a calculated Z* of +1. So, we see that our calculated Z* values provide only a rough estimate of the nuclear pull on electrons. The effect of distance between a shell and the nucleus also needs to be considered.

The combination of the increasing effective nuclear charge from left to right and the increasing number of shells from top to bottom creates a periodic trend in which the electrons in atoms at the upper right of the periodic table are held most strongly and the electrons in atoms at the lower left are held least strongly. This is reflected in **Figure 4.40**, which shows **ionization energy**, the amount of energy needed to pull an electron away from an atom. The greater the ionization energy, the greater the attraction between the nucleus and its outermost electrons.

How strongly an atomic nucleus is able to hold on to the outermost electrons in an atom plays an important role in determining the atom's chemical behavior. What do you suppose happens when an atom that holds its outermost electrons only weakly comes into contact with an atom that has a very strong pull on its outermost electrons? As we explore in Chapter 6, the atom that pulls strongly may remove one or more electrons from the other atom. The result is that the two atoms become chemically bonded. So, the shell model not only gives us insight into the workings of the periodic table, it also helps us to understand the heart of chemistry, which is the study of how new materials are created by the bonding of atoms.

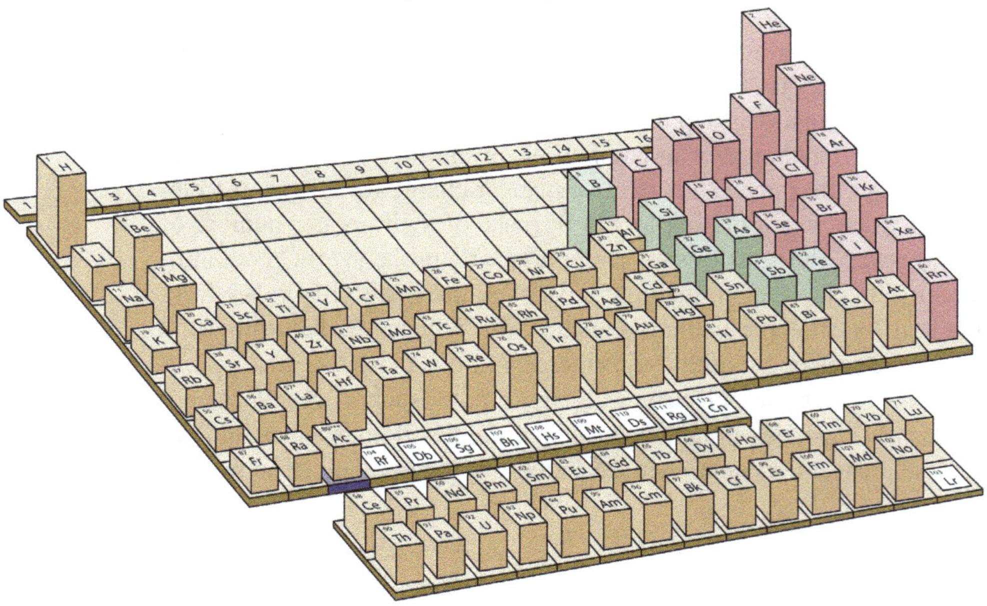

Figure 4.40 >
This chart shows the trends in ionization energy. The attraction an atomic nucleus has for its outermost electrons is indicated by height. Note that atoms at the upper right tend to have the greatest ionization energy and those at the lower left have the least.

CONCEPT **CHECK**

Which loses one of its outermost electrons more easily: a francium, Fr, atom (atomic number 87) or a helium, He, atom (atomic number 2)?

CHECK YOUR ANSWER

A francium, Fr, atom loses electrons much more easily than does a helium, He, atom. Why? Because a francium atom's outer electrons are not held as tightly by its nucleus, which is buried deep beneath many layers of shielding electrons.

4.10 Quantum Phenomena

You can measure the speed of a pitched baseball by having it fly through a pair of photogates that are a known distance apart, as shown in **Figure 4.41**. The ball is timed as it interrupts beams of light in the gates. The accuracy of the ball's measured speed has to do with uncertainties in the measured distance between the gates and in the timing mechanisms. Interactions between the macroscopic ball and the photons it encounters are insignificant.

In the case of measuring submicroscopic things like electrons we have a much different situation. Even a single photon bouncing off of an electron appreciably alters the motion of the electron, and in an unpredictable way. If you wish to observe an electron and determine its whereabouts with light, the wavelength of the light would have to bee very short. You fall into a dilemma. Light of a short wavelength, which can "see" the tiny electron better, corresponds to a large quantum of energy, which ,in turn, greatly alters the electron's state of motion. If, on the other hand, you use a long wavelength that corresponds to a smaller quantum of energy, the change you induce will be smaller, but the determination of its position by means of the coarser wave will be less accurate. The act of probing something as tiny as an electron produces a considerable uncertainty in either its position or its motion.

The uncertainty of measurement in the atomic domain, which was first stated mathematically by the German physicist Werner Heisenberg, is called the uncertainly principle. It is a fundamental principle in quantum mechanics. Heisenberg found that when the uncertainties in the measurement of the momentum and position of a particle are multiplied, the product must be equal to our greater than Planck's constant, h, divided by 2π:

$$(\Delta p)(\Delta x) \geq h/2\pi$$

Here, the Δ symbol means "uncertainty of": Δp is the uncertainty of momentum (the symbol for momentum is conventionally p), and Δx is the uncertainty of position. The product of these two uncertainties must be equal to or greater than ($\geq$) the size of $h/2\pi$.

The significance of the uncertainly principle is that, even in the best of conditions, the lower limit of uncertainly is $h/2\pi$. This means that if we wish to know the momentum of an electron with great accuracy (small Δp), the corresponding uncertainly in position will be large. Or, if we wish to know the position with great accuracy (small Δx), the corresponding uncertainly in momentum will be large. The shaper one of these quantities is, the less sharp is the other.

The uncertainly principle is relevant only to quantum phenomena. The inaccuracies in measure the position and momentum of a baseball due to quantum effects are completely negligible. But the inaccuracies in measuring the position and momentum of an electron are far from negligible. This is because the uncertainties in the measurements of these subatomic quantities are comparable to the magnitude of the quantities themselves.

Quantum Computing

A computer is a system that allows for computations. The abacus is an early form of a computer. It allows for the storage of information based upon the position of the beads. Also, the combination of various positions allows for computations to be completed in an efficient manner.

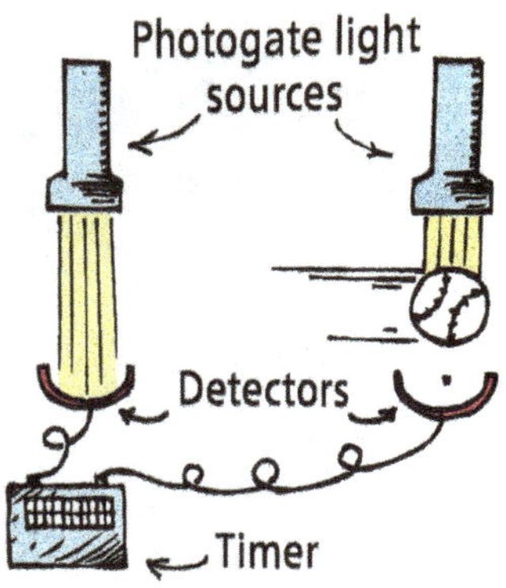

Figure 4.41
The ball's speed is measured by dividing the distance between the photogates by the time difference between crossing the two light paths. Photons hitting the ball alter its motion much less than the motion of an oil supertanker being altered by a few fleas bumping into it.

Werner Heisenberg (1901 - 1976)

fyi FOR YOUR INFORMATION

Planck's constant, h, is a fundamental constant of nature that serves to set a lower limit on the smallness of things. It ranks with the velocity of light and Newton's gravitational constant as a basic constant of nature. The equation $E = hf$ gives the smallest amount of energy that can be converted to light with frequency f. The radiation of light is not emitted by an atom continuously but is emitted as a sequence of photons, with each photon throbbing at a frequency f and carrying an energy hf.

Abacus

For the modern computer, storage of information and the processing of that information is achieved by using an electronic device called the *transistor*. Simply put, a transistor can hold information as either a 1 or a 0. A single transistor by itself is no big deal. But billions of them working together squeezed into a tiny integrated circuit is a very big deal. They form the basis of modern computers and many other modern electronic devices. They can hold massive amounts of information, and this includes programs that allow us to do amazing things with that information—drawing programs, writing programs, virtual reality programs, or any app you might be using on your smart phone.

There have been many advances in transistor technology over the past several decades. Yet transistors face a fundamental limit when it comes to the speed at which they can operate. Most notably, because they are electronic circuits, they run on electricity, which necessarily means they get hot. The higher the speed, the hotter (and less efficient) they get.

Enter quantum computers. While a traditional computer stores bits of information within miniature electronic gadgets, a quantum computer stores bits of information within what are known as "quantum bits", or qubits. Here's the key: while a transistor can be set to a "0" or a "1", a quantum bit can be set to both. This is a purely quantum phenomenon known as coherent superposition where two complementary states can exist simultaneously. What this means is that the qubit can hold twice the information. While a system of 10 transistors can be described by 10 bits of information, a system of 10 qubits would be described by $2^{10} = 10^{24}$ bits of information, which is explosively larger. The memory capacity of a quantum computers would thus also be explosively larger.

Also impressive is the speed at which calculations can be performed. While the transistors of a traditional computer are connected electronically, the qubits of a quantum computer are connected by a process known as entanglement, which is also a purely quantum phenomenon. Two entangled qubits respond to each other instantaneously, regardless of the distance between them. Einstein called entanglement "spooky action at a distance.". Because entangled qubits act together instantaneously, calculations made using groups of entangled quibits are exceedingly fast.

READING CHECK

Is Heisenberg's uncertainty principle applicable to the practical case of using a thermometer to measure the temperature of a glass of water?

Quantum Computing

Unfortunately, quantum computers are still at a very basic research level. There are many technical hurdles to a viable quantum computer. But it's something to keep an eye upon over the upcoming decades. That said, aspects of quantum computing are already here with us today when it comes to protecting highly sensitive data being sent from one place to another. Photons within fiber optic networks can be entangled. This allows for cryptographic keys that make it impossible for a third party to "peek in" on a transfer of private data. The moment they look, is the moment that the quantum entanglement is destroyed, which cuts off access to the data.

Concept Review

Chapter 4

Summary of Terms

Atomic Mass The total mass of an atom. The atomic mass of each element presented in the periodic table is the *average* atomic mass of the various isotopes of that element occurring in nature.

Atomic Nucleus The dense, positively charged center of every atom.

Atomic Number The number of protons in the atomic nucleus of each atom of a given element.

Atomic Spectrum The pattern of frequencies of electromagnetic radiation emitted by the atoms of an element, considered to be an element's "fingerprint."

Conceptual Model A representation of a system that helps us predict how the system behaves.

Effective Nuclear Charge The nuclear charge experienced by outer-shell electrons, diminished by the shielding effect of inner-shell electrons and also by the distance from the nucleus.

Electromagnetic Spectrum The complete range of waves, from radio waves to gamma rays.

Electron An extremely small, negatively charged subatomic particle found outside the atomic nucleus.

Electron Configuration The arrangement of an atom's electrons within orbitals.

Energy-level diagram A schematic drawing used to arrange atomic orbitals in order of increasing energy levels.

Inner-Shell Shielding The tendency of inner-shell electrons to partially shield outer-shell electrons from the attractive pull exerted by the positively charged nucleus.

Ionization Energy The amount of energy needed to pull an electron away from an atom.

Isotope Any member of a set of atoms of the same element whose nuclei contain the same number of protons but different numbers of neutrons.

Mass Number The number of nucleons (protons plus neutrons) in the atomic nucleus. Used primarily to identify isotopes.

Neutron An electrically neutral subatomic particle found in atomic nuclei.

Nucleon Any subatomic particle found in an atomic nucleus. Another name for either proton or neutron.

Physical Model A representation of an object on some convenient scale.

Proton A positively charged subatomic particle in atomic nuclei.

Quantum A small, discrete packet of energy.

Quantum Number An integer that specifies the quantized energy level within an atom.

Shell A graphic representation of a collection of orbitals of comparable energy in a multielectron atom. A shell can also be viewed as a region of space about the atomic nucleus within which electrons may reside.

Spectroscope A device that uses a prism or diffraction grating to separate light into its color components and measure their frequencies.

Review Questions

4.1 Physical and Conceptual Models

1. If a baseball were the size of the Earth, about how large would its atoms be?

2. What is the difference between a physical model and a conceptual model?

4.2 Discovering of the Electron

3. Why is a cathode ray deflected by a nearby electric charge or magnet?

4. What did Thomson discover about the electron?

5. What did Millikan discover about the electron?

4.3 Discovering the Atomic Nucleus

6. What did Rutherford discover about the atom?

7. To Rutherford's surprise, what was the fate of a tiny fraction of alpha particles in the gold-foil experiment?

8. What kind of force prevents atoms from squishing into one another?

4.4 Protons and Neutrons

9. What role does atomic number play in the periodic table?

10. Distinguish between atomic number and mass number.

11. Distinguish between mass number and atomic mass.

4.5 Light Is a Form of Energy

12. Does visible light constitute a large or small portion of the electromagnetic spectrum?

13. What does a spectroscope do to the light coming from an atom?

4.6 Atomic Spectra and the Quantum

14. What causes an atom to emit light?

15. Why do we say atomic spectra are like fingerprints of the elements?

16. What was Planck's quantum hypothesis?

17. Did Bohr think of his planetary model as an accurate representation of what an atom looks like?

4.7 Electrons Exhibit Wave Properties

18. Who first proposed that electrons exhibit the properties of a wave?

19. About how fast does an electron travel around the atomic nucleus?

20. How does the speed of an electron change its fundamental nature?

4.8 Orbitals and Energy-Level Diagrams

21. How many electrons can reside in a single atomic orbital?

22. What two elements are represented by these two energy-level diagrams?

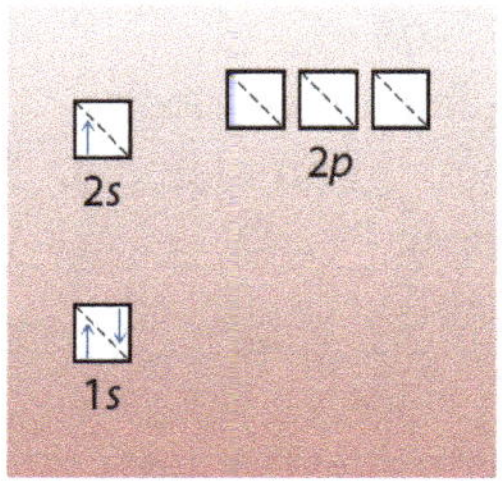

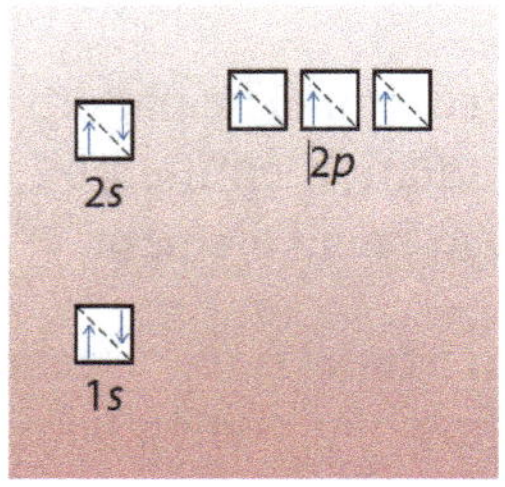

23. What element has the electron configuration: $1s^22s^22p^3$?

24. What atomic orbitals comprise the third shell?

4.9 The Shell Model and Periodic Table

25. Which electrons are most responsible for the properties of an atom?

26. What is the relationship between the number of electrons each shell can hold and the number of elements in each period of the periodic table?

27. What is effective nuclear charge?

28. How would you know from the periodic table that oxygen, O (number 8), molecules are smaller than nitrogen, N (number 7), molecules?

29. What happens to the strength of the electric force with increasing distance?

4.10 Quantum Phenomena

30. What is uncertain in the uncertainty principle?

31. What is a fundamental limit of the transistor?

Quantitative Questions

32. A class of 20 students takes an exam and every student scores 80 percent. What is the class average? Would the class average be slightly less, the same for slightly more if one of the students instead scored 100 percent? How is this similar to how we derived the atomic masses of elements?

33. The isotope lithium-7 has a mass of 7.0160 atomic mass units, and the isotope lithium-6 has a mass of 6.0151 atomic mass units. Given the information that 92.58 percent of all lithium atoms found in nature are lithium-7 and 7.42 percent are lithium-6, show that the atomic mass of lithium, Li (atomic number 3) is 6.941 amu

34. The element bromine, Br (atomic number 35), has two major isotopes of similar abundance, both around 50 percent. The atomic mass of bromine is reported in the periodic table as 79.904 atomic mass units. Choose the most likely set of mass numbers for these two bromine isotopes:

(a) ^{80}Br, ^{81}Br (b) ^{79}Br, ^{80}Br (c) ^{79}Br, ^{81}Br.

Solutions (Odd-Numbered)

1. The atoms in the baseball would be the size of ping pong balls if the baseball were the size of Earth.
3. The ray itself is negatively charged since it is a stream of electrons.
5. Millikan discovered the fundamental increment of all electrical charge to be 1.60×10^{-19} Coulombs.
7. Rutherford found that a few of the alpha particles were scattered backwards.
9. Elements are listed in the periodic table in order of increasing atomic number.
11. Mass number is the count of the number of nucleons in an isotope. Atomic mass is a measure of the total mass of an atom.
13. A spectroscope separates the light into color components whose frequencies can then be measured.
15. The atoms of each element emit only select frequencies of light. The pattern of these frequencies is unique to that element.
17. No, Bohr's model merely illustrated the different energy levels of an electron in an atom.
19. An electron moves around the nucleus at around 2 million meters per second.
21. Two.
23. Nitrogen
25. The electrons in the outer most shell of an atom are the ones most responsible for the properties of an atom.
27. Inner-shell electrons diminish the attraction outer-shell electrons have for the nucleus. The strength of the nuclear charge is also diminished for outer-shell electrons because they are farther away from the nucleus. This diminished nuclear charge experienced by outer-shell electrons is called the effective nuclear charge.
29. The electric force weakens with increasing distance.
31. A limiting factor for a transistor is the heat it generates.
33.

	Mass (amu)		Fraction of Abundance		
Li-6	6.0151	×	0.0742	=	0.446
Li-7	7.0160	×	0.9258	=	6.495
					6.941 amu

Solutions to Calculation Corner

Calculating Atomic Mass

Atomic mass of chlorine

	Contributing Mass of ^{35}Cl	**Contributing Mass of ^{37}Cl**
Fraction of Abundance	0.7553	0.2447
Mass (amu)	× 34.97	× 36.95
	26.41	9.04

atomic mass = 26.41 + 0.94 = 35.48

Mass of hydrogen atom in grams
(1.0079 ~~amu~~)(1.661×10^{-24} g / 1 ~~amu~~) = 1.674×10^{-24} g

Mass of water molecule in grams

Two hydrogen atoms: 3.348×10^{-24} g
One oxygen atom: 2.657×10^{-23} g

One water molecule: 2.992×10^{-23} g

Number of water molecules in 1 gram
(1.000 ~~g~~ H_2O)(1 molecule H_2O / 2.992×10^{-23} ~~g~~) = 3.342×10^{22}

Chapter 4: Contextual Chemistry Essay

Forensic Chemistry

The methods and tools of science can be used to discover the mysteries of the atom, but also the mysteries arising from crime. For example, "How did a murder victim die?" or "Who was the murderer?" The application of science to solving crimes is called *forensic science*, which can be subdivided into the various areas of science. Forensic medicine, for example, employs the methods and tools of medicine, such as autopsies, to determine a cause of death. Similarly, forensic chemistry employs the methods and tools of chemistry, such as the analysis of materials, to identify criminal suspects or, perhaps, criminal intent.

The pioneer who laid the cornerstone of modern forensic science was the early 20th-century criminologist Dr. Edmond Locard (1877–1966). Known as the Sherlock Holmes of France, Dr. Locard established the first police laboratory in Lyon, France, in 1910. His most widely recognized contribution has come to be known as the Locard Principle, which can be summarized as follows: "With contact between two items, there will be an exchange." According to this principle, a burglar cannot enter a house without leaving some trace of his presence, such as fingerprints or, perhaps, bits of hair. Knowing this, the burglar wears gloves and a cap. But the exchange of materials goes far beyond fingerprints and bits of hair. Consider the tiny grains of soil from the burglar's shoes or microfibers that shed from the burglar's clothing, including the gloves and cap! Furthermore, the burglar also carries bits of the crime scene, such as fibers from a carpet or furniture, with him as he leaves. In the words of Dr. Locard, "Wherever he steps, whatever he touches, whatever he leaves, even unconsciously, will serve as a silent witness against him."

Dr. Edmond Locard

The trained forensic chemist collects these bits of matter from the scene of the crime or from the suspect. The matter is then identified by assessing its properties, such as chemical composition or melting point. Once identified, this physical evidence may be used to support or refute the guilt of the suspect.

Samples of hair, skin, or semen left at the scene of the crime can be examined for DNA content. As we describe in Chapter 13, DNA is the biomolecule that holds a person's genetic information, which is unique for each individual. The amount of DNA recovered from a crime scene is often quite small. The tools of genetic engineering, however, allow the forensic chemist to use a minuscule amount of collected DNA as a template for the production of much larger amounts of identical DNA. This larger quantity of DNA can then be analyzed for patterns that may be identified as belonging (or not belonging) to the criminal suspect.

Modern analytical tools allow chemists to detect and identify chemicals at ultralow concentrations. One of the most widely used and sensitive analytical instruments is the mass spectrometer. Within the mass spectrometer, a sample molecule is subjected to harsh energy, which breaks the molecule into fragments. These fragments are then given an electric charge and accelerated via magnets down the length of a tube. The fragments separate from each other because more massive fragments travel slower while the lighter ones travel faster. The fragments, sorted by mass, produce a pattern that is characteristic of the original molecule. An unknown molecule can thus be identified by comparing its fragmentation pattern to a catalog of known fragmentation patterns. The wonder of the mass spectrometer is its great sensitivity—if you can see the sample injected into a mass spectrometer, it is way too much!

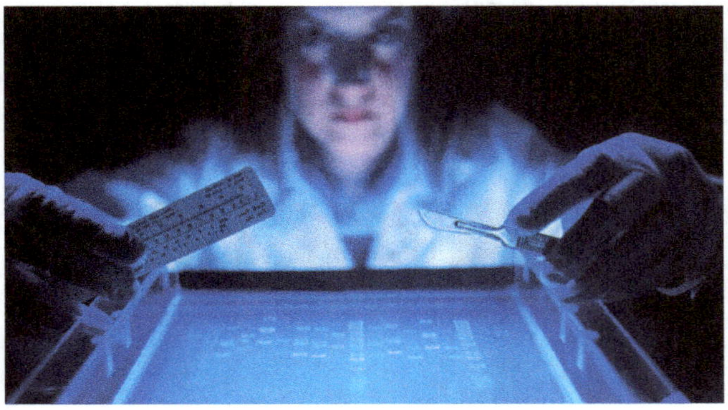

DNA can be fragmented and the fragments then separated to yield a pattern characteristic of an individual.

The mass spectrometer is often used in tandem with other analytical tools, such as the gas chromatograph, which volatilizes mixtures and separates them into their individual components. Each component is then analyzed with the mass spectrometer. These are the tools of choice for testing bodily fluids for ingested compounds, such as illicit drugs or sports-enhancing steroids. Instruments at the International Olympics, for example, are standardized to detect hundreds of different agents that are prohibited for use by Olympic athletes.

A type of mass spectrometer is also employed at airports to check for compounds that may be used as explosives. The technician rubs a swab inside a piece of luggage and then places the swab within the highly sensitive spectrometer, which tests for a wide assortment of potentially dangerous compounds. For luggage, the spectrometer is used in conjunction with an X ray machine that identifies the density of the contents. Many explosives have a density comparable to water, which is a reason why passengers are discouraged from packing water or liquid toiletries into their luggage. These X ray machines are also able to assess the average atomic number of the atoms within the luggage. This is helpful because the chemicals of explosives tend to be made from nitrogen (atomic number 7). A region of the luggage containing an average atomic number of around 7 likely contains explosive materials.

Airport Security

Modern technology used by forensic scientists is not foolproof or without its limitations. Collected material evidence needs to be taken in context and weighed against other factors, such as witness accounts and the possibility that physical evidence has been tampered with— intentionally or unintentionally—prior to being collected. That said, modern technology is a very powerful tool for the forensic scientists whose primary goal is the accurate reconstruction of criminal events as they occurred in the past with the hope that these events can be deterred from occurring again in the future.

Think and Discuss

1. A dog trained to sniff out fuel is brought to the remains of a building suspected to have been burned down by an arsonist. The dog barks excitedly in evidence of residual fuel at one and only one location. Does this suggest arson? What if the dog found residual fuel in two locations? Why are cases of arson frequently difficult to solve?

2. A woman is dropped off at her apartment by her boyfriend after a night of intimate romance. Entering her bedroom, she surprises a burglar, who then attacks and strangles her to death before fleeing. DNA evidence found on the woman implicates her boyfriend as being guilty of both rape and murder. How can the boyfriend prove beyond a reasonable doubt that he is innocent? How might the outcome of this case have been different if it had occurred 100 years ago?

3. According to the Innocence Project, a group that uses DNA testing to right wrongful convictions, police lineups

and similar forms of eyewitness identification are the leading cause of wrongful convictions across all DNA exonerated cases. Why might this be so? What might be done to improve the reliability of police lineups?

4. The homicide rate in states with the death penalty is almost double the rate in states without the death penalty. Is this because the homicide rates in these states are naturally high and the citizens seek the death penalty as a deterrent? Or is it because the death penalty cheapens the value of life and causes a higher homicide rate?

5. A deductive argument asserts that a conclusion necessarily follows from the truth of a premise. For example, all cats are mortal. Fluffy is a cat. Therefore, Fluffy is mortal. An inductive argument asserts that a conclusion follows, not necessarily but probably, from the truth of the premise. For example, all the cats you have ever seen are black. Therefore, all cats are black. Which of these forms of argument is used more often in a court of law? Which is used more often in science?

CONCEPT **CHECK**

Would it be a good idea for a burglar to own a type of dog that sheds?

CHECK YOUR ANSWER

For the burglar this would be a bad idea, because fur from the dog could easily end up at the scene of the crime, in accordance with Locard's principle. For the greater society, however, if this dog fur led to the just conviction of the burglar's crime, then the burglar's owning the dog would be a good thing.

Author Responses to Think and Discuss

1. Finding two sources of residual fuel most strongly implies arson because the arsonist is likely wanting to ensure that the building catches fire and catches fire as fast as possible. Of course, the burning destroys much of the physical evidence that would have otherwise been available to help solve the crime.

2. Ideally the boyfriend has an alibi for the time of the murder. Otherwise, it is in his interest for the investigator to find physical evidence of a third person, the burglar, at the scene of the crime. Character witnesses would also be vital. It would be bad news if the couple had a reputation for arguing with each other. Perhaps 100 years ago, without the benefit of careful forensic science, the boyfriend would have been found guilty.

3. Cognitive science shows that our ability to remember past events accurately is severely limited. More typical than remembering the past is remembering a time you thought you were remembering the past. For example, say a few years ago someone showed you an old photo of yourself as an 8 year old climbing your favorite tree. Do you then later remember climbing that tree? Or are you actually remembering your imagination as you gazed upon that photo from a few years ago? With each recollection, we have a tendency to fill-in any missing pieces. As time progresses, the accuracy of our memories diminishes. All that remains is a colored feeling. The assumption of a police line-up is that the witness has perfect memory and is not subject to hidden biases and is also not influenced by the body language of attending officers.

4. Why some areas have high rates of homicide and other do not is a complex issue with complex causes. It is interesting to note, however, that the homicide rate in Canada has declined since the death penalty was banned in 1976 while the overall conviction rate for first-degree murders doubled, suggesting juries are more willing to convict for murder when it does not involve a life-and-death decision.

5. Deductive arguments attempt to conclude with certainty, which makes them the preferred mode of argument within a court of law. As described in Chapter 1, however, science works primarily by a third mode of argument known as retroduction, which is the process of studying known facts and devising a theory to explain those facts. This is similar to induction in that a generalization is made based upon observed facts. Retroduction, however, takes it a step further by seeking a testable hypothesis for why that generalization is true. For example, all calico cats you have ever seen are female. The reason this is true is because the calico gene is located on the female-determining X chromosome. Unlike simple induction, retroduction provides a means of predicting new observations. For example, the fur of a calico cat consists of many different colors. The fur of the son of a calico cat will always consist of one and only one of these colors.

Chapter 5

The Atomic Nucleus

THE MAIN IDEA

> The atomic nucleus is the source of a tremendous amount of energy.

Λ What do volcanos and the Sun have in common? Both their energy sources are nuclear.

The atomic nucleus is the source of nuclear energy, which comes in different forms. Radioactivity is nuclear energy that occurs naturally in the air we breathe, in the rocks around us, and even in the food we eat. Some nuclear energy is human-controlled and is an important part of medical diagnostics and radiation therapy. Another source of nuclear energy is nuclear power plants, which transform nuclear energy into electricity. Of course, a nuclear bomb is another form of nuclear energy. Our most significant source of nuclear energy, however, is the Sun. We'll see in this chapter how solar energy comes from thermonuclear fusion and how it relates to Einstein's famous equation, $E = mc^2$. In this nuclear age it makes good sense to have a basic understanding of the atomic nucleus and the nuclear energy that touches every aspect of everyday life. Let's begin.

5.1 Unstable Nuclei

As we learned in Chapter 4, atoms are made of electrons, neutrons, and protons. The neutrons and protons are at the highly dense center of each atom—its nucleus. Most atoms have stable nuclei, which means they hold together well. This stability is due to an optimal balance of the number of protons with the number of neutrons. These stable nuclei remain unchanged.

Some atoms, however, have nuclei that are unstable because they contain "off-balance" numbers of protons and neutrons. They may either contain too many neutrons and not enough protons or vice versa. Sooner or later, these nuclei spontaneously change to a more stable composition. For example, excess neutrons may transform into protons to provide a better nuclear balance. During these transformations, the nucleus ejects highly energetic particles while also emitting highly energetic electromagnetic radiation.

A material containing unstable nuclei undergoing these sorts of transformations is said to be **radioactive**. The high-energy particles and radiation emitted by a radioactive substance is called **radioactivity**. Notably, a radioactive nucleus disintegrates as it emits radioactivity. In other words, after emitting the radioactivity, the radioactive nucleus is no longer what it was.

We say that the nucleus has "decayed," which is why the process of emitting radioactivity is often called radioactive decay.

Radioactive atoms emit three distinct types of radiation, named by the first three letters of the Greek alphabet, α, β, γ —alpha, beta, and gamma. Alpha rays carry a positive electric charge, beta rays carry a negative charge, and gamma rays carry no charge. The three rays can be separated by placing a magnetic field across their paths (**Figure 5.1**).

Alpha radiation is a stream of alpha particles. An **alpha particle** is the combination of two protons and two neutrons (in other words, it is the nucleus of the helium atom, atomic number 2). Alpha particles are relatively easy to shield against due to their relatively large size and their double positive charge (+2). For example, they do not normally penetrate such lightweight materials as paper or clothing. Because of their great kinetic energies, however, alpha particles can cause significant damage to the surface of a material, especially living tissue. When traveling through only a few centimeters of air, alpha particles pick up electrons and become nothing more than harmless helium. As a matter of fact, that's the origin of the helium in a child's balloon—practically all Earth's helium atoms were at one time energetic alpha particles.

Beta radiation is a stream of beta particles. A **beta particle** is merely an electron ejected from a radioactive nucleus. We will discuss in Section 5.3 how an electron arises from the nucleus. For now, just keep in mind that a beta particle is a fast-flying electron. A beta particle is normally faster than an alpha particle, and it carries only a single negative charge (–1). Beta particles are not as easily stopped as alpha particles, and they are able to penetrate light materials such as paper or clothing. They can

READING CHECK

What types of radiation are emitted by the nuclei of radioactive elements?

FOR YOUR INFORMATION

Most helium atoms produced within the Earth find their way to the surface and then upward to outer space. Some helium, however, collects within natural gas deposits, which can contain as much as 7 percent helium. Most helium used in the world is isolated from the natural gas reserves of the Great Plains of the United States. Prior to World War II, the United States stopped supplying helium to the Germans, who then needed to use combustible hydrogen gas to fill their large zeppelin airships. One such airship, the Hindenburg, famously exploded in 1937.

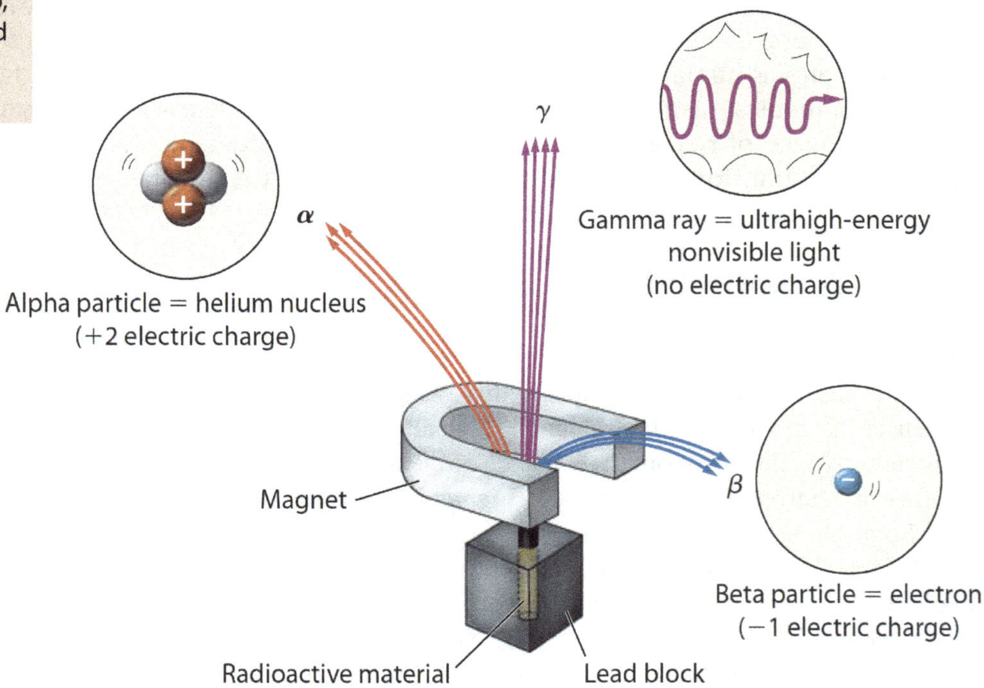

Figure 5.1

In a magnetic field, alpha rays bend one way, beta rays bend the other way, and gamma rays don't bend at all. Note that the alpha rays bend less than the beta rays. This occurs because alpha particles have more inertia (mass) than beta particles. The combined beam comes from a radioactive material placed at the bottom of a hole drilled in a block of lead.

penetrate fairly deeply into skin, where they have the potential to harm or kill living cells, but they are unable to penetrate even thin sheets of denser materials, such as aluminum. Beta particles, once stopped, simply become a part of the material they are in, like any other electron.

Gamma rays are the high-frequency electromagnetic radiation emitted by radioactive nuclei. Like photons of visible light, a gamma ray is pure energy. The amount of energy in a gamma ray, however, is much greater than that in visible light, ultraviolet light, or even X rays (Figure 5.2). Because they have no mass or electric charge, and because of their high energies, gamma rays are able to penetrate through most materials. However, they cannot easily penetrate very dense materials, such as lead. Lead is commonly used as a shielding material in laboratories or hospitals to combat against gamma radiation exposure. Delicate molecules inside cells throughout our bodies that are zapped by gamma rays suffer structural damage. Hence, gamma rays generally cause more damage to our cells than does alpha or beta radiation.

Figure 5.3 shows the relative penetrating power of the three types of radiation. Figure 5.4 shows an interesting practical use for gamma radiation.

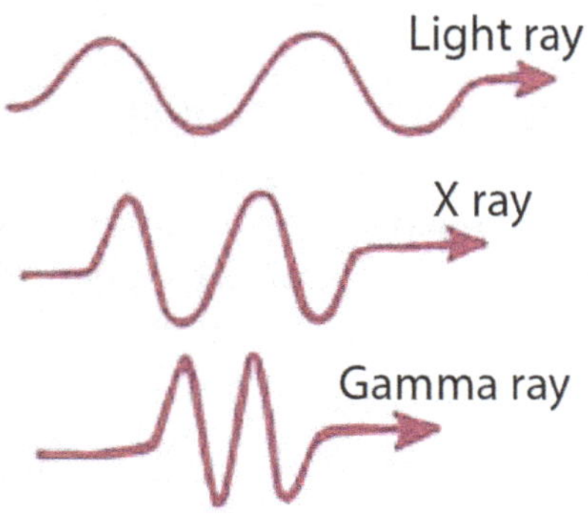

∧ Figure 5.2

A gamma ray is simply electromagnetic radiation, much higher in frequency and energy than light and X rays.

∧ Figure 5.4

The shelf life of fresh strawberries and other perishables is markedly increased when the food is subjected to gamma rays from a radioactive source, which kills the microorganisms that normally lead to spoilage. The strawberries are only a receiver of radiation and are not transformed into an emitter of radiation, as can be confirmed with a radiation detector.

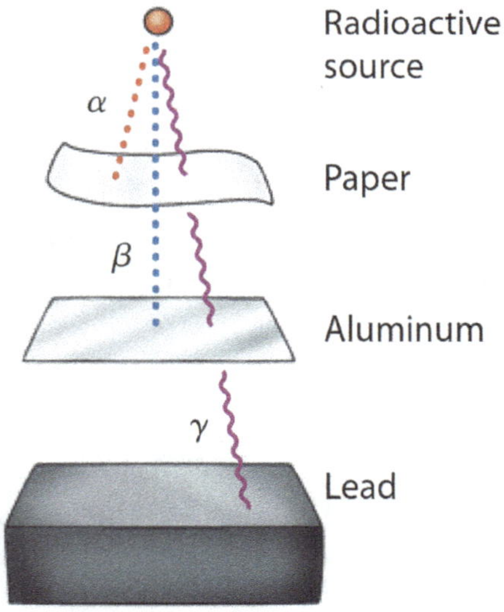

∧ Figure 5.3

Alpha particles are the least penetrating and can be stopped by a few sheets of paper. Beta particles will readily pass through paper but not through a sheet of aluminum. Gamma rays penetrate several centimeters into solid lead.

CONCEPT CHECK

Pretend that you are given three radioactive rocks. One is an alpha emitter, one is a beta emitter, and one is a gamma emitter, and you know which is which. You can throw away one, but of the remaining two, you must hold one in your hand and place one in your pocket. What can you do to minimize your exposure to radiation?

CHECK YOUR ANSWER

Throw away the gamma emitter, because it would penetrate your body from any of these locations. Hold the alpha emitter in your hand, because the skin on your hand is enough to shield you. Put the beta emitter in your pocket, because beta particles will likely be stopped by the combined thickness of your clothing and skin. (Ideally, of course, you should distance yourself as much as possible from all of the rocks.)

5.2 Radioactivity Is Natural

It is a common misconception that radioactivity is something new in the environment. Actually, radioactivity has been around far longer than humans. It has always been in the soil we walk on and in the air that we breathe, and it warms Earth's interior. In fact, radioactive decay in Earth's interior heats the water that spurts from a geyser or wells up from a natural hot spring.

As **Figure 5.5** shows, most of the radiation we encounter is natural background radiation that originates from the Earth, from the Sun, and from other stars. At sea level, the protective blanket of the atmosphere reduces this radiation, but radiation is more intense at higher altitudes, where the air is thinner. In Denver, the "Mile-High City," a person receives more than twice as much cosmic radiation as he or she does at sea level.

READING CHECK

What is a common source of radiation arising from Earth?

A common source of radiation of an Earthly origin is radon-222, an inert gas arising from uranium, which is widely found at low levels within all rock, soil, and water. Radon is a heavy gas that tends to accumulate in basements after it seeps up through cracks in the floor. Levels of radon vary from region to region, depending upon local geology. You can check the radon level in your home with a radon detector kit (**Figure 5.6**). If levels are high, corrective measures, such as sealing the basement floor and walls and maintaining adequate ventilation, should be taken. Radon gas poses a serious health risk.

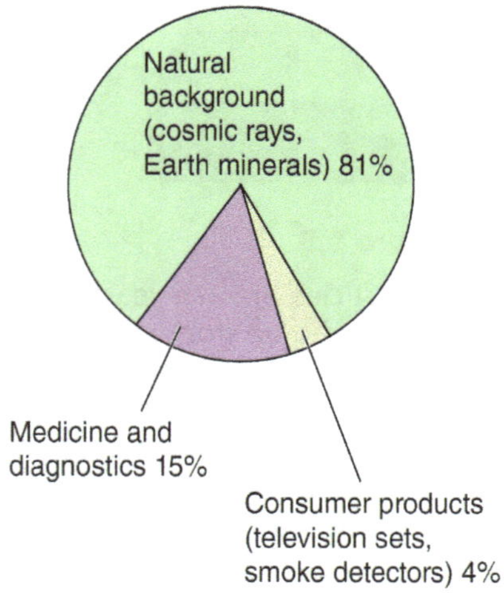

Figure 5.5

Origins of radiation exposure for an average individual in the United States.

About 20 percent of our annual exposure to radiation comes from sources outside of nature, primarily medical procedures. Fallout from nuclear testing and the coal and nuclear power industries are also contributors. Interestingly, the coal industry outranks the nuclear power industry as a source of radiation. The global combustion of coal annually releases over 13,000 tons of radioactive thorium and uranium into the atmosphere (in addition to other environmentally damaging molecules, including greenhouse gases). Both of these elements are found naturally in coal deposits, so their release is a natural consequence of burning coal. Nuclear power plants also produce radioactive by-products. Worldwide, the nuclear power industries generate about 10,000 tons of radioactive waste each year (about 2000 tons from the United States alone). Most of this waste, however, is contained and not released into the environment.

When radiation encounters our intricately structured cells, it can create chaos. Cells are able to repair most kinds of damage caused by radiation if the damage is not too severe. A cell can survive an otherwise lethal dose of radiation if the dose is spread over a long period of time to allow intervals for healing. When radiation is sufficient to kill cells, the dead cells can be replaced by new ones. Sometimes a radiated cell will survive with damaged DNA. This can alter the genetic information contained in a cell, producing one or more mutations. Although the effects of most mutations are inconsequential in terms of a person's health, some mutations affect the functioning of cells. Genetic mutations are the cause of most cancers, for example. In addition, mutations that occur in an individual's reproductive cells can be passed to the individual's offspring. In this case, the mutation will be present in every cell in the offspring organism's body—and may well have an effect on the functioning of the organism.

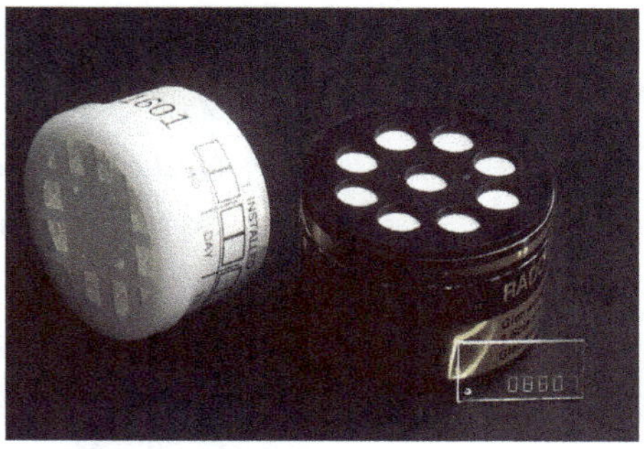

Figure 5.6

A commercially available radon test kit for the home. The canister is unsealed in the area to be sampled. Radon seeping into the canister is adsorbed by activated carbon within the canister. After several days, the canister is resealed and sent to a laboratory that determines the radon level by measuring the amount of radiation emitted by the adsorbed radon.

Rems Are Units of Radiation

We measure the ability of radiation to cause harm to living tissue in **rems**. Lethal doses of radiation begin at 500 rems. A person has about a 50 percent chance of surviving a dose of this magnitude received over a short period of time. During radiation therapy, a patient may receive localized doses in excess of 200 rems each day for a period of weeks (**Figure 5.7**).

Throughout our lives all the radiation we receive from natural sources and medical procedures is only a fraction of 1 rem. For convenience, the smaller unit *millirem* is used; 1 millirem (mrem) is 1/1000 of a rem.

The average person in the United States is exposed to about 360 millirems a year, as **Table 5.1** indicates. About 80 percent of this radiation comes from natural sources, such as cosmic rays (radiation from our sun as well as other stars) and the Earth. A typical diagnostic X ray exposes a person to between 5 and 30 millirems (0.005 and 0.030 rem), less than 1/10,000 of the lethal dose.

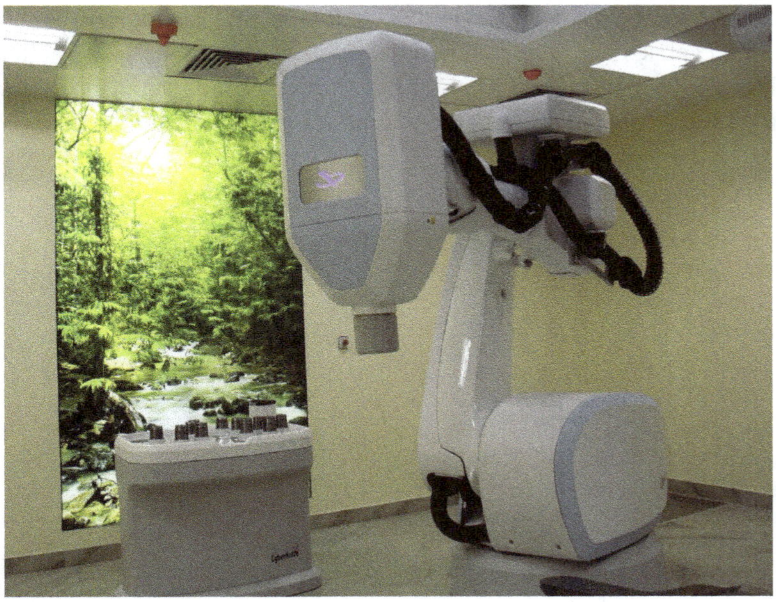

Figure 5.7

Nuclear radiation is focused on harmful tissue, such as a cancerous tumor, to selectively kill or shrink the tissue in a technique known as radiation therapy. This application of nuclear radiation has saved millions of lives—a clear-cut example of the benefits of nuclear technology. The CyberKnife System shown here uses robotics to deliver precise doses with extreme accuracy.

TABLE 5.1 Annual Radiation Exposure

SOURCE	TYPICAL AMOUNT RECEIVED IN 1 YEAR (MILLIREMS)
Natural Origin	
Cosmic radiation	26
Ground	33
Air (radon-222)	198
Human tissues (potassium-40; radium-226)	35
Human Origin	
Medical procedures	
Diagnostic X rays	40
Nuclear medicine	15
Television tubes, other consumer products	11
Weapons-test fallout	1

Source: U.S. Nuclear Regulatory Commission, http://www.nrc.gov/reading-rm/doc-collections/fact-sheets/bio-effects-radiation.html

Radioactive Tracers and Medical Imaging

Radioactive isotopes can be incorporated into molecules whose location can then be traced by the radiation they emit. When used in this way, radioactive isotopes are called *tracers*, and **Figure 5.8** shows one use for them. To check the action of a fertilizer, researchers incorporate radioactive isotopes into the molecules of the fertilizer and then apply the fertilizer to plants. The amount taken up by the plants can be measured with radiation detectors. From such measurements, scientists can tell farmers how much fertilizer to use, because fertilizer uptake is a physical and chemical process that is not affected by the radioactivity of the materials involved.

Tracers are also used in industry. Motor oil manufacturers can quantify the lubricating qualities of their products by running oil in engines containing small but measurable amounts of radioactive isotopes. As the engine runs and the pistons rub against the inner chambers, some of the metal from the engine invariably makes its way into the motor oil, and this metal carries with it the embedded radioactive isotopes. The better the lubricating qualities of a motor oil, the fewer radioactive isotopes it will contain after running in the engine for a given length of time.

fyi FOR YOUR INFORMATION

The deeper you go below the Earth's surface, the hotter it gets. At a depth of merely 30 kilometers the temperature is over 500°C. At greater depths it is so hot that rock melts into magma, which can rise to the Earth's surface to escape as lava. Superheated subterranean water can escape violently to form geysers or more gently to form a soothing natural hot spring. The main reason it gets hotter down below is that the Earth contains an abundance of radioactive isotopes and is heated as it absorbs the radiation from these isotopes. So, volcanoes, geysers, and hot springs are all powered by radioactivity. Even the drifting of continents (plate tectonics) is a consequence of the Earth's internal radioactivity.

Figure 5.8 >

Tracking fertilizer uptake with a radioactive isotope.

Fertilizer with radioactive isotope applied to crop

Radioactivity detected in plant

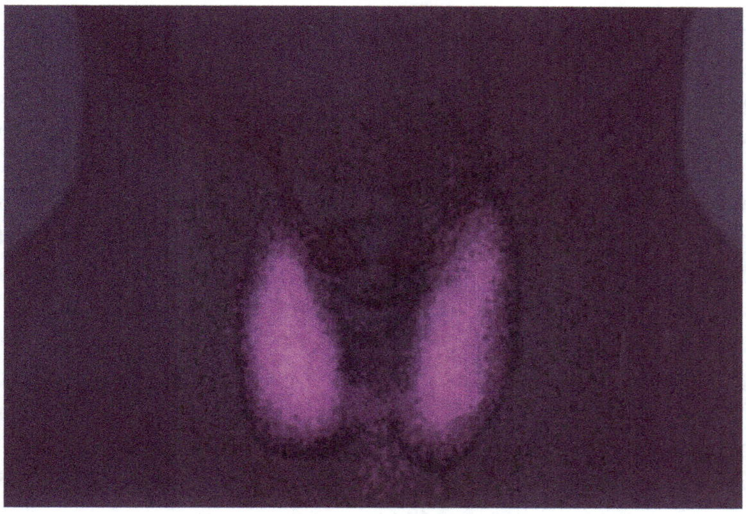

< Figure 5.9

The thyroid gland, located in the neck, absorbs much of the iodine that enters the body through food and drink. Images of the thyroid gland, such as the one shown here, can be obtained by giving a patient the radioactive isotope iodine-131. These images are useful in diagnosing metabolic disorders.

In a technique known as *medical imaging*, isotopes are used for the diagnosis of internal disorders. Small amounts of a radioactive material such as sodium iodide, NaI, which contains the radioactive isotope iodine-131, are administered to a patient and traced through the body with a radiation detector. The result, shown in **Figure 5.9,** is an image that shows how the material is distributed in the patient's body. This technique works because the path the tracer material takes is influenced only by its physical and chemical properties, not by its radioactivity. The tracer may be introduced alone or along with some other chemical, known as a *carrier compound*, that helps target the isotope to a particular type of tissue in the body.

TABLE 5.2 Uses for Various Radioactive Isotopes

ISOTOPE	USAGE
Fluorine-18	Medical PET scans
Calcium-47	The study of bone formation in mammals
Californium-252	Inspection of airline luggage for explosives
Hydrogen-3 (tritium)	Drug-metabolism studies
Iodine-131	Treatment of thyroid disorders
Iridium-192	Testing the integrity of metal parts
Thallium-201	Cardiology and for tumor detection
Xenon-133	Lung-ventilation and blood-flow studies

Source: The Regulation and Use of Radioisotopes in Today's World (NUREG/BR-0217, Revision 1), U.S. Nuclear Regulatory Commission

fyi FOR YOUR INFORMATION

The only element beyond uranium to find a commercial application is americium, Am, which is a key component of nearly all household smoke detectors. This element completes an electric circuit by ionizing air within a chamber. Smoke particles interfere with this ionization, thus breaking the circuit and triggering the alarm.

5.3 An Imbalance of Forces

We know that electric charges of the same charge repel one another. So how is it possible that positively charged protons in the nucleus stay clumped together? This question led to the discovery of an attraction called the **strong nuclear force**, which acts between all nucleons. This force is very strong but only over extremely short distances. Repulsive electric forces, on the other hand, are relatively long-ranged. Figure 5.10 suggests a comparison of the strengths of these two forces over distance.

Because the strong nuclear force decreases over distance, a large nucleus is not as stable as a small one, as shown in Figure 5.11. For protons that are close together, as in small nuclei, the attractive strong nuclear force easily overcomes the repulsive electric force. But for protons that are far apart, like those on opposite edges of a large nucleus, the attractive strong nuclear force may be weaker than the repulsive electric force.

READING CHECK

Are protons on opposite sides of a large nucleus attracted to each other?

A large atomic nucleus is more susceptible to the repulsive forces among protons. This means there is a limit to the size of the atomic nucleus. As evidence of this, we find that all nuclei having more than 83 protons are radioactive. Furthermore, the superheavy elements, such as those above uranium, atomic number 92, are not found in nature. These superheavy elements are difficult to make in the laboratory. When they are produced, they exist for only fractions of a second.

Neutrons serve as the "nuclear cement" holding an atomic nucleus together. Protons attract both protons and neutrons by the strong nuclear force. Protons also repel other protons by the electric force. Neutrons, on the other hand, have no electric charge and so attract protons and other neutrons only by the strong nuclear force. Therefore, the presence of neutrons adds to the attraction among nucleons and helps hold the nucleus together (Figure 5.12).

Nuclei with larger numbers of protons require larger numbers of neutrons to help balance the repulsive electric forces. For light elements, it is sufficient to have about as many neutrons as protons. For instance, the most

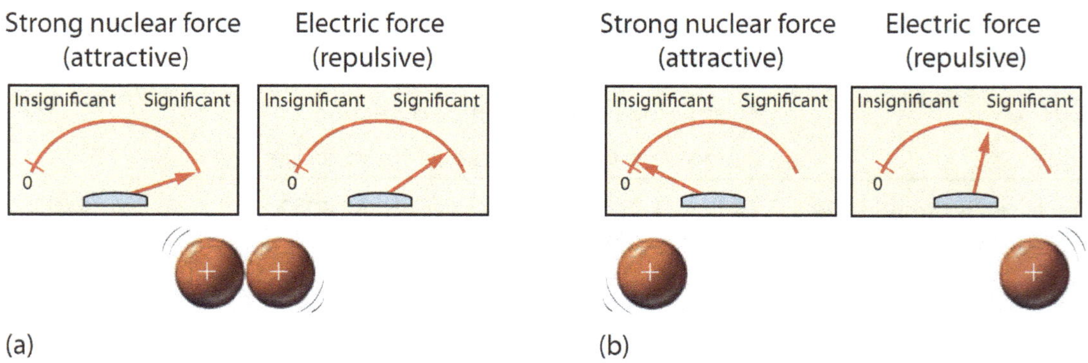

Figure 5.10

(a) Two protons near each other experience both an attractive strong nuclear force and a repulsive electric force. At this tiny separation distance, the strong nuclear force overcomes the electric force, resulting in their staying together. (b) When the two protons are relatively far from each other, the electric force is more significant. The protons repel each other. This proton–proton repulsion in large atomic nuclei reduces nuclear stability.

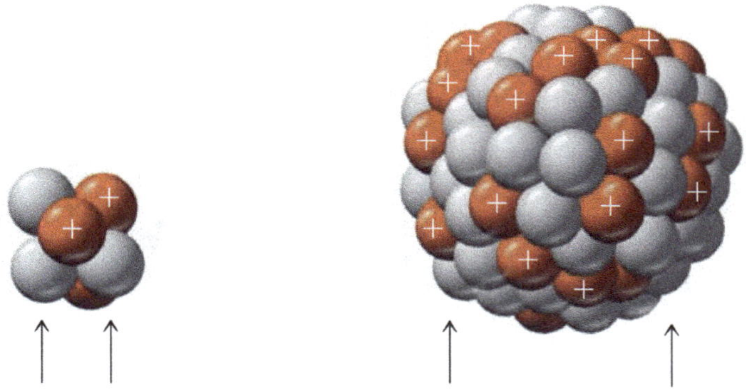

< Figure 5.11

All nucleons in a small atomic nucleus are close to one another; hence, they experience an attractive strong nuclear force. (b) Nucleons on opposite sides of a larger nucleus are not as close to one another, so the attractive strong nuclear forces holding them together are much weaker. The result is that the large nucleus is less stable.

common isotope of carbon, C-12, has equal numbers of each—six protons and six neutrons. For large nuclei, more neutrons than protons are needed. Because the strong nuclear force diminishes rapidly over distance, nucleons must be practically touching in order for the strong nuclear force to be effective. Nucleons on opposite sides of a large atomic nucleus are not so attracted to one another. The electric force, however, does not diminish by much across the diameter of a large nucleus and so begins to win out over the strong nuclear force. To compensate for the weakening of the strong nuclear force across the diameter of the nucleus, large nuclei have more neutrons than protons. Lead nuclei, for example, have about one-and-a-half times as many neutrons as protons.

So, we see that neutrons are stabilizing and large nuclei require an abundance of them. But neutrons are not always successful in keeping a nucleus intact. Interestingly, neutrons are not stable without protons. A lone neutron is radioactive and spontaneously transforms to a proton and an electron (**Figure 5.13a**). A neutron seems to need protons around to keep this from happening. After the size of a nucleus reaches a certain point, the neutrons so outnumber the protons that there are not sufficient protons in the mix to prevent the neutrons from turning into protons. As neutrons in a nucleus change into protons, the stability of the nucleus decreases because the repulsive electric force becomes increasingly significant. The result is that pieces of the nucleus fragment away in the form of radiation, as indicated in **Figure 5.13b**. This sets an upper limit on the mass and size of the atomic nucleus, which explains why the periodic table is limited in its number of elements.

CONCEPT CHECK

Two protons in the atomic nucleus repel each other, but they are also attracted to each other. Why?

CHECK YOUR ANSWER

While two protons repel each other by the electric force, they also attract each other by the strong nuclear force. These forces act simultaneously. So long as the attractive strong nuclear force is stronger than the repulsive electric force, the protons will remain together. Under conditions in which the electric force overcomes the strong nuclear force, however, the protons fly apart from each other.

Figure 5.12 >

(a) The presence of neutrons helps hold the nucleus together by increasing the effect of the strong nuclear force, represented by the single-headed arrows.

(b) The total strength of the attractive nuclear force, therefore, exceeds that of the repulsive electric force occurring between the two protons, represented by the double-headed arrow

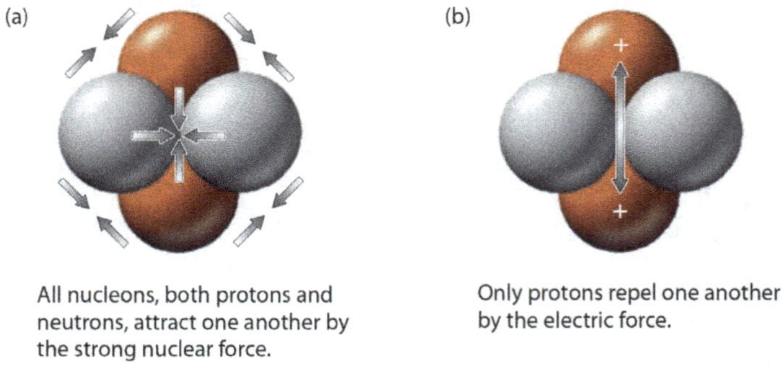

Figure 5.13 >

(a) A neutron without an adjacent proton is unstable and decays to a proton by emitting an electron.

(b) Large nuclei have more neutrons than protons, which means that some of these neutrons don't have a sufficient number of adjacent protons. One of these extra neutrons may transform into a proton. Destabilized by an increase in the number of protons, the nucleus begins to shed fragments, such as alpha particles.

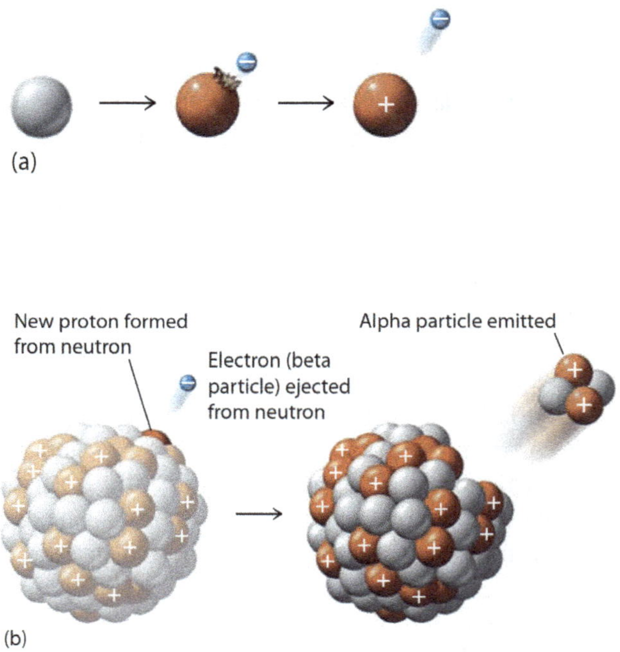

CONCEPT CHECK

What role do neutrons serve in the atomic nucleus? What is the fate of a neutron when alone or distant from one or more protons?

CHECK YOUR ANSWER

Neutrons serve as a nuclear cement in nuclei and add to nuclear stability. But when alone or away from protons, a neutron becomes radioactive and spontaneously transforms to a proton and an electron.

5.4 Transmutation

When a radioactive nucleus emits an alpha or beta particle, the atomic number of the nucleus is changed. It becomes another element.

The changing of one element to another is called **transmutation**. Consider a uranium-238 nucleus, which contains 92 protons and 146 neutrons. When an alpha particle is ejected, the nucleus loses two protons and two neutrons. Since an element is defined by the number of protons in its nucleus, the 90 protons and 144 neutrons remaining no longer constitute a uranium atom. What we have now is the nucleus of a different element—thorium, Th. This transmutation can be written as a nuclear equation:

$$^{238}_{92}\mathrm{U} \longrightarrow \, ^{234}_{90}\mathrm{Th} + \, ^{4}_{2}\mathrm{He}$$

We see that uranium-238 transmutes to the two elements written to the right of the arrow. When this transmutation occurs, energy is released, partly in the form of the kinetic energy of the alpha particle, He-2, partly in the kinetic energy of the thorium atom, and partly in the form of gamma radiation. In this and all such equations, the mass numbers at the top balance (238 = 234 + 4) and the atomic numbers at the bottom also balance (92 = 90 + 2).

Thorium-234, the product of this reaction, is also radioactive. When it decays, it emits a beta particle. Because the formation of a beta particle results in an additional proton, the atomic number of the resulting nucleus is increased by 1. So, after beta emission by thorium with 90 protons, the resulting element has 91 protons. It is no longer thorium but the element protactinium, Pa. Although the atomic number has increased by 1 in this process, the mass number (protons + neutrons) remains the same. The nuclear equation is

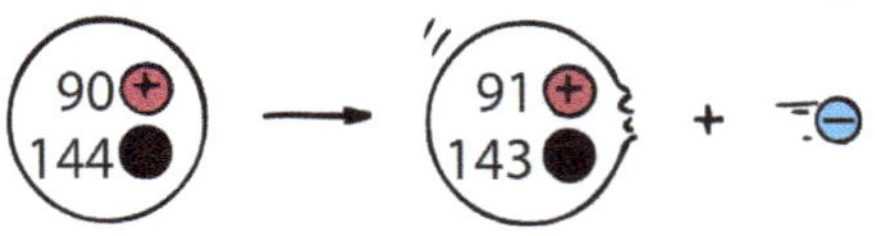

$$^{234}_{90}\mathrm{Th} \longrightarrow \, ^{234}_{91}\mathrm{Pa} + \, ^{0}_{-1}\mathrm{e}$$

We write an electron with a superscript 0 indicating that the electron's mass is insignificant relative to that of protons and neutrons. The subscript –1 is the electric charge of the electron.

So, we see that when an element ejects an alpha particle from its nucleus, the mass number of the remaining atom is decreased by 4 and its atomic number is decreased by 2. The resulting atom is an atom of the element two spaces back in the periodic table, because this atom has two fewer protons. When an element ejects a beta particle from its nucleus, the mass of the atom is practically unaffected, meaning there is no change in its mass number but its atomic number increases by 1. The resulting atom becomes the element one space forward in the periodic table because it has one more proton.

The decay of uranium-238 to lead-206 is shown in **Figure 5.14.** Each blue arrow shows an alpha decay, and each red arrow shows a beta decay.

READING CHECK

What is the term for the changing of one element into another? Hint: It is the same word used by the alchemists.

FOR YOUR INFORMATION

Beta emission is also accompanied by the emission of a neutrino, which is a neutral particle with nearly zero mass that travels at about the speed of light. Neutrinos are hard to detect because they interact very weakly with matter—a piece of lead about eight light-years thick would be needed to stop half the neutrinos produced in typical nuclear decays. Thousands of neutrinos are flying through you every second of every day, because the universe is filled with them. Only occasionally, one or two times a year, does a neutrino interact with an atom of your body.

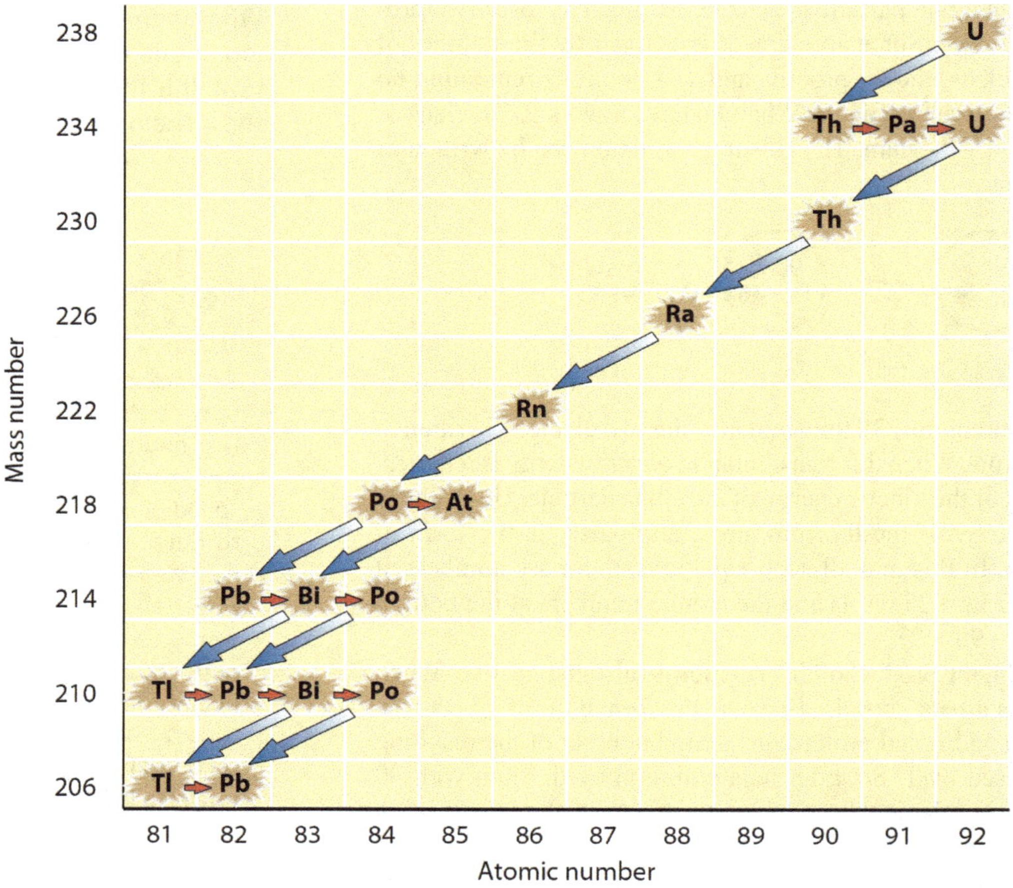

∧ Figure 5.14

U-238 decays to Pb-206 through a series of alpha and beta decays.

CONCEPT **CHECK**

What finally becomes of all the uranium that undergoes radioactive decay?

CHECK YOUR ANSWER

All uranium ultimately becomes lead. Along the way, it exists as the elements indicated in Figure 5.14.

5.5 Radioactive Half-Life

Radioactive isotopes decay at different rates. The radioactive decay rate is measured in terms of a characteristic time, the **half-life**.

The half-life of a radioactive material is the time required for half of the radioactive atoms to decay. Radium-226, for example, has a half-life of 1620 years. This means that half of any given specimen of Ra-226 decays by the end of 1620 years. In the next 1620 years, half of the remaining radium decays, leaving only one-fourth the original number of radium atoms. The other three-fourths convert, by a succession of decays, to lead. After 20 half-lives, an initial quantity of radioactive atoms is diminished to about one-millionth of the original quantity (**Figure 5.15**).

READING CHECK

How is radioactive half-life defined?

Half-lives are remarkably constant and not affected by external conditions. Some radioactive isotopes have half-lives that are less than a millionth of a second, while others have half-lives of more than a billion years. For example, uranium-238 has a half-life of 4.5 billion years. This means that in 4.5 billion years, half the uranium in the earth today will be lead-206.

It is not necessary to wait through the duration of a half-life in order to measure it. The half-life of an element can be accurately estimated by measuring the rate of decay of a known quantity of the element. This is easily done using a radiation detector. In general, the shorter the half-life of a substance, the faster it disintegrates and the more radioactivity per minute is detected. **Figures 5.16** and **5.17** show two ways of detecting radiation.

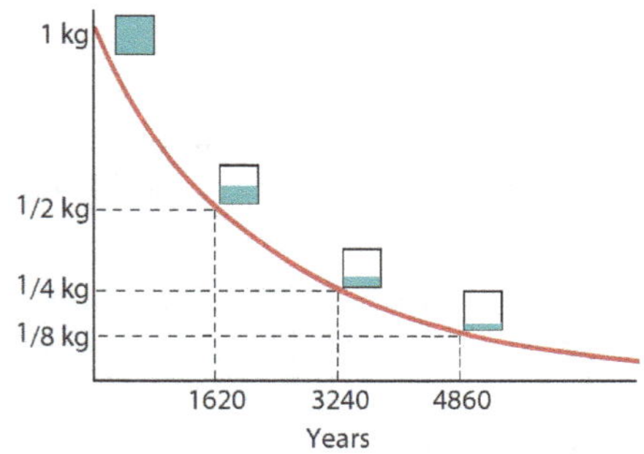

Λ Figure 5.15

Radium-226 has a half-life of 1620 years, meaning that every 1620 years the amount of radium decreases by half as it trans- mutes to other elements.

The half-life of radium-226 is 1620 years. Does this mean that an individual Ra-226 nucleus has to wait 1620 years before it decays? The answer is no. Some Ra-226 nuclei will decay within a few minutes. Others won't decay for tens of thousands of years. But if all the nuclei of a radioactive element decay at different times, then why is the half-life of a radioactive substance always the same? Half-life applies only to macroscopic quantities of a radioactive substance. In such quantities, there are billions upon billions of nuclei. Half-life is a statistical measure of what happens to all these nuclei *on average*. If, on average, the nuclei decay quickly, then the half-life will be short, which tells us that these nuclei are not very stable. Conversely, if, on average, it takes a long time for the nuclei to decay, then the half-life will be long, which tells us that these nuclei are more stable.

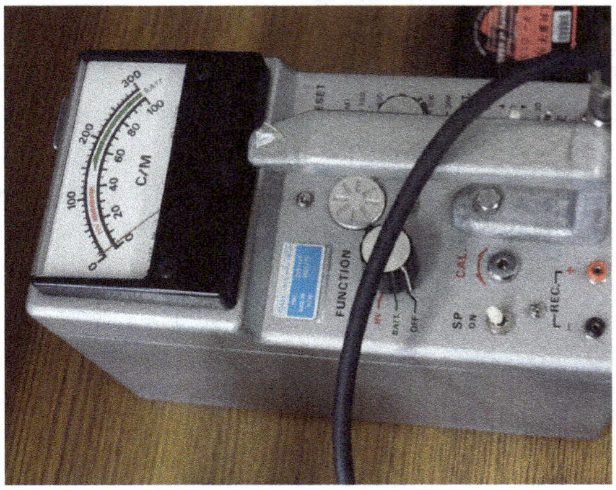

< Figure 5.16

The handheld tube of a Geiger counter contains a gas that gets ionized by incoming nuclear radiation. This completes an electric circuit, which generates audible clicks that increase with increasing radiation.

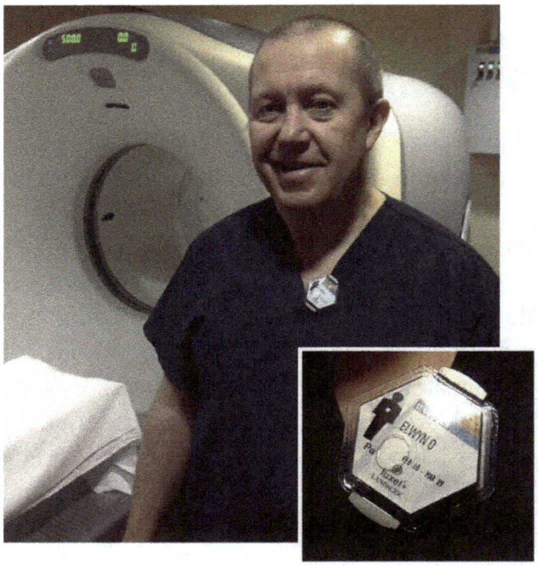

Figure 5.17 >

The film badge worn by Elwyn contains audible alerts for radiation surges and accumulated exposure. Information from the individualized badges is periodically downloaded to a database for analysis.

CONCEPT **CHECK**

1. If a sample of radioactive isotopes has a half-life of 1 day, how much of the original sample will remain at the end of the second day? The third day?

2. Which will give a higher counting rate on a radiation detector— radioactive material that has a short half-life or a long half-life?

CHECK YOUR ANSWER

1. In the second day, one-fourth of the original sample will be left. The three-fourths that underwent decay becomes a different element altogether. At the end of 3 days, one-eighth of the original sample will remain.

2. The material with the shorter half-life is more active and will show a higher counting rate on a radiation detector.

5.6 Isotopic Dating

Cosmic rays continually bombard the Earth's atmosphere. This bombardment causes many atoms in the upper atmosphere to transmute. These transmutations result in many protons and neutrons being "sprayed out" into the environment. Most of the protons are stopped as they collide with the atoms of the upper atmosphere. These protons strip electrons from the atoms they collide with and thus become hydrogen atoms. The neutrons, however, continue for longer distances because they have no electric charge and therefore do not interact electrically with matter. Eventually, many of them collide with atomic nuclei in the lower atmosphere. A nitrogen atom that captures a neutron, for instance, becomes an isotope of carbon by emitting a proton:

$$ {}^{1}_{0}\text{n} + {}^{14}_{7}\text{N} \longrightarrow {}^{14}_{6}\text{C} + {}^{1}_{1}\text{H} $$

This carbon-14 isotope, which makes up less than one-millionth of 1 percent of the carbon in the atmosphere, is radioactive and has eight neutrons. (The most common isotope, carbon-12, has six neutrons and is not radioactive.) Because both carbon-12 and carbon-14 are forms of carbon, they have the same chemical properties. Both of these isotopes, for example, chemically react with oxygen to form carbon dioxide, which is consumed by plants through the process of photosynthesis. This means that all plants contain a tiny quantity of radioactive carbon-14.

All animals eat either plants or plant-eating animals; therefore, all animals, including humans, have a little carbon-14 in them. So, we see why all living things on Earth contain some carbon-14. This radioactive isotope decays back into nitrogen as it emits a beta particle:

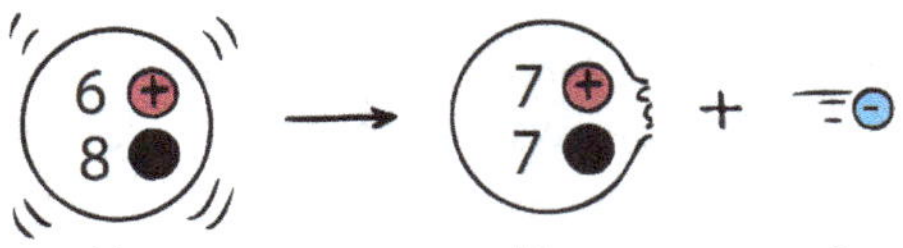

$$^{14}_{6}\mathrm{C} \longrightarrow {}^{14}_{7}\mathrm{N} + {}^{0}_{-1}\mathrm{e}$$

Because plants absorb carbon dioxide while alive, any carbon-14 that decays is immediately replenished with fresh carbon-14 from the atmosphere. In this way, a radioactive equilibrium is reached in which there is a constant ratio of about 1 carbon-14 atom to every 100 billion carbon-12 atoms. When a plant dies, replenishment of carbon-14 ends. Then the percentage of carbon-14 decreases at a constant rate determined by its half-life. The amount of carbon-12, however, does not change, because this isotope does not undergo radioactive decay. The longer a plant or other organism is dead, the less carbon-14 it contains relative to the constant amount of carbon-12.

The half-life of carbon-14 is about 5730 years. This means that half of the carbon-14 atoms now present in a plant or animal that dies today will decay in the next 5730 years. Half of the remaining carbon-14 atoms will then decay in the following 5730 years, and so on.

With this knowledge, scientists are able to calculate the age of carbon containing artifacts, such as wooden tools or the skeleton shown in **Figure 5.18**. Their age is measured by their current level of radioactivity. This process, known as **carbon-14 dating**, enables investigators to probe as far as 50,000 years into the past. Beyond this time span, there is too little carbon-14 remaining for accurate analysis.

Because of fluctuations in cosmic ray bombardment rates over the centuries, carbon dating has an uncertainty of about 15 percent. This means, for example, that the straw of an old adobe brick dated to be 500 years old may really be only 425 years old on the low side or 575 years old on the high side. For many purposes, this is an acceptable level of uncertainty.

Artifacts derived from nonliving materials can also be dated based on the radioactive minerals they contain. The naturally occurring mineral isotopes uranium-238 and uranium-235, for example, decay very slowly and ultimately become lead—but not the common isotope lead-208. Instead, as was shown in **Figure 5.14,** uranium-238 decays to lead-206. Uranium-235, on the other hand, decays to lead-207. Thus, the lead-206 and lead-207 that now exist in a uranium-bearing rock were at one time uranium. Older rocks contain higher percentages of these trace isotopes.

READING CHECK

Plants contain a little carbon-14, but why do humans also contain carbon-14?

FOR YOUR INFORMATION

A 1 g sample of carbon from recently living matter contains about 50 trillion billion (5×10^{22}) carbon atoms. Of these carbon atoms, about 65 billion (6.5×10^{10}) are the radioactive C-14 isotope. This gives the carbon a beta disintegration rate of about 13.5 decays per minute.

CHEMICAL CONNECTIONS

How is the radioactivity of your body connected to the stars?

Figure 5.18

The amount of radioactive carbon-14 in the skeleton is reduced by one-half every 5730 years. The result is that the same skeleton today contains only a trace amount of the original carbon-14. The arrows represent the relative amounts of carbon-14.

If you know the half-lives of uranium isotopes and the percentage of lead isotopes in some uranium-bearing rock, you can calculate the date of the rock's formation. Rocks dated in this manner have been found to be as much as 3.7 *billion* years old. Samples from the Moon have been dated at 4.2 billion years, which is close to the estimated age of our solar system: 4.6 billion years.

Figure 5.19

Carbon-14 dating was developed by the American chemist Willard F. Libby (1908–1980) at the University of Chicago in the 1950s. For this work, he received the Nobel Prize in Chemistry in 1960.

CONCEPT **CHECK**

Suppose that an archaeologist extracts 1 gram of carbon from an ancient ax handle and finds that it is one-fourth as radioactive as 1 gram of carbon extracted from a freshly cut tree branch. About how old is the ax handle?

CHECK YOUR ANSWER

Assuming the ratio of C-14/C-12 was the same when the ax was made, the ax handle is as old as two half-lives of C-14, or about 11,460 years old

5.7 Nuclear Fission

Biology students know that living tissue grows by the division of cells. The splitting in half of living cells is called fission. In a similar way, the splitting of an atomic nucleus into two smaller halves is called **nuclear fission**.

Nuclear fission involves the delicate balance between two forces within the nucleus. One force is the *strong nuclear force*, which is a force that holds all the nucleons together. The second force is the repulsive *electric force* occurring among all the like-sign protons. In most nuclei the nuclear strong force dominates. In uranium, however, this domination is weak. If the uranium nucleus is stretched into an elongated shape (**Figure 5.20**), the nuclear force quickly weakens and the electric forces may push it into an even more elongated shape. If the elongation passes a critical point, the electric forces overwhelm the strong nuclear forces, and the nucleus splits. This splitting process is called nuclear fission.

The absorption of a neutron by a uranium nucleus supplies enough energy to cause such an elongation. The resulting fission process may produce many different combinations of smaller nuclei. More significantly, the energy released by fission is enormous—about 7 million times that of a TNT molecule explosion. This energy is mainly in the form of the kinetic energy of the fission fragments, which fly apart from one another. A much smaller amount of energy is released as gamma radiation. Here is a pictorial with the equation for a typical uranium fission reaction:

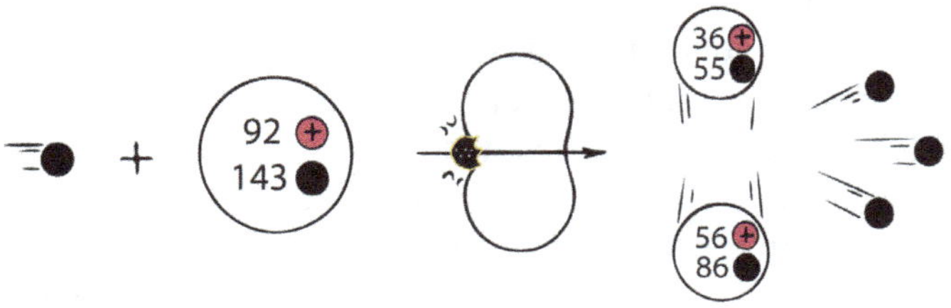

$$ {}^{1}_{0}n + {}^{235}_{92}U \longrightarrow {}^{91}_{36}Kr + {}^{142}_{56}Ba + 3({}^{1}_{0}n) $$

Note that in this reaction 1 neutron starts the fission of a single uranium nucleus, which produces nuclear fragments and 3 neutrons. These 3 neutrons can cause the fission of 3 more uranium atoms, releasing 9

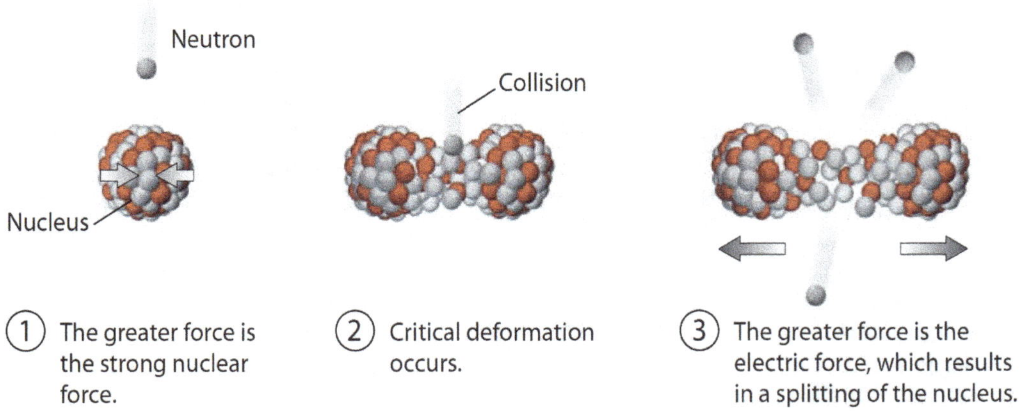

Figure 5.20

An elongation of the nucleus may result in the repulsive electric force's overcoming the attractive strong nuclear force, in which case fission occurs.

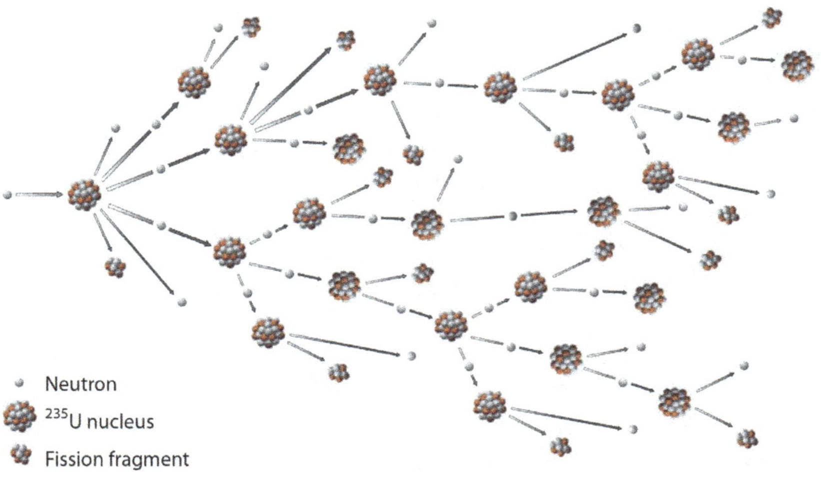

Figure 5.21 >
A chain reaction.

more neutrons. If each of these 9 neutrons succeeds in splitting a uranium atom, the next step in the reaction produces 27 neutrons, and so on. Such a sequence, illustrated in **Figure 5.21**, is called a **chain reaction**. A chain reaction is a self-sustaining reaction in which the products of one reaction event initiate further reaction events.

READING CHECK
What exactly is a chain reaction?

Why do chain reactions not happen in naturally occurring uranium ore deposits? They would if all uranium atoms fissioned so easily. Fission occurs mainly in the rare isotope U-235, which makes up only 0.7 percent of the uranium in pure uranium metal, as shown in **Figure 5.22**. When the more abundant isotope U-238 absorbs neutrons created by fission of U-235, the U-238 typically does not undergo fission. So, any chain reaction is snuffed out by the neutron-absorbing U-238, as well as by the rock in which the ore is imbedded. The rock is slightly warmed.

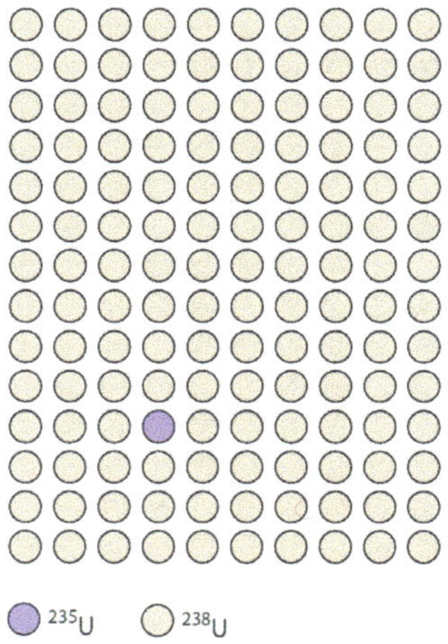

∧ Figure 5.22
Only 1 part in 140 of naturally occurring uranium is U-235.

Chain reactions are more effective in large chunks of uranium than in smaller chunks. In smaller chunks, neutrons easily find the surface and escape, as shown in **Figure 5.23**. As the neutrons escape, the chain reaction no longer builds up.

If two small chunks of uranium are suddenly pushed together, they make a larger chunk. Within this larger chunk, neutrons are no longer able to escape as easily. Instead, they continue the chain reaction, which becomes sustainable. The minimum size of a chunk needed for a sustainable chain reaction is called the **critical mass**. Any chunk at or above the critical mass produces a steady release of energy. With the correct engineering, this release of energy can be controlled, which is what happens within a nuclear power plant. A different sort of engineering, as shown in **Figure 5.24**, produces an explosion of energy, which is the basis of a nuclear fission bomb

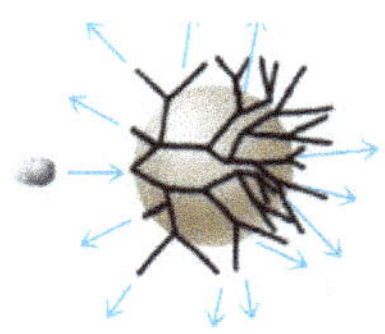

Neutrons escape small lump of uranium-235.

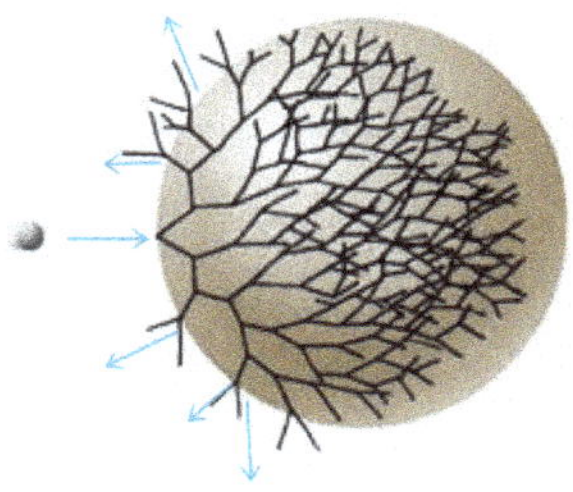

Neutrons trigger more reactions within large lump of uranium-235.

Figure 5.23 >
This exaggerated view shows that a chain reaction in a small piece of pure uranium-235 runs its course before it can cause a large explosion because neutrons leak from the surface too soon. The surface area of the small piece is large relative to its mass. In a larger piece, more uranium and less surface are presented to the neutrons.

Constructing a fission bomb is a formidable task. The difficulty is in separating enough uranium-235 from the more abundant uranium-238. Separation is difficult because, except for their slightly different masses, the two isotopes have the same physical and chemical properties. Scientists took more than 2 years to extract enough of the 235 isotope from uranium ore to make the bomb that was detonated at Hiroshima in 1945. To this day, uranium isotope separation, also known as uranium enrichment, remains a difficult process.

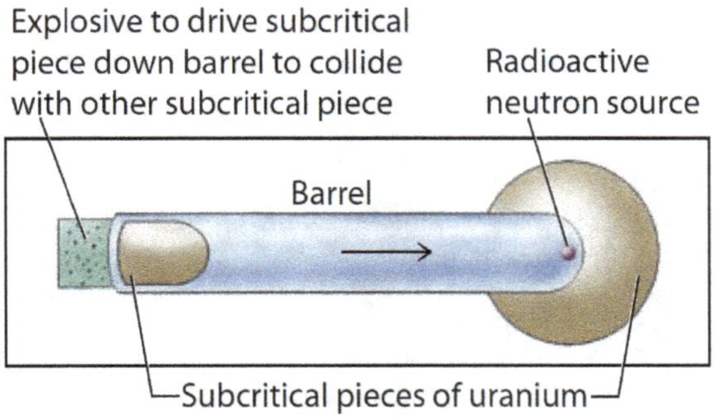

< Figure 5.24

Simplified diagram of a uranium-fission bomb. Two smaller masses of uranium are initially apart from each other. The bomb is ignited when high explosives (gunpowder) force the two masses to come together into a single lump with a critical mass.

Nuclear Fission Reactors

The awesome energy of nuclear fission was introduced to the world in the form of nuclear bombs, and this violent image still colors our thinking about nuclear power, making it difficult for many people to recognize its potential usefulness. Currently, about 20 percent of electric energy in the United States is generated by *nuclear fission reactors* (whereas most electric power is nuclear in some other countries—about 75 percent in France). These reactors are simply nuclear furnaces. They, like fossil fuel furnaces, do nothing more elegant than boil water to produce steam for a turbine (**Figure 5.25**). The greatest practical difference is the amount of fuel involved: a mere 1 kilogram of uranium fuel, less than the size of a baseball, yields more energy than 30 freight-car loads of coal.

A fission reactor contains four components: nuclear fuel, control rods, moderator (to slow the velocity of the neutrons, making them effective for the fission process), and a liquid (usually water) to transfer heat from the reactor to the turbine and generator. The nuclear fuel is primarily U-238, plus about 3 percent U-235. Because the U-235 isotopes are so highly diluted with U-238, an explosion like that of a nuclear bomb is not possible. The reaction rate, which depends on the number of neutrons that initiate the fission of other U-235 nuclei, is controlled by rods inserted into the reactor. The control rods are made of a neutron-absorbing material, usually metals like cadmium or boron.

Heated water around the nuclear fuel is kept under high pressure to keep it at a high temperature without boiling (see Section 8.4). It transfers heat to a second, lower-pressure water system, which operates the turbine and electric generator in a conventional fashion. In this design, two separate

FOR YOUR INFORMATION

A nuclear power plant "meltdown" occurs when the fissioning nuclear fuels are no longer submerged within a cooling fluid, such as water. The temperature rises to the point that the solid nuclear fuel and the reaction vessel itself melt into a liquid phase that has the potential of penetrating through the floor of the containment building.

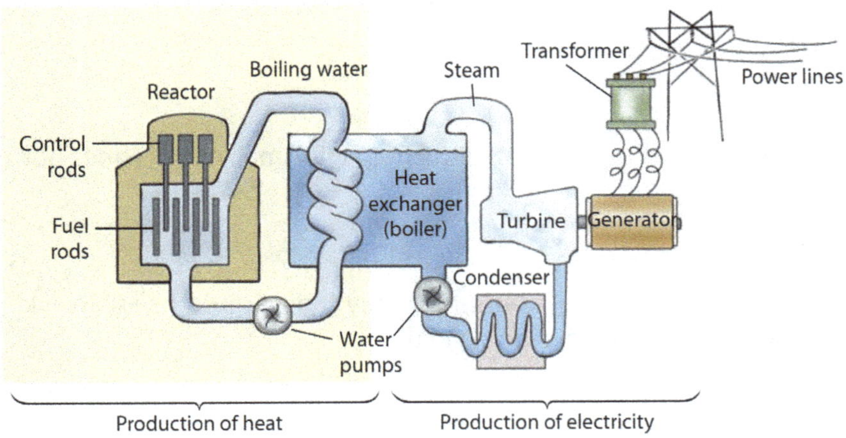

< Figure 5.25

Diagram of a nuclear fission power plant. Note that the water in contact with the fuel rods is completely contained and that radioactive materials are not involved directly in the generation of electricity.

Figure 5.26 >

The nuclear reactor is housed within a dome-shaped containment building designed to prevent the release of radioactive isotopes in the event of an accident. The Soviet-built Chornobyl nuclear power plant that reached meltdown in 1986 had no such containment building, leading to massive amounts of radiation being released into the surrounding environment.

water systems are used so that no radioactivity reaches the turbine or the outside environment. The entire setup resides inside a building like the one shown in **Figure 5.26**.

Fission power has benefits and drawbacks. The benefits include plentiful electricity, conservation of fossil fuels that presently are turned to heat and smoke (and in the long run may be far more precious as sources of organic molecules than as sources of heat), eliminating the megatons of sulfur oxides and other poisons as well as the greenhouse gas carbon dioxide released into the air each year by the burning of these fuels.

The drawbacks include the low-level release of radioactive materials into the air and groundwater, and, most important, the risk of an accidental release of large amounts of radioactivity. Then there's the problem of storing radioactive wastes. Some countries, such as France, dispose of wastes in deep ground repositories. Waste disposal nevertheless remains a problem.

Reasoned judgment requires not only that we examine fission power's benefits and drawbacks but that we compare its benefits and drawbacks with those of other power sources. We speak of "trade-offs," a comparison of benefits and drawbacks. The issue is not whether fission power is safe, but whether its dangers are greater or less than the dangers of supplying the same electric power in other ways. So, going nuclear isn't a long-term solution, but rather is a reasoned compromise.

fyi FOR YOUR INFORMATION

Sustainable fission chain reactions are a prime source of heat found within Earth's core, which is to say that Earth itself is a planetary-sized nuclear reactor. Know nukes before you say "No nukes"!

CONCEPT CHECK

A 52-kilogram ball of uranium-235 has critical mass, but the same ball broken up into small chunks does not. Explain.

CHECK YOUR ANSWER

The small chunks have more combined surface area than the ball from which they came. Neutrons escape via the surface of each small chunk before a sustained chain reaction can build up.

5.8 Mass and Energy

Clearly, a lot of energy comes from every gram of nuclear fuel when it undergoes fission. What is the source of this energy? As we will see, it comes from nucleons losing mass through nuclear reactions.

In the early 1900s, Albert Einstein discovered that mass is actually "congealed" energy. Mass and energy are two sides of the same coin, as stated in his celebrated equation, $E = mc^2$. In this equation, E stands for the energy contained by any mass when at rest, m stands for mass, and c is the speed of light. This relationship between energy and mass is the key to understanding why and how energy is released in nuclear reactions.

How easy might it be to pull a nucleon out of a nucleus? (Remember, "nucleon" is the generic name for either a proton or a neutron.) To do this, you would have to fight against the strong nuclear force, which holds the nucleon to the nucleus. So, a lot of energy would be required to pull this nucleon out of the nucleus. What we learn from Einstein's equation is that the energy you contribute to pull the nucleon out is not lost. Instead, this energy is absorbed by the nucleon, which thus becomes more massive as you pull it out. For example, if the mass of the nucleon were 1.00000 while in the nucleus, its mass might be a slightly greater 1.00728 after it has been pulled out of the nucleus. The energy you put into pulling the nucleon out was converted into mass. Energy and mass are two sides of the same coin. In other words, energy can become mass and mass can become energy as dictated by the equation $E = mc^2$.

So, the mass of a nucleon depends upon where it is. In general, a nucleon's mass is greatest when it is free by itself outside of the nucleus. The nucleon's mass is smallest when it is tightly bound within the nucleus.

Not all nuclei are the same. In one nucleus, for example, a nucleon might find itself more tightly bound than in another. The mass of the nucleon, therefore, also depends upon which nucleus it is in. As illustrated in the graph of **Figure 5.27**, a nucleon has its greatest mass when in the hydrogen nucleus and its smallest mass when in the iron nucleus. The mass of a nucleon then gradually increases as it enters heavier nuclei, such as that of uranium.

The graph of Figure 5.27 is the key to understanding the energy released in nuclear processes. Uranium, being toward the right-hand side of the graph, is shown to have a relatively large amount of mass per nucleon. When the uranium nucleus splits in half, however, smaller nuclei of lower

READING CHECK

What do the symbols represent in the equation $E = mc^2$?

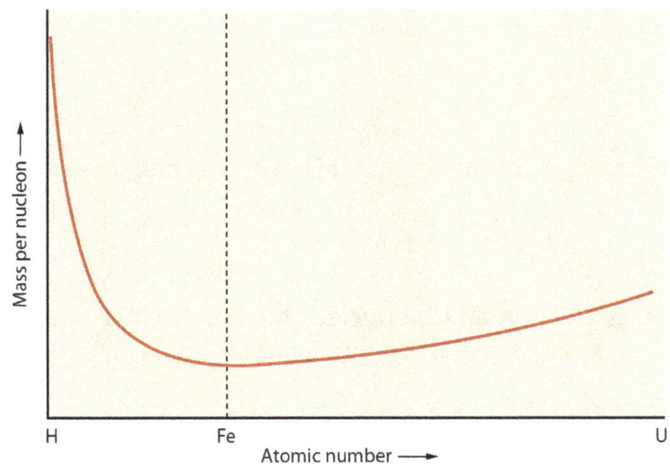

< Figure 5.27

This graph shows that the average mass of a nucleon depends on which nucleus it is in. A nucleon has the most mass when in the lightest (hydrogen) nucleus. The nucleon has its least mass when in an iron nucleus and an intermediate-mass when in the heaviest (uranium) nucleus.

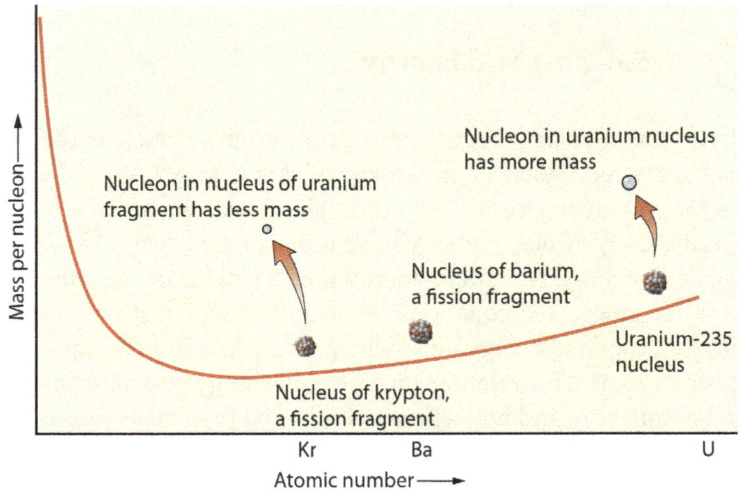

Figure 5.28 >

The mass of each nucleon in a uranium nucleus is greater than the mass of each nucleon in any one of its nuclear fission fragments. This lost mass is mass that has been transformed into energy, which is why nuclear fission is an energy-releasing process.

atomic numbers are formed. As shown in **Figure 5.28**, these nuclei are lower on the graph than uranium, which means that they have a smaller amount of mass per nucleon. Thus, nucleons lose mass in their transition from being in a uranium nucleus to being in one of its fragments. When this decrease in mass is multiplied by the speed of light squared (c^2 in Einstein's equation), the product is equal to the energy yielded by each uranium nucleus as it undergoes fission.

Interestingly, Einstein's mass–energy relationship applies to chemical reactions as well as to nuclear reactions. For nuclear reactions, the energies involved are so great that the change in mass is measurable, corresponding to about 1 part in 1000. In chemical reactions, the energy involved is so small that the change in mass, about 1 part in 1,000,000,000, is not detectable. This is why chemistry's law of mass conservation is best stated as follows: There is no *detectable* change in the total mass of materials as they chemically react to form new materials. In truth, there are changes in the mass of atoms during a chemical reaction. These changes, however, are too small to be of any concern to the working chemist.

We can think of the mass-per-nucleon graph as an energy valley that starts at hydrogen (the highest point) and slopes steeply to iron (the lowest point), then slopes gradually up to uranium. Iron is at the bottom of the energy valley and has the most stable nucleus. It also has the most tightly bound nucleus; more energy per nucleon is required to separate nucleons from iron's nucleus than from any other element's nucleus. In this sense, iron, Fe, is a very special element.

All nuclear power today is produced by nuclear fission. A more promising long-range source of energy is to be found on the left side of the energy valley, in a process known as nuclear fusion.

fyi FOR YOUR INFORMATION

Work is required to pull a nucleon from an atomic nucleus. This work increases the energy and hence the mass of the nucleon outside the nucleus.

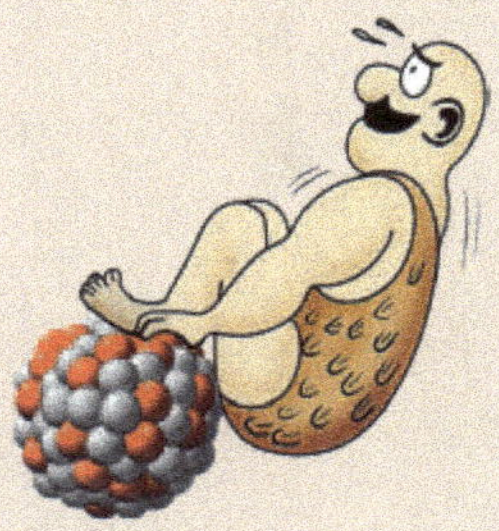

CONCEPT **CHECK**

A nucleon has the least mass when in an iron nucleus. Does this mean that it is relatively easy to pull a nucleon out of an iron nucleus?

CHECK YOUR ANSWER

No! Just the opposite. The nucleon has the least mass when in an iron nucleus because that is where it is most tightly bound. To pull the nucleon from the iron nucleus would be much harder than pulling it from any other nucleus.

CONCEPT **CHECK**

Correct the following incorrect statement: when a heavy element such as uranium undergoes fission, there are fewer nucleons after the reaction than before.

CHECK YOUR ANSWER

When a heavy element such as uranium undergoes fission, there aren't fewer nucleons after the reaction. Instead, the combined mass of the nucleons is less than was originally the case.

5.9 Nuclear Fusion

In the graphs of Figures 5.27 and 5.28, we see that the steepest part of the energy valley goes from hydrogen to iron. Energy is gained by the nucleons as light nuclei combine. This combining of nuclei is called *nuclear fusion*, and it is the opposite of nuclear fission. We can see from **Figure 5.29** that as we move along the list of elements from hydrogen to iron, the average mass per nucleon decreases. Thus, when two small nuclei fuse—for example, a pair of hydrogen isotopes—the mass of the resulting helium nucleus is less than the mass of the two small nuclei before fusion. The mass difference is released in the form of energy (**Figure 5.30**).

READING CHECK

How does the mass of a pair of hydrogen isotopes about to fuse compare with the mass of the resulting helium nucleus?

For a fusion reaction to occur, the nuclei must collide at a very high speed in order to overcome their mutual electric repulsion. The required speeds correspond to the extremely high temperatures found in the Sun and other stars. Fusion brought about by high temperatures is called **thermonuclear fusion**. In the high temperatures of the Sun, approximately 657 million tons of hydrogen is converted into 653 million tons of helium each second. The missing 4 million tons of mass is converted to energy—a tiny bit of which reaches our planet as sunshine. So thermonuclear fusion is the energy source of our Sun, which is, in turn, the ultimate energy source of life on Earth (**Figure 5.31**).

Such reactions are, quite literally, nuclear burning. Thermonuclear fusion is analogous to ordinary chemical combustion. In both chemical and nuclear burning, a high temperature starts the reaction; the release of

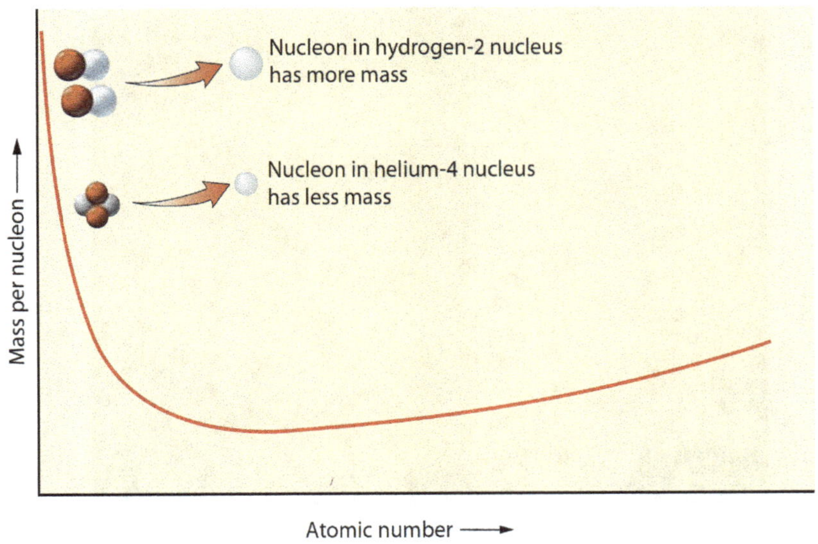

< Figure 5.29

The mass of each nucleon in a hydrogen-2 nucleus is greater than the mass of each nucleon in a helium-4 nucleus, which results from the fusion of two hydrogen-2 nuclei. This lost mass is mass that has been converted to energy, which is why nuclear fusion is a process that releases energy.

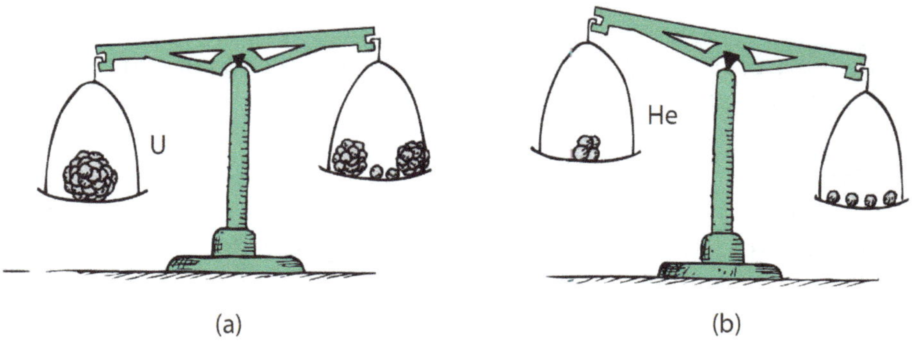

Figure 5.30

The mass of a nucleus is not equal to the sum of the mass of its parts. (a) The fission fragments of a uranium nucleus are less massive than the uranium nucleus. (b) Two protons and two neutrons are more massive in their free states than when they are combined to form a helium nucleus. Can you relate this to the graphs of Figure 5.28 and 5.29?

energy by the reaction maintains a high enough temperature to spread the fire. The net result of the chemical reaction is the combination of atoms into more tightly bound molecules. In nuclear fusion reactions, the net result is more tightly bound nuclei. In both cases, mass decreases as the corresponding amount of energy is released.

Prior to the development of the atomic bomb, the temperatures required to initiate nuclear fusion on Earth were unattainable. When researchers found that the temperature inside an exploding atomic bomb is four to five times the temperature at the center of the Sun, the thermonuclear bomb was but a step away. This first thermonuclear bomb, a hydrogen bomb, was detonated in 1952. Whereas the critical mass of fissionable material limits the size of a fission bomb (atomic bomb), no such limit is imposed on a fusion bomb (thermonuclear, or hydrogen, bomb). Just as there is no limit to the size of an oil-storage depot, there

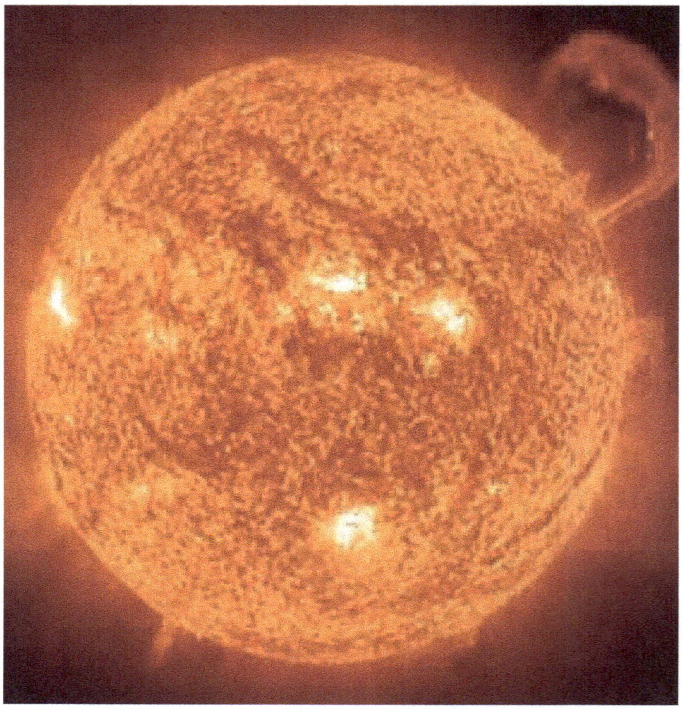

Figure 5.31

Thermonuclear fusion takes place in stars, such as the Sun. Some day, humans may produce vast quantities of energy through thermonuclear fusion, as the stars have always done.

is no theoretical limit to the size of a fusion bomb. Like the oil in the storage depot, any amount of fusion fuel can be stored with safety until ignited. Although a mere match can ignite an oil depot, nothing less energetic than an atomic bomb can ignite a thermonuclear bomb. We can see that there is no such thing as a "baby" hydrogen bomb. A typical thermonuclear bomb stockpiled by the United States today, for example, is about 1000 times more destructive than the atomic bomb detonated over Hiroshima at the end of World War II.

The hydrogen bomb is another example of a discovery used for destructive rather than constructive purposes. The potential constructive possibility is the controlled release of vast amounts of clean energy.

CHEMICAL CONNECTIONS

How are the atoms of your body connected to sunlight?

CONCEPT **CHECK**

1. Fission and fusion are opposite processes, yet each releases energy. Isn't this contradictory?
2. To get a release of nuclear energy from the element iron, should iron undergo fission or fusion?

CHECK YOUR ANSWER

1. No, no, no! This is contradictory only if the same element is said to release energy by both fission and fusion. Only the fusion of light elements and the fission of heavy elements result in a decrease in nucleon mass and a release of energy.

2. Neither, because iron is at the very bottom of the "energy valley." Fusing a pair of iron nuclei produces an element to the right of iron on the curve, in which the mass per nucleon is higher. If you split an iron nucleus, the products will lie to the left of iron on the curve and again have a higher mass per nucleon. So, no energy is released. For energy release, "decrease mass" is the name of the game—any game, chemical or nuclear.

Controlling Fusion

Carrying out fusion reactions under controlled conditions requires temperatures of millions of degrees. There are a variety of techniques for attaining high temperatures. No matter how the temperature is produced, a problem is that all materials melt and vaporize at the temperatures required for fusion. One solution to this problem is to confine the reaction in a nonmaterial container.

A nonmaterial container is a magnetic field, which can exist at any temperature and can exert powerful forces on charged particles in motion. "Magnetic walls" of sufficient strength provide a kind of magnetic straitjacket for hot ionized gases called plasmas. Magnetic compression further heats the plasma to fusion temperature.

Although there are no nuclear fusion power plants currently in operation, an international project now exists whose goal is to prove the feasibility of nuclear fusion power in the near future. This fusion power project is the International Thermonuclear Experimental Reactor (ITER) (**Figure 5.32**). The reactor will house electrically charged hydrogen gas (plasma) heated to about 150 million degrees Celsius, which is about 10 times hotter than the center of the Sun. In addition to producing about 500 MW of power, the reactor could be the energy source for the creation of hydrogen, H_2, which could be used to power fuel cells.

If people are one day to dart about the universe in the same way we jet about the Earth today, their supply of fuel is assured. The fuel for fusion—hydrogen—is found in every part of the universe, not only in the stars but also in the space between them. About 90 percent of the atoms in the universe are estimated to be hydrogen. For people of the future, the supply of raw materials is also assured, because all the elements known to exist result from the fusing of more and more hydrogen nuclei. Future humans might synthesize their own elements and produce energy in the process, just as the stars have always done.

fyi FOR YOUR INFORMATION

Elements are created in stars as smaller nuclei fuse to form larger nuclei. This process is energy releasing only up to iron. The manufacture of elements heavier than iron cannot be sustained within a shining star. So, where do heavier elements, such as gold, come from? The final stage of certain very large stars involves a mighty collapse, called a supernova. The energy of a supernova can outshine an entire galaxy, and it is this energy that makes the heavier-than-iron elements. The gold atoms in your jewelry were created using the abundant energy of a supernova that exploded very long ago.

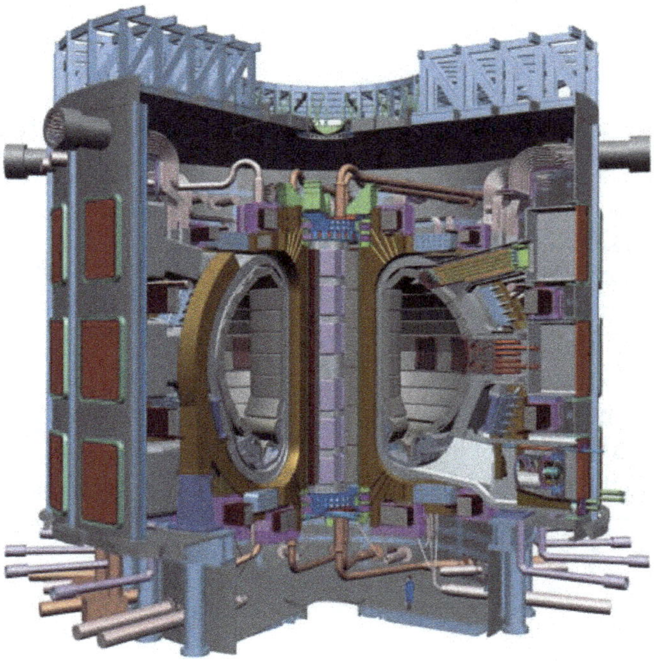

Figure 5.32 >

Cross-sectional views of the ITER (rhymes with "fitter") now underway in Cadarache, France.

Concept Review

Chapter 5

Summary of Terms

Alpha Particle A subatomic particle consisting of the combination of two protons and two neutrons ejected by a radioactive nucleus. The composition of an alpha particle is the same as that of the nucleus of a helium atom.

Beta Particle An electron emitted during the radioactive decay of a radioactive nucleus.

Carbon-14 Dating The process of estimating the age of once-living material by measuring the amount of radioactive carbon-14 present in the material.

Chain Reaction A self-sustaining reaction in which the products of one reaction event initiate further reaction events.

Critical mass The minimum mass of fissionable material needed for a sustainable chain reaction.

Gamma Rays High-frequency electromagnetic radiation emitted by radioactive nuclei.

Half-Life The time required for half the atoms in a sample of a radioactive isotope to decay.

Nuclear Fission The splitting of the atomic nucleus into two smaller halves.

Nuclear Fusion The combining of nuclei of light atoms to form heavier nuclei.

Radioactive A descriptive term for a material that contains nuclei that emit energetic particles and radiation.

Radioactivity The phenomenon exhibited by the spontaneous emission of energetic particles and radiation by an atomic nucleus.

Rem A unit for measuring the ability of radiation to harm living tissue.

Strong Nuclear Force The attractive force between all nucleons, effective only at very short distances.

Thermonuclear Fusion Nuclear fusion brought about by high temperatures.

Transmutation The changing of an atomic nucleus of one element into an atomic nucleus of another element through a decrease or increase in the number of protons.

Review Questions

5.1 Unstable Nuclei

1. Which type of radiation—alpha, beta, or gamma—results in the greatest change in mass number? The greatest change in atomic number?

2. Which of the three rays—alpha, beta, or gamma— has the greatest penetrating power?

5.2 Radioactivity Is Natural

3. What is the origin of most of the natural radiation we encounter?

4. Which produces more radioactivity in the atmosphere, coal-fired power plants or nuclear power plants?

5. Is radioactivity on the Earth something relatively new? Defend your answer.

5.3 An Imbalance of Forces

6. Why doesn't the repulsive electric force of protons in the atomic nucleus cause the protons to fly apart?

7. Which have more neutrons than protons, large nuclei or small nuclei?

8. What role do neutrons play in the atomic nucleus?

5.4 Transmutation

9. In what form is most of the energy released by atomic transmutation?

10. What change in atomic number occurs when a nucleus emits an alpha particle? A beta particle?

11. What is the long-range fate of all the uranium that exists in the world today?

5.5 Radioactive Half-Life

12. What is meant by the half-life of a radioactive sample?

13. What is the half-life of uranium-238?

5.6 Isotopic Dating

14. What happens to a nitrogen atom in the atmosphere that captures a neutron?

15. Why is there more carbon-14 in living bones than in once-living ancient bones of the same mass?

16. Why is lead found in all deposits of uranium ores?

5.7 Nuclear Fission

17. What happens to the uranium-235 nucleus when it is stretched out?

18. Is a chain reaction more likely to occur in two separate pieces of uranium-235 or in the same pieces stuck together?

19. How is a nuclear reactor similar to a conventional fossil-fuel power plant? How is it different?

20. Nuclear power reactors that are relatively safe but have yet to be developed are those based upon the fission of what isotope?

5.8 Mass and Energy

21. Who discovered that energy and mass are two different forms of the same thing?

22. In which atomic nucleus do nucleons have the least mass?

23. How does the mass per nucleon in uranium compare with the mass per nucleon in the fission fragments of uranium?

5.9 Nuclear Fusion

24. How does the mass of a pair of atoms that have fused compare to the sum of their masses before fusion?

25. What kind of containers are used to contain plasmas at temperatures of millions of degrees?

26. What kind of nuclear power is responsible for sunshine?

Quantitative Questions

27. Radiation from a point source follows an inverse-square law where the amount of radiation received is proportional to $1/d^2$, where d is distance. If a Geiger counter that is 1 meter away from a small source reads 100 counts per minute, what will be its reading 2 meter from the source? 3 meters from it?

28. Consider a radioactive sample with a half-life of one week. How much of the original sample will be left at the end of the second week? The third week? The fourth week?

29. A radioisotope is placed near a radiation detector, which registers 80 counts per second. Eight hours later, the detector registers 5 counts per second. What is the half-life of the radioactive isotope?

30. Uranium-238 absorbs a neutron and then emits a beta particle. Show that the resulting nucleus is neptunium-239.

Solutions (Odd-Numbered)

1. Alpha radiation decreases the atomic number of the emitting element by 2 and the atomic mass number by 4. Beta radiation increases the atomic number of an element by 1 and does not affect the atomic mass number. Gamma radiation does not affect the atomic number or the atomic mass number. So alpha radiation results in the greatest change in atomic number, and hence charge, and mass number as well.

3. Most of the radiation we encounter is natural background radiation that originated in earth and in space.

5. No. It has been around longer than the human race and has been a part of the environment as long as the sun, rain and soil have been.

7. A large nucleus requires more neutrons to help overcome the repulsions among the many protons.

9. Most is kinetic energy of the ejected particle and some is the kinetic energy of the recoiling nucleus.

11. All uranium will eventually decay to lead.

13. 4.5 billion years.

15. Carbon-14 lost to decay is replenished with carbon-14 from the atmosphere. When a living thing dies, replenishment stops and the percentage of carbon-14 decreases at a constant rate.

17. An elongated uranium-235 nucleus splits in half because the strong nuclear force between distant nucleons quickly diminishes.

19. A nuclear reactor and a fossil fuel power plant both boil water to produce steam for turbines. The main difference between them is the amount of fuel involved. Nuclear reactors are much more efficient—one kilogram of uranium-235 yields more than 30 freight car loads of coal.

21. Albert Einstein.

23. Mass per nucleon is greater in uranium than the mass per nucleon of fission fragments of uranium.

25. Magnetic containers.

27. Intensity decreases with distance by the inverse-square law. Twice the distance is 1/4 the intensity and 1/4 the reading. Three times as far is 1/9 the intensity and 1/9 the reading.

29. Count: It's decreased to 40 cps (one half life), 20 cps (2 half-lives), 10 cps (3 half-lives), 5 cps, which takes 4 half lives. If 4 half-lives equals 8 hours, then a single half-life equals $8/4 = 2$ hours.

Chapter 5: Contextual Chemistry Essay

Safer Fission Reactors

The designs for nuclear power plants have progressed over the years. The earliest designs, from the 1950s through the 1980s, are called the Generation I and II reactors. The safety systems of these reactors are "active," in that they rely on a series of active measures, such as water pumps, that come into play to keep the reactor core cool in the event of an accident. The problem with this is that the active measure itself may fail in the event of a crisis.

Notably, active safety measures failed when the Generation II Fukushima Daiichi nuclear plant of Japan was hit by the powerful tsunami of 2011. In this case, the water pumps could not operate because the electric generators designed to power these pumps where destroyed by seawater.

The Generation III reactors, built in the 1990s, are more economical to build, operate, and maintain. These reactors, however, still rely on active safety measures. While not yet operational, the latest Generation IV nuclear reactors have fundamentally different reactor designs. Importantly, they incorporate passive safety measures that will cause the reactor to cool down by itself in the event of an emergency. For example, as the fuel gets hotter, it expands, which leads to a decrease in the rate of fission.

The fuel source for a Generation IV reactor may be the depleted uranium stockpiled from earlier-generation reactors. Designs will also allow for the formation of hydrogen fuel from water. Furthermore, these reactors can be built as small modular units generating between 150 and 600 megawatts of power rather than the 1500 megawatts that is the usual output of today's reactors. Smaller reactors are easier to manage and can be used in tandem to build a generating capacity suited to the community being served.

The Generation IV International Forum (gen-4.org) aims to have Generation IV power plants operating by the 2030s. Some of these Generation IV power plants may employ the element thorium as the primary fuel.

^ Figure A

(a) Aerial view of the Fukushima Daiichi nuclear power plant destroyed by the 2011 tsunami. (b) International inspectors in 2013 reviewing Japan's plans to decommission the facility.

The Thorium Reactor

It is well known that uranium and plutonium have fissionable isotopes, U-235 and Pu-239. These isotopes are historic in that they comprised the two nuclear bombs that ended World War II, and were favored in nuclear power reactors that followed. But there is a third isotope that is fissionable, which is the less well-known U-233.

U-233 does not occur naturally on Earth, but is produced in specially designed reactors where thorium-232 captures a neutron to form Th-233, which beta decays into U-233. Thorium is about four times more abundant in Earth's crust than uranium, and most thorium is the needed Th-232 isotope. So, the world already has an abundant supply of this fuel.

The thorium is first blended with fluoride salts to bring the melting point to a relatively cool 360°C. To initiate the fission chain reaction, neutron-releasing isotopes of uranium or plutonium are mixed into a molten blend of thorium and fluoride salts. This leads to the formation of U-233, which upon fission releases more neutrons making the chain reaction self-sustaining while also creating more U-233 from Th-232, as shown in Figure 1. With a higher temperature, the liquid mixture flows through a heat exchanger where it causes a turbine to generate electricity. This system is called a Liquid Fluoride Thorium Reactor, LFTR.

Nuclear power plants using U-235 or Pu-239 as a fuel are prone to explosive accidents because they operate at extremely high pressures. The LFTR, by contrast, operates near atmospheric pressure, which translates to greater safety. And whereas the fuel in conventional nuclear power plants invariable leads to the creation of plutonium, production of this toxic element in the LFTR is low enough to be retained and controlled within the molten mixture. The LFTR also has inherent passive safety mechanisms that prevent a meltdown scenario.

Conventional nuclear power plants have to be shut down once every 18 months for replacement of solid fuel pellets. The LFTR, however, can function continuously for years using liquid fuel in which by-products can be removed by distillation or electrolysis. Furthermore, most of the by-products isolated from the LFTR have short half-lives that decay within a matter of hours or days. Its longest-lived isotope Cs-137 requires only decades of safe storage (rather than the hundreds of millennia with plutonium reactors).

Bomb technology has historically avoided the fissionable isotope U-233, mainly because a dangerous by-product is the thallium-208 isotope, which is a powerful gamma ray emitter. Nuclear bombs built of U-233 were impracticable because the gamma rays from residual thallium-208 would fry the detonation circuitry making a bomb "temperamental." The resultant Tl-208 isotope produced within the LFTR, however, can simply remain within the molten mixture to provide additional energy. And indeed, the presence of the pesky Tl-208 isotope provides a great deterrent against anyone trying to extract U-233 for clandestine nuclear bomb making.

With a growing awareness of the dangers posed by greenhouse gases, more attention is being given to the

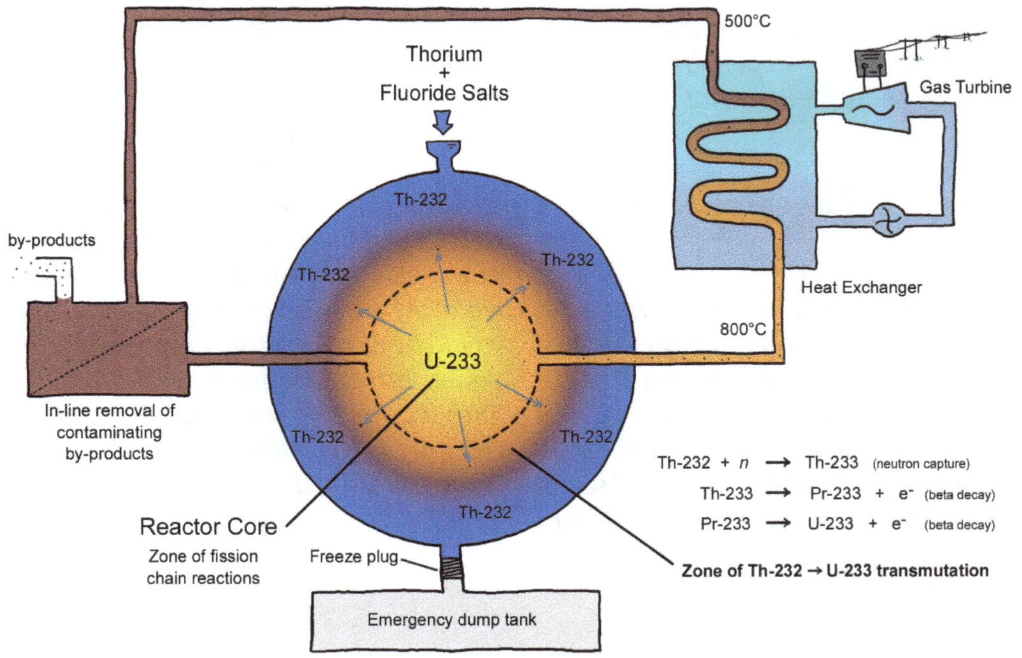

∧ Figure B

Diagram of a thorium reactor. Fissionable U-233 is produced as the thorium is exposed to neutrons. Because of its liquid phase, the fuel mixture itself can be passed through the heat exchanger and then treated for the removal of by-products.

potential benefits of next-generation nuclear technology. Japan, China, the UK and India are all sponsoring research into the LFTR as well as private companies in the US, Czech Republic, Canada, and Australia. Check online for the latest developments.

CONCEPT CHECK

Thorium-232 is not fissionable. How then can this common isotope be used as a fuel for nuclear fission?

CHECK YOUR ANSWER

The Th-232 isotope transforms into Th-233 upon capturing a neutron. The Th-233 isotope then undergoes two beta decays to form the fissionable U-233 isotope.

Think and Discuss

1. The United States has an abundant supply of coal and natural gas. To what extent does this affect the reluctance of Americans to accept nuclear power as a means of generating electricity?

2. Rank the following issues in order of importance for the implementation of nuclear power: (a) Sustainable supply of fuel, (b) Management of nuclear wastes, (c) Remaining economically competitive, (d) Safe operation of the nuclear power plant, (e) Reducing risk of weapons proliferation.

3. Describe how each of these might be a threat or salvation to human existence: nuclear-fired power plants, coal-fired power plants, no power plants.

Author Responses to Think and Discuss

1. The reluctance of Americans toward nuclear energy has many causes, most notably safety issues and the association of nuclear technology with nuclear bombs. Americans can supply most all of their electrical needs through coal and gas fired powered plants, which is to say that nuclear power plants are not so essential. In all, only about 20% of electricity in the United States is generated by nuclear power. France, by contrast, relies on imported fossil fuels. In all, over 70% of electricity in France is generated by nuclear power

2. It is vitally important that each one of these issues be addressed.

3. The fissionable isotopes from a nuclear power plant could potentially be concentrated sufficiently to make a nuclear weapon. If such a nuclear weapon were used, it could potentially set off a nuclear war in which already made nuclear bombs are set off. In addition to killing millions of people, a "prolonged" nuclear war could set off a "nuclear winter" in which residuals from these bombs reflect sunlight creating a feedback loop in which the planet plunges into an ice age. Nuclear power plants, by contrast, generate no greenhouse gases. If they were used to substitute for coal-fired power plants, then this would dramatically lower our carbon dioxide emissions. A great disadvantage of coal-fired power plants is the they produce significant amounts of carbon dioxide, a potent greenhouse gas. Excessive reliance on coal-fired power plants could lead to a devastating global climate change. Coal, however, is abundant and therefore relatively cheap. The energy we get from coal-fired power plants helps our economy. With a strong economy we're in a better position to offer assistance in times of crises as they happen. If all power plants were to suddenly shut down, so would our electricity dependent society leading to much chaos. There would be a strong push, however, toward decentralized power, such as solar panels or wind turbines at every home. Such technology could easily be exported to developing countries who already lack centralized power. This, in turn, would minimize the global output of greenhouse gases.

Chapter 6

How Atoms Bond

THE MAIN IDEA

Atoms bond by exchanging or sharing electrons

∧ These large cubic salt crystals grew many years ago from an evaporating sea in what is now Colorado.

Salt crystals, as shown in the opening photograph, are cubic because that's the way sodium and chlorine join together to form sodium chloride. Smash these crystals with a hammer and you'll merely get smaller cubes! Why are metals opaque to light, and why do they conduct both electricity and heat so well? Again, the answer has to do with how their atoms are bonded. You know carbon dioxide to be a gas at room temperature and water to be a liquid. Would it surprise you that, molecule for molecule, carbon dioxide is over twice as heavy as water? As we explore in this chapter, carbon dioxide molecules are non-sticky, which allows them to float away from each other into a gaseous phase. Conversely, the relatively light water molecules are sticky, which keeps them together in a liquid phase. The reason has to do with, you guessed it, how the atoms within these molecules are bonded. So, if we are to understand why materials behave as they do, it's important that we have some understanding of how it is that their atoms bond.

6.1 Electron-Dot Structures

An atomic model is needed to help us understand how atoms bond. We begin this chapter with a brief overview of the shell model presented in Section 4.8. You may recall how electrons are arranged around an atomic nucleus. Rather than moving in neat orbits like planets around the Sun, electrons are wavelike entities that hover in various volumes of space called *shells*.

As was shown in Figure 4.32, there are seven shells available to the electrons in an atom, and the electrons fill these shells in order from innermost to outermost. Furthermore, the maximum number of electrons allowed in the first shell is 2, and for the second and third shells it is 8. The fourth and fifth shells can each hold 18 electrons, and the sixth and seventh shells can each hold 32 electrons.* These numbers match the number of elements in each period (horizontal row) of the periodic table. **Figure 6.1** shows how this model applies to the first three elements of group 18.

* As a point of reference for advanced physics students reading this text, these are shells of orbitals grouped by similar energy levels rather than by principal quantum number. They are the "argonian" shells developed by Linus Pauling in the 1930s to explain chemical bonding and the organization of the periodic table. This is an old atomic model, but it works well for a simple description of chemical bonding.

Electrons in the outermost occupied shell of any atom play a significant role in that atom's chemical properties, including its ability to form chemical bonds. To indicate their importance, these outermost electrons are called **valence electrons** (from the Latin valentia, "strength"), and the shell they occupy is called the **valence shell**. Valence electrons can be conveniently represented as a series of dots surrounding an atomic symbol. This notation is called an **electron-dot structure** or, sometimes, a *Lewis dot symbol* in honor of the American chemist Gilbert N. Lewis, who first proposed the concepts of shells and valence electrons (**Figure 6.2**).

READING CHECK

Electron-dot structures are needed to help us understand what kinds of chemical bonds?

The electron-dot structures shown in Figure 6.3 help us to understand ionic and covalent bonds. Electron-dot structures, however, are not so useful in describing metallic bonds. The reason is because within a metallic bond, the bonding electrons readily flow from one atom to the next, as is discussed further in Section 6.4. This is why the metallic groups 3–12 are not included in **Figure 6.3**.

When you look at the electron-dot structure of an atom, you immediately know two important things about that element. You know how many valence electrons it has and how many of these electrons are *paired*. Chlorine, for example, has three pairs of electrons and one unpaired electron, and carbon has four unpaired electrons:

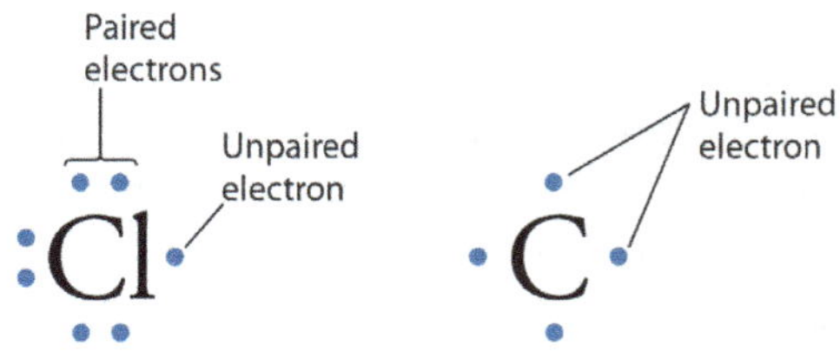

Paired valence electrons are relatively stable with little need to form bonds with other atoms. For this reason, electron pairs in an atom's electron-dot structure are called **nonbonding pairs**. (Do not take this term literally, however, because under the right conditions even "nonbonding" pairs can form a chemical bond. This is explored further in Chapter 10.)

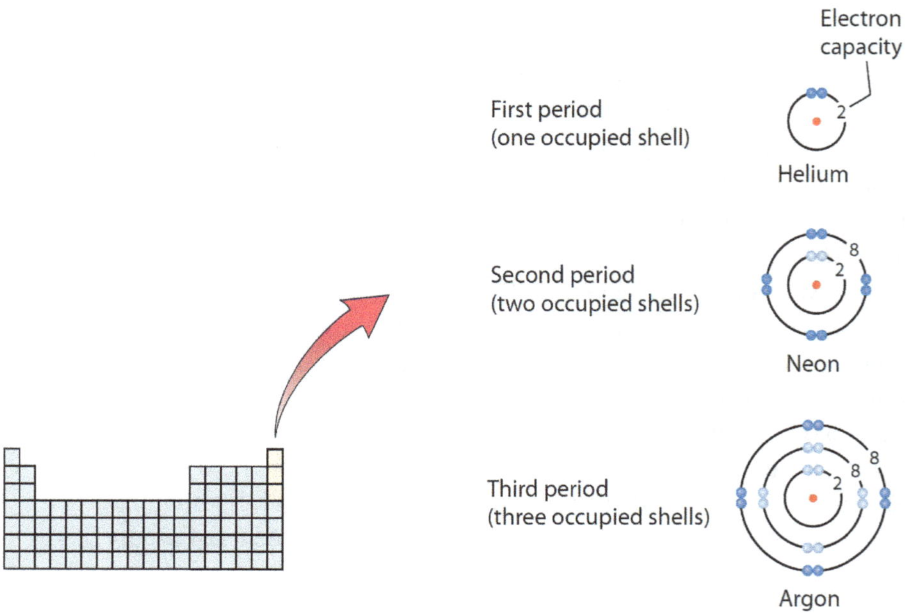

Figure 6.1 >

Occupied shells in the group 18 elements helium through argon. Each of these elements has a filled outermost shell. Note that the number of electrons in each shell (2, 8, 8, and so on) corresponds to the number of elements in the periods of the periodic table. For more detail, please review Sections 4.8 and 4.9.

Valence electrons that are *unpaired*, by contrast, have a strong tendency to participate in chemical bonding. By doing so, they become paired with an electron from another atom. The ionic and covalent bonds discussed in this chapter all result from either a transfer or a sharing of unpaired valence electrons.

So why do electrons like to pair up? As was discussed in Chapter 4, electrons have a property called *spin*. Spin can have two states, and these states are analogous to a ball's spinning either clockwise or counterclockwise, as was shown in Figure 4.28. Because an electron has an electric charge, the spinning generates a tiny magnetic field. Two electrons spinning in opposite directions have oppositely aligned magnetic fields, which allows them to come together as a pair.

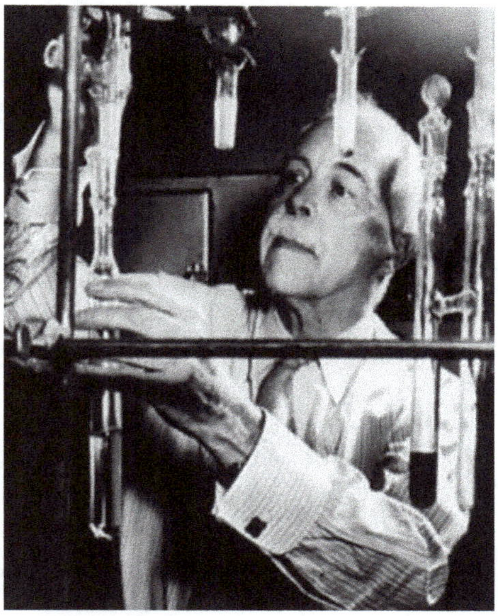

Figure 6.2

Gilbert Newton Lewis (1875–1946) revolutionized chemistry with his theory of chemical bonding, which he published in 1916. He worked most of his life in the chemistry department of the University of California, Berkeley, where he was not only a productive researcher but also an exceptional teacher. Among his teaching innovations was the idea of providing students with problem sets (homework) as a follow-up to lectures and readings.

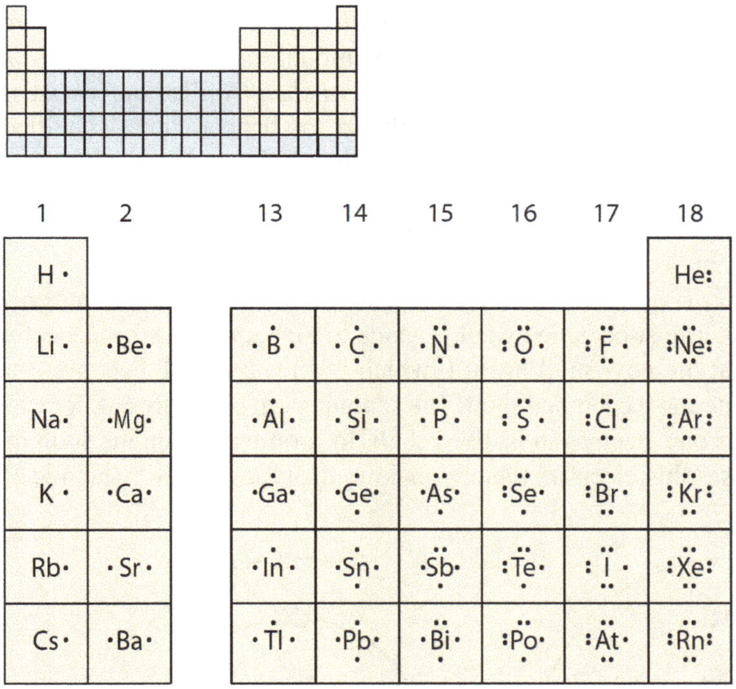

Figure 6.3

The valence electrons of an atom are shown in its electron-dot structure. Note that the first three periods here parallel Figure 4.33. Also note that for larger atoms, not all the electrons in the valence shell are valence electrons. Krypton, Kr, for example, has 18 electrons in its valence shell, but only 8 of these are classified as valence electrons. You can learn the reason for this detail, which involves sub-orbitals, in a follow-up course on advanced chemistry.

CONCEPT **CHECK**

Where are valence electrons located, and why are they important?

CHECK YOUR ANSWER

Valence electrons are located in the outermost occupied shell of an atom. They are important because they play a leading role in determining the chemical properties of the atom.

6.2 Ion Formation

When the number of protons and electrons in an atom are equal, the charges balance and the atom is electrically neutral. If electrons are lost or gained, then the balance is lost and the atom takes on a net electric charge. Any atom having a net electric charge is an **ion**. When electrons are lost, protons outnumber electrons and the ion has a positive net charge, as shown in **Figure 6.4**. A positively charged ion is sometimes referred to as a *cation*. When electrons are gained, electrons outnumber protons and the ion has a negative net charge, as shown in **Figure 6.5**. A negatively charged ion is sometimes referred to as an *anion*.

Chemists use a superscript to the right of the atomic symbol to indicate the strength and sign of an ion's charge. Thus, as shown in Figures 6.4 and 6.5, the positive ion formed from the sodium atom is written Na^{1+} and the negative ion formed from the fluorine atom is written F^{1-}. Usually the numeral 1 is omitted when indicating either a 1+ or 1- charge. Hence, these two ions are most frequently written Na^{+} and F^{-}. Two more examples: a calcium atom that loses two electrons is written Ca^{2+}, and an oxygen atom that gains two electrons is written O^{2-}.

We can use the shell model to deduce the type of ion an atom tends to form. According to this model, *atoms tend to lose or gain electrons to create a filled outermost shell.* A filled outermost shell is stable because there's not much that can happen there. The filled shell can't gain electrons because it's already filled. Conversely, the filled shell can't lose electrons because these electrons are held tightly by the positive charge of the nucleus. Let's take a moment to consider this point, looking to Figures 6.4 and 6.5 as visual guides.

If an atom has only one electron or only a few electrons in its outermost occupied shell, it tends to give up (lose) these electrons so that the next shell inward, which is already filled, becomes the new outermost occupied shell. The sodium atom of Figure 6.4, for example, has one electron in its third shell. In forming an ion, the sodium atom loses this electron, thereby making the second shell, which is already

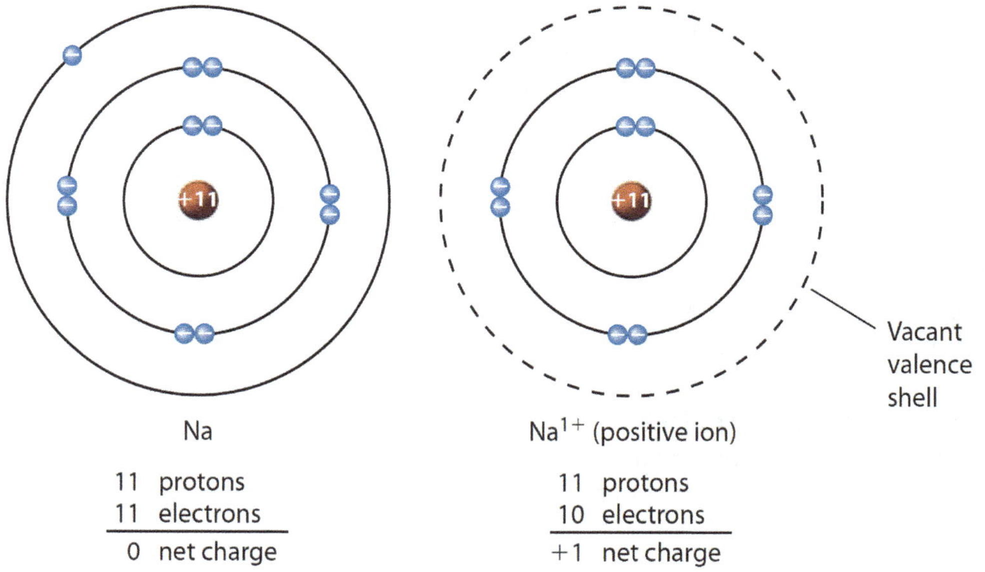

Figure 6.4 >

An electrically neutral sodium atom contains 11 negatively charged electrons surrounding the 11 positively charged protons of the nucleus. When this atom loses an electron, the result is a positive ion.

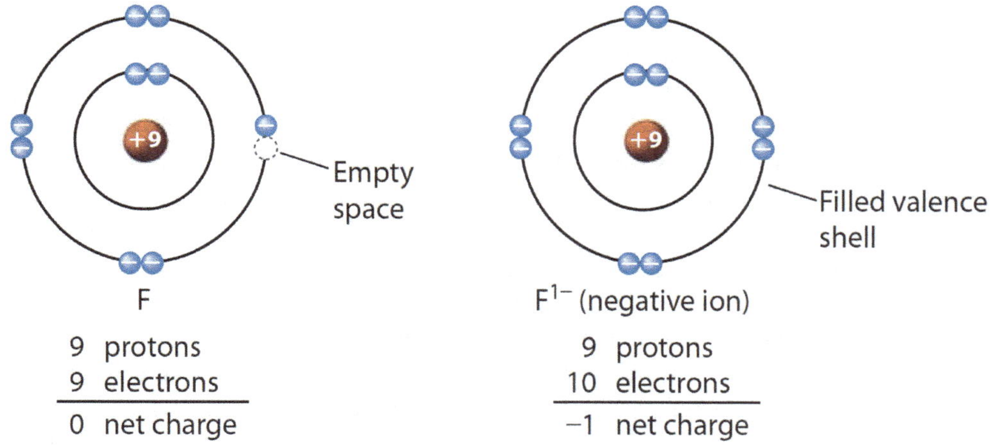

< Figure 6.5

An electrically neutral fluorine atom contains nine protons and nine electrons. When this atom gains an electron, the result is a negative ion.

filled to capacity, the new outermost occupied shell. Because the sodium atom has only one valence electron to lose, it tends to form only the 1+ ion. It is a *positive* ion, because the number of protons (11+) now exceeds the number of electrons (10–).

If the outermost shell of an atom is *almost* filled, that atom attracts electrons from another atom and so forms a negative ion. The fluorine atom of Figure 6.5, for example, has available space for only one additional valence electron. After this additional electron is gained, the fluorine atom achieves a filled outermost shell. Fluorine therefore tends to form the 1– ion. Remember, electrons are negatively charged. So, gaining an electron results in a *negative* ion.

The periodic table tells us the type of ion each atom tends to form. As **Figure 6.6** shows, each atom of any group 1 element, for example, has only one valence electron and so tends to form the 1+ ion. Each atom of any group 17 element has room for one additional electron in its valence shell and therefore tends to form the 1– ion. Atoms of the noble gas elements tend not to form ions of any type because their valence shells are already filled to capacity.

As is indicated by color in Figure 6.6, the attraction between an atom's nucleus and its valence electrons is weakest for elements on the left in the periodic table and strongest for elements on the right. From

READING CHECK

Why do you NOT need to memorize the type of ion each atom tends to form?

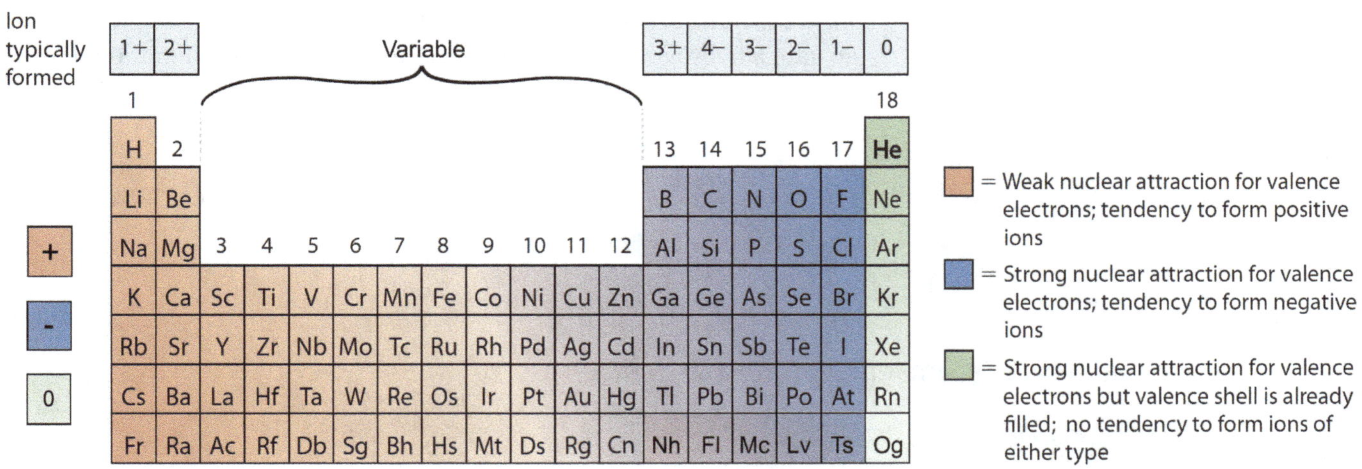

Λ Figure 6.6

The periodic table is your guide to the types of ions that atoms tend to form within an ionic compound.

> **fyi FOR YOUR INFORMATION**
>
> What do the ions of the following elements have in common: calcium, Ca; chromium, Cr; cobalt, Co; copper, Cu; iodine, I; iron, Fe; magnesium, Mg; manganese, Mn; molybdenum, Mo; nickel, Ni; phosphorus, P; potassium, K; selenium, Se; sodium, Na; sulfur, S; and zinc, Zn? They are all dietary minerals that are essential for good health but that can be harmful, even lethal, when consumed in excessive amounts.

sodium's position in the table, we can see that a sodium atom's single valence electron is not held very strongly, which explains why it is so easily lost. The attraction the sodium nucleus has for its second-shell electrons, however, is much stronger, which is why the sodium atom rarely loses more than one electron.

On the other side of the periodic table, the nucleus of a fluorine atom strongly holds onto its valence electrons, which explains why the fluorine atom tends not to lose any electrons to form a positive ion. Instead, fluorine's nuclear pull on the valence electrons is strong enough to accommodate even an additional electron taken from some other atom.

The nucleus of a noble gas atom pulls so strongly on its valence electrons that they are very difficult to remove, so noble gas atoms generally do not lose electrons. However, because their outermost shells are already filled to capacity, they generally don't gain electrons either. Thus, these atoms tend not to form ions of any sort.

Using our shell model to explain the formation of ions works well for groups 1 and 2 and groups 13 through 18. However, this model is too simplified to work well for the transition metals of groups 3 through 12 and the inner transition metals. In general, these metal atoms tend to form positive ions, but the number of electrons lost varies. For example, depending on conditions, an iron atom may lose two electrons to form the Fe^{2+} ion or lose three electrons to form the Fe^{3+} ion.

CONCEPT CHECK

What type of ion does the magnesium atom, Mg, tend to form?

CHECK YOUR ANSWER

The magnesium atom (atomic number 12) is found in group 2 and has two valence electrons to lose (see Figure 6.6). Therefore, it tends to form the 2+ ion.

Molecules Can Form Ions

We have seen that atoms form ions by losing or gaining electrons. Interestingly, molecules can also become ions. In most cases, this occurs whenever a molecule loses or gains a proton—equivalent to the hydrogen ion, H^+. (Recall that a hydrogen atom is a proton together with an electron. The hydrogen ion, H^+, therefore, is simply a proton.) For example, a water molecule, H_2O, can gain a hydrogen ion, H^+ (a proton), to form the hydronium ion, H_3O^+:

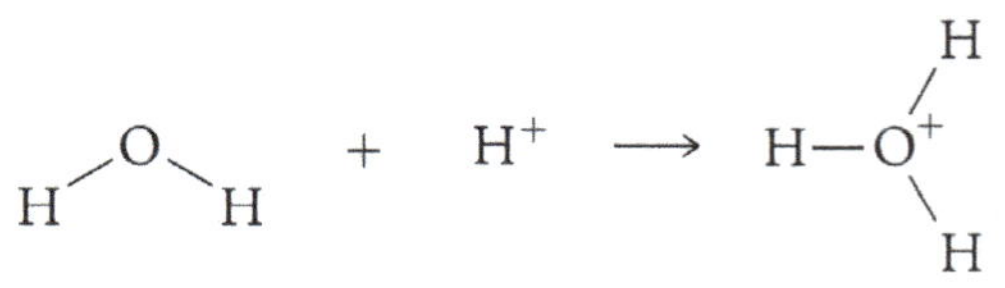

Water — Hydrogen ion (proton) — Hydronium ion

CONCEPT CHECK

Why does the magnesium atom tend to form a 2+ ion?

CHECK YOUR ANSWER

Magnesium is on the left in the periodic table, so atoms of this element do not hold onto their two valence electrons very strongly. Because these electrons are not held very tightly, they are easily lost, which is why the magnesium atom tends to form the 2+ ion.

Similarly, the carbonic acid molecule, H_2CO_3, can lose two protons to form the carbonate ion, CO_3^{2-}:

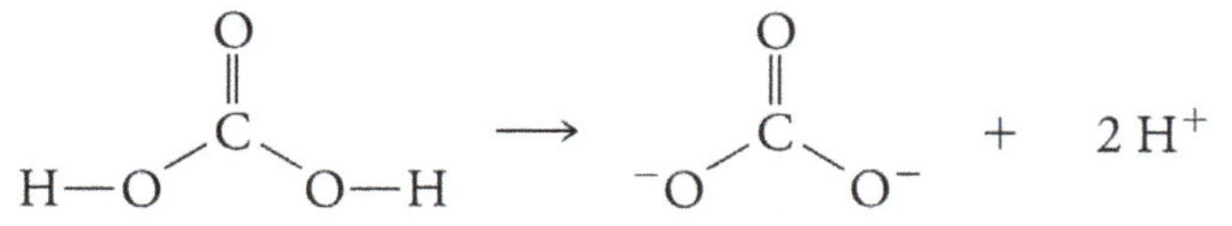

Carbonic acid — Carbonate ion — Hydrogen ions (protons)

How these reactions occur will be explored in later chapters. For now, you should understand that the hydronium and carbonate ions are examples of **polyatomic ions**, which are molecules that carry a net electric charge. Table 6.1 lists some commonly encountered polyatomic ions. Interestingly, these common polyatomic ions consist of all non-metal atoms. The reason for this is because of the tendency of non-metal atoms to form molecules, which we discuss later in this chapter.

TABLE 6.1 Common Polyatomic Ions

NAME	FORMULA
Hydronium ion	H_3O^+
Ammonium ion	NH_4^+
Bicarbonate ion	HCO_3^-
Acetate ion	$CH_3CO_2^-$
Nitrate ion	NO_3^-
Cyanide ion	CN^-
Hydroxide ion	OH^-
Carbonate ion	CO_3^{2-}
Sulfate ion	SO_4^{2-}
Phosphate ion	PO_4^{3-}

6.3 Ionic Bonds

When an atom that tends to lose electrons is placed in contact with an atom that tends to gain them, the result is an electron transfer and the formation of two oppositely charged ions. This occurs when sodium and chlorine are combined. As shown in Figure 6.7, the sodium atom loses one of its electrons to the chlorine atom, resulting in the formation of a positive sodium ion and a negative chloride ion. The two oppositely charged ions are attracted to each other by the electric force, which holds them close together. This electric force of attraction between two oppositely charged ions is called an **ionic bond**.

A sodium ion and a chloride ion together make the chemical compound sodium chloride, commonly known as table salt. This and all other chemical compounds containing ions are referred to as **ionic compounds**. All ionic compounds are completely different from the elements from which they are made. As discussed in Section 3.4, sodium chloride is not sodium, nor is it chlorine. Rather, it is a collection of sodium and chloride ions that form a unique material having its own physical and chemical properties.

As Figure 6.8 shows, ionic compounds typically consist of elements that are found on opposite sides of the periodic table. Also, because of how the metals and nonmetals are organized in the periodic table, positive ions are generally derived from metallic elements and negative ions are generally derived from nonmetallic elements.

READING CHECK

What type of force gives rise to an ionic bond?

CONCEPT CHECK

Is the transfer of an electron from a sodium atom to a chlorine atom a physical change or a chemical change?

CHECK YOUR ANSWER

Recall from Chapter 3 that only a chemical change involves the formation of new material. The transfer of an electron results in the formation of a new substance, which makes this a chemical change.

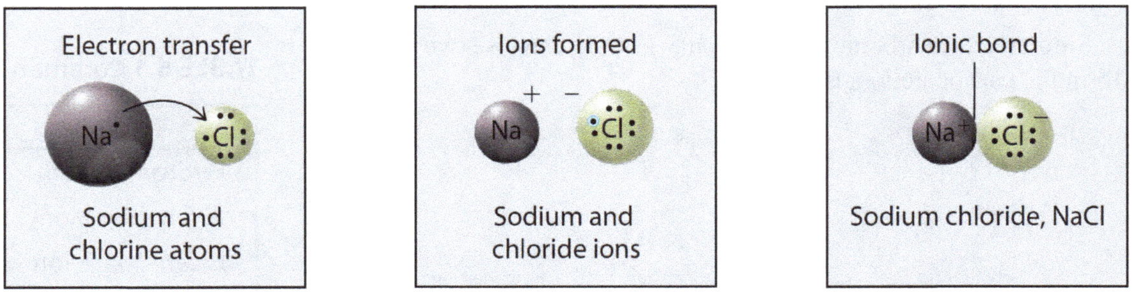

(a) (b) (c)

∧ Figure 6.7

An electrically neutral sodium atom loses its valence electron to an electrically neutral chlorine atom. (b) This electron transfer results in two oppositely charged ions. (c) The ions are then held together by an ionic bond. The spheres drawn around these and subsequent illustrations of electron-dot structures indicate the relative sizes of the atoms and ions.

For all ionic compounds, positive and negative charges must balance. In sodium chloride, for example, there is one sodium 1+ ion for every chloride 1– ion. Charges must also balance in compounds containing ions that carry various charges. The calcium ion, for example, carries a charge of 2+, but the fluoride ion carries a charge of only 1–. Because two fluoride ions are needed to balance each calcium ion, the formula for calcium fluoride is CaF_2, as **Figure 6.9** illustrates.

An aluminum ion carries a 3+ charge, and an oxide ion carries a 2– charge. Together, these ions make the ionic compound aluminum oxide, Al_2O_3, the main component of such gemstones as rubies and sapphires. **Figure 6.10** illustrates the formation of aluminum oxide. The three oxide ions in Al_2O_3 carry a total charge of 6–, which balances the total 6+ charge of the two aluminum ions. Rubies and sapphires differ in color because of the impurities they contain. Rubies are red because of minor amounts of chromium ions, and sapphires are blue because of minor amounts of iron and titanium ions.

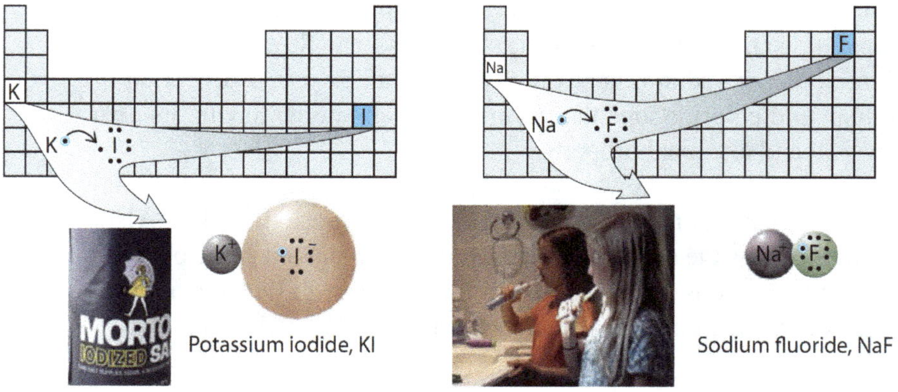

(a) (b)

∧ Figure 6.8

The ionic compound potassium iodide, KI, is added in minute quantities to commercial salt because the iodide ion, I^-, that it contains is an essential dietary mineral. The ionic compound sodium fluoride, NaF, is often added to toothpastes because it is a good source of the tooth-strengthening fluoride ion, F^-.

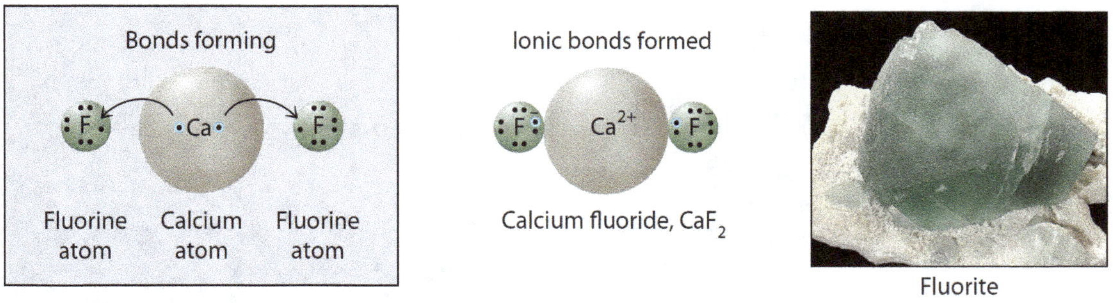

Figure 6.9

A calcium atom loses two electrons to a pair of fluorine atoms. In the process, the calcium atom becomes a calcium ion, Ca^{2+}, and the fluorine atoms become fluoride ions, F^-. The oppositely charged ions join to form the ionic compound calcium fluoride, CaF_2, which occurs naturally as the mineral fluorite.

An ionic compound typically contains a multitude of ions grouped together in a highly ordered 3-dimensional array. In sodium chloride, for example, each sodium ion is surrounded by six chloride ions and each chloride ion is surrounded by six sodium ions (**Figure 6.11**). Overall, there is one sodium ion for each chloride ion, but there are no identifiable sodium–chloride pairs. Such an orderly array of ions is known as an *ionic crystal*. As mentioned at the onset of this chapter, on the atomic level, the crystalline structure of sodium chloride is cubic, which is why macroscopic crystals of table salt are also cubic. Smash a large cubic sodium chloride crystal with a hammer, and what do you get? Smaller cubic sodium chloride crystals! Similarly, the crystalline structures of other ionic compounds, such as calcium fluoride and aluminum oxide, are a consequence of how the ions pack together.

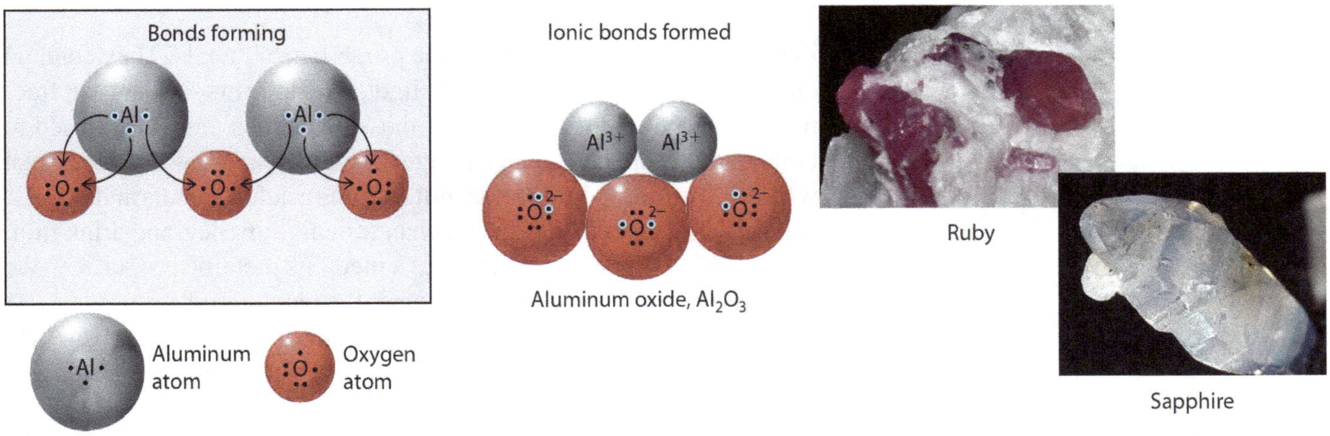

Figure 6.10

Two aluminum atoms lose a total of six electrons to three oxygen atoms. In the process, the aluminum atoms become aluminum ions, Al^{3+}, and the oxygen atoms become oxide ions, O^{2-}. The oppositely charged ions join to form the ionic compound aluminum oxide, Al_2O_3. Certain gemstones are crystalline aluminum oxide with trace amounts of impurities, such as chromium, which makes ruby, and titanium, which makes sapphire.

Figure 6.11

Sodium chloride, as well as other ionic compounds, forms ionic crystals in which every internal ion is surrounded by ions of the opposite charge. (For simplicity, only a small portion of the ion array is shown here. A typical NaCl crystal involves billions and billions of ions.) (b) A view of crystals of table salt through a microscope shows their cubic structure. The cubic shape is a consequence of the cubic arrangement of sodium and chloride ions.

CONCEPT **CHECK**

What is the chemical formula for the ionic compound magnesium oxide?

CHECK YOUR ANSWER

Because magnesium is a group 2 element, you know a magnesium atom must lose two electrons to form a Mg^{2+} ion. Because oxygen is a group 16 element, an oxygen atom gains two electrons to form an O^{2-} ion. These charges balance in a one-to-one ratio, so the formula for magnesium oxide is MgO.

6.4 Metallic Bonds

In Chapter 3, you learned about the properties of metals. They conduct electricity and heat, are opaque to light, and deform—rather than fracture—under pressure. Because of these properties, metals are used to build homes, appliances, cars, bridges, airplanes, and skyscrapers. Metal wires across the landscape transmit communication signals and electric power. We wear metal jewelry, exchange metal currency, and drink from metal cans. Yet what is it that gives a metal its metallic properties? We can answer this question by looking at the behavior of its atoms.

The outer electrons of most metal atoms tend to be weakly held to the atomic nucleus. Consequently, these electrons are easily dislodged, leaving behind positively charged metal ions. The many electrons dislodged from a large group of metal atoms flow freely through the resulting metal ions, as depicted in **Figure 6.12**. This "fluid" of electrons holds the positively charged metal ions together in the type of chemical bond known as a **metallic bond**.

READING CHECK

What holds the positively charged metal ions together within a metallic bond?

The mobility of electrons in a metal accounts for the metal's significant ability to conduct electricity and heat. Also, metals are opaque and shiny because the free electrons easily vibrate to the oscillations of

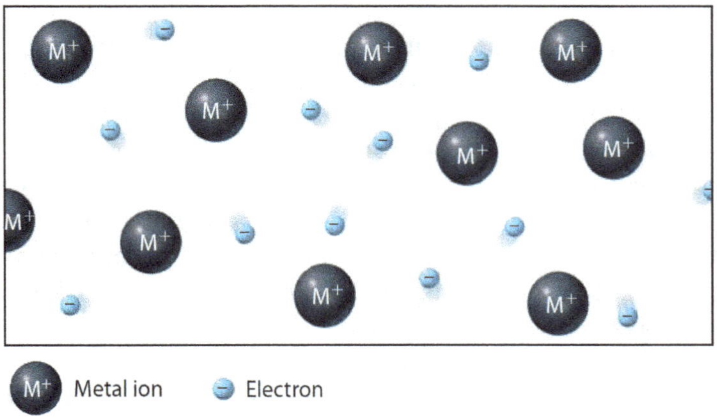

< **Figure 6.12**

Metal ions are held together by freely flowing electrons. These loose electrons form a kind of "electronic fluid," which flows through the lattice of positively charged ions.

any light falling on them, reflecting most of it. Furthermore, the metal ions are not rigidly held to fixed positions, as ions are in an ionic crystal. Rather, because the metal ions are held together by a "fluid" of electrons, these ions can move into various orientations relative to one another, which explains why a metal can be pounded, pulled, or molded into a different shape.

Two or more different metals can be bonded to each other by metallic bonds. This occurs, for example, when molten gold and molten palladium are blended to form the homogeneous solution known as white gold. The quality of the white gold can be modified simply by changing the proportions of gold and palladium. White gold is an example of an alloy, which is any mixture composed of two or more metallic elements. By playing around with proportions, metalworkers can readily modify the properties of an alloy. For example, in designing the Sacagawea dollar coin, shown in **Figure 6.13**, the U.S. Mint needed a metal having a gold color—so that it would be popular—and having the same electrical characteristics as the Susan B. Anthony dollar coin—so that the new coin could substitute for the Anthony coin in vending machines.

Only a few metals—gold and platinum are two examples—appear in nature in metallic form. Deposits of these natural metals, also known as *native metals*, are quite rare. For the most part, metals found in nature are chemical compounds. Iron, for example, is most frequently found as iron oxide, Fe_2O_3, and copper as chalcopyrite, $CuFeS_2$. Geologic deposits containing relatively high concentrations of metal-containing compounds are called **ores**. The metals industry mines these ores from the ground, as shown in **Figure 6.14**, and then chemically processes them into metals, as

∧ **Figure 6.13**

The gold color of the Sacagawea U.S. dollar coin is achieved by an outer surface made of an alloy of 77% copper, 12% zinc, 7% manganese, and 4% nickel. The interior of the coin is pure copper.

< **Figure 6.14**

The world's biggest open-pit mine is the copper mine at Bingham Canyon, Utah.

discussed in Chapter 11. Although metal-containing compounds occur just about everywhere, only ores are concentrated enough to make the extraction of the metal profitable.

FOR YOUR INFORMATION

Metal ores contain ionic compounds in which the metal atoms have lost electrons to become positive ions. As we discuss in Chapter 11, converting the ores to metals requires that electrons be given back to the metal ions. This is done by heating the ore with electron-releasing materials, such as carbon, in hot furnaces that reach about 1500°C. The metal emerges in a molten state that can be cast into a variety of useful shapes.

Because our planet is chock-full of metal-containing compounds, it is difficult to imagine how we could ever incur a shortage of metals. Experts suggest, however, that if we continue with our present rate of consumption, such shortages will occur within the next two centuries. The problem is not a shortage of metal-containing compounds but rather a shortage of ores from which these compounds can be extracted *at a reasonable cost.*

Consider the recovery of gold. All the gold in the world isolated from nature so far could form a single cube 18 meters on a side, which would have a mass of about 130,000 tons. This includes all the naturally occurring elemental gold we have mined plus all the gold purified from gold-containing ores. Because the rate of gold production is steadily decreasing, one might think we have already isolated a significant portion of the Earth's total gold reserves. Our oceans, however, are laden with gold—as much as 2 milligrams per ton of seawater. Given that there is about 1.5×10^{18} tons of seawater on the planet, our oceans contain 3.4 billion tons of gold! As yet, however, no method has been found for recovering gold from seawater profitably—this gold is simply too dilute (**Figure 6.15**).

Figure 6.15

Natural resources are unavailable to us when the energy required to collect them far exceeds the resource's inherent value. For example, most of the world's gold is found in the oceans, but this gold is too dilute for extraction to be worthwhile.

Like the gold in the ocean, most of the metal-containing compounds in the Earth's crust are finely mixed with other stuff, which is to say the compounds are diluted. Ores are, by definition, parts of the Earth's crust where, for geologic reasons, the compounds have been concentrated. High-grade ores, those containing relatively large concentrations of compounds, are the first to be mined. After these are depleted, we move on to lower-grade ores, which have lower yields that translate into greater costs. Eventually, a nation's ore supplies are depleted, as are the aluminum oxide ores in the United States, as described in **Figure 6.16**. The nation is forced to import metals or their ores from other countries, which also have finite ore resources.

Interestingly, high-grade ore nodules discovered on the ocean floor are a potential new source of metals, as are metal-rich asteroids in space. Rather than investing in new ore resources, however, it is far cheaper to produce metals from recycled products. Perhaps one day in the future our recycling programs will be so strong worldwide that mining from such unusual places will not be necessary.

Figure 6.16

An open-pit aluminum mine in Australia. Aluminum ore is no longer mined in the United States because the reserves have dwindled to the point where it is less expensive to import aluminum from other countries, including Australia.

6.5 Covalent Bonds

Imagine two children playing together and sharing their toys. Perhaps a force that keeps the children together is their mutual attraction to the toys they share. In a similar fashion, two atoms can be held together by their mutual attraction for electrons they share. A fluorine atom, for example, has a strong tendency to acquire an additional electron, which fills the outer shell. As shown in **Figure 6.17**, a fluorine atom can add an additional electron to its outer shell by grabbing onto the unpaired valence electron of another fluorine atom. This results in a situation in which the two fluorine atoms are mutually attracted to the same two electrons. This type of electrical attraction in which atoms are held together by their mutual attraction for shared electrons is called a **covalent bond**, where co- signifies sharing and -valent refers to the fact that it is valence electrons that are being shared.

A substance composed of atoms held together by covalent bonds is a **covalent compound**. The fundamental unit of most covalent compounds is a **molecule**, which we can now formally define as any group of atoms held together by covalent bonds. **Figure 6.18** uses the element fluorine to illustrate this principle.

When writing electron-dot structures for covalent compounds, chemists often use a straight line to represent the two electrons involved in a covalent bond. In some representations, the nonbonding electron pairs are ignored. This occurs in instances where these electrons play no significant role in the process being illustrated. Here are two frequently used ways of showing the electron-dot structure for a fluorine molecule without using spheres to represent the atoms:

FOR YOUR INFORMATION

Spectroscopic studies of interstellar dust within our galaxy have revealed the presence of more than 120 kinds of molecules, such as hydrogen chloride, HCl; water, H_2O; acetylene, H_2C_2; formic acid, HCO_2H; methanol, CH_3OH; methyl amine, NH_2CH_3; acetic acid, CH_3CO_2H; and even the amino acid glycine, $NH_2CH_2CO_2H$. Notably, about half of these interstellar molecules are carbon-based organic molecules. As discussed in Chapter 5, the atoms originated from the nuclear fusion of ancient stars. How interesting that these atoms then join together to form molecules even in the deep vacuum of outer space.

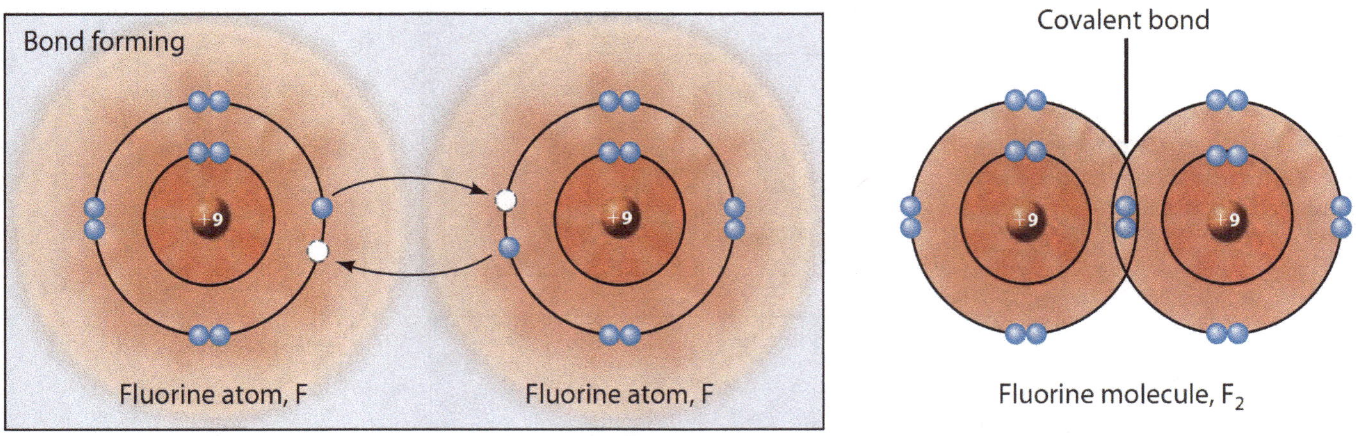

Figure 6.17

The positive nuclear charge (represented by red shading) of a fluorine atom can cause the fluorine atom to become attracted to the unpaired valence electron of a neighboring fluorine atom. In this way, each fluorine atom achieves a filled valence shell and the two atoms are held together in a fluorine molecule by the attraction they both have for the two shared electrons. The atoms are said to be held together by a covalent bond.

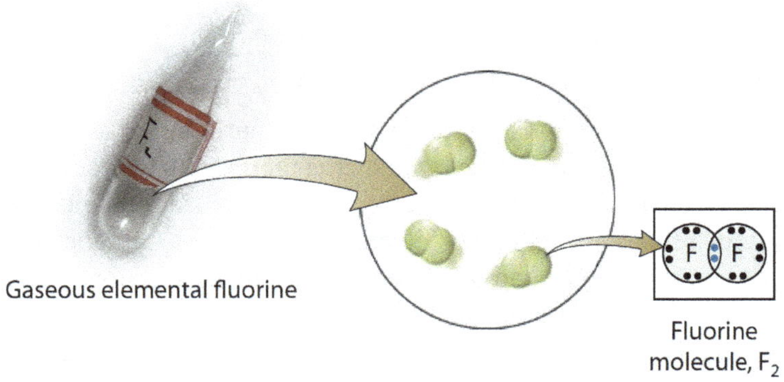

Figure 6.18 >

Molecules are the fundamental units of the gaseous covalent compound fluorine, F2. Notice that in this model of a fluorine molecule, the spheres overlap, whereas the spheres shown earlier for ionic compounds do not. Now you know that this difference in representation is because of the difference in bond types.

Remember—the straight line in both versions represents two electrons, one from each atom. Thus, we now have two types of electron pairs to keep track of. The term *nonbonding pair* refers to any pair that exists in the electron-dot structure of an individual atom, and the term *bonding pair* refers to any pair that results from formation of a covalent bond. In a nonbonding pair, both electrons originate in the same atom; in a bonding pair, one electron comes from each atom participating in the bond.

Recall from Section 6.3 that an ionic bond is formed when an atom that tends to lose electrons makes contact with an atom that tends to gain them. A covalent bond, by contrast, is formed when two atoms that *both* tend to gain electrons are brought into contact with each other. Atoms that tend to form covalent bonds are therefore primarily atoms of the nonmetallic elements, in the upper right corner of the periodic table (with the exception of the noble gas elements, which are very stable and tend not to form bonds).

Hydrogen tends to form covalent bonds because, unlike the other group 1 elements, it has a fairly strong attraction for an additional electron. Two hydrogen atoms, for example, covalently bond to form a hydrogen molecule, H_2, as shown in **Figure 6.19**.

The number of covalent bonds an atom commonly forms is equal to the number of additional electrons it can attract, which is the number of electrons needed to fill its valence shell. Hydrogen attracts only one additional electron, so it forms only one covalent bond. Oxygen, which attracts two additional electrons, finds them when it encounters two hydrogen atoms and reacts with them to form water, H_2O, as **Figure 6.20** shows. In water,

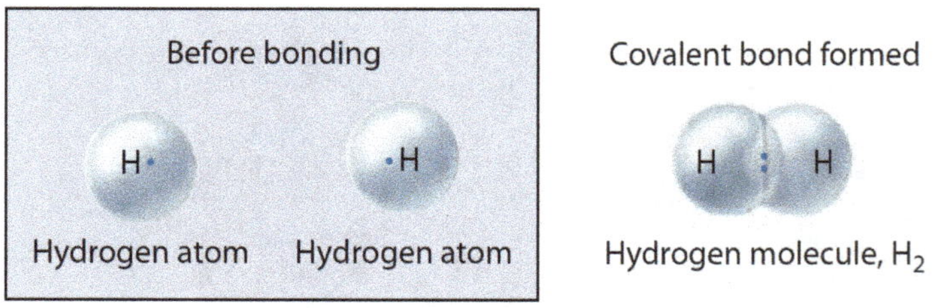

Figure 6.19 >

Two hydrogen atoms form a covalent bond as they share their unpaired electrons.

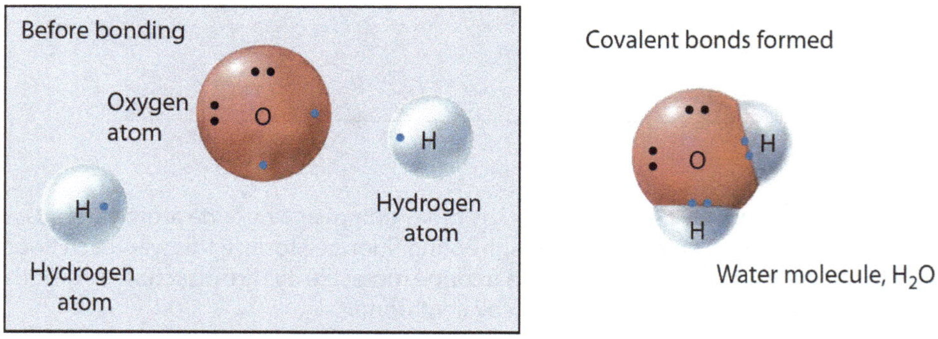

Figure 6.20 >

The two unpaired valence electrons of oxygen pair with the unpaired valence electrons of two hydrogen atoms to form the covalent compound water.

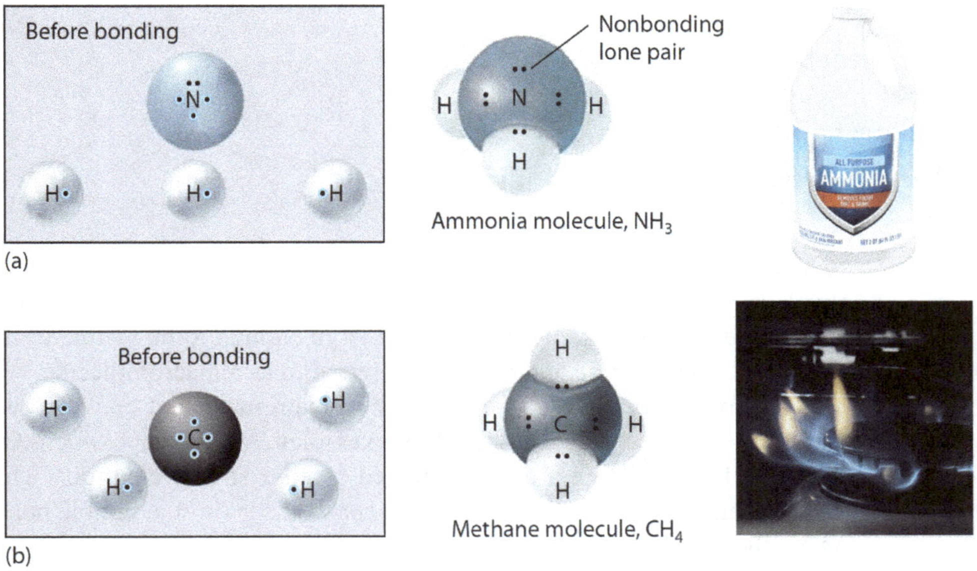

< Figure 6.21

(a) A nitrogen atom attracts the three electrons in three hydrogen atoms to form ammonia, NH_3, a gas that can dissolve in water to make an effective cleanser. (b) A carbon atom attracts the four electrons in four hydrogen atoms to form methane, CH_4, the primary component of natural gas. In these and most other cases of covalent-bond formation, the result is a filled valence shell for all the atoms involved.

not only does the oxygen atom have access to two additional electrons by covalently bonding to two hydrogen atoms, but each hydrogen atom has access to an additional electron by bonding to the oxygen atom. Each atom thus achieves a filled valence shell.

Nitrogen attracts three additional electrons and is thus able to form three covalent bonds, as occurs in ammonia, NH_3, shown in **Figure 6.21**. Likewise, a carbon atom can attract four additional electrons and is thus able to form four covalent bonds, as occurs in methane, CH_4, also shown in Figure 6.21. Note that the number of covalent bonds formed by these and other nonmetallic elements parallels the negative charge these ions tend to form (see Figure 6.6). This makes sense, because covalent-bond formation and negative-ion formation are both applications of the same concept: nonmetallic atoms tend to gain electrons until their valence shells are filled.

READING CHECK

What do covalent bonds have in common with the negative ions formed by nonmetals?

Diamond is a most unusual covalent compound consisting of carbon atoms covalently bonded to one another in four directions. The result is a *covalent crystal*, which, as shown in **Figure 6.22**, is a highly ordered, 3-dimensional network of covalently bonded atoms. This network of carbon atoms forms a very strong and rigid structure, which is why diamonds are

< Figure 6.22

The crystalline structure of diamond is nicely illustrated with sticks to represent the covalent bonds. The molecular nature of a diamond is responsible for its extreme hardness. Interestingly, most naturally occurring diamonds formed deep within the Earth, not millions, but billions of years ago.

CONCEPT CHECK

How many electrons make up a covalent bond?

CHECK YOUR ANSWER

One covalent bond consists of two electrons.

so hard. Also, because a diamond is a group of atoms held together only by covalent bonds, it can be characterized as a single molecule. Unlike most other molecules, a diamond molecule is large enough to be visible to the naked eye, so it is more appropriately referred to as a macromolecule.

It is possible to have more than two electrons shared between two atoms, and **Figure 6.23** shows a few examples. Molecular oxygen, O_2, consists of two oxygen atoms connected by four shared electrons. This arrangement is called a *double covalent bond* or, for short, a double bond. As another example, the covalent compound carbon dioxide, CO_2, consists of two double bonds connecting two oxygen atoms to a central carbon atom.

Some atoms can form *triple covalent bonds*, in which six electrons—three from each atom—are shared. One example is molecular nitrogen, N_2. Any double or triple bond is often referred to as a multiple bond. Multiple bonds higher than these, such as the quadruple covalent bond, are not commonly observed.

Figure 6.24 shows electron-dot structures, also sometimes called *Lewis structures*, for some additional molecules. Note from these structures that each element consistently forms the same number of bonds. Hydrogen, for example, forms one bond, while carbon forms four bonds, nitrogen forms three bonds, and oxygen forms two bonds. As discussed earlier, this number

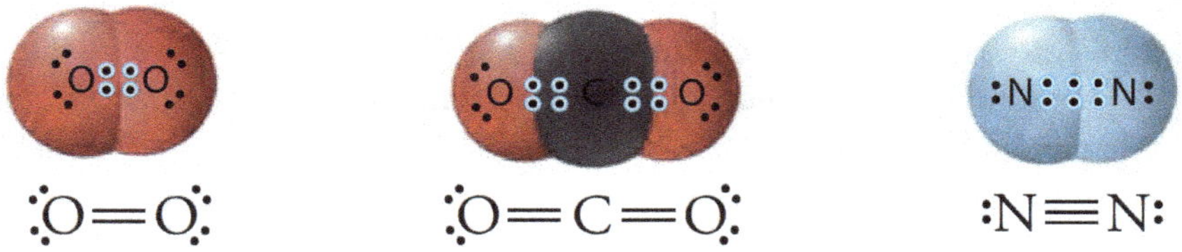

Oxygen, O_2 | Carbon dioxide, CO_2 | Nitrogen, N_2

Figure 6.23 >

Double covalent bonds in molecules of oxygen, O_2, and carbon dioxide, CO_2, and a triple covalent bond in a molecule of nitrogen, N_2.

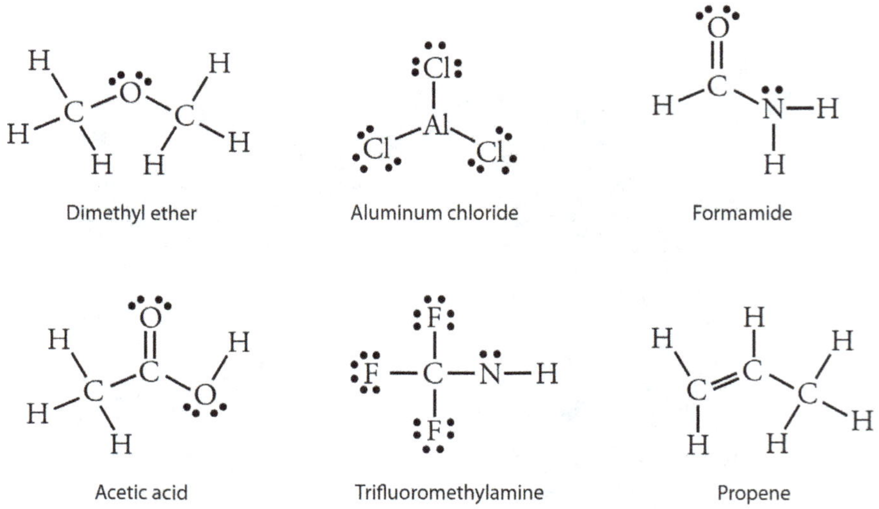

Figure 6.24 >

Additional electron-dot structures. Note that each carbon atom always forms four bonds. In formamide, two of these bonds are single bonds and the other two are part of a double bond. A double bond, therefore, counts as two bonds, and a triple bond counts as three bonds.

of bonds is equal to the number of additional electrons the atom is able to attract as indicated by its position in the periodic table. Also note from these structures that the number of nonbonding lone pairs on atoms is also consistent. Hydrogen, aluminum, and carbon show no lone pairs. Nitrogen has one lone pair, oxygen has two, and chlorine and fluorine have three. These numbers of lone pairs allow each atom to have a filled shell of electrons.

The pattern is a bit different for the charged atoms within a polyatomic ion. A positively charged atom within a polyatomic ion has one additional covalent bond and one fewer lone pair. For example, the nitrogen atom of an ammonia molecule, NH_3, has three bonds and one lone pair, as shown in Figure 6.21a. Add a hydrogen ion, H+, to ammonia and you create the ammonium ion, NH_4^+, as shown in **Figure 6.25a**. The nitrogen of the ammonium ion has four bonds and no lone pairs. (As we discuss in Chapter 10, the reason for this is that the lone pair of ammonia combined with a hydrogen ion to form the fourth bond.)

Similarly, a negatively charged atom within a polyatomic ion has one fewer covalent bond and one additional lone pair. For example, the oxygen of a water molecule, H_2O, has two bonds and two lone pairs, as shown in Figure 6.20. Remove a hydrogen ion, H^+, from a water molecule and you create the hydroxide ion, OH^-, as shown in **Figure 6.25b**. The oxygen of the hydroxide ion has only one bond but three lone pairs. (The additional lone pair was created as the hydrogen ion left the water molecule without an electron.)

FOR YOUR INFORMATION

Astronomers have recently discovered an expired star that has a solid core made of diamond. This star-sized diamond measures about 4000 kilometers wide, which amounts to about 10 billion trillion trillion carats. It has been named "Lucy," after The Beatles song "Lucy in the Sky with Diamonds." In about 7 billion years, our own star, the Sun, is also likely to crystallize into a huge diamond ball.

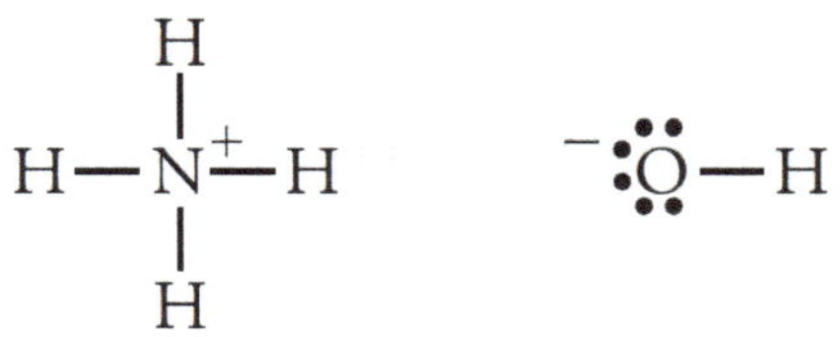

Ammonium ion

(a)

Hydroxide ion

(b)

< Figure 6.25

The electron-dot structures for the ammonium and hydroxide ions.

CONCEPT CHECK

Shown below is the electron-dot structure for the polyatomic bicarbonate ion. Why is one of the oxygen atoms shown with only one covalent bond?

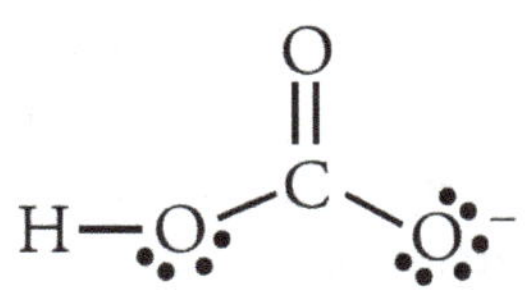

Bicarbonate ion

CHECK YOUR ANSWER

In a neutral molecule, one would expect each oxygen atom to have two bonds. The oxygen seen to the right in this electron-dot structure is bonded only once because it is carrying a negative charge. Note that although it is bonded only once, it has three lone pairs, which means the valence shell is filled.

6.6 Molecular Shape

READING CHECK
What do electron pairs, either bonding or nonbonding, tend to do?

Molecules are 3-dimensional entities and therefore best depicted in three dimensions. We can translate the 2-dimensional electron-dot structure representing a molecule into a more accurate 3-dimensional rendering by using the model known as **valence-shell electron-pair repulsion**, also called VSEPR (pronounced ves-per). According to this model, electron pairs in a valence shell arrange themselves to get as far away as possible from all other electron pairs in the shell. This includes nonbonding pairs and any bonding pairs or groups of bonding pairs held together in a double or triple bond. This behavior is the result of simple electrostatic repulsions between the electron pairs.

The 2-dimensional electron-dot structure for methane, CH_4,

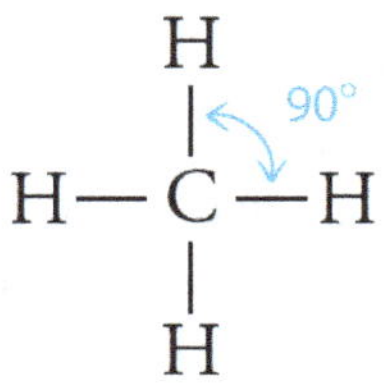

In this structure, the bonding electron pairs (shown as straight lines representing two electrons each) are set 90° apart because that is the farthest apart they can be shown in two dimensions. When we extend the view to three dimensions, however, we can create a more accurate rendering, in which the four bonding pairs are 109.5° apart:

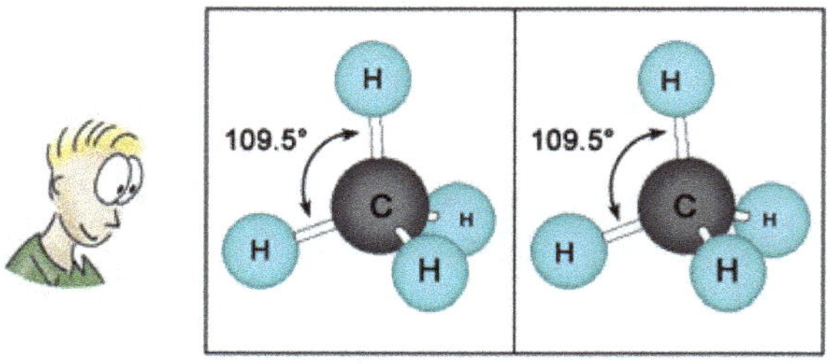

Stereo image

These two renderings of methane are *stereo images*—you can see them in three dimensions by looking at them cross-eyed so that they appear to overlap. To get the images to overlap, you can also try touching your nose to the page and then slowly pulling the book away from your face. If neither of these techniques works, try building these structures with gumdrops and toothpicks.

You can think of this 3-dimensional structure as follows: the central carbon atom has one hydrogen atom sticking out of its top and is supported on a tripod whose legs are formed by the three lower C–H bonds.

Draw the four triangles defined by the hydrogen atoms in the above stereo image of CH_4 (one triangle being the base, the other three being the three upright faces) and you'll see that the shape of the methane molecule is a pyramid that has a triangular base supporting three other triangles that meet at the pyramid's apex. In geometry, a pyramid that has a triangular base is given the special name *tetrahedron*, and so chemists say that the methane molecule is tetrahedral:

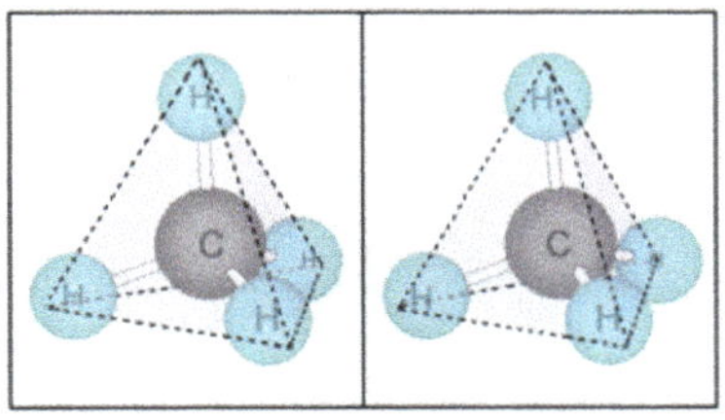

Stereo image of tetrahedral methane molecule

The VSEPR model allows us to use electron-dot structures to predict the 3-dimensional geometry of simple molecules. This geometry is determined by considering the number of atoms or nonbonding electron pairs surrounding the centrally located atom. Chemists refer to any atom or nonbonding electron pair surrounding a central atom as a **substituent**. They use this fancy term the same way people commonly use the word appendage to refer to either an arm or a leg. Thus, to simplify our discussion, we define a substituent as any atom or nonbonding pair of electrons surrounding a centrally located atom. For example, the carbon of a methane molecule has four substituents—the four hydrogen atoms. The oxygen atom of a water molecule also has four substituents—two hydrogen atoms and two nonbonding pairs of electrons:

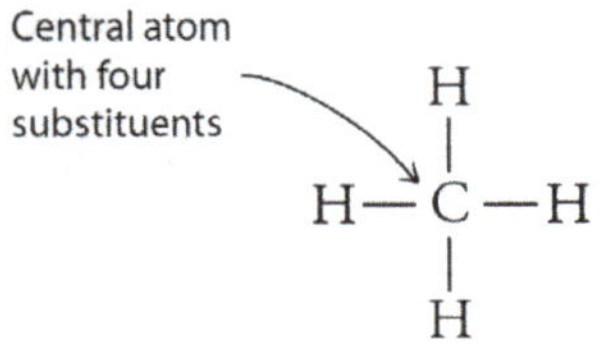

Methane, CH_4

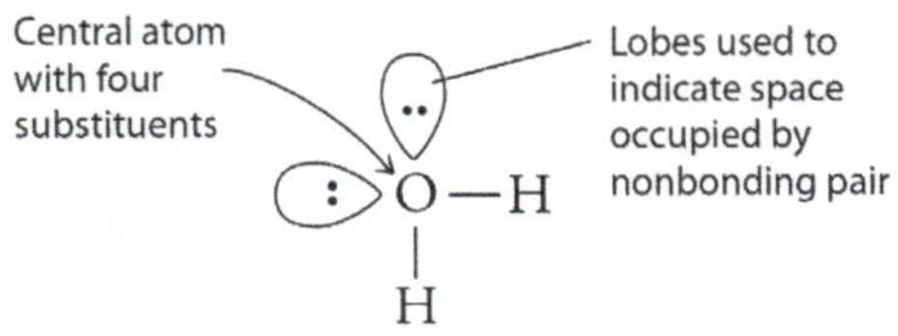

Water, H_2O

As shown in **Table 6.2**, when a central atom has only two substituents, the geometry of the molecule is linear, meaning that a single straight line may be drawn passing through both substituents and the central atom. Three substituents arrange themselves in a triangle—the plane of which passes through the central atom—so this geometry is called *triangular planar*. Four substituents form a tetrahedron, as already discussed. Why these geometries? Simply put, these are the geometries that allow the maximum distance between substituents.

CONCEPT CHECK

Why are the two oxygen atoms in carbon dioxide, CO_2, spaced 180° apart?

CHECK YOUR ANSWER

If the two oxygen atoms were on the same side of the carbon atom, the bonding electrons would be relatively close to each other. Because electron pairs repel one another, this is not a stable situation. Instead, the oxygen atoms position themselves so that the bonding pairs of the two double bonds are as far from each other as possible, which is on opposite sides of the carbon atom, 180° apart, as shown in Table 6.2.

TABLE 6.2 Molecular Geometries

NUMBER OF SUBSTITUENTS	3-DIMENSIONAL GEOMERTY	EXAMPLES		
2	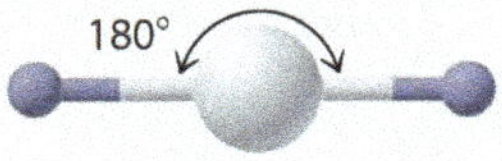 Linear	H—Be—H BeH_2	O=C=O CO_2	H—C≡N HCN
3	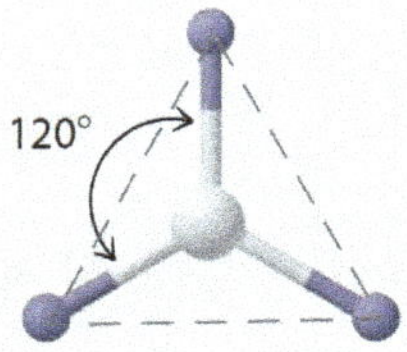 Triangular planar	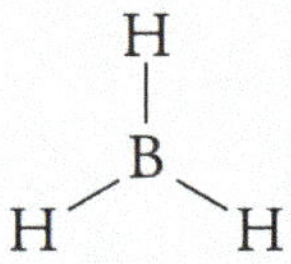 BH_3	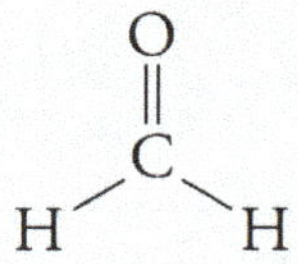 H_2CO	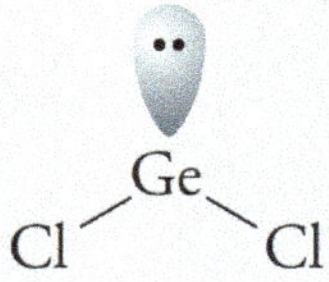 $GeCl_2$
4	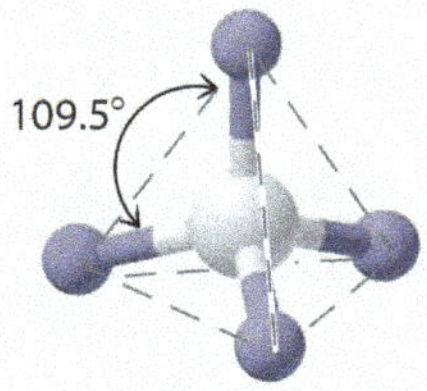 Tetrahedral	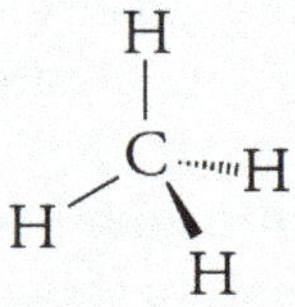 CH_4	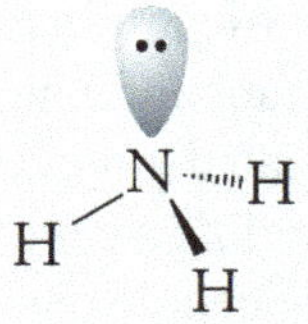 NH_3	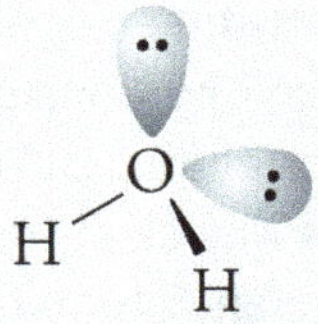 H_2O

Molecular Shape Is Defined by the Substituent Atoms

When chemists talk about the *shape of a molecule*, they are talking about the relative positions of the atoms of the molecule. How atoms position themselves is spelled out by VSEPR. But from VSEPR, we see that the nonbonding electrons (lone pairs) also affect how atoms within a molecule get positioned.

Figuring out the shape of a molecule is a two-step process. The first step is to use VSEPR to position all substituents, both atoms and nonbonding pairs, around a central atom. The second step is to ignore all nonbonding pairs and decide what 3-dimensional shape the atoms form. Let's work through a few examples from Table 6.2 to see what all this means.

In any molecule in which there are no nonbonding pairs around the central atom, the molecular shape is the same as the VSEPR geometry. Thus, to use the examples from Table 6.2, all the molecules of the first row (BeH_2, CO_2, HCN) have a linear shape. Two molecules of the second row (BH_3 and H_2CO) have a triangular planar shape, and the first molecule of the third row (CH_4) has a tetrahedral shape.

Now let's look at molecules that have nonbonding pairs, beginning with germanium chloride, $GeCl_2$, and its one nonbonding pair. The VSEPR geometry is triangular planar, but to get the shape of the molecule, we focus only on the atoms. This reveals the germanium and two chlorine atoms held together at an angle—a shape known as *bent*. Similarly, focusing only on the atoms, a water molecule is also seen to have a bent shape. Now you know why water molecule are always depicted with the two hydrogen atoms close to each other, like a set of mouse ears, rather than as far apart as possible on opposite sides of the oxygen atom—two nonbonding pairs are pushing them into this orientation.

Focusing only on the atoms of the ammonia molecule, NH_3, in Table 6.2 means the shape is not tetrahedral, because in a tetrahedron, all four corners must be equally distant from the central atom. Ammonia's shape, as shown in **Figure 6.26**, is typically described as *triangular pyramidal*.

fyi FOR YOUR INFORMATION

"Dry cleaning" is the process of washing clothes without water. The most common dry cleaning solvent is perchloroethylene, C_2Cl_4. (Can you deduce its chemical structure?) The advantage of dry cleaning is that dirt, grime, and stains are typically more soluble in the dry cleaning solvent, which is also less harsh on the clothing and can do a full load in under 10 minutes. After a washing cycle, the solvent is centrifuged out of the machine, filtered, distilled, and recycled for the next load. Clothes come out of the machine already dried and ready for folding. Perchloroethylene, also known as perc, is mildly carcinogenic and can cause dizziness in those who work with it. An up-and-coming alternative to perc is carbon dioxide, CO_2, which at super high pressures forms an unusual liquid-like and gas-like medium, called supercritical fluid, that has remarkable cleaning properties.

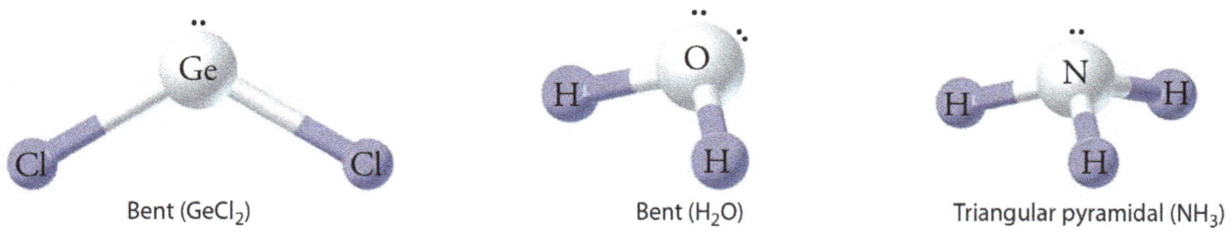

∧ Figure 6.26

The shapes of molecules from Table 6.2.

The rules for determining the molecular shape of a small molecule are summarized in **Figure 6.27**. The first step is to count the electron pairs surrounding the central atom including both bonding and nonbonding. This determines the geometry. You then count the number of surrounding atoms and this will give you the molecular shape.

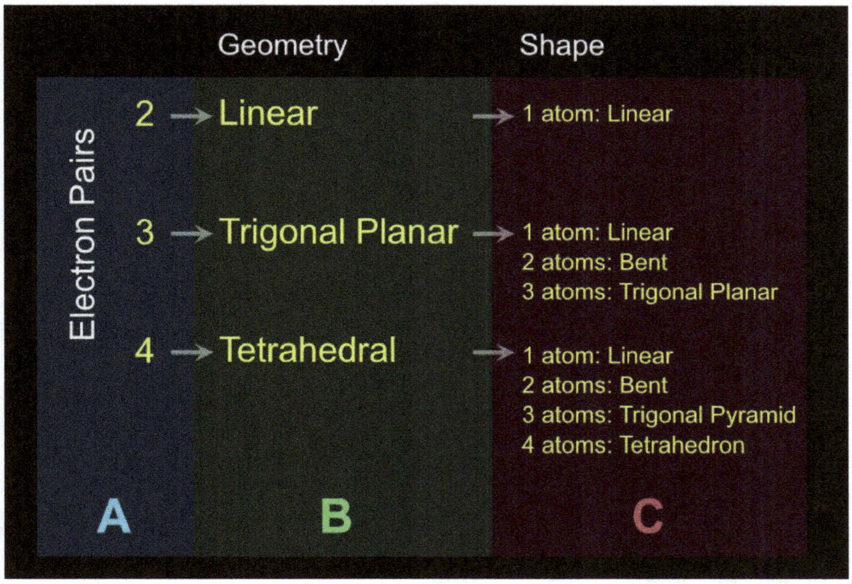

< Figure 6.27

A flow chart for determining the shape of a small molecule.

CONCEPT **CHECK**

What is the shape of a chlorine trifluoride molecule, ClF_3?

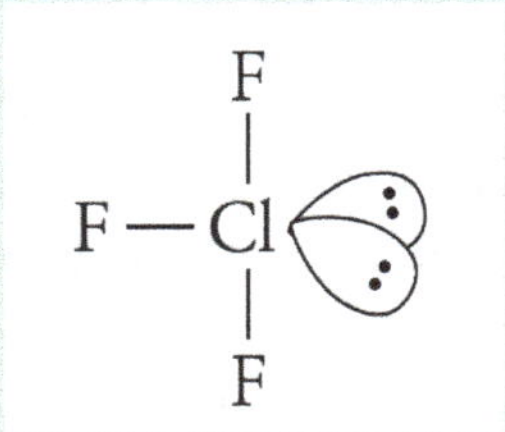

CHECK YOUR ANSWER

Describe what you see as you focus only on the atoms (ignoring the nonbonding pairs). For chlorine trifluoride, the shape has all four atoms in the same plane. They form a triangle having a fluorine atom at each corner and the chlorine atom sitting at the midpoint on one side. Call it what you like—most chemists simply call it T-shaped. If you're curious how chlorine is able to form three covalent bonds, you should consider taking a follow-up course in general chemistry.

6.7 Polar Covalent Bonds

If the two atoms in a covalent bond are identical, as shown below, their nuclei have the same positive charge. The electrons shared between these two identical atoms are shared evenly. We can represent these electrons being shared evenly using an electron-dot structure with the electrons situated exactly halfway between the two atomic symbols. Alternatively, we can draw a cloud in which the positions of the two bonding electrons over time are shown as a series of dots. Where the dots are most concentrated is where the electrons have the greatest probability of being located:

H : H

In a covalent bond between nonidentical atoms, the nuclear charges are different, and consequently, the bonding electrons may be shared *unevenly*. This occurs in a hydrogen–fluorine bond, in which electrons are more attracted to fluorine's greater nuclear charge:

H : F

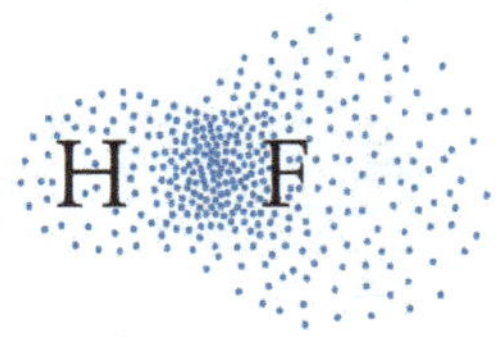

The bonding electrons spend more time around the fluorine atom. For this reason, the fluorine side of the bond is slightly negative, and because the bonding electrons have been drawn away from the hydrogen atom, the hydrogen side of the bond is slightly positive. This separation of charges is called a **dipole** (pronounced die-pole) and is represented either by the characters ∂− and ∂+ (read "slightly negative" and "slightly positive," respectively) or by a crossed arrow pointing to the negative side of the bond:

$\overset{\delta+}{H}-\overset{\delta-}{F}$

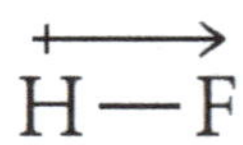

READING CHECK
What is an atom with great electronegativity able to do?

So, atoms forming a chemical bond engage in a tug-of-war for electrons. How strongly an atom is able to tug on bonding electrons has been measured experimentally and quantified as the atom's **electronegativity**. The range of electronegativities runs from 0.7 to 3.98, as **Figure 6.28** shows. The greater an atom's electronegativity, the greater its ability to pull electrons toward itself when bonded. Thus, in hydrogen fluoride, fluorine has a greater electronegativity, or pulling power, than hydrogen.

Electronegativity is greatest for elements at the upper right of the periodic table and lowest for elements at the lower left. Noble gases are not considered in electronegativity discussions because, as previously mentioned, they rarely participate in chemical bonding.

H 2.2																	He —
Li 0.98	Be 1.57											B 2.04	C 2.55	N 3.04	O 3.44	F 3.98	Ne —
Na 0.93	Mg 1.31											Al 1.61	Si 1.9	P 2.19	S 2.58	Cl 3.16	Ar —
K 0.82	Ca 1.0	Sc 1.36	Ti 1.54	V 1.63	Cr 1.66	Mn 1.55	Fe 1.83	Co 1.88	Ni 1.91	Cu 1.90	Zn 1.65	Ga 1.81	Ge 2.01	As 2.18	Se 2.55	Br 2.96	Kr —
Rb 0.82	Sr 0.95	Y 1.22	Zr 1.33	Nb 1.6	Mo 2.16	Tc 1.9	Ru 2.2	Rh 2.28	Pd 2.20	Ag 1.93	Cd 1.69	In 1.78	Sn 1.96	Sb 2.05	Te 2.1	I 2.66	Xe —
Cs 0.79	Ba 0.89	La 1.10	Hf 1.3	Ta 1.5	W 2.36	Re 1.9	Os 2.2	Ir 2.20	Pt 2.8	Au 2.54	Hg 2.00	Tl 2.04	Pb 2.33	Bi 2.02	Po 2.0	At 2.2	Rn —
Fr 0.7	Ra 0.9	Ac 1.1	Rf —	Db —	Sg —	Bh —	Hs —	Mt —	Ds —	Rg —	Cn —	Nh —	Fl —	Mc —	Lv —	Ts —	Og —

Λ Figure 6.28

The experimentally measured electronegativities of elements.

When the two atoms in a covalent bond have the same electronegativity, no dipole is formed (as is the case with H_2) and the bond is classified as a **nonpolar bond**. When the electronegativities of the atoms differ, a dipole may form (as with HF) and the bond is classified as a **polar bond**. Just how polar a bond is depends on the difference between the electronegativity values of the two atoms—the greater the difference, the more polar the bond. An electronegativity difference between two atoms greater than about 1.7 indicates an ionic bond. A difference of less than 1.7 but greater than about 0.4 indicates a polar covalent bond. Nonpolar covalent bonds have electronegativity differences of less than 0.4.

As can be seen in Figure 6.28, the greater the distance between two atoms in the periodic table, the greater the difference in their electronegativities, and hence the greater the polarity of the bond between them. So, a chemist can predict which bonds are more polar than others without reading the electronegativities. Bond polarity can usually be inferred by looking at the relative positions of the atoms in the periodic table—the farther apart they are, especially when one is at the lower left and one is at the upper right, the greater the polarity of the bond between them.

The magnitude of bond polarity is sometimes indicated by the size of the crossed arrow or the $\partial-$ and $\partial+$ symbols used to depict a dipole, as shown in **Figure 6.29**. Note that the electronegativity difference between atoms in an ionic bond is relatively large. For example, the bond in NaCl has an electronegativity difference of 2.23, far greater than the difference of 1.43 shown for the C–F bond.

What is important to understand here is that there is no black-and-white distinction between ionic and covalent bonds. Rather, there is a gradual change from one to the other as the atoms that bond are located farther and farther apart in the periodic table. This continuum is illustrated in **Figure 6.30**. Atoms on opposite sides of the periodic table have great differences in electronegativity, and hence the bonds between them are highly polar—in other words, ionic. Nonmetallic atoms of the same element have the same electronegativities, so their bonds are nonpolar covalent. The polar covalent bond, with its uneven sharing of electrons and slightly charged atoms, lies between these two extremes

fyi FOR YOUR INFORMATION

The term electronegativity was coined by Linus Pauling in his famous 1932 publications "The Nature of the Chemical Bond, Parts I–IV." Through these articles, Pauling was the first to formally describe the periodic trends of ionic, polar covalent, and nonpolar covalent bonding. The term electronegativity refers to the negative charge that an atom in a molecule picks up due to the electrons it may attract. There are other electronegativity scales, but Pauling's original scale has remained the most commonly used.

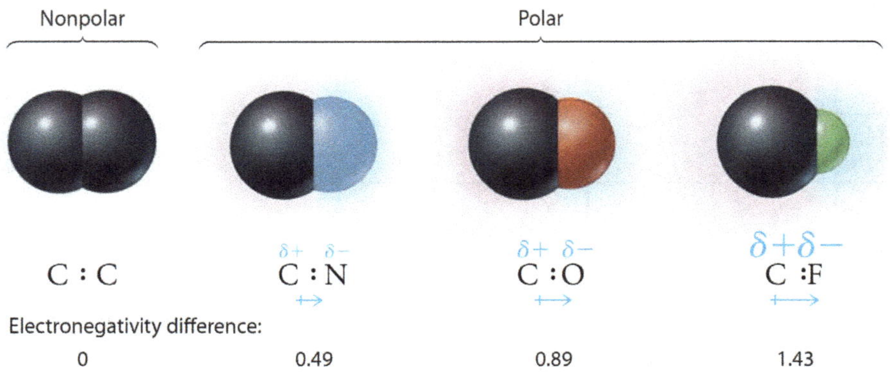

Figure 6.29 >

These bonds are in order of increasing polarity from left to right, a trend indicated by the larger and larger crossed arrows and ∂−/∂+ symbols. Which of these pairs of elements are farthest apart in the periodic table?

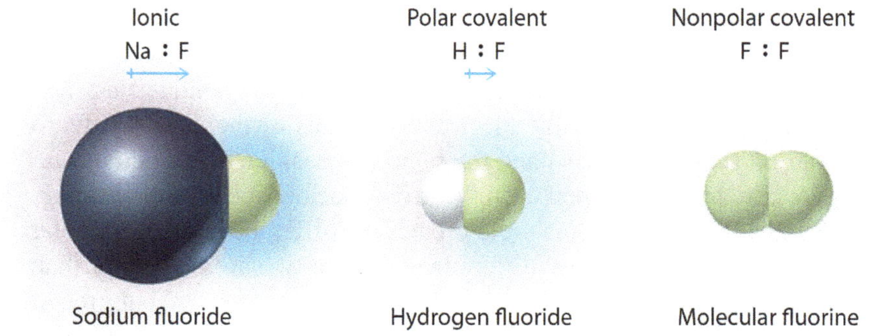

Figure 6.30 >

The ionic bond and the nonpolar covalent bond represent the two extremes of chemical bonding. The ionic bond involves a transfer of one or more electrons, and the nonpolar covalent bond involves the equitable sharing of electrons. The character of a polar covalent bond falls between these two extremes

CONCEPT CHECK

List these bonds in order of increasing polarity: P–F, S–F, Ga–F, Ge–F.

(least polar) ______, ______, ______, ______. (most polar)

CHECK YOUR ANSWER

If you answered the question, or attempted to, before reading this answer, hooray for you! You're doing more than reading the text—you're learning chemistry. The greater the difference in electronegativities between two bonded atoms, the greater the polarity of the bond; so, the order of increasing polarity is S–F, P–F, Ge–F, Ga–F.

Note that this answer can be obtained merely by looking at the relative positions of these elements in the periodic table rather than by calculating the differences in their electronegativities.

6.8 Molecular Polarity

When all the bonds in a molecule are nonpolar, the molecule as a whole is also nonpolar—as is the case with H_2, O_2, and N_2. When a molecule consists of only two atoms and the bond between them is polar, the polarity of the molecule is the same as the polarity of the bond—as with HF and HCl.

Complexities arise when assessing the polarity of a molecule containing more than two atoms. Consider carbon dioxide, CO_2, shown in **Figure 6.31**. The cause of the dipole in either of the carbon–oxygen bonds is oxygen's greater pull on the bonding electrons (because oxygen is more electronegative than carbon). At the same time, however, the oxygen atom on the opposite side of the carbon pulls those electrons back to the carbon. The net result is an even distribution of bonding electrons around the entire molecule. So, dipoles that are of equal strength but pull in opposite directions in a molecule effectively cancel each other, with the result that the molecule as a whole is nonpolar.

Figure 6.32 illustrates a similar situation in boron trifluoride, BF_3, where three fluorine atoms are oriented 120° from one another around a central boron atom. Because the angles are all the same and because each fluorine atom pulls on the electrons of its boron–fluorine bond with the same force, the resulting polarity of this molecule is zero.

As is discussed further in the next chapter, nonpolar molecules have only relatively weak attractions to other nonpolar molecules. This lack of attraction between nonpolar molecules explains the low boiling points of many nonpolar substances. Recall from Section 2.7 that boiling is a process wherein the molecules of a liquid separate from one another as they go into the gaseous phase. When there are only weak attractions between the molecules of a liquid, less heat energy is required to liberate the molecules from one another and allow them to

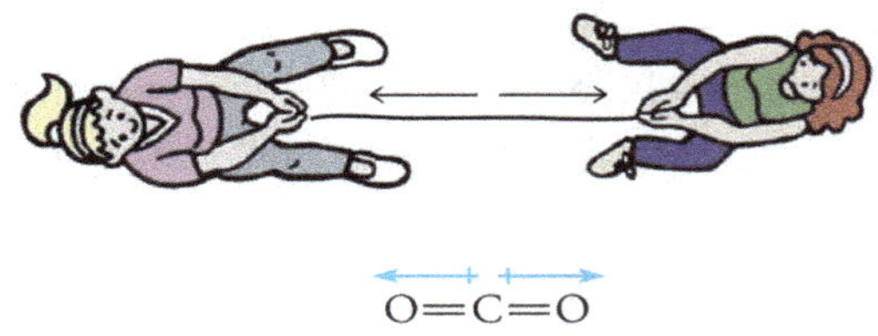

< Figure 6.31

There is no net dipole in a carbon dioxide molecule, so the molecule is nonpolar. This is analogous to two people in a tug-of-war. As long as they pull with equal force but in opposite directions, the rope remains stationary.

< Figure 6.32

The three dipoles of a boron trifluoride molecule oppose one another at 120° angles, which makes the overall molecule nonpolar. This is analogous to three people pulling with equal force on ropes attached to a central ring. As long as they all pull with equal force and all maintain the 120° angles, the ring will remain stationary.

FOR YOUR INFORMATION

Polar interactions between molecules are many times weaker than the attractions between ions within an ionic bond. The polar attractions described here and the ionic bonds described earlier are both electrical attractions. The main difference between them is one of magnitude.

enter the gaseous phase. This translates into a relatively low boiling point for the liquid, as shown for nitrogen, N_2 in **Figure 6.33**. The boiling points of hydrogen, H_2; oxygen, O_2; carbon dioxide, CO_2; and boron trifluoride, BF_3, are also quite low for the same reason.

There are many instances, however, in which the dipoles of different bonds in a molecule do *not* cancel each other. Reconsider the rope analogy of Figure 6.32. As long as everyone pulls equally, the ring stays put. Imagine, however, that one person begins to ease off on the rope. Now the pulls are no longer balanced, and the ring begins to move away from the person who is slacking off, as **Figure 6.34** shows. Likewise, if one person began to pull harder, the ring would move away from the other two people.

A similar situation occurs in molecules in which polar covalent bonds are not equal and opposite. Perhaps the most relevant example is water, H_2O. Each hydrogen–oxygen covalent bond has a relatively large dipole because of the great electronegativity difference. Because of the bent shape of the molecule, however, the two dipoles, shown in blue in **Figure 6.35,** do not cancel each other the way the C–O dipoles in Figure 6.31 do. Instead, the dipoles in the water molecule work together to give an overall dipole, shown in purple, for the molecule.

CHEMICAL CONNECTIONS

How is a glacier connected to a lone pair of electrons?

Figure 6.36 illustrates how polar molecules electrically attract one another and, as a result, are relatively difficult to separate. In other words, polar molecules can be thought of as being "sticky," which is why it takes more energy to separate them—to change the substance's phase. For this reason, substances composed of polar molecules typically have higher boiling points than substances composed of nonpolar molecules, as **Table 6.3** shows. Water, for example, boils at 100°C, whereas carbon dioxide boils at –79°C. This 179°C difference is quite dramatic when you consider that a carbon dioxide molecule is more than twice as massive as a water molecule.

Because molecular "stickiness" can play a lead role in determining a substance's macroscopic properties, molecular polarity is a central concept of chemistry. **Figure 6.37** helps bring to mind, in a tragic way, the fact that oil (both petroleum-based oil and cooking oil) and water don't mix. It's not, however, that oil and water repel each other. Rather, water molecules are so attracted to themselves because of their polarity that they pull themselves

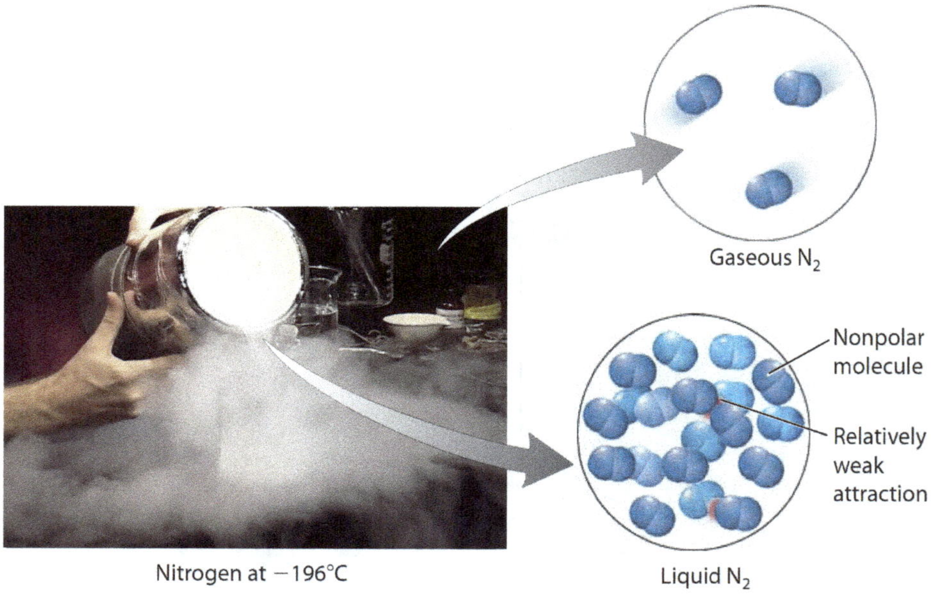

Figure 6.33 >

Nitrogen is a liquid at temperatures below its chilly boiling point of −196°C. Nitrogen molecules are not very attracted to one another because they are nonpolar. As a result, the small amount of heat energy available at −196°C is enough to separate them and allow them to enter the gaseous phase.

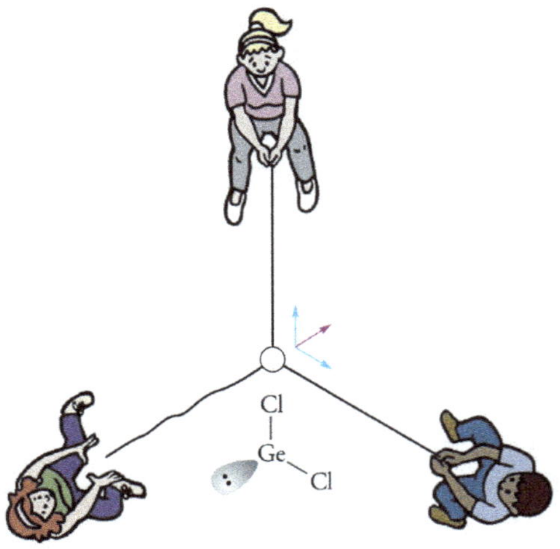

< Figure 6.34

The two strongly electronegative chlorine atoms in $GeCl_2$ pull germanium's lone pair inward. By analogy, if one person eases off in a three-way tug-of-war but the other two continue to pull, the ring moves in the direction of the purple arrow.

together. The nonpolar oil molecules are thus excluded and left to themselves. We see this in a bottle of oil and vinegar salad dressing. After the bottle is shaken vigorously, the water molecules of the vinegar cling together, excluding the oil molecules, which separate into their own layer. Being less dense than water, the oil rises to the top.

The shape of a molecule, as determined by VSEPR, plays a major role in determining the polarity of that molecule. A molecule's polarity, in turn, has a great influence on macroscopic behavior. Consider what the world would be like if the oxygen atom in a water molecule did not have its two nonbonding pairs of electrons. Instead of being bent, each water molecule would be linear, much like carbon dioxide. The dipoles of the two hydrogen–oxygen bonds would cancel each other, which would make water a nonpolar substance and give it a relatively low boiling point. Water would not be a liquid at the ambient temperatures of our planet, and we in turn would not be here discussing these concepts. We should be thankful that the oxygen within water has those two nonbonding pairs. That's good chemistry.

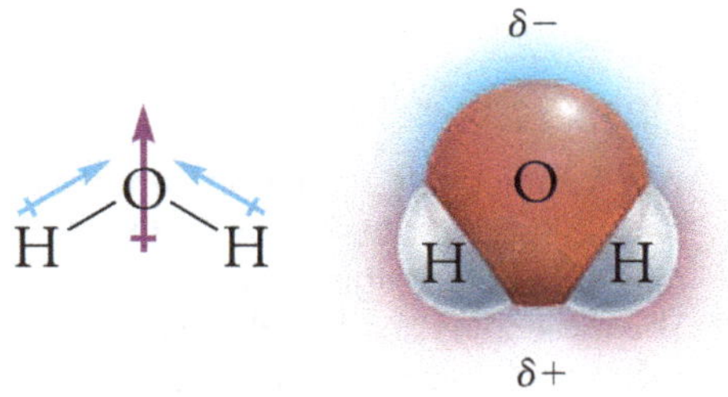

(a) (b)

∧ Figure 6.35

(a) The individual dipoles in a water molecule add together to give a large overall dipole for the whole molecule, shown in purple. (b) The region around the oxygen atom is therefore slightly negative, and the region around the two hydrogen atoms is slightly positive.

CONCEPT CHECK

Which of these molecules is polar, and which is nonpolar?

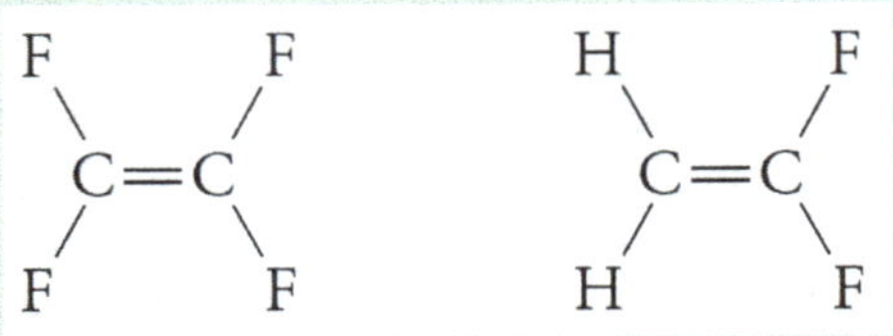

CHECK YOUR ANSWER

Symmetry is often the greatest clue for determining polarity. Because the molecule on the left is symmetrical, the dipoles on the two sides cancel each other. This molecule is therefore nonpolar.

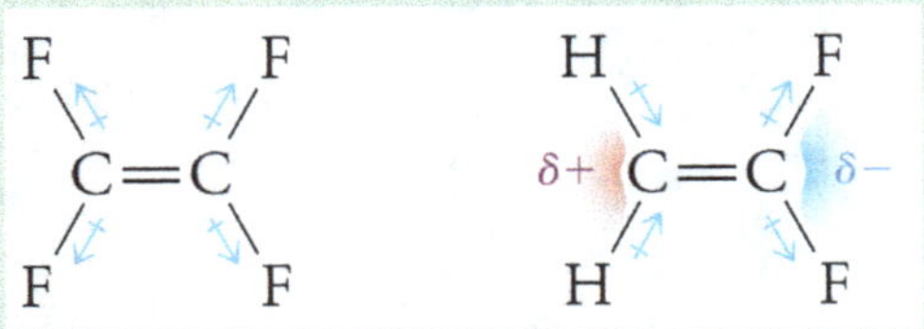

Because the molecule on the right is less symmetrical (more "lopsided"), it is a polar molecule.

TABLE 6.3 Boiling Points of Some Polar and Nonpolar Substances

SUBSTANCE	BOILING POINT (°C)
Polar	
Hydrogen fluoride, HF	20
Water, H_2O	100
Ammonia, NH_3	–33
Nonpolar	
Hydrogen, H_2	–253
Oxygen, O_2	–183
Nitrogen, N_2	–196
Boron trifluoride, BF_3	–100
Carbon dioxide, CO_2	–79

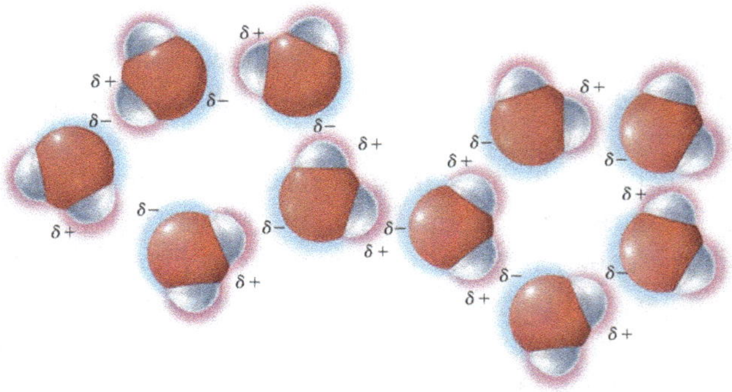

Figure 6.36>

Water molecules attract one another because each contains a slightly positive side and a slightly negative side. The molecules position themselves in such a way that the positive side of one faces the negative side of a neighbor.

∧ Figure 6.37

Oil and water are difficult to mix, as is evident from these photo of the Deepwater Horizon oil spill of 2010 in the Gulf of Mexico—the largest accidental oil spill in human history. The oil came from a sea-floor oil gusher that released an estimated 53,000 barrels every day for three months before it was finally capped. Petroleum-based oil molecules are not able to compete with the attraction water molecules have for themselves, which is why the oil and water don't mix.

Figure 6.38>

All species on our planet depend upon water, H_2O, for survival.

Concept Review

Chapter 6

Summary of Terms

Alloy A mixture of two or more metallic elements.

Covalent bond A chemical bond in which atoms are held together by their mutual attraction for two or more electrons they share.

Covalent compound A substance, such as an element or chemical compound, in which atoms are held together by covalent bonds.

Dipole A separation of charge that occurs in a chemical bond because of differences in the electronegativities of the bonded atoms.

Electron-dot structure A shorthand notation of the shell model of the atom, in which valence electrons are shown around an atomic symbol. The electron-dot structure for an atom or ion is sometimes called a Lewis dot symbol while the electron-dot structure of a molecule or polyatomic ion is sometimes called a Lewis structure.

Electronegativity The ability of an atom to attract a bonding pair of electrons to itself when bonded to another atom.

Ion An atom having a net electrical charge because of either a loss or gain of electrons.

Ionic bond A chemical bond in which there is an electric force of attraction between two oppositely charged ions.

Ionic compound A chemical compound containing ions.

Metallic Bond A chemical bond in which positively charged metal ions are held together within a "fluid" of loosely held electrons.

Molecule The fundamental unit of a chemical compound, which is a group of atoms held tightly together by covalent bonds.

Nonbonding pairs Two paired valence electrons that are not participating in a chemical bond.

Nonpolar Said of a chemical bond or molecule that has no dipole. In a nonpolar bond or molecule, the electrons are distributed evenly.

Ore A geologic deposit containing relatively high concentrations of one or more metal-containing compounds.

Polar Said of a chemical bond or molecule that has a dipole. In a polar bond or molecule, electrons are congregated to one side. This makes that side slightly negative while the opposite side (lacking electrons) becomes slightly positive.

Polyatomic ion An ionically charged molecule.

Substituent A term used to describe an atom or nonbonding pair of electrons surrounding a centrally located atom.

Valence electron The electrons in the outermost occupied shell of an atom.

Valence shell The outermost occupied shell of an atom.

Valence-shell electron-pair repulsion A model, also known as VSEPR (pronounced ves-per), that explains molecular geometries in terms of electron pairs striving to be as far apart from one another as possible.

Review Questions

6.1 Electron-Dot Structures

1. How many electrons can occupy the first shell? How many can occupy the second shell?

2. Which electrons are represented by an electron-dot structure?

3. How do the electron-dot structures of elements in the same group in the periodic table compare with one another?

4. Which are more stable: paired electrons or unpaired electrons?

6.2 Ion Formation

5. How does an ion differ from an atom?

6. To become a negative ion, does an atom lose or gain electrons?

7. Why does the fluorine atom tend to gain only one electron?

8. Why does the sodium atom to lose only one electron?

9. What is a polyatomic ion?

6.3 Ionic Bonds

10. Which elements tend to form ionic bonds?

11. Suppose an oxygen atom gains two electrons to become an oxygen ion. What is its net electric charge?

12. What is an ionic crystal?

13. In the formation aluminum oxide, Al_2O_3, who gains electrons and who loses them?

6.4 Metallic Bonds

14. Do metal atoms more readily gain or lose electrons?

15. What is an alloy?

16. What is a native metal?

17. What are two potential sources of metals that have yet to be exploited?

6.5 Covalent Bonds

18. Which elements tend to form covalent bonds?

19. How many electrons are shared in a double covalent bond?

20. Within a neutral molecule, how many covalent bonds does an oxygen atom form?

21. Within a polyatomic ion, how many covalent bonds does a negatively charged oxygen form?

6.6 Molecular Shape

22. What does vsepr stand for?

23. What is meant by the term substituent?

24. How many substituents does the oxygen atom in a water molecule have?

6.7 Polar Covalent Bonds

25. What is a dipole?

26. Which element in the periodic table has the greatest electronegativity? Which has the least electronegativity?

27. Which is more polar: a carbon–oxygen bond or a carbon–nitrogen bond?

6.8 Molecular Polarity

28. How can a molecule be nonpolar when it consists of atoms that have different electronegativities?

29. Why do nonpolar substances boil at relatively low temperatures?

30. Which has a greater degree of symmetry: a polar molecule or a nonpolar molecule?

31. Why don't oil and water mix?

Quantitative Questions

32. Ores of manganese, Mn, sometimes contain the mineral rhodochrosite, $MnCO_3$, which is an ionic compound of manganese ions and carbonate ions. How many electrons has each manganese atom lost to make this compound?

33. What is the electric charge on the calcium ion in calcium chloride, $CaCl_2$?

34. Magnesium ions carry a 2+ charge, and chloride ions carry a 1- charge. What is the chemical formula for the ionic compound magnesium chloride?

35. Barium ions carry a 2+ charge, and nitrogen ions carry a 3- charge. What would be the chemical formula for the ionic compound barium nitride?

36. Rank the following bonds in order of increasing polarity:

    a) C—H b) O—H c) N—H

37. Rank the following compounds in order of increasing boiling point:

    a) Fluorine, F_2 b) Hydrogen fluoride, HF
    c) Hydrogen chloride, HCl

38. Rank the following in order of increasing symmetry:

    a) CH_4 b) NH_3 c) H_2O

Solutions (Odd-Numbered)

1. Two electrons fit into the first shell. Eight electrons fit into the second shell.

3. Electron dot structures of elements in the same group have the same number of valence electrons.

5. An ion has a charge and an atom does not.

7. Fluorine only has room for one more electron in its outermost shell.

9. A polyatomic ion is a molecule that carries a charge. A polyatomic ion is formed when ever the molecule loses or gains a hydrogen ion, which is the same thing as a proton.

11. The charge on the O atoms is –2.

13. Each aluminum loses three electrons while each oxygen gains two electrons.

15. An alloy is a mixture composed of two or more metallic elements.

17. High-grade ore on the ocean floor and asteroids.

19. Four electrons are shared within a double covalent bond.

21. The negatively charged oxygen within a polyatomic ion forms only one bond and has three lone pairs as a consequence.

23. A substituent is any atom or lone pair of electrons emanating from a central atom.

25. A dipole is an uneven distribution of electrons in a bond caused by a difference in electronegativity between two atoms.

27. A carbon-oxygen bond is more polar.

29. The molecules of a nonpolar substance tends to have weak attractions to themselves, which causes a low boiling point.

31. Oil and water do not mix because water molecules are more attracted to themselves than to oil molecules.

Quantitative Questions

33. The electronic charge on Ca is +2 in $CaCl_2$.

35. Ba_3N_2 A shortcut way of solving these sorts of problems is to take the charge of one ion and make it the subscript of the opposite ion. For example, take the 2+ charge of the barium and make it the subscript on the nitrogen. Then take the 3- charge on the nitrogen and make that the subscript of the barium: Ba3N2.

37. Fluorine, F_2 (-188°C) $<$ Hydrogen chloride, HCl (-85°C) $<$ Hydrogen fluoride, HF (+19°C)

    Explanation: The fluorine is a nonpolar compound and so should have the lowest boiling point—the attraction between fluorine molecules is so weak. The fluorine has a greater electronegativity than the chlorine, which means that the H—F bond should be more polar than the H—Cl bond. That HF is the most polar suggests that it should also have the highest boiling point.

Chapter 6: Contextual Chemistry Essay

Water Fluoridation

In the early 1900s, a young dentist, Frederick McKay, opened a dental practice in Colorado Springs, Colorado. He soon discovered that many local residents had brown hole-pocked teeth that, despite being unsightly, had a resistance to cavity formation. Following years of investigation, McKay and his colleagues determined that this condition was caused by unusually high levels of naturally occurring calcium fluoride in the drinking water. The condition became known as dental fluorosis. By the 1930s, the U.S. Public Health Service began to study the idea of adding fluoride ions to drinking water. These studies indicated that dental fluorosis could be avoided but resistance to cavity formation maintained at fluoride ion concentrations of about 1 part per million (ppm), which equals 1 mg per liter.

< Figure A
What are the pros and cons of adding fluoride ions to drinking water?

The first municipal fluoridation test programs began in the 1940s, soon after World War II, which was a time when the general public held great trust in the chemical industry as well as government. In Grand Rapids, Michigan, for example, sodium fluoride, once widely regarded as rat poison, was to be added to the municipal drinking water for 15 years, after which time the program would be evaluated for effectiveness. Also, the upstate New York towns of Newburgh and Kingston, which had similar demographic and water profiles, were chosen for a double study in which the water of only one town (Newburgh) would be fluoridated. After 10 years, the rates of dental decay in each city would be compared. Halfway through the Newburgh/Kingston trial, however, the U.S. Public Health Service announced a preliminary 65 percent decrease in dental decay in Newburgh. Word of this and other trials quickly spread, and soon many municipalities were requesting water fluoridation, well before any of these initial trials could be completed.

There were many who advocated against the fluoridation of public drinking water. They argued that fluoride is not a natural nutrient, but a drug and industrial pollutant, and that fluoridation is unethical because individuals are not being asked for their informed consent prior to being given medication. As any physician knows, what is good for one patient is not necessarily good for another. A counterargument to this was that fluoridation of public water provides a benefit to the whole community, including the underprivileged, and not just those who can afford good health care. Opponents to fluoridation, however, also pointed out that the long-term side effects of fluoride were not known. Of particular concern were the risks of skeletal fluorosis, bone cancer, and hypothyroidism. Believing these risks to be minimal compared to the great dental benefits, the government, motivated in large part by professional dental associations, continued to push for fluoridation. Today, about 145 million people in the United States regularly drink fluoridated water.

As a quick Internet search reveals, the fluoridation debate continues to be most vigorous. For example, the American Dental Association, www.ada.org, publishes a well-known booklet that clearly advocates the great benefits and minimal risks of fluoridating municipal drinking water. This is countered by a review article published by the National Center for Biotecnhology Information (NCBI) and entitled: "The Fluoride Debate: The Pros and Cons of Fluoridation." Of note is the fact that drinking water in most of Europe is no longer fluoridated. Most European nations, as well as Japan, halted water fluoridation in the 1970s, after which their rates of dental caries continued to decline, as within the United States. Furthermore, the U.S. Centers for Disease Control and Prevention have noted the lack of evidence supporting the effectiveness of ingested fluoride. Rather, teeth appear to benefit by surface exposure, as occurs when the teeth are brushed with fluoridated toothpaste.

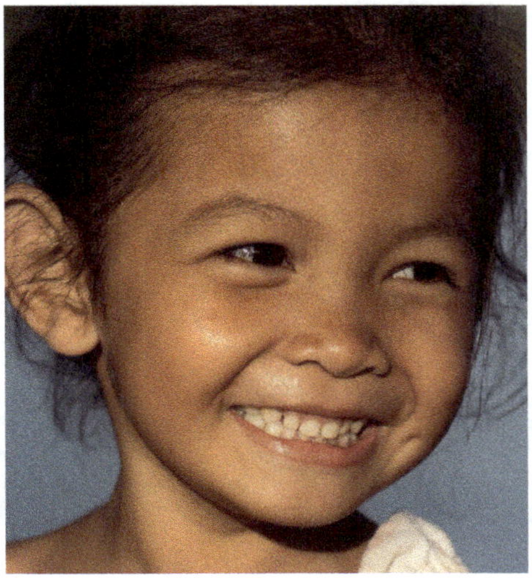

Λ Figure B
Evidences suggests that small amounts of fluoride help growing children to become resistant to tooth decay. Should this fluoride be ingested through public drinking water?

In 1987, the U.S. government–sponsored National Institute of Dental Research (NIDR) examined the teeth of about 39,000 school-children aged 5 to 17 from 84 geographical areas, with and without fluoridated water. Analysis of this data by John Yiamouyiannis, a biochemist and ardent opponent of fluoridation, showed no meaningful statistical differences between children growing up in fluoridated and non-fluoridated areas, except that 5- to 6-year-olds drinking fluoridated water tended to keep their baby teeth longer. Analysis by the NIDR of the same data indicated that children who have always lived in fluoridated areas have 18 percent fewer decayed surfaces than those who have never lived in fluoridated areas.

As the debate continues, one thing remains certain: those familiar with the basic concepts of chemistry and methodologies of science are at a great advantage for understanding the issues and for being able to recognize well-informed arguments and decisions.

Think and Discuss

1. In 1962, the Kettering Laboratory of the University of Cincinnati conducted a study in which dogs were exposed to calcium fluoride dust at a rate simulating human occupational exposure. The results of the study showed significant damage to the dogs' lungs and lymph nodes. This study was funded by an industry group that was seeking evidence to help counter worker claims of crippling skeletal fluorosis. Consequently, the study remained unpublished. Should there be laws that prohibit the suppression of unfavorable research data? Try answering this question from the point of view of a company that might have contributed millions of dollars to the research.

CONCEPT CHECK

Teflon® is a carbon-based molecule with lots of fluorine. Might flossing with Teflon® coated dental floss be a good way of introducing additional fluoride to your teeth?

CHECK YOUR ANSWER

Sources of fluoride ions include the ionic compounds sodium fluoride, NaF, and stannous fluoride, SnF_2. The fluorine within Teflon® is not ionic, rather as a molecule (covalent compound) where fluorine atoms are covalently bound to the carbon atoms. These fluorine atoms cannot escape the molecule to join the structure of tooth enamel.

2. What kind of pressures might a scientist face if she discovered evidence suggesting that fluoridated water had neurotoxic effects in rats? Assume she works for the government, for a government-funded university, for a private research firm not related to dental health, or for a professional dental association.

3. The most common form of fluoride added to municipal drinking water today is hydrofluosilicic acid, H_2SiF_6, which is obtained from the antipollution smokestack scrubbers of the phosphate fertilizer industry. What are the advantages and disadvantages of this system? If water fluoridation were banned, what might become of this hydrofluosilicic acid?

4. How could the Newburgh/ Kingston trials have been designed to ensure that the results were not affected by the bias of pro-fluoridation dentists and government officials? Were government officials justified when they published the preliminary results?

5. Vitamin B12 deficiency affects about one-quarter of the U.S. population and is more com- mon in the elderly. This deficiency is often undetected and can lead to devastating and irreversible complications. Should vitamin B12 be added to municipal drinking water? Why or why not?

6. A town claims that over the course of 10 years, it'll save at least \$5 million and at most \$32 million by not fluoridating its water. How might this town use this saved money to help protect the dental health of all its citizens, including children of under- employed families? Should the saved money instead be returned to the citizens in the form of lower taxes?

7. In many countries lacking elaborate waterworks systems, fluoride is provided to the general population by adding it to commercial table salt. What might be some

of the advantages and disadvantages of this sort of fluoride delivery system?

8. People have been told that the fluoride in their drinking water helps to protect their teeth. To what extent does this prompt them to ignore good dental health habits such as eating healthfully, avoiding sweets, and brushing and flossing regularly?

9. Studies indicate that males exposed to fluoridated drinking water throughout their childhood experience a drop in their intelligence by 3 to 4 IQ points. If this is the case, who should be liable?

Author Responses to Think and Discuss

1. As of the writing of this edition, there are still no laws governing the suppression of privately funded unfavorable research results.

2. This discussion question reflects the real-life case of Phyllis Mullenix, a neurotoxicologist hired by the Forsyth Dental Center in Boston. Mullenix's research showed that, with chronic exposure, fluoride ions could cross the blood-brain barrier and serve as a neurotoxin in rats. She was told not to publish for fear that Forsyth would lose funding from the NIDR. She published anyway within Neurotoxicology & Teratology, and within a few days her contract with Forsyth was not renewed.

3. Initially, the aluminum industry was the leading fluoride polluter turned municipal fluoride provider. With the downturn of domestic aluminum production, the phosphate fertilizer industry became the leading fluoride provider. Phosphate rock is treated with sulfuric acid, which reacts with the fluorite impurities of the phosphate rock to produce silicon tetrafluoride and hydrogen fluoride, HF. These volatile fluoride compounds are sequestered in the scubber as they react with water to form hydroflusilicic acid, also known as fluorosilicic acid or fluosilicic acid. This is a highly corrosive acid that is delivered typically as a 23% aqueous solution. The average cost for a year supply of hydroflusilicic acid is about 15 cents per person. Trace amounts of cadmium, lead, arsenic, and other heavy metals are usually found within this industrial grade hydroflusilicic acid, but their concentrations are determined to be negligible.

4. The general consensus among fluoridation opponents is that these early trials were seriously flawed by the bias of both the dentists and government officials involved. One possible way of improving the reliablity of the results would have been to upgrade the experiment from a double study to a double-blind study where neither the dentists nor the government officials would know which of the two cities was being fluoridated. However, what would prevent an overly-enthusiastic administrator from secretly analyzing the tap water for its fluoride content? The government officials were about to lose their jobs with the end of the Truman presidency. Little wonder that they were tempted to announce positive results only halfway through the trial.

5. Sources of fluoride are cheap and abundant. Not so with vitamin B-12. Furthermore, vitamin B-12 is a reactive organic molecule that would not well survive the trip from the water treatment plant to the faucet, especially with chlorinated water. These technical problems aside, would it be ethical to enforce vitamin B-12 supplements on everyone? It is interesting how different the public attitudes and perceptions were back in the 1940s and 1950s. It is from this very era that municipalities embraced the addition of fluoride, a known toxin, to their drinking water.

6. This money could be used to provide all children with dental sealants, which are far more effective at preventing tooth decay than water fluoridation. Good dental health also comes from a healthy diet consisting of many fruits and vegetables. This money, therefore, could be used to help promote healthy diets. Alternatively, where do we draw the line when it comes to decisions regarding personal health? Some would rather it was not dictated by governmental agencies. Others point out that governmental agencies are very effective at programs designed to inform the general public.

7. Regulating how much fluoridated salt each community should received would be difficult. Also, many people are told to avoid salt because of high blood pressure. Encouraging these people to eat a lot of salt to protect their teeth, therefore, could cause more harm than good.

8. The extent will vary. What everyone should understand is that one cannot rely solely on the fluoride in drinking water to protect teeth from cavities. For good dental health, eating healthy, avoiding sweets, and brushing and flossing regularly are essential.

9. To learn more, search Google Scholar for fluoride as a neurotoxin. Here's a link to a news article concerning this issue: https://www.cnn.com/2019/08/19/health/fluoride-neurotoxin-canada-study/index.html

Chapter 7

How Molecules Mix

THE MAIN IDEA

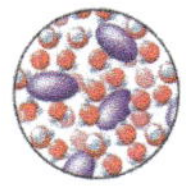

Molecules are "sticky."

Λ Water in these hot springs is loaded with calcium carbonate, a mineral whose solubility decreases with increasing temperature, resulting in these spectacular pools of Pamukkale, Turkey.

In Chapter 6, we focused on how atoms bond together to form ionic or covalent compounds. You understand from earlier chapters that the formation of these compounds is a chemical change. In this chapter, we look at how the ions of ionic compounds and the molecules of covalent compounds—once created—interact to form mixtures, which, as you know, is a physical change.

So, when you stir sugar into water, the sugar crystals disappear, but where do they go? What does it mean to say that one solution is more concentrated, while another is more dilute? Why does a warm soda fizzle in your mouth more than a cold one does? How does soap remove water-resistant grime? What is "hard water," and why does it hinder soap from working? How is water made safe for drinking, and why is it so difficult to create fresh water from ocean water? The answers to these sorts of questions involve an understanding of mixtures.

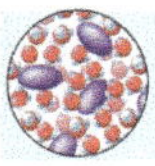

7.1 Dipole Attractions

Recall from Chapter 6 that compounds can have a *dipole*. This means the electrons tend to congregate preferentially to one side of the compound. An extreme case occurs within an ionic compound such as sodium chloride, NaCl, where the bonding electrons spend most all of their time with the chlorine, which becomes the negatively charged chloride ion, Cl^-, while the sodium becomes a positively charged sodium ion, Na^+. Sodium chloride, NaCl, has a rather high melting point of 801°C and boiling point of 1465°C because of the strong attractions among all the highly charged sodium and chloride ions.

A milder form of a dipole occurs with water where the oxygen side of the water molecule slightly negative, while the hydrogen side is slightly positive. A molecule with such a dipole is said to be a *polar molecule.* Water has a relatively high boiling point of 100°C because of the electrical attractions among all the polar water molecules.

We begin this chapter by describing molecular attractions involving dipoles, as shown in **Table 7.1**. Each is electrical in nature, involving the attraction between positive and negative charges.

TABLE 7.1 Molecular Attractions

ATTRACTION	RELATIVE STRENGTH
Ion–dipole	Strongest
Dipole–dipole	↑
Dipole–induced dipole	
Induced dipole–induced dipole	Weakest

Ion–Dipole Attractions

What happens to polar molecules, such as water molecules, when they are near an ionic compound, such as sodium chloride? The opposite charges electrically attract one another. A positive sodium ion attracts the negative side of a water molecule, and a negative chloride ion attracts the positive side of a water molecule. This phenomenon is illustrated in **Figure 7.1.** Such an attraction between an ion and the dipole of a polar molecule is called an ion–dipole attraction.

Ion–dipole attractions are much weaker than ionic bonds. However, a large number of ion–dipole attractions can act collectively to disrupt ionic bonds. This is what happens to sodium chloride in water. Attractions exerted by the water molecules break the ionic bonds and pull the ions away from one another. The result, represented in **Figure 7.2**, is a solution of sodium chloride in water. (A solution in water is called an *aqueous solution*.)

Dipole–Dipole Attractions

An attraction between two polar molecules is called a *dipole–dipole attraction*. An unusually strong dipole–dipole attraction is the **hydrogen bond**. This attraction occurs between molecules that have a hydrogen atom covalently bonded to a small, highly electronegative atom, usually nitrogen, oxygen, or fluorine. Recall from Section 6.7 that the electronegativity of an atom describes how well that atom is able to pull bonding electrons toward itself. The greater the atom's electronegativity, the better it can attract bonding electrons and become negatively charged.

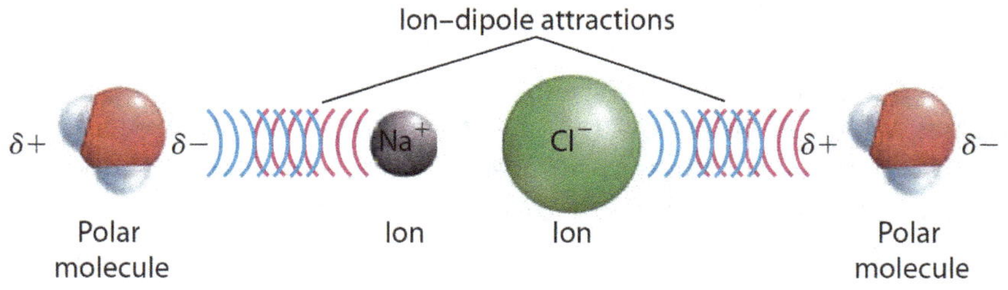

Figure 7.1 >

Electrical attractions are shown as a series of overlapping arcs. The blue arcs indicate negative charge, and the red arcs indicate positive charge.

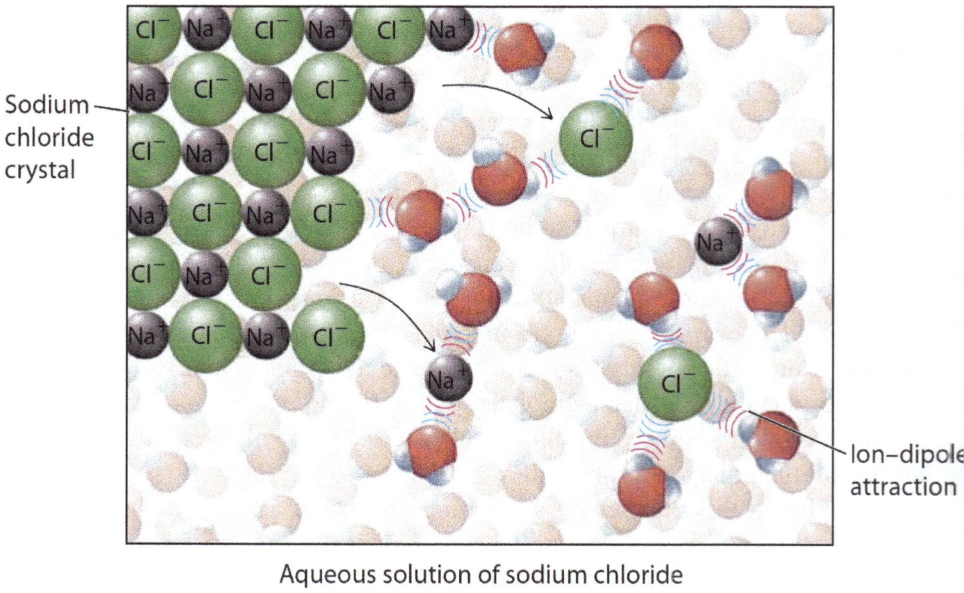

Figure 7.2 >

Sodium and chloride ions tightly bound in a crystal lattice are separated from one another by the collective attraction exerted by many water molecules to form an aqueous solution of sodium chloride.

Look at **Figure 7.3** to see how hydrogen bonding works. The hydrogen side of a polar molecule (water, in this example) has a partial positive charge because the more electronegative oxygen atom pulls more strongly on the electrons of the covalent bond. The hydrogen is therefore electrically attracted to a pair of non-bonding electrons on the partially negatively charged atom of another molecule (in this case, another water molecule). This mutual attraction between hydrogen and the negatively charged atom of another molecule is a hydrogen bond.

Even though the hydrogen bond is much weaker than any covalent or ionic bond, the effects of hydrogen bonding can be very pronounced. As we explore in Chapter 8, water owes many of its properties to hydrogen bonds. The hydrogen bond is also of great importance in the chemistry of the large molecules, such as DNA and proteins, which we discuss in Chapter 13.

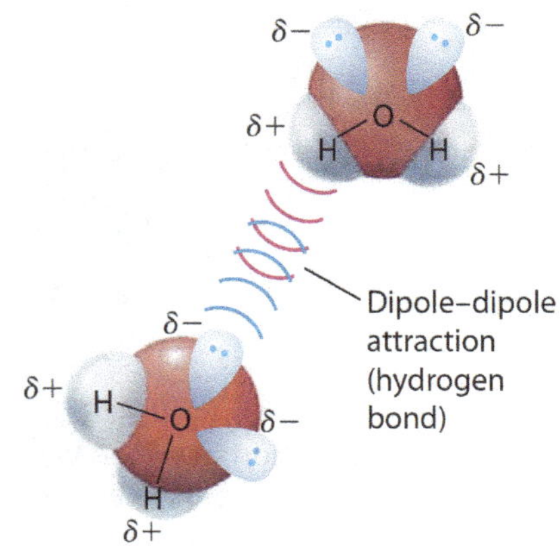

Λ Figure 7.3

The dipole–dipole attraction between two water molecules is a hydrogen bond because it involves hydrogen atoms bonded to highly electronegative oxygen atoms.

Dipole–Induced Dipole Attractions

In many molecules, the electrons are distributed evenly, so there is no dipole. The oxygen molecule, O_2, is an example. Such a nonpolar molecule can be induced to become a *temporary dipole*, however, when it is brought close to a water molecule (or to any other polar molecule), as **Figure 7.4** illustrates. The slightly negative side of the water molecule pushes the electrons in the oxygen molecule away. The result is a temporarily uneven distribution of electrons called an **induced dipole**. The resulting attraction between the permanent dipole (water) and the induced dipole (oxygen) is a *dipole–induced dipole* attraction.

Remember, induced dipoles are only temporary. If the water molecule in **Figure 7.4b** were removed, the oxygen molecule would return to its normal, non-polar state. In general, dipole–induced dipole attractions are much weaker than dipole–dipole attractions. But dipole–induced dipole attractions are strong enough to hold relatively small quantities of oxygen dissolved in water, as depicted in **Figure 7.5**. This attraction between water and molecular oxygen is vital for fish and other forms of aquatic life that rely on molecular oxygen dissolved in water.

Dipole–induced dipole attractions are also responsible for holding plastic wrap to glass, as shown in **Figure 7.6**. These wraps are made of very long non- polar molecules that are induced to have dipoles when placed in contact with glass, which is highly polar. As we will discuss next, the molecules of a nonpolar material, such as plastic wrap, can also induce dipoles among themselves. This explains why plastic wrap sticks not only to polar materials such as glass but also to itself.

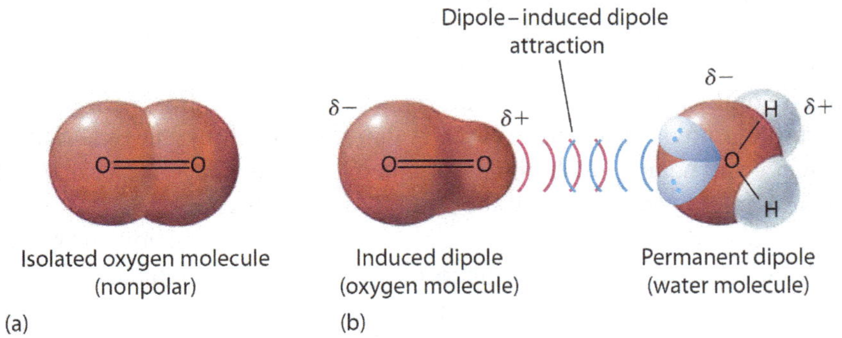

< Figure 7.4

(a) An isolated oxygen molecule has no dipole; its electrons are distributed evenly. (b) An adjacent water molecule induces a redistribution of electrons in the oxygen molecule. (The slightly negative side of the oxygen molecule is shown as larger than the slightly positive side because the slightly negative side contains more electrons.)

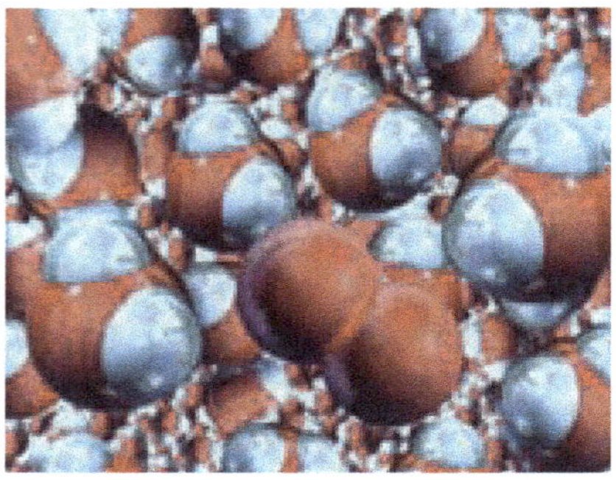

Figure 7.5

The electrical attraction between water and oxygen molecules is relatively weak, which explains why not much oxygen is able to dissolve in water. For example, water fully aerated at room temperature contains only about 1 oxygen molecule for every 200,000 water molecules. The gills of a fish, therefore, must be highly efficient at extracting this molecular oxygen from water.

Figure 7.6

Temporary dipoles induced in the normally nonpolar molecules in plastic wrap make it stick to glass.

> **CONCEPT CHECK**
> How does the electron distribution in an oxygen molecule change when the hydrogen side of a water molecule is nearby?
>
> **CHECK YOUR ANSWER**
> Because the hydrogen side of the water molecule is slightly positive, the electrons in the oxygen molecule are pulled toward the water molecule, inducing a temporary dipole in the oxygen molecule.

READING CHECK
Is it possible for the electrons of an atom or a nonpolar molecule to be bunched to one side?

Induced Dipole–Induced Dipoles Attractions

Individual atoms and nonpolar molecules, on average, have a fairly even distribution of electrons. Because of the randomness of electron distribution, however, at any given moment, the electrons in an atom or a nonpolar molecule may be bunched to one side. The result is a temporary dipole, as shown in **Figure 7.7**.

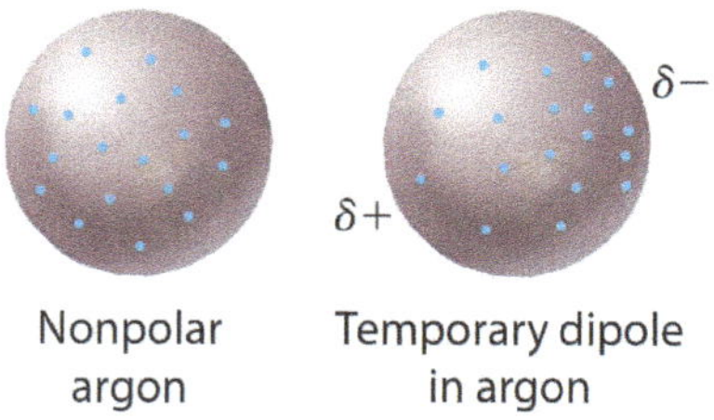

Figure 7.7

The electron distribution in an atom is normally even. At any given moment, however, the electron distribution may be somewhat uneven, resulting in a temporary dipole.

Just as the permanent dipole of a polar molecule can induce a dipole in a nonpolar molecule, a temporary dipole can do the same thing. This gives rise to the relatively weak *induced dipole–induced dipole* attraction, illustrated in **Figure 7.8**.

Induced dipole–induced dipole attractions (sometimes called *dispersion forces*) help explain why natural gas is a gas at room temperature but gasoline is a liquid. The major component of natural gas is methane, CH_4, and one of the major components of gasoline is octane, C_8H_{18}. We can see in **Figure 7.9** that the number of induced dipole–induced dipole attractions between two methane molecules is appreciably less than the number between two octane molecules. You know that two small pieces of Velcro are easier to pull apart than two long pieces. Like short pieces of Velcro, methane molecules can be pulled apart with little effort. That's why methane has a low boiling point, –161°C, and is a gas at room temperature. Octane molecules, like long strips of Velcro, are relatively difficult to pull apart because of the larger number of induced dipole–induced dipole attractions. The boiling point of octane, 125°C, is therefore much higher than that of methane, and octane is a liquid at room temperature. (The greater mass of octane also plays a role in making its boiling point higher.)

Induced dipole–induced dipole attractions also explain how the gecko can race up a glass wall and support its entire body weight with only a single toe. A gecko's feet are covered with billions of microscopic hairs called *spatulae*, each of which is about 1/300 as thick as a human hair. The

CONCEPT **CHECK**

Distinguish between a dipole–dipole attraction and a dipole–induced dipole attraction.

CHECK YOUR ANSWER

The dipole–dipole attraction is stronger and involves two permanent dipoles. The dipole–induced dipole attraction is weaker and involves a permanent dipole and a temporary one.

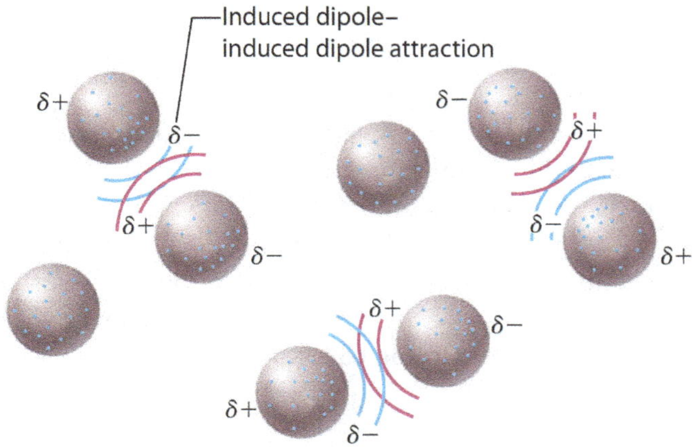

< Figure 7.8

Because the normally even distribution of electrons in atoms can momentarily become uneven, atoms can be attracted to one another through induced dipole–induced dipole attractions.

force of attraction between these hairs and the wall is the weak induced dipole–induced dipole attraction. But because there are so many hairs, the surface area of contact is relatively great; hence, the total force of attraction is enough to prevent the gecko from falling (**Figure 7.10**). Look for synthetic dry glue and dry, yet sticky, fabrics (allowing humans to stick to walls) based on the principle of gecko adhesion.

Temporary dipoles are more significant for larger atoms. A reason for this is that their outermost electrons are relatively far from the nucleus as well as from each other. As a result, these electrons are easily "pushed around," as described in **Figure 7.11.** Thus, larger atoms are sometimes described as "soft"—more like a marshmallow than a marble. The technical term for this property is *polarizability*. We say that a larger atom is more "polarizable," which means it can form an induced dipole more easily.

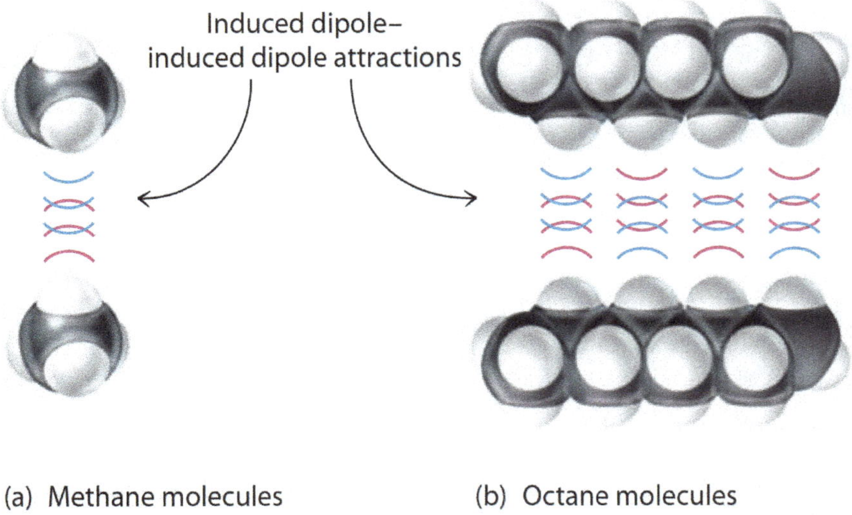

< Figure 7.9

Two nonpolar methane molecules are attracted to each other by induced dipole–induced dipole attractions, but there is only one attraction per molecule. (b) Two nonpolar octane molecules are similar to methane molecules, but they are longer. The number of induced dipole–induced dipole attractions between these two molecules is therefore greater.

FOR YOUR INFORMATION

Dipole–induced dipole attractions are sometimes called Debye forces, while induced dipole–induced dipole attractions are sometimes called London dispersion forces, or more simply London forces. Each of these are the last names of the 20th-century scientists who first described them.

Figure 7.10 >

If the gecko's foot is so sticky, how does the gecko keep its feet clean? Answer: The gecko's foot is extremely nonpolar. Dirt may stick to it briefly, but after a few steps, the dirt sticks better to the surface upon which the gecko walks.

Fluorine is one of the smallest atoms, which means it is a "hard" atom (like a marble) and not very polarizable. Nonpolar molecules made with fluorine atoms exhibit only very weak induced dipole–induced dipole attractions. This is the principle behind the Teflon® nonstick surface. The Teflon® molecule, part of which is shown in **Figure 7.12**, is a long chain of carbon atoms chemically bonded to fluorine atoms, and the fluorine atoms exert essentially no attractions on any material in contact with the Teflon® surface—scrambled eggs in a frying pan, for instance.

CONCEPT CHECK

Distinguish between a dipole–induced dipole attraction and an induced dipole–induced dipole attraction.

CHECK YOUR ANSWER

The dipole–induced dipole attraction is stronger and involves a permanent dipole and a temporary one. The induced dipole–induced dipole attraction is weaker and involves two temporary dipoles.

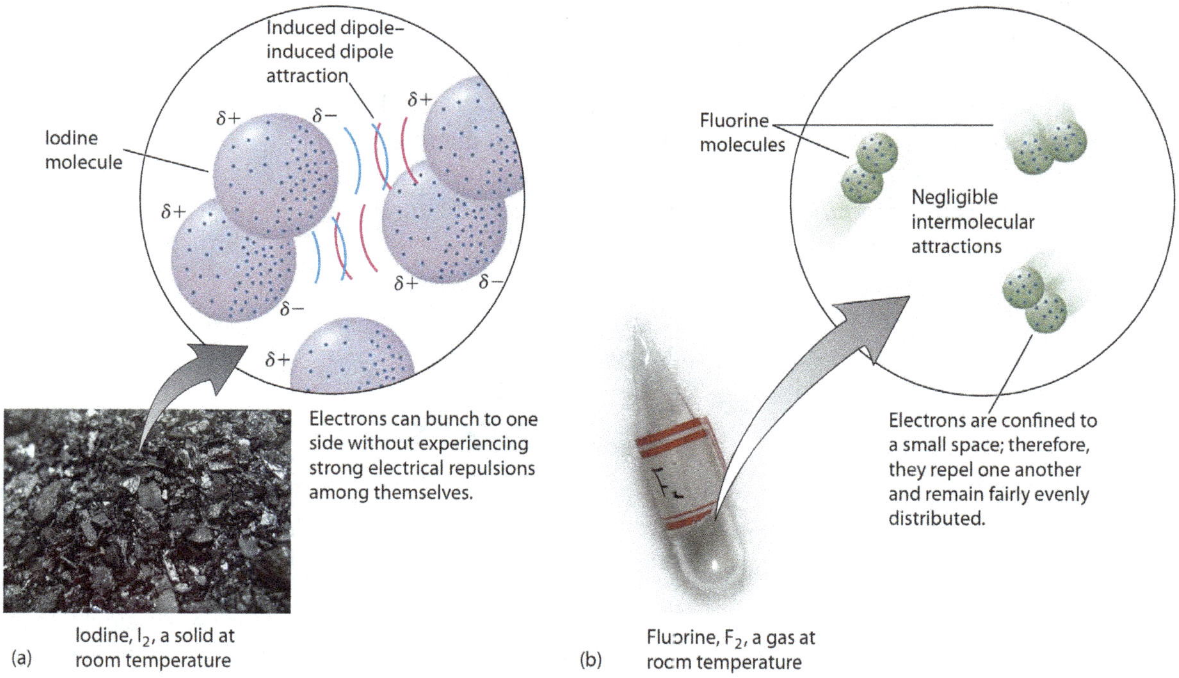

Λ Figure 7.11

(a) Temporary dipoles form more readily in larger atoms, such as those in an iodine molecule, because in larger atoms, electrons are more loosely held and can bunch to one side and still be relatively far apart from one another. (b) In smaller atoms, such as those in a fluorine molecule, electrons are tightly held and cannot bunch to one side so well because the repulsive electric force increases as the electrons get closer together.

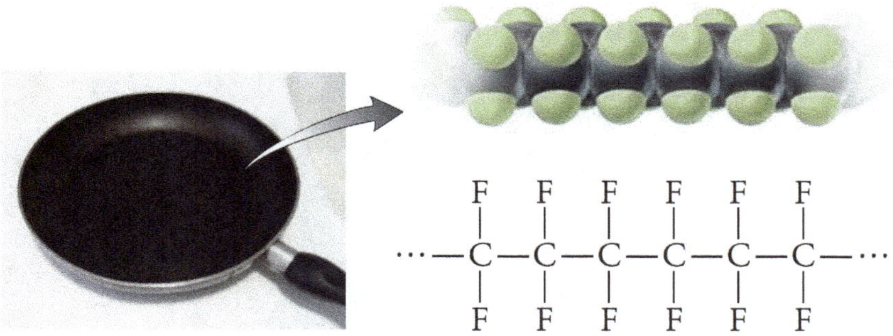

Figure 7.12

Few things stick to Teflon® because of the high proportion of fluorine atoms that it contains. The structure depicted here is only a portion of the full length of the molecule.

CONCEPT CHECK

Molecules of methanol, CH_3OH are not much larger than molecules of methane, CH_4. Methanol, however, is a liquid at room temperature while methane is a gas. Suggest why.

CHECK YOUR ANSWER

The polar oxygen–hydrogen covalent bond in each methanol molecule leads to hydrogen bonding between molecules. These relatively strong interparticle attractions hold methanol molecules together as a liquid at room temperature.

7.2 Solutions

What happens when table sugar, known chemically as sucrose, is stirred into water? Is the sucrose destroyed? We know it isn't because it sweetens the water and all we have to do is evaporate the water and the sugar reappears in its solid form. When sucrose is dissolved in water, does it disappear because it somehow ceases to occupy space or because it fits within the nooks and crannies of the water? Not so, for the addition of sucrose changes the volume. This may not be noticeable at first, but if you continue adding sucrose to a glass of water, you'll see that the water level rises, just as it would if you were adding sand.

Sucrose stirred into water loses its crystalline form. Each sucrose crystal consists of billions upon billions of sucrose molecules packed neatly together. When the crystal is exposed to water (as was first shown in Figure 3.26 and is shown again here in **Figure 7.13**), an even greater number of water molecules pull on the sucrose molecules via hydrogen bonds formed between the sucrose molecules and the water molecules. With a little stirring, the sucrose molecules soon mix throughout the water. In place of sucrose crystals and water, we have a homogeneous mixture of sucrose molecules in

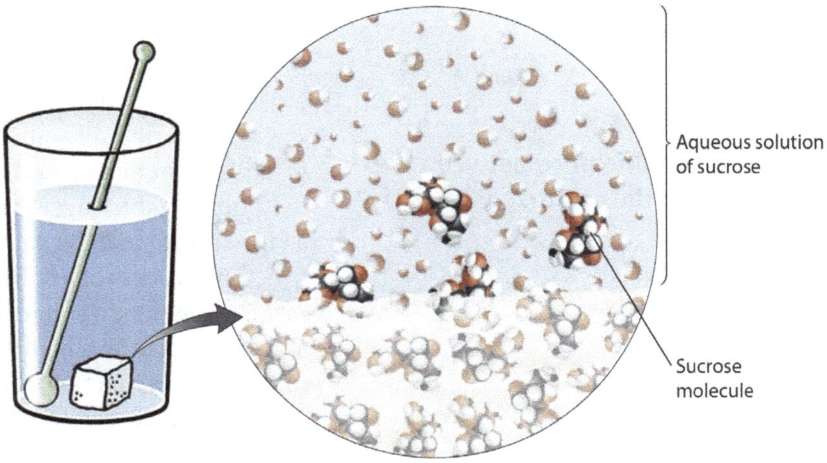

Figure 7.13

Water molecules pull the sucrose molecules in a sucrose crystal away from one another. This pulling away does not, however, affect the covalent bonds within each sucrose molecule, which is why each dissolved sucrose molecule remains intact as a single molecule.

(a) (b) (c)

∧ Figure 7.14

Solutions may occur in (a) the solid phase, (b) the liquid phase, or (c) the gaseous phase.

water. As discussed earlier, *homogeneous* means that a sample taken from any part of a mixture is the same as a sample taken from any other part. In our sucrose example, this means that the sweetness of the first sip of the solution is the same as the sweetness of the last sip.

Recall that a homogeneous mixture consisting of a single phase is called a *solution*. Sugar water is a solution in the liquid phase. Solutions aren't always liquids, however. They can also be solid or gaseous, as **Figure 7.14** shows. Gemstones are solid solutions. A ruby, for example, is a solid solution of trace quantities of red chromium compounds in transparent aluminum oxide. A blue sapphire is a solid solution of trace quantities of light green iron compounds and blue titanium compounds in aluminum oxide. Another important example of solid solutions is metal alloys, which are mixtures of different metallic elements. The alloy known as brass is a solid solution of copper and zinc. Similarly, the alloy stainless steel is a solid solution of iron, chromium, nickel, and carbon.

An example of a gaseous solution is the air we breathe. By volume, this solution is 78 percent nitrogen gas, 21 percent oxygen gas, and 1 percent other gaseous materials. The air we exhale is a gaseous solution of 75 percent nitrogen, 14 percent oxygen, 5 percent carbon dioxide, and around 6 percent water vapor.

READING CHECK

What is the difference between the solvent and the solute?

In describing solutions, the component present in the largest amount is the **solvent** and any other components are **solutes**. For example, when a teaspoon of table sugar is mixed with 1 liter of water, we identify the water as the solvent and the sugar as the solute.

The process of a solute's mixing with a solvent is called **dissolving**. To make a solution, a solute must dissolve in a solvent; that is, the solute and solvent must form a homogeneous mixture. Whether one material dissolves in another is a function of the electrical attractions between their molecules. The stronger these electrical attractions are, the greater the likelihood that dissolving will occur.

There is a limit to how much of a given solute can be dissolved in a given solvent, as **Figure 7.15** illustrates. We know that when you add sugar to a glass of water, for example, the sugar rapidly dissolves. As you continue to add sugar, however, there comes a point when it no longer dissolves. Instead, it collects at the bottom of the glass, even after stirring. At this point, the water is saturated with sugar, meaning that the water cannot accept any more sugar. When this happens, we have what is called a **saturated solution**, defined as a solution in which no more solute can be dissolved. A solution that has not reached the limit of solute that will dissolve is called an **unsaturated solution.**

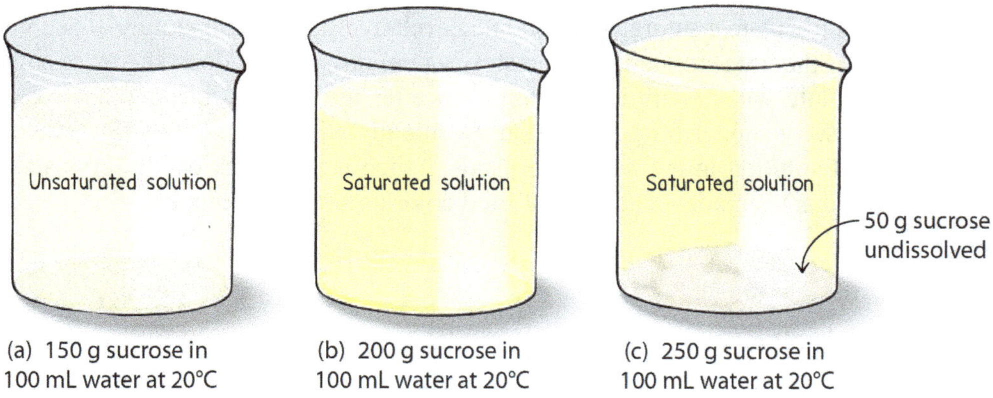

Λ Figure 7.15

A maximum of 200 grams of sucrose dissolves in 100 milliliters of water at 20°C. (a) Mixing 150 grams of sucrose in 100 milliliters of water at 20°C produces an unsaturated solution. (b) Mixing 200 grams of sucrose in 100 milliliters of water at 20°C produces a saturated solution. (c) If 250 grams of sucrose is mixed with 100 milliliters of water at 20°C, 50 grams of sucrose remains undissolved. (As we discuss later, the concentration of a saturated solution varies with temperature, which can impact volume.)

CONCEPT CHECK

What is the solvent in the gaseous solution we call air?

CHECK YOUR ANSWER

Nitrogen is the solvent, because it is the component present in the greatest quantity.

There are some special cases where more solute can be dissolved beyond the point of saturation. Such a solution is said to be **supersaturated**. Interestingly, it's fairly easy to produce a supersaturated solution of sugar (sucrose) in water. For sugar, the point of saturation is highly dependent upon the temperature. Consider that while about 200 g of sucrose can dissolve in 100 mL of water at 20°C, at a higher temperature of about 100°C, it's possible to dissolve almost 500 g in 100 mL of water, as can be seen in sugar's solubility curve:

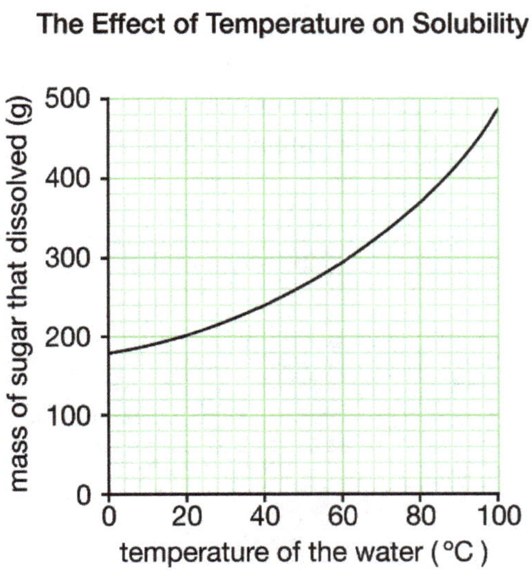

The first step to making a supersaturated solution of sugar in water is to create a saturated solution at a relatively high temperature. Then you let the hot saturated solution slowly cool down. As it does, the extra sugar solute will remain in solution even after the solution has cooled—you'll have a supersaturated solution. Eventually, that extra sugar will come out of solution in the form of crystals, which is how rock candy (crystals of sugar) is made—add a bit of food coloring for further appeal.

FOR YOUR INFORMATION

Be it large or small, an individual sugar crystal is transparent. A teaspoon of table sugar appears white because of the way light gets scattered as it passes in and out of the numerous tiny crystals at numerous different angles. Technically speaking, when table sugar dissolves in water, it is this scattering effect that disappears, not the sugar itself.

Two other categories of solutions include electrolyte and nonelectrolyte. An **electrolyte** is any dissolved ion, such as a sodium ion, Na^+, or a chloride ion, Cl^-, that allows for the flow of an electric current through the solution. The greater the concentration of electrolytes, the stronger the ability of the solution to conduct an electric current. Think of this in reverse: If you find that a solution is able to conduct electricity, evidenced by the brightness of a light bulb, then you can conclude it has electrolytes. The concentration of those electrolytes, in turn, can be deduced based upon how brightly the light bulb shines. But it need not be a light bulb in the circuit. Rather it can be a meter of some sort, which can be calibrated to measure the concentration of the electrolyte solution.

Electrolyte solutions tend to be those having water as the solvent. Why? Because water's polarity allows for the easy dissolving of various ions. An *electrolyte solution* is any solution containing electrolytes. Your blood is an example of an electrolyte solution containing electrolytes such as ions of sodium, calcium, and potassium. For your body to remain healthy, the concentration of these electrolytes needs to be just right—not too high and not too low. After a vigorous aerobic workout, you may be sweating. As you sweat you're losing not only water but the electrolytes dissolved in that water, which is why many sports drinks are fortified with electrolytes.

Lastly, a nonelectrolyte solution is one were there are no electrolytes to conduct electricity. Pure water is mostly nonelectrolytic—it won't allow the passage of an electric current. However, obtaining absolutely pure water is rather difficult (to impossible) because so many things dissolve in water quite readily, including the air. And, as we discuss in Chapter 10, absolutely pure water, by its nature, has an electrolyte concentration of 0.0000001 moles per liter. But ignoring these technicalities, nicely purified water can be regarded as nonelectrolytic. So, let's add some sugar to this purified water. Will it then readily conduct electricity? The answer is no. Why not? Because when sugar dissolves in water the sugar molecules remain sugar molecules—they don't dissociate into ions like a salt. A *nonelectrolytic solution* is any solution not containing electrolytes. Sugar dissolved in purified water is a good example. More common in laboratory research are solutions where water is not the solvent, but some non-polar solvent, such as methylene chloride, CH_2Cl_2. Organic compounds dissolved in such a solvent are fully nonelectrolytic solutions.

CONCEPT **CHECK**

What is an electrolyte?

CHECK YOUR ANSWER

And electrolyte is any dissolved ion, such as a sodium ion, Na^+, or a chloride ion, Cl^-, that allows for the flow of an electric current through the solution.

7.3 Concentration and the Mole

The quantity of solute dissolved in a solution is described in mathematical terms by the solution's **concentration**, which is the amount of solute dissolved per amount of solution:

READING CHECK

What does concentration measure?

$$\text{concentration} = \frac{\text{amount of solute}}{\text{amount of solution}}$$

For example, a sucrose–water solution may have a concentration of 1 gram of sucrose for every liter of solution. This can be compared with concentrations of other solutions. A sucrose–water solution containing 2 grams of sucrose per liter of solution, for example, is more concentrated, and one containing only 0.5 gram of sucrose per liter of solution is less concentrated, or more dilute.

Chemists are often more interested in the number of solute particles in a solution than in the number of grams of solute. Submicroscopic particles, however, are so very small that the number of them in any observable sample is incredibly large. To avoid awkwardly large numbers, scientists use a unit called the *mole*. One **mole** of any type of particle is equal to 6.02×10^{23} particles. (This super large number is 602 billion trillion, or 602,000,000,000,000,000,000,000 particles.) One mole of gold atoms, for example, is 6.02×10^{23} gold atoms, and 1 mole of sucrose molecules is 6.02×10^{23} sucrose molecules. Interestingly, the term mole is derived from the Latin word *moles*, meaning heap, mass, or pile.

Even if you've never heard the term mole before now, you are already familiar with the basic idea. Saying "one mole" is just a shorthand way of saying "six point oh two times ten to the twenty-third particles." Just as "a couple of" means two of something and "a dozen of" means 12 of something, "a mole of" means 6.02×10^{23} of some elementary unit, such as atoms, molecules, or ions. It's as simple as that:

- a couple of coconuts = 2 coconuts
- a dozen donuts = 12 donuts
- a mole of molecules = 6.02×10^{23} molecules

One mole of gold atoms, for example, is 6.02×10^{23} gold atoms, and 1 mole of sucrose molecules is 6.02×10^{23} sucrose molecules. A stack containing "1 mole" of pennies would reach a height of about 97,000 light years, which is roughly equal to the diameter of our galaxy, the Milky Way. And "1 mole" of marbles would be enough to fill the volume of all Earth's oceans.

But sucrose molecules are so very small that 6.02×10^{23} of them are in only 342 grams of sucrose, which is about a cupful. Thus, because 342 grams of sucrose contains 6.02×10^{23} molecules of sucrose, we can use our shorthand wording and say that 342 grams of sucrose contains 1 mole of sucrose. As **Figure 7.16** shows, an aqueous solution that has a concentration of 342 grams of sucrose per liter of solution has a concentration of 6.02×10^{23} sucrose molecules per liter of solution. This would be a concentration of 1 mole of sucrose per liter of solution. While the number of *grams* tells you the mass of solute in a given solution, and the number of *moles* indicates the actual number of molecules.

A common unit of concentration used by chemists is **molarity**, which is the solution's concentration expressed in moles of solute per liter of solution:

$$\text{molarity} = \frac{\text{number of moles of solute}}{\text{liters of solution}}$$

A solution that contains 1 mole of solute per liter of solution is a 1-molar solution, which is often abbreviated 1 *M*. A 2-molar (2 *M*) solution contains 2 moles of solute per liter of solution.

fyi FOR YOUR INFORMATION

It would take 11.6 days of nonstop counting to count to a million. To count to a billion would take 31.7 years. To count to a trillion would take 31,700 years! Counting to a trillion over and over 602 billion times would take about 2 million times the estimated age of the universe. In short, 602 billion trillion, 6.02×10^{23}, is an inconceivably large number.

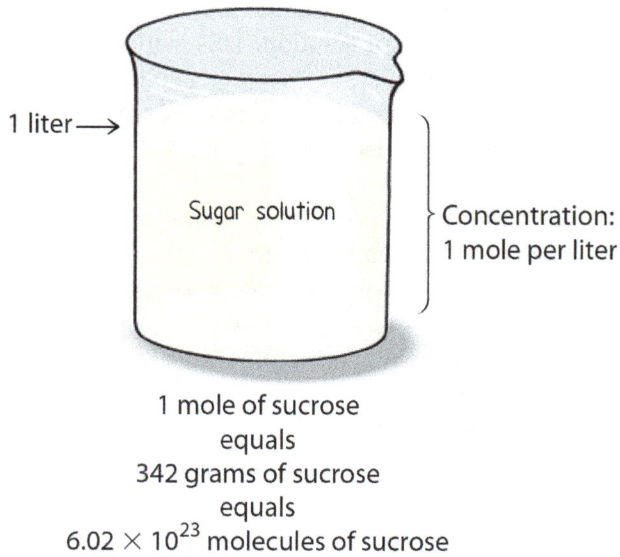

Figure 7.16 >

An aqueous solution of sucrose that has a concentration of 1 mole of sucrose per liter of solution contains 6.02×10^{23} sucrose molecules (342 grams) in every liter of solution.

The difference between referring to the number of molecules of solute and referring to the number of grams of solute can be illustrated by the following question. A saturated aqueous solution of sucrose contains 200 grams of sucrose and 100 grams of water. Which is the solvent: sucrose or water?

As shown in **Figure 7.17**, there are 3.5×10^{23} molecules of sucrose in 200 grams of sucrose, but there are almost 10 times as many molecules of water in 100 grams of water—3.3×10^{24} molecules. As defined earlier, the solvent is the component present in the largest amount, but what do we mean by amount? If amount means number of molecules, then water is the solvent. If amount means mass, then sucrose is the solvent. So, the answer depends on how you look at it. From a chemist's point of view, *amount* typically means the number of molecules, so water is the solvent in this case.

Component	Mass	Number of molecules
Sucrose	200 g	3.5×10^{23}
Water	100 g	3.3×10^{24}

< Figure 7.17

Although 200 grams of sucrose is twice as massive as 100 grams of water, there are about 10 times as many water molecules in 100 grams of water as there are sucrose molecules in 200 grams of sucrose. How can this be? Each water molecule is about one-twentieth as massive (and much smaller) than each sucrose molecule

CONCEPT **CHECK**

1. How many moles of sucrose are in 0.5 liter of a 2-molar solution? How many molecules of sucrose is this?

2. Does 1 liter of a 1-molar solution of sucrose in water contain 1 liter of water, less than 1 liter of water, or more than 1 liter of water?

CHECK YOUR ANSWER

First, you need to understand that 2-molar means 2 moles of sucrose per liter of solution. To obtain the amount of solute, you should multiply solution concentration by amount of solution:

(2 moles/L)(0.5 L) = 1 mole, which is the same as 6.02×10^{23} molecules.

The definition of molarity refers to the number of liters of solution, not to the number of liters of solvent. When sucrose is added to a given volume of water, the volume of the solution increases. So, if 1 mole of sucrose is added to 1 liter of water, the result is more than 1 liter of solution. Therefore, 1 liter of a 1-molar solution requires less than 1 liter of water.

Calculation Corner: Concentrating on Solutions

From the formula for the concentration of a solution, we can derive equations for the amount of solute and the amount of solution:

$$\text{concentration of solution} = \frac{\text{amount of solute}}{\text{volume of solution}}$$

$$\text{amount of solute} = \text{concentration of solution} \times \text{volume of solution}$$

$$\text{volume of solution} = \frac{\text{amount of solute}}{\text{concentration of solution}}$$

In solving for any of these values, the units must always match. If concentration is given in grams per liter of solution, for example, the amount of solute must be in grams and the amount of solution must be in liters. Note that these equations are set up for calculating the volume of solution rather than the volume of solvent. The volume of solution is greater than the volume of solvent because, in addition to containing the solvent, the solution also contains the solute. As discussed at the beginning of Section 7.2, for example, the volume of an aqueous solution of sucrose depends not only on the volume of water but also on the volume of dissolved sucrose.

EXAMPLE 1

How many grams of sucrose are in 3 liters of an aqueous solution that has a concentration of 2 grams of sucrose per liter of solution?

ANSWER 1

This question asks for amount of solute, so you should use the second of the three formulas given above:

$$\text{amount of solute} = \frac{2\text{ g}}{1\text{ L}} \times 3\text{ L} = 6\text{ g}$$

EXAMPLE 2

A solution you are using in an experiment has a concentration of 10 grams of solute per liter of solution. If you pour enough of this solution into an empty laboratory flask so that the flask contains 5 grams of the solute, how many liters of the solution have you poured into the flask?

ANSWER 2

This question asks for amount of solution, and you will want to use the third formula:

$$\text{volume of solution} = \frac{5\text{ g}}{10\text{ g/L}} = 0.5\text{ L}$$

YOUR TURN

1. At 20°C, a saturated solution of sodium chloride in water has a concentration of about 380 grams of sodium chloride per liter of solution. How many grams of sodium chloride are required to make 3 liters of a saturated solution?

2. A student is told to use 20 grams of sodium chloride to make an aqueous solution with a concentration of 10 grams of sodium chloride per liter of solution. How many liters of solution does she end up with?

Dilutions

It happens fairly often that a working chemist needs to take a concentrated solution and dilute it to some known lesser concentration of a given volume. For that working chemistry the following formula is their great friend:

$$(before)\quad C_1 \times V_1 = C_2 \times V_2 \quad (after)$$

Where C_1 is the concentration and V_1 is the volume before the dilution and C_2 is the concentration and V_2 is the volume after the dilution.

Here is the key thing to understand with any dilution: The amount of solute remains the same. This is what the above equation is saying. You should recognize that concentration times volume gives the amount of solute. For example, (moles/liter)(liter) = moles. Thus, the amount of solute before equals the amount of solute after the dilution. This is what the equation is telling us with that equals sign. So let's put this equation to practical use.

Say you have a 6.0 *M* aqueous solution of hydrochloric acid, HCl. From this you need to create 0.50 L of a 2.0 *M* HCl solution. How do you proceed? How much 6.0 *M* HCl should you use? How much water should you add? And how exactly do you add it? The equation comes to your rescue telling you exactly how much of each. Start by writing down all you know and don't know:

What you have:

$C_1 = 6.0\ M\ HCl$

$V_1 = x$

What you need:

$C_2 = 2.0\ M\ HCl$

$V_2 = 0.50\ L$

With three known values you can calculate the fourth unknown value:

$$(before)\quad C_1 \times V_1 = C_2 \times V_2 \quad (after)$$

$$(before)\quad (6.0\ M\ HCl)(x\ L) = (2.0\ M\ HCl)(0.50\ L)\quad (after)$$

Solve for "*x*" and you'll find you need 0.42 liters of this 6.0 *M* solution. To make your diluted solution, you add 0.42 liters of 6.0 *M* HCl to a beaker that you then fill with water to the 0.50 liter mark. You will thus be diluting a small portion of the 6.0 *M* HCl solution to create the 0.50 liters of 2.0 *M* HCl that you need.

Safety Note: How you mix solutions can make a big difference! HCl is a strong acid and for safety reasons it's important to "do what you oughta and add acid to water". In this way, any splashes will be primarily water and not the strong acid. There's also heat released when concentrated HCl and water are mixed. Add the water to the HCl and the HCl gets hot. Not

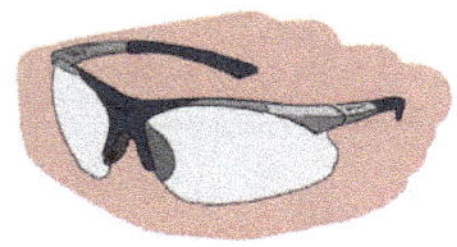

good! Far better to add the HCl to the water so that the water absorbs this heat. So, to make your dilution, you would first add a good portion of the water to the beaker, then carefully pour in the 0.42 liters of 6 M HCl. Only then do you carefully pour in the additional water to bring it to the 0.50 liter mark. And, of course, one always wears eye protection when working with chemicals.

CONCEPT **CHECK**

You need 0.75 liters of a 1 *M* sodium chloride, NaCl, solution. On hand is a solution of 3 *M* NaCl. Describe how you would make this solution. Note: A sodium chloride solution is much safer to work with than a concentrated hydrochloric acid solution. But still exercise caution!

CHECK YOUR ANSWER

You need 0.75 liters of a 1.0 *M* sodium chloride, NaCl, solution. On hand is a solution of 3.0 *M* NaCl. Describe how you would make this solution. Note: A sodium chloride solution is much safer to work with than a concentrated hydrochloric acid solution. But still exercise caution!

$$(before)\quad C_1 \times V_1 = C_2 \times V_2 \quad (after)$$

$$(before)\quad (3.0\ M\ \text{NaCl})(x\ \text{L}) = (1.0\ M\ \text{NaCl})(0.75\ \text{L}) \quad (after)$$

Solve for "x" and you'll find you need 0.25 liters of the concentrated 3.0 *M* solution. To make your diluted solution, you add 0.25 liters of 3.0 *M* NaCl to a beaker that you then fill with water to the 0.75 liter mark. You will thus be diluting a small portion of the 3.0 *M* NaCl solution to create the 0.75 liters of 1.0 *M* NaCl that you need.

Calculation Corner: Concentrating on Dilutions

$$(before)\quad C_1 \times V_1 = C_2 \times V_2 \quad (after)$$

YOUR TURN

1. Available to you is a bottle of 6.00 *M* sodium hydroxide, NaOH. How much of this 6.00 *M* NaOH solution is required to make 465 mL of a 2.00 *M* solution?

2. If you're given a dilution problem with all volume data provided in milliliters, why is it not necessary to first convert to liters?

3. You have 55.0 mL of a 3.00 *M* potassium bromide, KBr, solution. You need to dilute this to a 1.00 *M* solution. How many milliliters of water do you add?

7.4 Solubility

The **solubility** of a solute is its ability to dissolve in a solvent. As can be expected, solubility mainly depends on the attractions between the fundamental particles of the solute and solvent. If a solute has any appreciable solubility in a solvent, then that solute is said to be **soluble** in that solvent.

Solubility also depends on attractions of solute particles for one another and attractions of solvent particles for one another. As shown in **Figure 7.18**, for example, there are many polar hydrogen–oxygen bonds in a sucrose molecule. Sucrose molecules, therefore, can form multiple hydrogen bonds with one another. These hydrogen bonds are strong enough to make sucrose a solid at room temperature and to give it the relatively high melting point of 185°C. In order for sucrose to dissolve in water, the water molecules must first pull sucrose molecules away from one another. This puts a limit on the amount of sucrose that can dissolve in water— eventually, a point is reached at which there are not enough water molecules to separate the sucrose molecules from one another. As we discussed in Section 7.2, this is the point of saturation and any additional sucrose added to the solution does not dissolve.

READING CHECK

Solubility depends upon what?

When the molecule-to-molecule attractions among solute molecules are comparable to the molecule-to-molecule attractions among solvent molecules, there is no practical point of saturation. As shown in **Figure 7.19**, for example, the hydrogen bonds among water molecules are about as strong as those between ethanol molecules. These two liquids therefore mix together quite well in just about any proportion. We can even add ethanol to water until the ethanol, rather than the water, can be considered the solvent.

A solute that has no practical point of saturation in a given solvent is said to be *infinitely soluble* in that solvent. Ethanol, for example, is infinitely soluble in water. Also, all gases are generally infinitely soluble in other gases, because they can be mixed together in just about any proportion.

Let's now look at the other extreme of solubility, where a solute has very little solubility in a given solvent. An example is oxygen, O_2, in water. In contrast to sucrose, which has a solubility of 200 grams per 100 milliliters of water, only 0.004 gram of oxygen can dissolve in 100 milliliters of water. We can account for oxygen's low solubility in water by noting that the only attractions that occur between oxygen molecules and water molecules are relatively weak dipole–induced dipole attractions. More important,

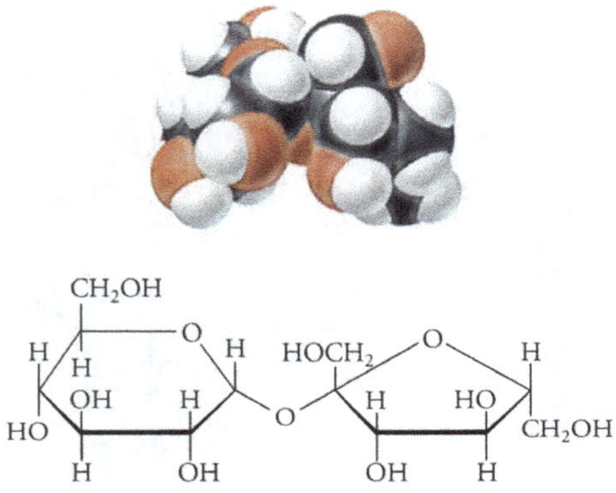

Figure 7.18

A sucrose molecule contains many hydrogen–oxygen covalent bonds, in which the hydrogen atoms are slightly positive and the oxygen atoms are slightly negative. These dipoles in any given sucrose molecule result in the formation of hydrogen bonds with neighboring sucrose molecules.

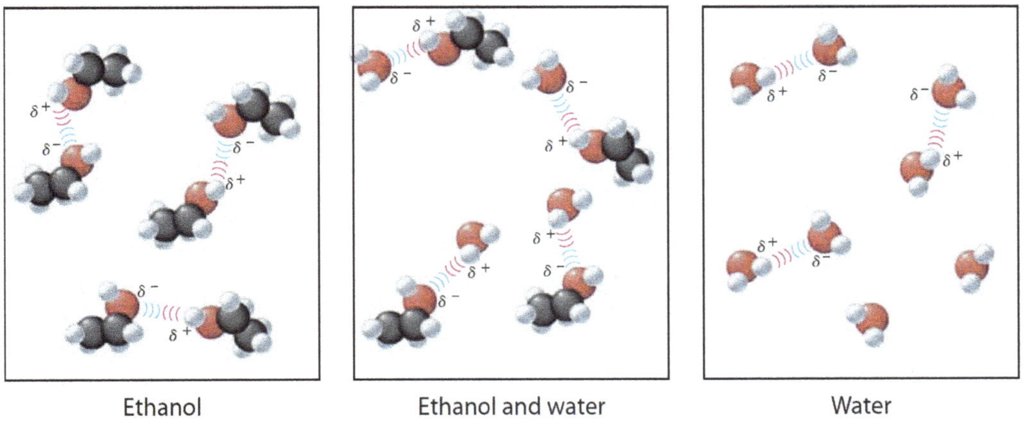

Figure 7.19

Ethanol and water molecules are about the same size, and they both form hydrogen bonds. As a result, ethanol and water will readily mix with each other.

∧ **Figure 7.20**

Glass is frosted by dissolving its outer surface in hydrofluoric acid.

however, is the fact that the stronger attraction of water molecules for one another—through the hydrogen bonds that the water molecules form with one another—effectively excludes oxygen molecules from intermingling with them.

A material that does not dissolve in a solvent to any appreciable extent is said to be insoluble in that solvent. We find many substances to be insoluble in water, including sand and glass. Just because a material is not soluble in one solvent, however, does not mean it won't dissolve in another. Sand and glass, for example, are soluble in hydrofluoric acid, HF, which is used to give glass the decorative frosted look shown in **Figure 7.20**. Also, although Styrofoam™ is insoluble in water, it is partially soluble in acetone, a solvent used in fingernail-polish remover. Pour a little acetone into a Styrofoam™ cup, and the acetone soon causes the Styrofoam™ to deform, as demonstrated in **Figure 7.21**.

∧ **Figure 7.21**

Is this cup melting or dissolving?

Solubility Changes with Temperature

You probably know from experience that water soluble solids usually dissolve better in hot water than in cold water. A highly concentrated solution of sucrose in water, for example, can be made by heating the solution almost to the boiling point. This is how syrup and hard candy are made.

Solubility increases with increasing temperature because hot water molecules have greater kinetic energy and therefore are able to collide with the solid solute more vigorously. The vigorous collisions facilitate the disruption of particle-to-particle attractions in the solid.

Although the solubilities of many solid solutes—sucrose, to name just one example—are greatly increased by rises in temperature, the solubilities of other solid solutes, such as sodium chloride, are only mildly affected, as **Figure 7.22** shows. This difference involves a number of factors, including the strength of the chemical bonds in the solute molecules and the way those molecules are packed together. Some chemicals, such as calcium carbonate, $CaCO_3$, actually become less soluble as the water temperature increases. This explains why the inner surfaces of tea kettles are often coated with calcium carbonate residues.

When a sugar solution saturated at a high temperature is allowed to cool, some of the sugar usually comes out of solution and forms what is

Figure 7.22 >

The solubility of many water-soluble solids increases with temperature, while the solubility of others is only very slightly affected by temperature.

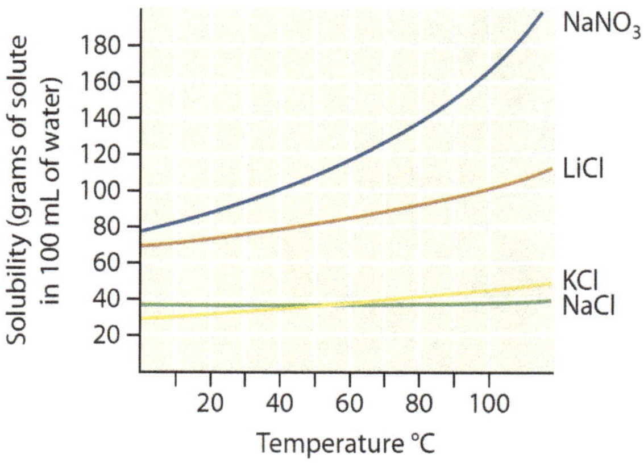

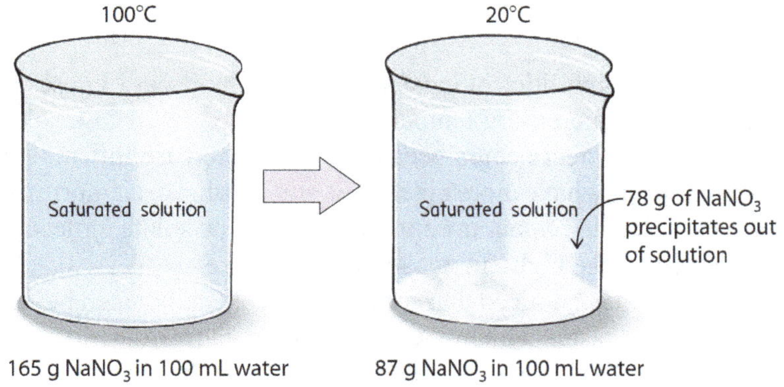

< Figure 7.23

The solubility of sodium nitrate is 165 grams per 100 milliliters of water at 100°C but only 87 grams per 100 milliliters at 20°C. Cooling a 100°C saturated solution of $NaNO_3$ to 20°C causes 78 grams of the solute to precipitate.

called a **precipitate**. When this occurs, the solute, sugar in this case, is said to have precipitated from the solution.

Let's put on our quantitative thinking caps and consider another example. At 100°C, the solubility of sodium nitrate, $NaNO_3$, in water is 165 grams per 100 milliliters of water. As we cool this solution, the solubility of $NaNO_3$ decreases, as shown in Figure 7.22, and this change in solubility causes some of the dissolved $NaNO_3$ to precipitate (come out of solution). At 20°C, the solubility of $NaNO_3$ is only 87 grams per 100 milliliters of water. So, if we cool the 100°C solution to 20°C, 78 grams (165 grams – 87 grams) precipitates, as shown in **Figure 7.23**.

CONCEPT **CHECK**

Why isn't sucrose infinitely soluble in water?

CHECK YOUR ANSWER

The attraction between two sucrose molecules is much stronger than the attraction between a sucrose molecule and a water molecule. Because of this, sucrose dissolves in water only as long as the number of water molecules far exceeds the number of sucrose molecules. When there are too few water molecules to dissolve any additional sucrose, the solution is saturated.

Rates of Dissolving

The temperature of the solution can impact not only how much solute that solution can hold, but also the rate at which the solute dissolves. At a warmer temperature the solvent molecules have greater kinetic energy, which allows them to "attack" the solute in a quicker fashion. This is why you'll see sugar dissolve faster in a hot tea than within cold iced tea.

A large crystal of sugar takes longer to dissolve in water than the same mass of sugar crushed up into a fine powder. How so? Drop a large crystal of sugar into water and it's only at the surface of the crystal where the solvent water molecules can attack the sugar molecules—those sugar molecules deep within the crystal are hidden from the water. Take that same crystal and grind it into a powder before adding it to the water. In this case, there is a greater exposure of the sugar molecules to the water. We say that the *surface area* of the sugar has been increased. Thus, we find that the greater the surface area of the solute, the quicker it will dissolve.

Lastly, consider right along the surface of the solute where the solute molecules are dissolving. Right along the surface, solute molecules are going into solution. This means it's getting fairly concentrated right along the surface, which means 1) the solution might be saturated at that location, and 2) the number of water molecules is getting less. Both these factors can be overcome by stirring the solution, or providing some agitation to assure there is a movement of both the solvent and solute.

TABLE 7.2
Temperature-Dependent Solubility of Oxygen Gas in Water at a Pressure of 1 Atmosphere

TEMPERATURE (°C)	O_2 SOLUBILITY ($g\ O_2/L\ H_2O$)
0	0.0141
10	0.0109
20	0.0092
25	0.0083
30	0.0077
35	0.0070
40	0.0065

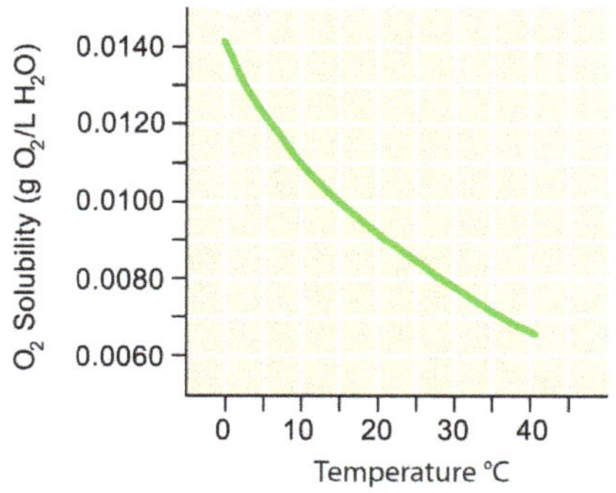

Solubility of Gases

In contrast to the solubilities of most solids, the solubilities of gases in liquids decrease with increasing temperature, as **Table 7.2** with a solubility curve shows. This effect occurs because with an increase in temperature, the solvent molecules have more kinetic energy. This makes it more difficult for a gaseous solute to remain in solution, because the solute molecules are literally ejected by the high-energy solvent molecules.

Perhaps you have noticed that compared to cold carbonated beverages, warm ones go flat faster. The higher temperature causes the molecules of carbon dioxide gas to leave the liquid solvent at a higher rate.

The solubility of a gas in a liquid also depends on the pressure of the gas immediately above the liquid. In general, a higher gas pressure above the liquid means more of the gas dissolves. A gas at a high pressure has many, many gas particles crammed into a given volume. The "empty" space in an unopened soft drink bottle, for example, is crammed with carbon dioxide molecules in the gaseous phase. With nowhere else to go, many of these molecules dissolve in the liquid, as shown in **Figure 7.24**. Alternatively, we might say that the great pressure forces the carbon dioxide molecules into solution. When the bottle is opened, the "head" of highly pressurized carbon dioxide gas escapes. Now the gas pressure above the liquid is lower than it was before. As a result, the solubility of the carbon dioxide drops, and the carbon dioxide molecules that were once squeezed into the solution begin to escape into the air above the liquid.

The rate at which carbon dioxide molecules leave an opened soft drink is relatively slow. You can increase the rate by pouring in granulated sugar, salt, or sand. The microscopic nooks and crannies on the surfaces of the grains serve as nucleation sites where carbon dioxide bubbles are able to form rapidly and then escape by buoyant forces. Shaking the beverage also increases the surface area of the liquid-to-gas interface, making it easier for the carbon dioxide to escape from the solution. Once the solution is shaken, the rate at which carbon dioxide escapes becomes so great that the beverage froths over. You also increase the rate at which carbon dioxide escapes when you pour the beverage into your mouth, which abounds in nucleation sites. You can feel the resulting tingly sensation.

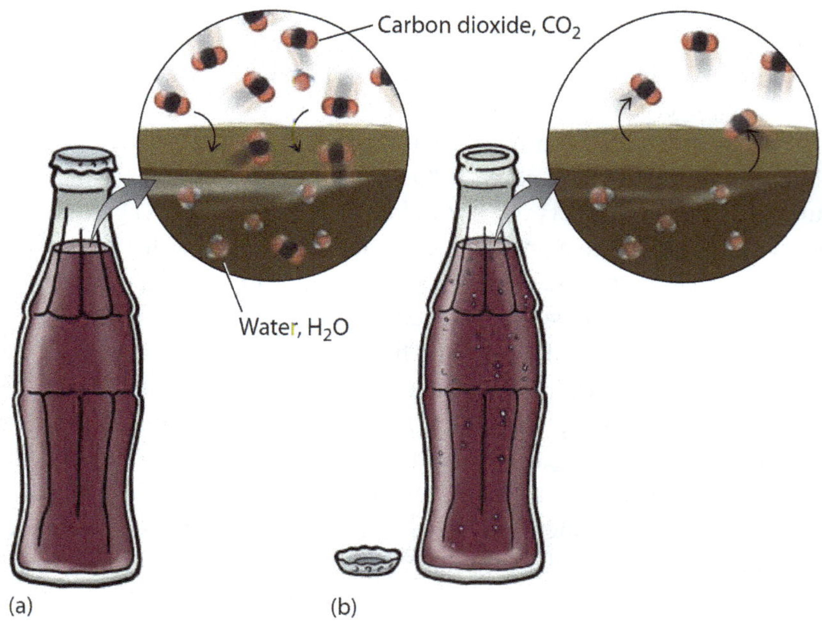

Figure 7.24 >

(a) The carbon dioxide gas above the liquid in an unopened soft drink bottle consists of many tightly packed carbon dioxide molecules that are forced by pressure into solution. (b) When the bottle is opened, the pressure is released and carbon dioxide molecules originally dissolved in the liquid can escape into the air.

What happens to the solubility of gases in water when salt is added? Consider what water molecules do when sodium and chloride ions are introduced to the mix. The water molecules are drawn to these highly charged ions, to the point of surrounding them forming a "sphere of hydration". In effect, there are now fewer water molecules available to interact with the nonpolar gas molecules. This can be any sort of gas, such as oxygen, O_2, or carbon dioxide, CO_2. Simply put, these gas molecules cannot compete with the salt for water's attention. As salt is introduced to water, the solubility of any gas dissolved in that water decreases. This is called a salting effect.

Thus, adding salt to a carbonated beverage is particularly effective at the formation of bubbles—there are the nucleation sites discussed earlier as well as this salting effect. This also explains why oxygen, O_2, is more soluble in freshwater streams and lakes than in seawater. A fresh water trout won't survive in seawater—its tissues are not accustomed to the high salinity, but also there is less oxygen, O_2, available for it to absorb through its gills.

fyi FOR YOUR INFORMATION

It is not just dipole–induced dipole attractions that keep carbon dioxide dissolved in water. As we'll discuss in Chapter 10, carbon dioxide reacts with water to form carbonic acid, which is much more soluble in water. When a can of carbonated soda is opened, much of this carbonic acid rapidly transforms back into water and carbon dioxide, which quickly bubbles out of solution because of its low solubility.

CONCEPT CHECK

You open two cans of soft drink, one from a warm kitchen shelf and the other from the coldest depths of your refrigerator. Which provides the greater fizz in your mouth?

CHECK YOUR ANSWER

The solubility of carbon dioxide in water decreases with increasing temperature. The warm drink, therefore, will fizz in your mouth more than the cold one will.

Precipitation Reactions

During a chemical reaction, atoms are changing partners to form new compounds. For example, consider the reaction between lead nitrate, $Pb(NO_3)_2$, and potassium iodide, KI, to form lead iodide, PbI_2, and potassium nitrate, KNO_3. We can represent this as follows:

$$Pb(NO_3)_2 + 2\,KI \rightarrow PbI_2 + 2\,KNO_3$$

What happens in this reaction is that the iodine, I, jumps from the potassium, K, to the lead, Pb, to form the lead iodide, PbI_2. The reactants, lead nitrate, $Pb(NO_3)_2$, and potassium iodide, KI, are both soluble in water. That's why this reaction can be performed within an *aqueous* solution, which means water is the solvent. As discussed further in Chapter 9, we can indicate this by placing an (*aq*) after each compound that's soluble in water:

$$Pb(NO_3)_2\,(aq) + 2\,KI\,(aq) \rightarrow PbI_2\,(s) + 2\,KNO_3\,(aq)$$

While the lead nitrate, $Pb(NO_3)_2$, potassium iodide, KI, and the potassium nitrate, KNO_3, product are all soluble in cold water, the lead iodide, PbI_2, product is not. The result is that the lead iodide, PbI_2, comes out of solution as it is formed—it *precipitates*. As a yellow solid, this particular "precipitation reaction" can be quite beautiful, especially if the solution is heated to dissolve the lead iodide, which then recrystallizes out of solution as the solution is slowly cooled, as shown on the next page.

Lead iodide, PbI_2, in cold water

There are many other types of precipitation reactions. What they all have in common is that one of the chemical compound products of the reaction is not soluble in the solvent. As a result, this product precipitates out of solution, which makes it relatively easy to collect by passing the solution through filter paper.

Solubility Rules

Lastly, chemists have found some general trends when it comes to the solubility of compounds within water. Not to be memorized but provided as a reference, some of them are as follows:

1. Salts containing Group I ions (Li^+, Na^+, K^+, Cs^+, Rb^+) are soluble . There are few exceptions to this rule. Salts containing the ammonium ion (NH_4^+) are also soluble.
2. Salts containing nitrate ion (NO_3^-) are generally soluble.
3. Salts containing Group 17 ions (Cl^-, Br^-, or I^-) are generally soluble. Important exceptions to this rule are halide salts of Ag^+, Pb^{2+}, and Hg^{2+}. Thus, AgCl, $PbBr_2$, and Hg_2Cl_2 are insoluble.
4. Most silver salts are insoluble. $AgNO_3$ and $Ag(C_2H_3O_2)$ are common soluble salts of silver; virtually all others are insoluble.
5. Most sulfate salts are soluble. Important exceptions to this rule include $CaSO_4$, $BaSO_4$, $PbSO_4$, Ag_2SO_4 and $SrSO_4$.
6. Most hydroxide salts are only slightly soluble. Hydroxide salts of Group I ions are soluble. Hydroxide salts of Group II ions (Ca, Sr, and Ba) are slightly soluble. Hydroxide salts of transition metals and Al^{3+} are insoluble. Thus, $Fe(OH)_3$, $Al(OH)_3$, $Co(OH)_2$ are not soluble.
7. Most sulfides of transition metals are highly insoluble, including CdS, FeS, ZnS, and Ag_2S. Arsenic, antimony, bismuth, and lead sulfides are also insoluble.
8. Carbonates are frequently insoluble. Group II carbonates ($CaCO_3$, $SrCO_3$, and $BaCO_3$) are insoluble, as are $FeCO_3$ and $PbCO_3$.
9. Chromates are frequently insoluble. Examples include $PbCrO_4$ and $BaCrO_4$.
10. Phosphates such as $Ca_3(PO_4)_2$ and Ag_3PO_4 are frequently insoluble.
11. Fluorides such as BaF_2, MgF_2, and PbF_2 are frequently insoluble.

Add sodium chloride, NaCl, (table salt) to water and stir. Through this investigation you should find truth to the first and third solubility rules. The ions of Group 1 elements are soluble in water (Rule 1) as are many salts containing the Group 17 ions (Rule 3).

All these rules are *empirical*, meaning they are based simply upon investigation and observation, rather than detailed theory. That said, the working chemist can use these rules to help predict whether any chemicals they produce might not be soluble in the solvent.

Might they memorize these rules? That would be an exercise in memorization, not in chemistry. Like the periodic table, the solubility rules can be easily looked up. So they need not be memorized. Also like the periodic table, the more you work with the solubility rules, the more familiar they become. Working chemists know these solubility rules not so much because the rules were memorized, but because they work with these rules on a regular basis.

Consider these two reactions shown earlier:

$$Pb(NO_3)_2\,(aq) + 2\,KI\,(aq) \rightarrow PbI_2\,(s) + 2\,KNO_3\,(aq)$$

$$AgNO_3\,(aq) + NaCl\,(aq) \rightarrow AgCl\,(s) + NaNO_3\,(aq)$$

As we'll be discussing later, these two reactions are known as double replacement reactions, which simply means the two reactants exchange an atom with each other. Look to see how the products lead iodide, PbI_2, and silver chloride, AgCl, are presented in the solubility rules. Rule 3 states that salts of lead tend and silver are insoluble. Thus we see the (*s*) for "solid" appearing after those two chemical formulas.

7.5 How Soap Works

Dirt and grease together make *grime*. Because grime contains many nonpolar components, it is difficult to remove from hands or clothing with water alone. To remove most grime, we can use a nonpolar solvent such as paint thinner, which dissolves the grime because of strong induced dipole–induced dipole attractions. Paint thinner is good for removing the grime left on hands after an activity such as changing a car's motor oil. But nonpolar solvents such as paint thinner are toxic, they have offensive odors, and they are harsh on the skin. Rather than washing our dirty hands and clothes with nonpolar solvents, we have a more pleasant alternative—soap and water. *Soap* works because soap molecules have both nonpolar and polar properties. A typical soap molecule has two parts: a long, nonpolar tail of carbon and hydrogen atoms and a polar head containing at least one ionic bond.

READING CHECK
What properties do soap molecules have?

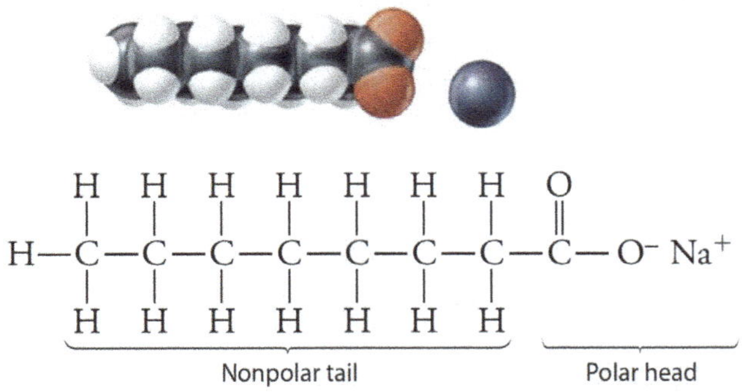

Because most of a soap molecule is nonpolar, it attracts nonpolar grime molecules via induced dipole–induced dipole attractions (dispersion forces) as **Figure 7.25** illustrates. In fact, grime quickly finds itself surrounded in three dimensions by the nonpolar tails of soap molecules. This attraction is usually sufficient to lift the grime away from the surface being cleaned. With the nonpolar tails facing inward toward the grime, the polar heads are all directed outward, where they are attracted to water molecules by relatively strong ion–dipole attractions. If the water is flowing, the whole conglomeration of grime and soap molecules flows with it, away from your hands or clothes and then down the drain.

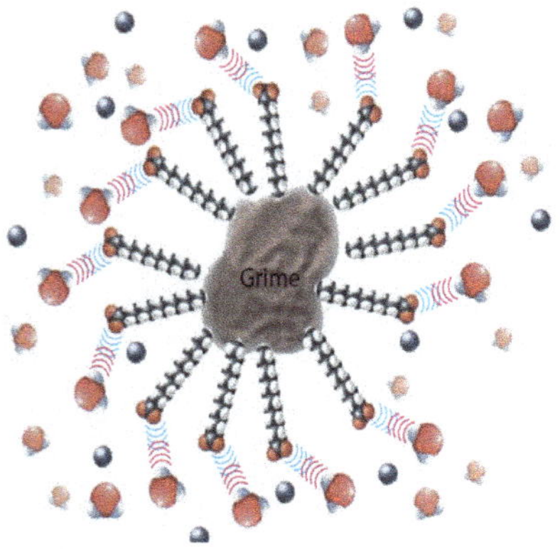

Figure 7.25
Nonpolar grime attracts and is surrounded by the nonpolar tails of soap molecules, forming what is called a micelle. The polar heads of the soap molecules are attracted by ion–dipole attractions to water molecules, which then carry away the soap–grime combination.

For the past several centuries, soaps have been prepared by treating animal fats with sodium hydroxide, NaOH, also known as caustic lye. In this reaction, which is still used today, each fat molecule is broken down into three *fatty acid* soap molecules and one glycerol molecule:

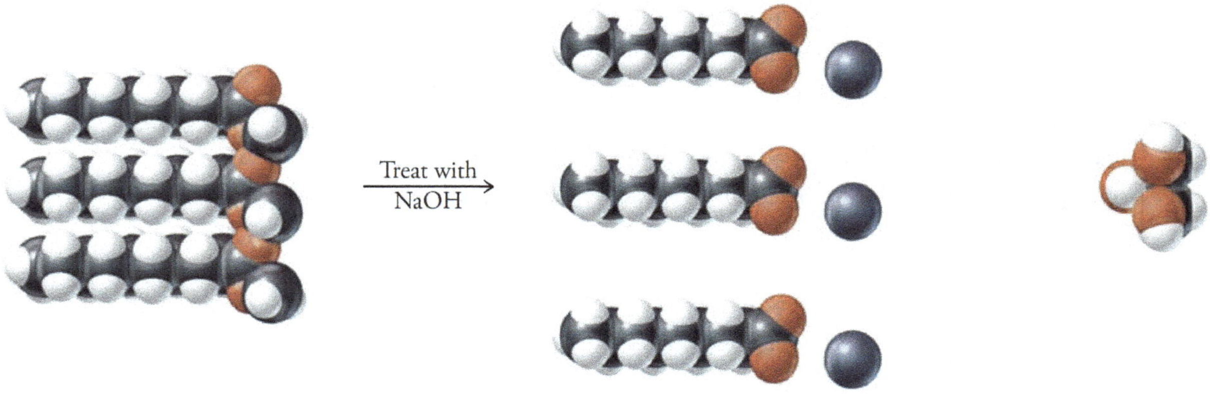

Fat molecule   Three fatty acid soap molecules   Glycerol molecule

In the 1940s, chemists began developing a class of synthetic soap-like compounds known as *detergents*, which offer several advantages over true soaps, such as stronger grease penetration and lower price.

The chemical structure of detergent molecules is similar to that of soap molecules in that both possess a polar head attached to a nonpolar tail. The polar head in a detergent molecule, however, typically consists of either a sulfate group, $-OSO_3^-$, or a sulfonate group, $-SO_3^-$, and the nonpolar tail can have an assortment of structures.

One of the most common sulfate detergents is sodium lauryl sulfate, a main ingredient of many toothpastes. A common sulfonate detergent is sodium dodecyl benzenesulfonate, also known as a linear alkylsulfonate, or LAS, often found in dishwashing liquids. Both of these detergents are biodegradable, which means that microorganisms can break down the molecules once they are released into the environment.

CONCEPT **CHECK**

What type of attractions hold soap or detergent molecules to grime?

CHECK YOUR ANSWER

If you haven't yet formulated an answer, why not back up and reread the question? You've got only four choices: ion–dipole, dipole–dipole, dipole–induced dipole, and induced dipole–induced dipole. The answer is induced dipole–induced dipole attractions, because the interaction is between two nonpolar entities—the grime and the nonpolar tail of a soap or detergent molecule.

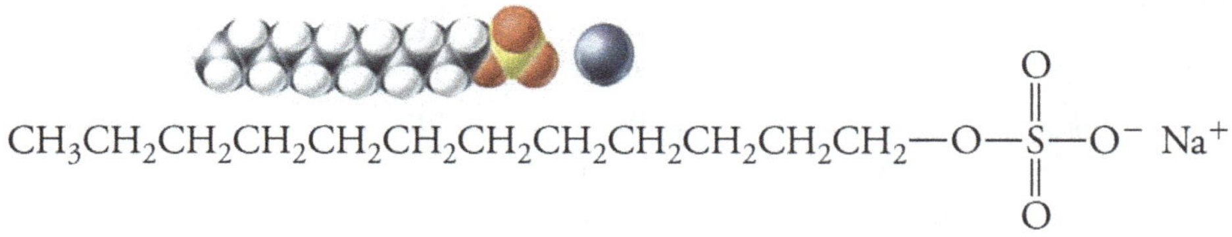

Sodium lauryl sulfate

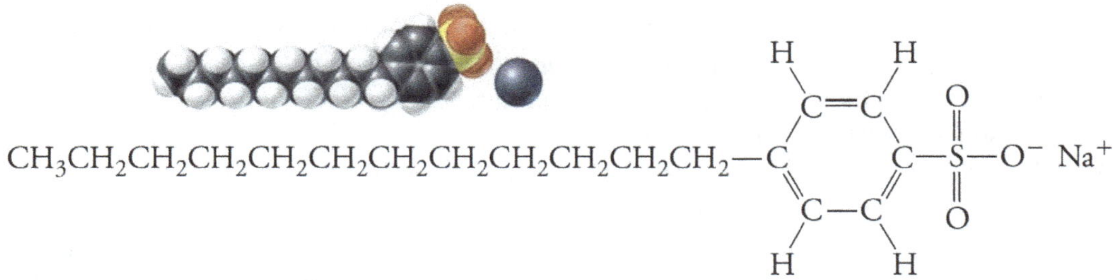

Sodium dodecyl benzenesulfonate

7.6 Softening Hard Water

Water containing large amounts of calcium and magnesium ions is said to be **hard water,** and it has many undesirable qualities. For example, when hard water is heated, its calcium and magnesium ions tend to bind with negatively charged ions also found in the water to form solid compounds, like those shown in **Figure 7.26**. These can clog water heaters and boilers. You'll also find coatings of these calcium and magnesium compounds on the inside surfaces of a well-used tea kettle (because the solubility of these compounds decreases with increasing temperature, as discussed earlier).

Hard water also inhibits the cleansing actions of soaps and, to a lesser extent, detergents. The sodium ions of soap and detergent molecules carry a 1+ charge, and calcium and magnesium ions carry a 2+ charge (note their positions in the periodic table). The negatively charged portion of the polar head of a soap or detergent molecule is more attracted to the double positive charge of calcium and magnesium ions than to the single positive charge of sodium ions. Soap or detergent molecules, therefore, give up their sodium ions to bind selectively with calcium or magnesium ions.

READING CHECK

Is the polar head of a soap molecule more attracted to calcium or sodium ions?

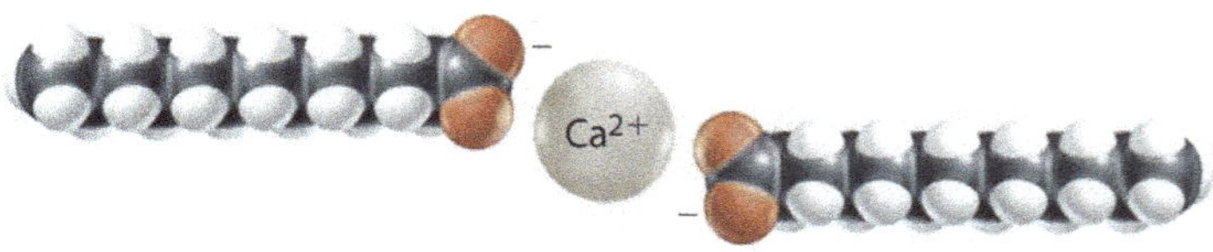

Figure 7.26

Hard water causes calcium and magnesium compounds to build up on the surfaces of pipes and fixtures, especially those used to carry hot water.

Soap or detergent molecules bound to calcium or magnesium ions tend to be insoluble in water. As they come out of solution, they form a scum, which can appear as a ring around the inside of your bathtub. Because the soap or detergent molecules are tied up with calcium and magnesium ions, more soap or detergent must be added to maintain cleaning effectiveness.

Many detergents today contain sodium carbonate, Na_2CO_3, commonly known as washing soda. The calcium and magnesium ions in hard water are more attracted to the carbonate ion with its two negative charges than they are to a soap or detergent molecule with its single negative charge. With the calcium and magnesium ions bound to the carbonate ion, as shown in **Figure 7.27**, the soap or detergent is free to do its job. Because it removes the ions that make water hard, sodium carbonate is known as a water-softening agent.

In some homes, the water is so hard that it must be passed through a *water-softening unit*. In a typical unit, illustrated in **Figure 7.28**, hard water is passed through a large tank filled with tiny beads of a water-insoluble resin known as an *ion-exchange resin*. The surface of the resin contains many negatively charged ions bound to positively charged sodium ions. As calcium and magnesium ions pass over the resin, they displace the sodium ions and thereby become bound to the resin. The calcium and magnesium ions are able to do this because their positive charge (2+) is greater than that of the sodium ions (1+). The calcium and magnesium ions therefore have a greater attraction for the negative sites on the resin. The net result is that for every one calcium or magnesium ion that binds, two sodium ions are set free. In this way, the resin exchanges ions. The water that exits from the unit is now free of calcium and magnesium ions, but it does contain sodium ions in their place.

Eventually, all the sites for calcium and magnesium on the resin are filled, and then the resin needs to be either discarded or recharged. It is recharged by flushing it with a concentrated solution of sodium chloride, NaCl. The abundant sodium ions displace the calcium and magnesium ions (ions are *exchanged* once again), freeing up the binding sites on the resin.

fyi FOR YOUR INFORMATION

Most modern water softeners are equipped with meters that let you know the rate at which you consume water. This is a great way to keep tabs on your water-conservation efforts.

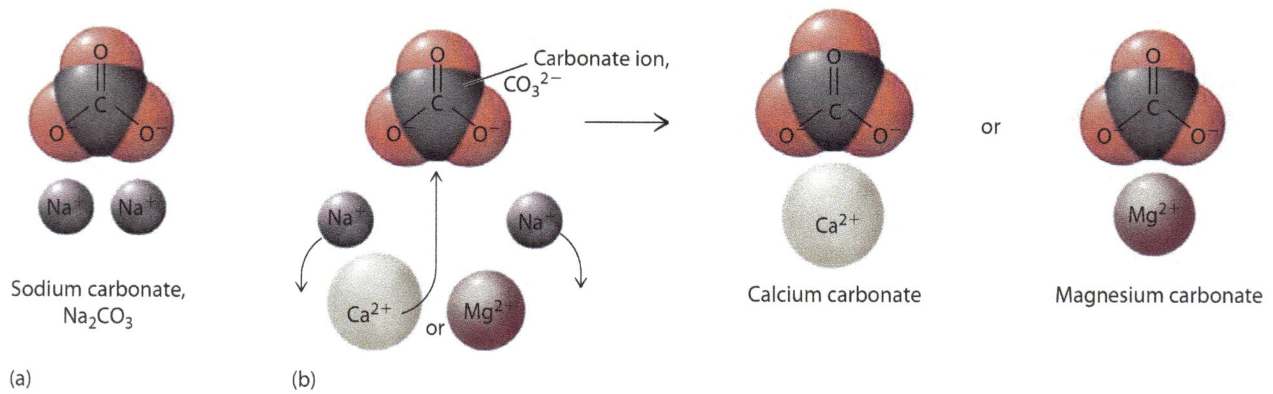

Figure 7.27

(a) Sodium carbonate is added to many detergents as a water-softening agent. (b) The doubly positive calcium and magnesium ions of hard water preferentially bind with the doubly negative carbonate ion, freeing the detergent to do its job.

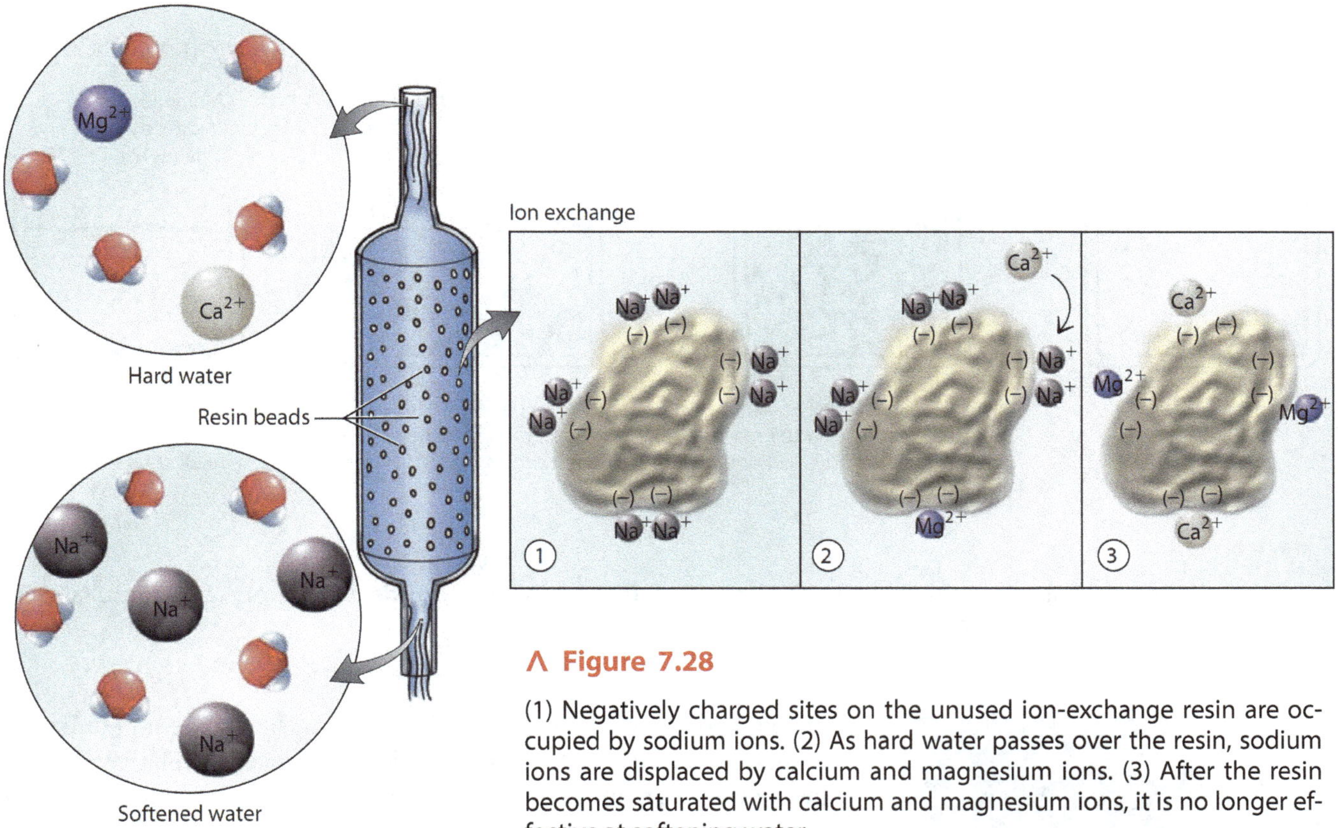

∧ Figure 7.28

(1) Negatively charged sites on the unused ion-exchange resin are occupied by sodium ions. (2) As hard water passes over the resin, sodium ions are displaced by calcium and magnesium ions. (3) After the resin becomes saturated with calcium and magnesium ions, it is no longer effective at softening water.

7.7 Purifying Drinking Water

As was discussed in Section 3.7, it is impossible to obtain 100 percent pure water. However, we are able to purify water to meet our needs. We do this by taking advantage of the differences in physical properties of water and the solutes or particulates it contains.

Water safe for drinking is said to be *potable*. In the United States, potable water is currently used for everything from cooking to flushing our toilets. The first step most public utilities take to produce potable water from natural sources is to remove any dirt particles or pathogens, such as bacteria. This is done by mixing the water with certain minerals, such as slaked lime and aluminum sulfate, which coagulate into a gelatinous material, that becomes interspersed throughout the water (**Figure 7.29**). This is done in a large settling basin. Slow stirring causes the gelatinous material to clump together and settle to the bottom of the basin. As these clumps form and settle, they carry with them many of the dirt particles and bacteria. The water is then filtered through sand and gravel. (**Figure 7.30**).

To improve the odor and flavor of the water, many treatment facilities also aerate the water by cascading it through a column of air. Aeration removes many unpleasant-smelling volatile chemicals, such as sulfur compounds. At the same time, air dissolves into the water, giving it a better taste—without dissolved air, the water tastes flat. As a final step, the water is treated with a disinfectant, usually chlorine gas, Cl_2, but sometimes ozone, O_3, and then stored in a holding tank that feeds into the city mains.

Developed countries have the technology and infrastructure to produce vast quantities of water suitable for drinking—with the result that many

READING CHECK

How do we purify our drinking water?

fyi FOR YOUR INFORMATION

In 1908, Jersey City, New Jersey, became the first American city to begin chlorinating its drinking water. By 1910, as disinfecting drinking water with chlorine became more widespread, the death rate from typhoid fever dropped from about 100 to 20 lives per 100,000. By 1935, the death rate fell to three lives per 100,000. In 1960, fewer than 20 persons in the entire United States died from typhoid fever.

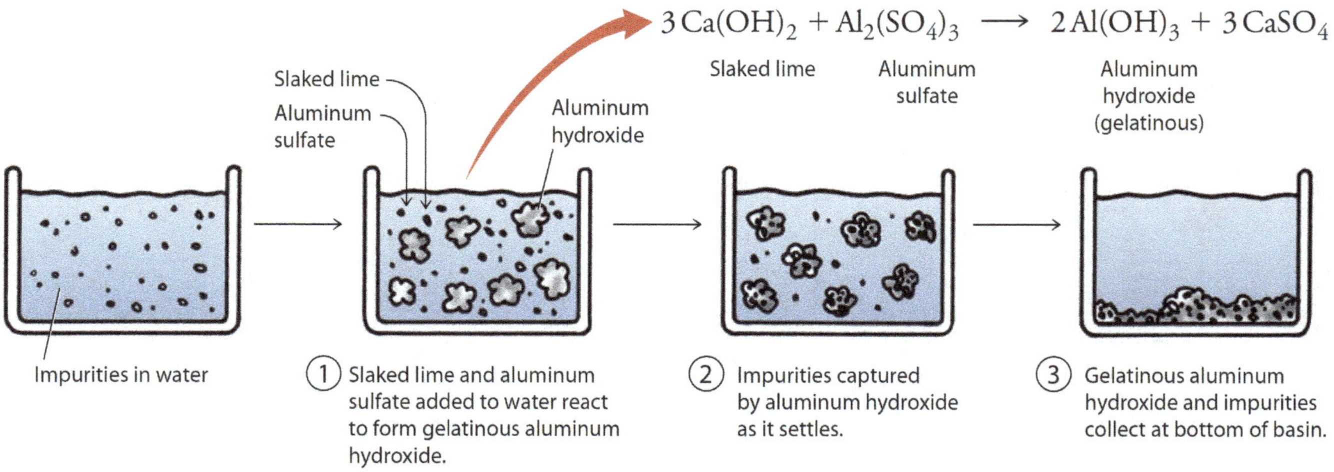

∧ Figure 7.29

Slaked lime, $Ca(OH)_2$, and aluminum sulfate, $Al_2(SO_4)_3$, react to form aluminum hydroxide, $Al(OH)_3$, and calcium sulfate, $CaSO_4$, which together form a gelatinous material.

∧ Figure 7.30

To become safe for drinking, water is filtered through layers of sand, gravel, and charcoal as shown in this display. The water can then be aerated to help remove volatile impurities. As a final step, chlorine, Cl_2, is sometimes added to fight against waterborne pathogens.

citizens take their drinking water for granted. The number of public water-treatment facilities in developing nations, however, is relatively small. In these locations, many people drink their water in the form of a hot beverage, such as tea, which is disinfected through boiling. Alternatively, disinfecting iodine tablets can be used.

Fuel for boiling and tablets for disinfecting, however, are not always available. As a result, more than 400 people in the world (mostly children) die every hour from preventable diseases or infections such as cholera, typhoid fever, dysentery, and hepatitis, which they contract by drinking contaminated water. In response, several American manufacturers have developed tabletop systems that bathe water with pathogen-killing ultraviolet light. One prototype model, shown in **Figure 7.31**, disinfects 15 gallons per minute, weighs about 15 pounds, and is powered by photovoltaic solar cells, which permit it to run unsupervised in remote locations.

Aside from pathogens, untreated water from wells or rivers may contain toxic metals that seep into the water supply from natural geologic formations. Many of the wells in Bangladesh, for example, are made very deep to avoid the pathogens that run rampant in the surface waters of the region. The water obtained from these deep wells, however, is highly contaminated with arsenic—a naturally occurring element in the Earth's crust. The arsenic is in the underlying rock, which formed from river sediments carried down from the Himalayas. Because this region is so densely populated, as many as 70 million people may be subject to some level of arsenic poisoning, which manifests itself as skin lesions and a higher susceptibility to cancer. Low-cost methods for removing arsenic from well water are greatly needed, as are worldwide recognition of this problem and the political, economic, and social support to overcome it.

Fresh Water Can Be Made from Salt Water

With the depletion of sources of natural fresh water in many regions, there has been growing interest in techniques for generating fresh water from the Earth's far larger reserves of seawater or from *brackish* (moderately salty)

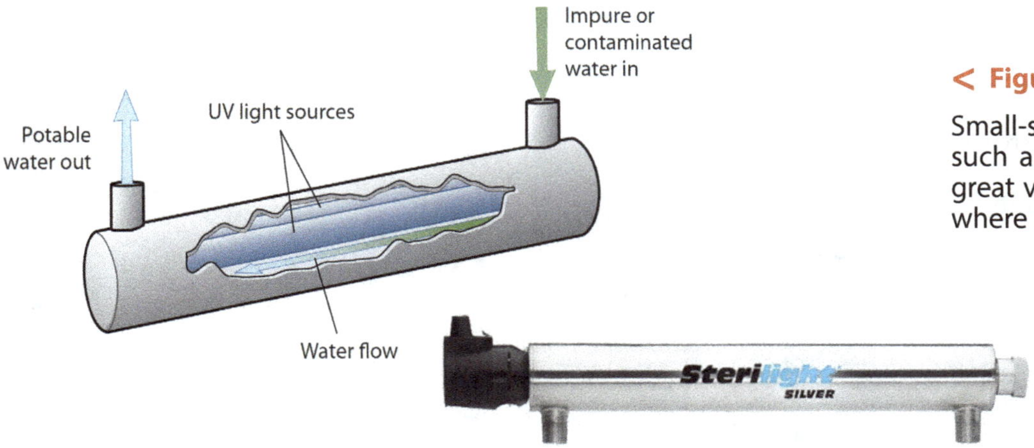

< **Figure 7.31**

Small-scale water-disinfecting units such as the one shown here hold great value in regions of the world where potable water is scarce.

CONCEPT **CHECK**

At a water-treatment facility, how does adding slaked lime and aluminum sulfate to water purify the water?

CHECK YOUR ANSWER

The water entering a water-treatment plant is usually a heterogeneous mixture containing suspended solids. Adding slaked lime and aluminum sulfate serves to capture these suspended solids, which then sink to the bottom, where they are easily removed.

groundwater. Worldwide, *desalination* plants operate in about 120 countries. In many areas of the Caribbean, North Africa, and the Middle East, desalinized water is the main source of municipal supply (**Figure 7.32**).

The two primary methods of removing salts from seawater or brackish water are *distillation* and *reverse osmosis*. These techniques are also highly effective in removing a host of other contaminants, such as pathogens, fertilizers, and pesticides. Distillation and reverse osmosis, therefore, are also used to purify naturally occurring fresh water. Many popular brands of bottled water, for example, contain fresh water that has been treated either by distillation or by reverse osmosis.

Distillation involves vaporizing water with heat and then condensing the vapors into purified liquid water (see Section 3.6). Earth's desalinized water is produced using this technique. Because water has such a high heat of vaporization, however, this technique is energy intensive. Solar distillers avoid the burning of fuels, but they require about 1 square meter of surface area to produce 4 liters of fresh water per day, as shown in **Figure 7.33**. For a single home or a small village, this surface-area requirement may be easily accommodated.

For many regions, *reverse osmosis* is a preferable method of water desalination. In order to understand reverse osmosis, you must first understand osmosis. Osmosis involves a **semipermeable membrane**. A semi-permeable membrane contains submicroscopic pores that allow the passage of water molecules but not of larger solute ions or solute molecules. When a body of fresh water is partitioned from a body of salt water by a semipermeable membrane, water molecules pass from the fresh water into the salt water at a higher rate than from the salt water into the fresh water.

fyi FOR YOUR INFORMATION

Hygroscopic is a term used to describe the ability of a material to absorb moisture from the air. Table salt, for example, is hygroscopic, which is why rice is sometime added to a salt shaker. The rice is even more hygroscopic than the salt. Thus it preferentially absorbs the moisture preventing the salt from clumping up. A company called Source has developed a device in which hygroscopic nanofibers absorb atmospheric moisture. With a system of fans run by photovoltaic cells, this moisture condenses into liquid water—up to 5 liters a day. Collecting moisture from dry desert areas requires more energy, but in such areas the solar energy is more abundant. To evaluate this scientific solution, visit Source.co.

Figure 7.32 >

Transforming sea water into fresh water, the Adelaide Desalination Plant of Australia has an operating capacity of 300 million liters per day.

The reason for this is the presence of more water molecules along the fresh water face of the membrane than along the salt water face. The result is a net movement of fresh water into the body of salt water, as illustrated in **Figure 7.34.** This net flow of water across a semipermeable membrane into a more concentrated solution is called **osmosis**.

The result of osmosis is a buildup in volume of the salt water and a decrease in volume of the fresh water. These changes in volume, in turn, allow a buildup in pressure, called *osmotic pressure*. For the system in **Figure 7.35a**, osmotic pressure is the consequence of the salt water's greater height and, therefore, weight bearing onto the semipermeable membrane. As osmotic pressure builds, the rate at which water molecules are able to pass from the salt water into the fresh water increases. The water molecules in the salt water are literally being squeezed back across the membrane by the osmotic pressure. Eventually, the rates of water molecules passing in both directions across the membrane are the same, and the system reaches equilibrium, as shown in **Figure 7.35b**. If external pressure is applied to the salt water, even more water molecules are squeezed across the membrane from the salt water into the fresh water, as shown in **Figure 7.35c**. Water forced across a semipermeable membrane into a less concentrated solution

CHEMICAL CONNECTIONS

How is a pickle connected to bottled water?

V Figure 7.33

These solar distillers are popular in the remote communities along the Texas–Mexico border, where the waters from the Rio Grande basin are saline and tainted by the runoff of agricultural chemicals from upstream irrigation.

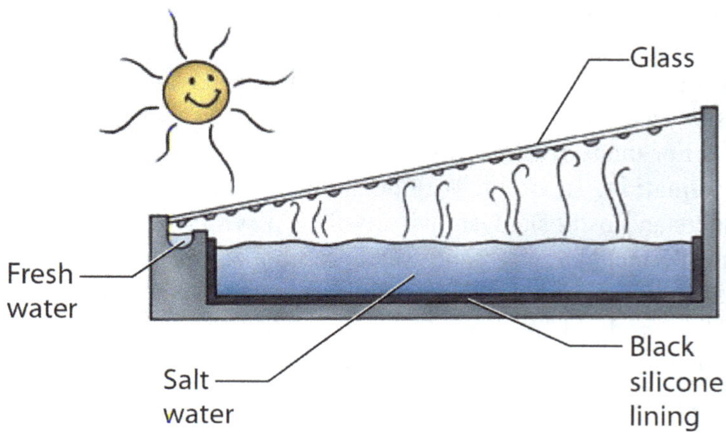

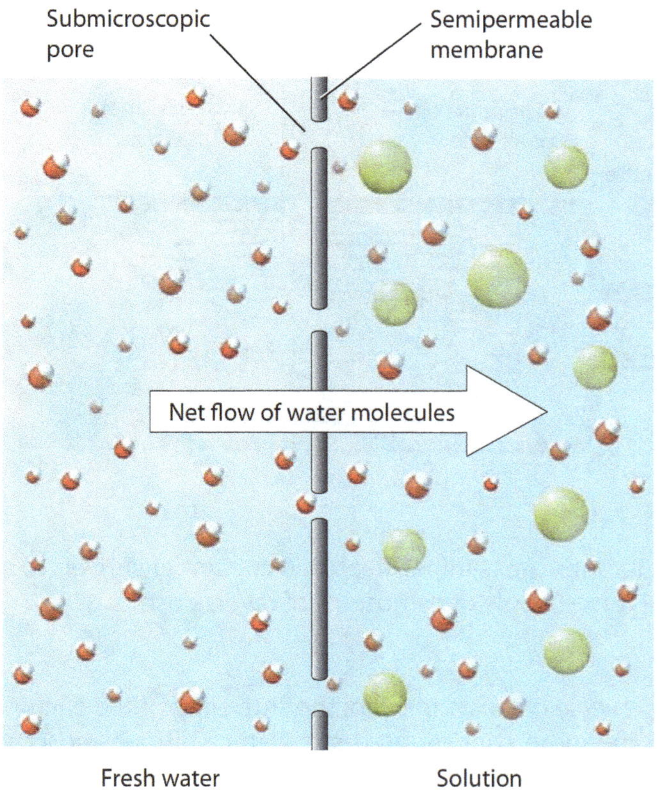

< Figure 7.34

Osmosis. The submicroscopic pores of a semipermeable membrane allow only water molecules to pass. Because there are more water molecules along the freshwater face of the membrane than along the solution face, more water molecules are available to migrate into the solution than are available to migrate into the fresh water.

is **reverse osmosis**. So, we see that reverse osmosis is a mechanism for generating fresh water from salt water.

The osmotic pressure for seawater, however, is an astounding 24.8 atmospheres (365 pounds per square inch). Generating pressures greater than this has its share of technical difficulties and is an energy-intensive process. Nonetheless, engineers have succeeded in building durable reverse osmosis units, shown in **Figure 7.36**, that can be networked together to generate fresh water from seawater at rates of millions of gallons per day. Reverse osmosis desalination facilities treating brackish water, which require much lower external pressures, are proportionately more economical.

fyi FOR YOUR INFORMATION

Lack of clean drinking water is one of the world's leading causes of death, especially among children in developing nations.

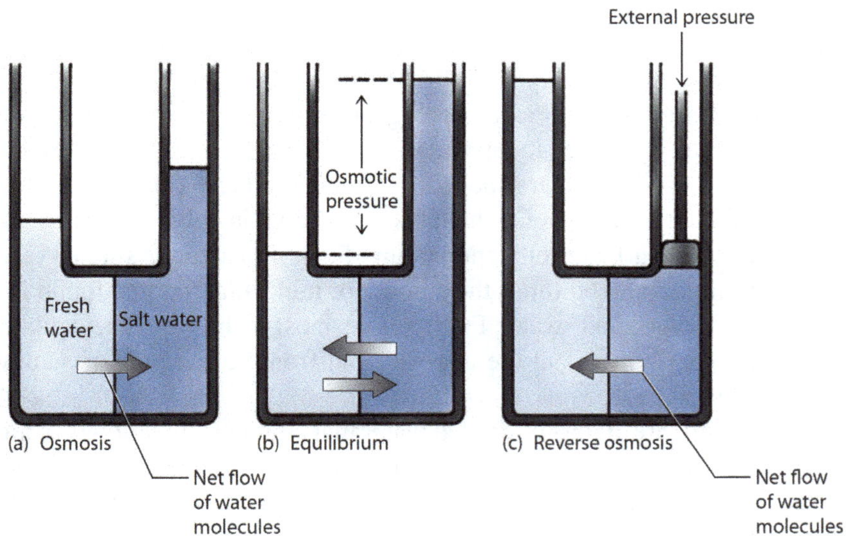

< Figure 7.35

(a) Osmosis results in a greater volume of salt water, which causes the pressure to increase on the salt side of the membrane.

(b) When the pressure on the salt side becomes high enough, equal numbers of water molecules pass in both directions.

(c) The application of external pressure forces water molecules to pass from the salt water to the fresh water so that now the salt-to-fresh rate exceeds the fresh-to-salt rate.

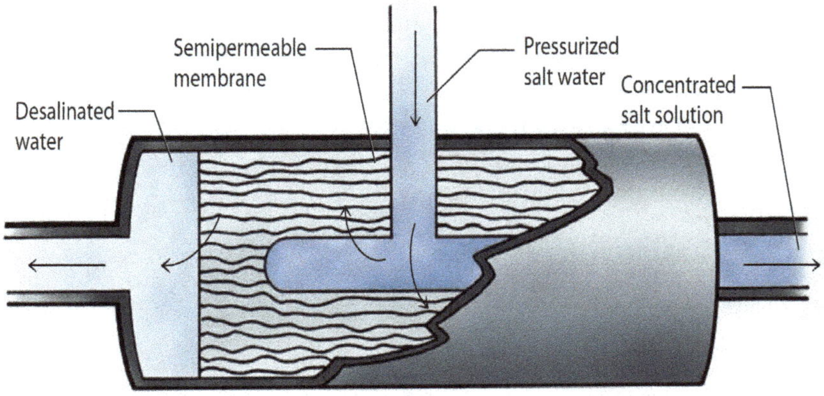

Figure 7.36 >

An industrial reverse osmosis unit consists of many semipermeable membranes packed around highly pressurized salt water. As desalinated water is pushed out one side, the remaining salt water, which is now even more concentrated, exits on the other side. A network of reverse osmosis units operating parallel to one another can produce enormous volumes of fresh water from salt water.

CONCEPT **CHECK**

Biological membranes, including cucumber membranes, are semipermeable. A cucumber shrivels to a smaller size when it is left in a solution of salt water. Is this an example of osmosis or of reverse osmosis?

CHECK YOUR ANSWER

No external pressure is involved, which rules out reverse osmosis. Instead, the shriveling of the cucumber tells us that the cucumber's cells are losing water to the more concentrated salt water. This is osmosis, whereby water molecules migrate across a semipermeable membrane into regions of higher salt concentrations. If you were to add a few other ingredients to the solution, such as spices and the right kinds of microorganisms, you would have a pickle.

∧ Figure 7.37

An economical way to great tasting drinking water is to use a carbon-based filter. As the water passes through the filter many bad tasting chemical compounds are absorbed by the carbon, including chlorine.

Is Bottled Water Worth the Price?

In the United States, natural sources of fresh water are relatively plentiful, allowing utilities to sell fresh water at rates of a fraction of a penny per liter. Nonetheless, consumers are still willing to purchase bottled water at a price that is about 2,000 times as expensive than drinking it straight from the tap. In 2020, Americans spend about $16 billion on bottled water, which was up from about $4 billion in 2010.

Bottled water also comes with environmental costs. In the United States, billions of water bottles are discarded each year. Of these, only about 20 percent are recycled—the rest end up primarily in landfills. Over 50 million barrels of oil are burned each year to meet the market demand for bottled water in the United States alone. To put this into perspective, according to the Earth Policy Institute, the volume of oil needed to produce a single plastic bottle is about one-quarter the volume of the bottle.

In an effort to overcome the ecological negatives of bottled water and a hefty price tag compared to tap water, many bottled-water marketers now focus on supposed peripheral benefits of their product. This includes the addition of dissolved oxygen, which, as was shown in Table 7.2, can be no more than 0.0083 g/L at room temperature. For comparison, a single breath of air contains about 100 times the amount of molecular oxygen found in a half liter of "oxygenated" water. Furthermore, most of the gases accumulated in your gut simply pass out the opposite end from your mouth, assuming you don't burp.

Worse still are claims of bottled water that contains "functional" water in which the structure of water has been modified using undetectable "subtle energy" to make the water more nutritious. Thank you for thinking critically about this assertion.

There is also water through which an electric current has been passed. The water companies claim that this creates two forms of water: alkaline and acidic. The alkaline water, they say, is good for you because it is "ionized, restructured, micro-clustered, activated, hydrogen–saturated, and oxidation–reduced." These sorts of misleading claims take advantage of the millions of people who don't understand that modifying the structure of water, H_2O, gives you something that is no longer water. Buyer beware! Water is H_2O. If it's not H_2O, it isn't water. Note: In Chapter 11 we'll be discussing what actually happens when an electric current passes through water.

Interestingly, about 25 percent of bottled water sold in the United States is simply municipal water that has been purified via reverse osmosis. Many homeowners are now discovering that rather than purchasing purified water, it is less expensive and more ecologically sound to install a small reverse osmosis unit in their own home. Simple carbon filters, such as Brita™, are also highly effective at purifying water of nonpolar contaminants (**Figure 7.37**). For fun, carbonators can also be purchased so that you can have your very own soda fountain, as shown in **Figure 7.38**.

Figure 7.38

Carbonating your own water is not only fun but also cheaper and more ecologically sound than purchasing your soda from a store.

Concept Review

Chapter 7

Summary of Terms

Concentration A quantitative measure of the amount of solute in a solution.

Dissolving The process of mixing a solute in a solvent to produce a homogeneous mixture.

Electrolyte Any dissolved ion that allows for the flow of an electric current through the solution.

Hard water Water containing large amounts of calcium and magnesium ions.

Hydrogen bond An unusually strong dipole-dipole attraction occurring between molecules that have a hydrogen atom covalently bonded to a small highly electronegative atom, usually nitrogen, oxygen, chlorine, or fluorine.

Induced dipole A temporarily uneven distribution of electrons in an otherwise nonpolar atom or molecule.

Insoluble Not capable of dissolving to any appreciable extent in a given solvent.

Molarity A common unit of concentration equal to the number of moles of a solute per liter of solution.

Mole A very large number equal to 6.02×10^{23} and usually used in reference to the number of atoms, ions, or molecules within a macroscopic amount of a material.

Osmosis The net flow (diffusion) of water across a semipermeable membrane from a region of low solute concentration to a region of high solute concentration.

Precipitate A solute that has come out of solution.

Reverse osmosis A technique for purifying water by forcing it through a semipermeable membrane into a region of lower solute concentration.

Saturated solution A solution containing the maximum amount of solute that will dissolve in its solvent.

Semipermeable membrane A membrane containing submicroscopic pores that allow passage of water molecules but not of larger solute ions or solute molecules.

Solubility The ability of a solute to dissolve in a given solvent.

Soluble Capable of dissolving to an appreciable extent in a given solvent.

Solute Any component in a solution that is not the solvent.

Solvent The component in a solution that is present in the largest amount.

Supersaturated Solution A solution in which more solvent is dissolved than is normal at a particular temperature.

Unsaturated solution A solution that is capable of dissolving additional solute.

Review Questions

7.1 Dipole Attractions

1. What is the primary difference between a chemical bond and an attraction between two molecules?

2. Which is stronger, the ion–dipole attraction or the induced dipole–induced dipole attraction?

3. What is a hydrogen bond?

4. Are induced dipoles permanent?

7.2 Solutions

5. What happens to the volume of a sugar solution as more sugar is dissolved in it?

6. Why is a ruby considered to be a solution?

7. Distinguish between a solute and a solvent.

7.3 Concentration and the Mole

8. What does it mean to say that a solution is concentrated?

9. Is 1 mole of particles a very large number of particles or a very small number?

10. Is concentration typically given with the volume of solvent or the volume of solution?

7.4 Solubility

11. Why does the solubility of a gas solute in a liquid solvent decrease with increasing temperature?

12. Why do sugar crystals dissolve faster when crushed?

13. Is sugar a polar or nonpolar substance?

14. Use the solubility rules to predict whether any of the products formed from these reactions will precipitate as a solid:

$$HCl\ (aq) + NaOH\ (aq) \rightarrow NaCl + H_2O$$

15. Use the solubility rules to predict whether any of the products formed from these reactions will precipitate as a solid:

$$Na_2SO_4\ (aq) + BaCl_2\ (aq) \rightarrow 2\ NaCl + BaSO_4$$

7.5 How Soap Works

16. Which portion of a soap molecule is nonpolar?

17. Water and soap are attracted to each other by what type of molecular attraction?

18. What is the difference between a soap and a detergent?

7.6 Softening Hard Water

19. What component of hard water makes it hard?

20. Why are soap molecules so attracted to calcium and magnesium ions?

21. Calcium and magnesium ions are more attracted to sodium carbonate than to soap. Why?

7.7 Purifying Drinking Water

22. Why is treated water sprayed into the air prior to being piped to users?

23. What are two ways in which people disinfect water in areas where municipal treatment facilities are not available?

24. What naturally occurring element has been contaminating the water supply of Bangladesh?

Quantitative Questions

25. Assume the total number of molecules in a solution is about 3 million trillion. One million trillion of these are molecules of some poison, while two million trillion of these are water molecules. What percentage of all the molecules in the glass are water?

26. Assume the total number of molecules in a solution is about 1,000,000 million trillion. One million trillion of these are molecules of some poison, while 999,999 million trillion of these are water molecules. What percentage of all the molecules in the glass are water?

27. You drink a small glass of water that is 99.9999% pure water and 0.0001% some poison. Assume the glass contains about a 1,000,000 million trillion molecules, which is about 30 mL. How many poison molecules did you just drink. Should you be concerned?

28. How much sodium chloride, in grams, is needed to make 15 L of a solution that has a concentration of 3.0 grams of sodium chloride per liter of solution?

29. If water is added to 1 mole of sodium chloride in a flask until the volume of the solution is 1 liter, what is the molarity of the solution? What is the molarity when water is added to 2 moles of sodium chloride to make 0.5 liter of solution?

30. A student is told to use 20.0 grams of sodium chloride to make an aqueous solution that has a concentration of 10.0 g/L (grams of sodium chloride per liter of solution). Assuming that 20.0 grams of sodium chloride has a volume of 7.50 milliliters, show that she will need about

1.99 liters of water to make this solution. In making this solution, should she add the solute to the solvent or the solvent to the solute?

31. Rank the following solutions in order of increasing concentration. Solution A: 0.5 moles of sucrose in 2.0 liters of solution. Solution B: 1.0 moles of sucrose in 3.0 liters of solution. Solution C: 1.5 moles of sucrose in 4.0 liters of solution.

32. List the following compounds in order of increasing boiling point: CI_4, CBr_4, CCl_4, CF_4.

33. Rank the following compounds in order of increasing solubility in water:

CH_3CH_2-OH
Ethanol

$CH_3CH_2CH_2CH_2$-OH
n-Butanol

$CH_3CH_2CH_2CH_2CH_2CH_2$-OH
n-Hexanol

34. Rank in order of increasing number of solute molecules:
    a. 7.0 L of 1.0 M sugar water
    b. 5.0 L of 2.0 M sugar water
    c. 2.0 L of 4.0 M sugar water

Solutions (Odd-Numbered)

1. A chemical bond is many times stronger than an attraction between molecules.

3. A hydrogen bond is a very strong dipole-dipole attraction involving a hydrogen atom bonded to a highly electronegative atom, such as oxygen.

5. The volume of a sugar solution gradually increases as more sugar is dissolved in it.

7. A solute is the component of lesser quantity in a solution (e.g., a pinch of salt in a glass of water). A solvent is the component of greater quantity in a solution (e.g., the water in a salt water solution).

9. A mole is a very large number: 6.02×10^{23}. For example, a mole of marbles would be enough to cover the entire land area of the United States to a depth greater than four meters.

11. The solubility of a gas decreases with increasing temperatures because the gas molecules will have more kinetic energy and be more likely to escape from solution. The greater kinetic energy of the solvent molecules also helps the gas molecules to escape.

13. Sugar is very polar as evidenced by its great solubility in water.

15. The barium sulfate, $BaSO_4$, should precipitate.

17. Water and soap are attracted to each other by ion-dipole attractions between each water molecule and the polar head of each soap molecule.

19. Large amounts of calcium and magnesium ions are found in hard water.

21. Sodium carbonate (Na_2CO_3) has a 2- charge in the carbonate ion $(CO_3)^{2-}$ to which calcium and magnesium are more attracted than to the 1- charge found in a molecule of soap. The hard water ions, calcium and magnesium, bind to the carbonate ions, which "softens" the water.

23. People disinfect their water by boiling it or adding disinfecting iodide tablets.

25. Divide the number of water molecules by the total number of molecules and multiply by 100 to get the percentage:

    2 million trillion/3 million trillion $\times$ 100 = 67%

27. Study the previous question and you'll find that you just ingested one million trillion poison molecules, along with 999,999 million trillion water molecules. Should you be concerned? Consider that the ratio of poison molecules to water molecules is about one to a million. This would be a concentration of 1 part per million (1 ppm). Whether this concentration is dangerous or not depends not only upon the poison, but also upon whether you drank this dose on a daily basis. Drinking this much arsenic daily would be very dangerous—arsenic poisoning becomes evident when the concentration is as little as 0.05 ppm! So think twice if you're ever tempted to buy water solely because it is touted to be 99.9999% pure. Water will always contain impurities. For water to be safe, these impurities need to be below or well-below toxic concentrations. Water containing 0.0000000001% arsenic, for example, is considered safe. Interestingly, however, an 8 oz glass

of such water, in most places, contains about 7 trillion arsenic atoms, which is a very, very tiny number compared to the 7×10^{24} water molecules.

29. a) 1 mole1 Liter = 1 Molar (1 M)
    b) 2 moles 0.5 Liters = 4 Molar (4 M)

31. Solution A = 0.25 M < Solution B = 0.33 M < Solution C = 0.375 M

33. The carbon-hydrogen structures are nonpolar. The structure for hexanol, therefore, is the most nonpolar of these three molecules, hence, it has the lowest solubility in water. Another way to look at this is as follows: the OH bond is a polar bond and this is what is needed to allow for good solubility in water. Only a small percentage of the *n*-hexanol molecule is made up of the OH bond. A greater percentage of the ethanol molecule (about a 33%) is made up of the OH bond. Water, therefore, has an easier time being attracted to an ethanol molecule than to a *n*-hexanol molecule. From left to right in order of increasing solubility in water:

$CH_3CH_2CH_2CH_2CH_2CH_2$-OH
n-Hexanol

$CH_3CH_2CH_2CH_2$-OH
n-Butanol

CH_3CH_2-OH
Ethanol

Solutions to Chapter 7
Calculation Corner

Concentrating on Solutions

1. Multiply the solution concentration by the final volume of the solution. This provides the amount of solute required: (380 g/L)(3 L) = 1140 g.

2. Divide the amount of solute by the solution concentration to obtain the amount of solution prepared: 20g/10 g/L = 2 L.

Concentrating on Dilutions

1. (6.00 *M* NaOH)(*x* mL) = (2.00 *M* NaOH)(465 mL)

   $x = 155$ mL

2. Units of volume cancel out within the math. Thus, so long as you have the same units of volume for before and after the solution, then the calculations will work.

3. (3.00 *M* KBr)(55.0 mL) = (1.00 *M*)(*x* mL)

   $x = 165$ mL

Chapter 7: Contextual Chemistry Essay

The Story of Superfund Act

Λ Figure A

A 1978 aerial view of the south end of the Love Canal district.

The story of one of the most significant environmental disasters involving chemical pollutants began in 1892. This was the year when developer William T. Love proposed the building of a canal linking the upper and lower portions of the Niagara River, which is the river feeding Niagara Falls in western New York. By 1910, however, the project had lost its financial backing and all that remained was a ditch some 60 feet wide and 3000 feet long. Located just east of the town of Niagara Falls, the ditch was purchased by the Hooker Chemical Corporation, which maintained it as a dump site for toxic chemical wastes, including poisonous heavy metals such as mercury and cancer-causing solvents such as benzene. By the early 1950s, Love Canal was completely filled and covered over with earth, creating a long vacant field.

At about this time, the population of the town of Niagara Falls was growing rapidly and more land was needed for the building of new communities. In 1953, a local school board, using the threat of eminent domain, convinced Hooker Chemical to sell the land above and around Love Canal. Hooker sold the land for $1, with a warning of the dangers of the toxic wastes and a disclaimer of all subsequent liability. An elementary school was soon built directly above the toxic wastes, along with 800 single-family homes and 240 apartments in the surrounding neighborhood.

Over the years, the buried canisters ruptured, spewing their toxic contents underground. Being close to Lake Erie, the water table of Love Canal is close to the surface. After heavy rains, the water table would rise to mix with the leaking toxic wastes, which would leach away toward the basements of neighboring houses. In some places, the wastes would leach directly to the surface. The residents complained about foul odors and health issues, but city and county officials did not know how to respond aside from installing fans in homes or covering pools of toxic wastes with dirt.

Finally, in 1978, health officials began to study the situation. They found numerous toxins in the surrounding soil and were able to correlate these toxins to reported health problems such as miscarriages and birth defects. The school was closed, and 239 families living on or immediately adjacent to Love Canal were evacuated. A fence was built around the canal to keep people away from the toxic wastes. The many families living just outside the fenced area, however, grew increasingly angry. They not only continued to face unusual health problems but also were unable to move away because their homes were no longer sellable. Their concerns captured little attention until on May 19, 1980, they held hostage two representatives of the Environmental Protection Agency (EPA), with the demand that the government buy their homes so that they could relocate. Within a few days, then-president Jimmy Carter promised the funding that allowed all Love Canal families to leave.

Love Canal turned the nation's attention to the problem of toxic wastes. It was soon recognized that Love Canal was just one of thousands of areas where toxic wastes were improperly disposed. To remedy this massive problem, in December 1980, Congress enacted the Comprehensive Environmental Response, Compensation, and Liability Act, which established a "Superfund" that would provide the money to clean up abandoned hazardous sites. This fund was to be supported by special taxes on industries related to the production of toxic chemicals Furthermore, in cases where the polluter could be identified, the act granted the government the authority to enforce a "polluter pays" policy. For example, the federal government spent $101 million from the Superfund to remedy the Love Canal toxic wastes. By 1995, federal courts ruled that the Hooker Corporation, despite its original disclaimer, was liable. Occidental Petroleum, which had bought Hooker in 1969, was thus required to reimburse the $101 million to the Superfund and to pay the $27 million in interest the Superfund would have earned had this money not been spent.

Over its history, however, the Superfund has received only a small fraction of its funding from the collection of fines from the polluters. The reason is that in many cases, the polluter cannot be identified or has since gone out of business or has no money to pay. Then in 1994, Congress decided that chemical industries should no longer pay the Superfund tax. As shown in **Figure C**, until that year, these taxes (shown in green) were the main source of funding. Since then, the government has had to dedicate increasing amounts of public money each year to keep the Superfund program afloat. For example, the 2009 economic stimulus program of the Obama administration directed $600 million to specific Superfund sites that were in danger of no longer being supported.

Since its enactment, the Superfund program has taken responsibility for about 1600 toxic-waste sites across the country. Cleanup efforts have been completed at about 300 of these sites. Of the still-active sites, another 700 or so are near completion. Most of these sites, however, are small sites where the cleanup has been relatively easy. A greater proportion of the remaining sites are the so-called "megasites," which will require decades of cleanup efforts and environmental monitoring.

Λ Figure B

Clean up at Love Canal involved the removal of easily accessible toxic-waste containers, along with polluted soil. It was deemed, however, that the bulk of the buried toxic wastes would be too dangerous to remove. Instead, the site was capped with a thick layer of water-resistant clay. Also, a trench system was dug around the site to direct groundwater to a specially designed water treatment facility.

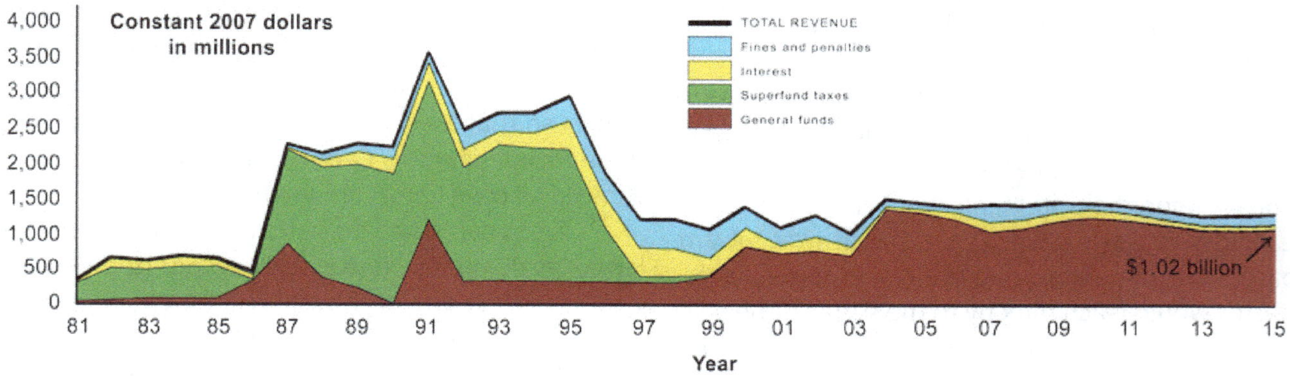

Λ Figure C

Partial contributions from various sources to the Superfund trust fund from 1981 through 2015. Note that in 1993, Superfund taxes accounted for about 70 percent of the trust revenue. More recently, the proportion of revenue coming in from fines has been small. Thus, the "polluter pays" ethic of the original Superfund Act has been essentially lost.

Λ Figure D

To find the Superfund site nearest to where you live and to learn the story of how it came to be, visit epa.gov/superfund.

CONCEPT **CHECK**

Why were heavy rains particularly dangerous for the former residents of Love Canal?

CHECK YOUR ANSWER

Heavy rains raised the water table, allowing it to mix with the hazardous wastes, which would then leach outward to the surrounding homes.

Think and Discuss

1. In a 1983 settlement, Occidental Petroleum paid \$20 million to 1328 residents of Love Canal and set aside another \$1 million for a medical trust fund. Do you think this was sufficient, too little, or too much? Why? Consider that the deeds of the homeowners contained clauses warning them of the health risks of the area.

2. Why did the town of Niagara Falls not have to pay into the Superfund, especially given that it pressured the Hooker Corporation into selling the land? Might the town have been liable had it taken the land against the will of Hooker by way of eminent domain?

3. In the 1940s, it was the accepted rule that if a company owned a parcel of land, the company had the right to use that land as a dump site for whatever materials it wanted to dispose, including hazardous wastes. Accordingly, Hooker had the right to use Love Canal as a toxic-waste repository. Was it fair, then, that the parent company of the Hooker Corporation had to pay for the cleanup of Love Canal?

4. Over many years, the Homeowners Association of Wide Beach Development, New York, along the shores of Lake Erie, sprayed thousands of gallons of waste oil onto area dirt roads to control dust. Unknown to the association, this oil contained dangerous cancer-causing chemicals known as PCBs. Should the Wide Beach Development qualify as a Superfund site, or should the Homeowners Association take care of the problem themselves?

5. The American Chemical Council, which represents chemical industries, argues that Superfund taxes unfairly punish all chemical manufacturers for the mistakes of a few. Furthermore, all consumers benefit from the products (from soaps to toys to computers) of chemical manufacturers, so supporting the Superfund with public taxpayer money is most appropriate. Do you agree or disagree? Why?

Author Responses to Think and Discuss

1. Despite the clause in the deeds, the Love Canal residents were largely unaware of the health risks of the area when they bought their homes. In fact, scientists were not fully aware of the health risks. They knew the health effects of when industry workers were accidentally exposed to large amounts of the hazardous chemicals over a single incident. But no one knew of the effects of prolonged exposure to much smaller amounts. Adjusting for inflation, $20 million in 1983 equals about $43 million in 2009. This amount was divided among the residents based upon need such that some individuals received more and other less. But if divided evenly, this would come to about $32,000 per individual (in 2009 dollars). This would correspond to about a year's salary, which would be a rather small compensation in cases where the resulting debilitation was lifelong.

2. In hindsight, we see that the local school boards and city officials were very wrong not to listen to the warnings of the Hooker Corporation. These entities representing a relatively small community, however, did not have the financial resources to tackle the hazardous waste problem. Keep in mind that their main function is to spend money, not make it. Occidental Petroleum, on the other hand, is a huge source of capital wealth, which is why, in part, it became the target of litigation. Had the school board pressed eminent domain, perhaps the litigation against Occidental would have taken longer or perhaps it might have been unsuccessful.

3. This sort of question was addressed by the Superfund Act, which boldly dictated that such previous "rights to pollute" were subservient to the greater harm caused by the release of toxic wastes into the environment.

4. Who has the best technical know-how and experience when it comes to cleaning up toxic wastes in the environment? Answer: Cleanup companies that have been contracted by way of the Superfund. Thus, an important advantage of the Superfund Act is that it has supported the training of hazardous waste clean-up experts. In fact, the Wide Beach Development was placed on the Superfund site roster in 1983. By 1992, contaminated soil had been removed, homes were decontaminated, roadways were paved, and a nearby wetland was restored.

5. If you drive a car, you take the risk of having an accident, which may or may not be your fault. Similarly, if you are a company that works with hazardous materials, you stand the chance of having these materials released dangerously into the environment. Perhaps, the Superfund tax should be re-worked to look more like a Superfund insurance premium. As companies at risk pay into this system, they are assured of protection in the event of some past, present, or future environmental disaster. Perhaps a company paying into this system would also be less liable for Superfund fines in the event it became the unwitting polluter.

Chapter 8

How Water Behaves

THE MAIN IDEA

Water is exceptional

∧ We are made of water, born into a world of water, and then forevermore dependent on water.

Water is so common in our lives that its many unusual properties easily escape our notice. Consider, for example, that water is the only chemical substance on our planet's surface that can be found abundantly in all three phases—solid, liquid, and gas. Another unique property of water is its great resistance to any change in temperature. As a result, the water in you moderates your body temperature, just as the oceans moderate global temperatures. Unlike most other liquids, which freeze from the bottom up, liquid water freezes from the top down. To the trained eye of a chemist, water is far from a usual substance. Rather, it's downright bizarre and exotic.

Almost all of the amazing properties of water are a consequence of the ability of water molecules to cling tenaciously to one another by way of electrical attractions. In this chapter, we explore the physical behavior of water while diving into the details and consequences of the "stickiness" of water molecules. We begin by exploring first the physical properties of solid water (ice); then those of liquid water; and finally those of gaseous water, also known as water vapor.

8.1 Open Structured Crystals

Experience tells us not to place a sealed glass jar of liquid water in the freezer, for we know that water expands as it freezes. Trapped in the jar, the freezing water expands outward with a force strong enough to shatter the glass into a hazardous mess or to pop the lid from the jar (**Figure 8.1**). This expansion occurs because when the water freezes, the water molecules arrange themselves in a six-sided crystalline structure that contains many open spaces. As **Figure 8.2** shows, a given number of water molecules in the liquid can get relatively close to one another and so occupy a certain volume. Water molecules in the crystalline structure of ice , however, occupy a greater volume than they do in liquid water. Consequently, ice is less dense compared to liquid water, which is why ice floats in water.

READING CHECK
Why does water expand as it freezes?

This property of expanding upon freezing is quite rare. The atoms or molecules of most frozen solids pack in such a way that the solid phase occupies a smaller volume than in the liquid phase (**Figure 8.3**).

Figure 8.1 >

Freezing water expands, which can damage the container holding it, be it a glass or metal pipe of outdoor plumbing.

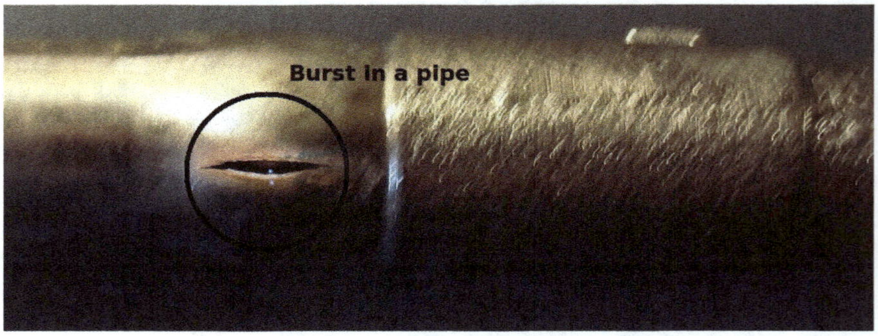

Figure 8.2 >

Water molecules in the liquid phase are arranged more compactly than are water molecules in the solid phase, in which they form an open crystalline structure.

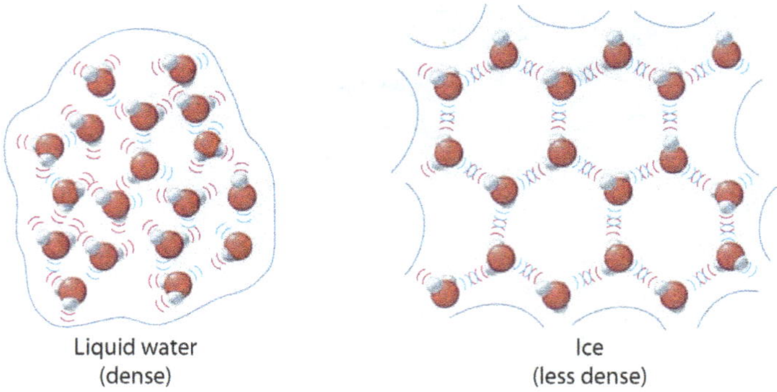

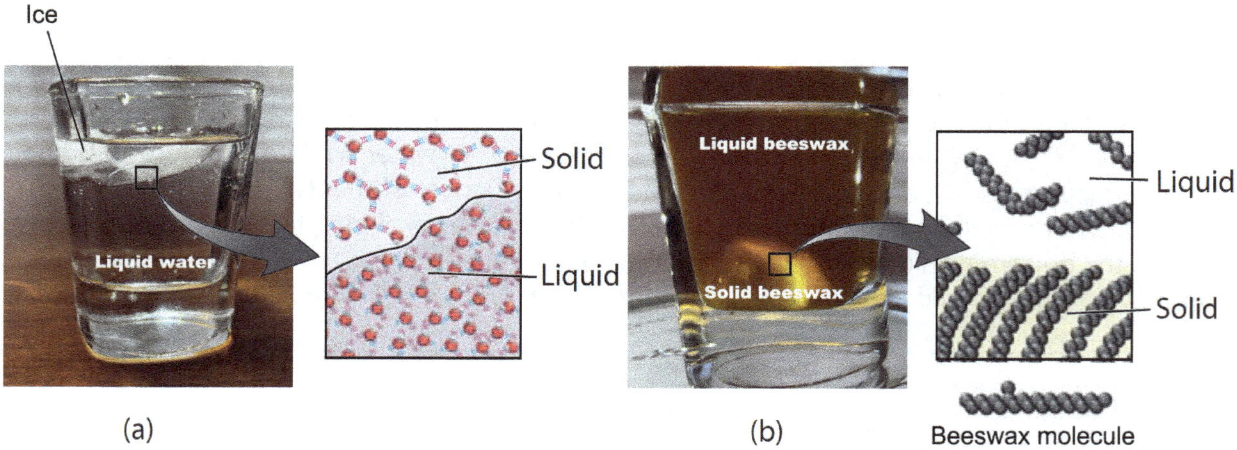

∧ Figure 8.3

Because water expands as it freezes, ice is less dense compared to liquid water and so floats in the water. (b) Like most other materials, beeswax is denser in its solid phase than in its liquid phase. Solid beeswax thus sinks in liquid beeswax.

CONCEPT **CHECK**

Are you interpreting correctly the illustration of water's open crystalline structure shown in Figure 8.2? If you are, you'll be able to answer this question: What's inside one of the open spaces?

a. air b. water vapor c. nothing

CHECK YOUR ANSWER

If air were in the spaces, the illustration would have to show the molecules that make up air, such as O_2 and N_2, which are comparable in size to water molecules. Any water vapor in the spaces would have to be shown as free-roaming water molecules spaced relatively far apart. The open spaces shown here represent nothing but empty space. The answer is c.

The hexagonal crystalline structure of H_2O molecules in ice has some interesting effects. Most snowflakes, like the one in **Figure 8.4**, share a similar hexagonal shape, which is the microscopic consequence of this molecular geometry. Also, applying great pressure to ice causes the open spaces to collapse. The result is the formation of liquid water. The great weight of a glacier, for example, causes the underside of the glacier to melt. With its wet underside, the glacier slowly slips down mountains toward the ocean, where it spawns icebergs.

Interestingly, ice is covered by a thin film of liquid water even at temperatures just below freezing. This is because ice's hexagonal structure requires support in three dimensions. At the surface, water molecules find nothing above them to cling to. The hexagonal structure at the surface is thus weakened to the point where it collapses into a thin film of liquid, which is what makes ice so slippery. An ice-skater's skate glides over this thin film of water, as shown in **Figure 8.5**.

So, what happens when the flat surfaces of two ice cubes are held together? The thin liquid films are no longer on a surface. Sandwiched between the two ice cubes, this liquid film freezes, effectively gluing the two ice cubes together. Lumps of snow are similarly "glued" together in the making of a snowball. Of course, if the weather is too cold, even the surfaces of the snow crystals are solid, which prevents the formation of snowballs no matter how hard you press the snow together. Skiers tend to prefer this "dry snow" because it behaves more like a powder than a sticky slush. An alternative to skiing in super cold weather is skiing at high altitudes, where lower atmospheric pressure favors thinner liquid films, and hence drier, more powder-like snow.

fyi FOR YOUR INFORMATION

Water expands upon freezing. By how much it expands is equal to the volume of ice that floats above the water's surface. This would be the tip of an iceberg floating in a freshwater lake.

∧ Figure 8.4

The six-sided geometry of ice crystals gives rise to the six-sided structure of snowflakes.

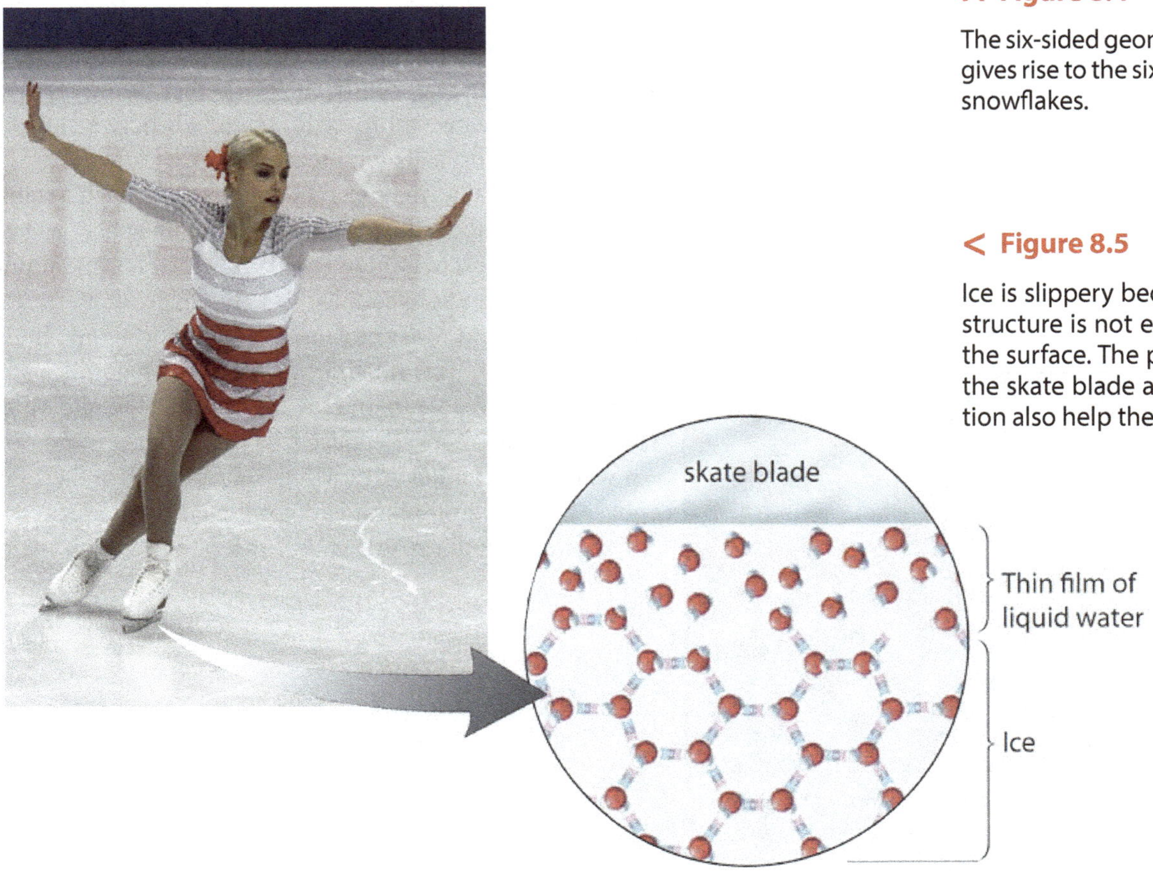

< Figure 8.5

Ice is slippery because its crystalline structure is not easily maintained at the surface. The pressure applied by the skate blade and the heat of friction also help the ice skater to glide.

8.2 Melting and Freezing

> **fyi FOR YOUR INFORMATION**
>
> If left undisturbed, chilled purified bottled water can become "supercooled," which means it remains liquid at temperatures below 0°C. This happens because crystal formation usually requires a substrate, such as a microparticle, upon which to start growing. Upon opening, the bottle of supercooled water will rapidly turn into an icy slush. Similarly, during an ice storm, raindrops formed in warm upper layers of the atmosphere become supercooled as they pass through colder lower layers. Upon hitting the ground, tree branches, and power lines, the supercooled raindrops, known as freezing rain, quickly freeze. The results can be beautiful but also quite hazardous, especially for aircraft.

As discussed in Section 2.7, *melting* occurs when a substance changes from solid to liquid, and *freezing* occurs when a substance changes from liquid to solid. When we view these processes from a molecular perspective, we see that melting and freezing occur simultaneously, as **Figure 8.6** illustrates.

The temperature 0°C is both the melting temperature and the freezing temperature of water. At this temperature, water molecules in the liquid phase are moving slowly enough that they tend to clump together to form ice crystals—they freeze. At this same temperature, however, water molecules in ice are vibrating with great commotion, much more than they vibrate at colder temperatures. Many thus break loose from the crystalline structure to form liquid water—they melt. Thus, melting and freezing occur simultaneously.

For water, 0°C is the special temperature at which the rate of ice formation equals the rate of liquid water formation. In other words, it is the temperature at which the opposite processes of melting and freezing counterbalance each other. This means that if a mixture of ice and liquid water is maintained at exactly 0°C, the two phases are able to coexist indefinitely.

Anytime we want a mixture of ice and liquid water at 0°C to freeze solid, we need to favor the rate of ice formation. This is accomplished by removing heat (cooling it down), a process that facilitates the formation of hydrogen bonds. As shown in **Figure 8.7**, when water molecules come together to form a hydrogen bond, heat energy is released. In order for the molecules to remain hydrogen-bonded, this heat energy must be removed—otherwise, it can be reabsorbed by the molecules, causing them to separate. The removal of heat energy therefore allows hydrogen bonds to remain intact after they have formed. As a result, there is the tendency for the ice crystals to grow.

Conversely, we can get a mixture of ice and liquid water at 0°C to melt completely by adding heat. This heat energy goes into breaking apart the hydrogen bonds that hold the water molecules together in the form of ice, as shown in **Figure 8.8**. Because more hydrogen bonds between water molecules are breaking, the ice crystals tend to melt.

Solutes tend to inhibit crystal formation. Anytime a solute, such as table salt or sugar, is added to water, the solute molecules take up space, as you learned in Section 7.2. When a solute is added to a mixture of ice and liquid water at 0°C, the solute molecules effectively decrease the number of liquid water molecules at the solid–liquid interface, as **Figure 8.9** illustrates. With fewer liquid water molecules available to join the ice crystals, the rate of ice formation decreases. Because ice is a relatively pure form of water, the number of molecules moving from the solid phase to the liquid phase is not affected by the presence of solute. The net result is that the rate at which water

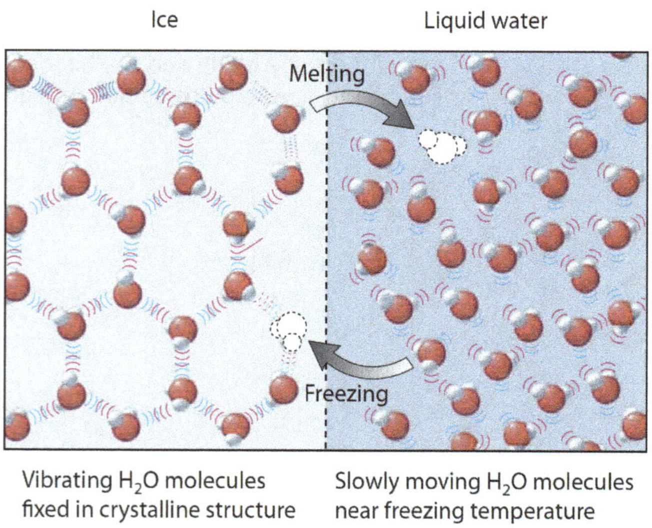

< Figure 8.6

At 0°C, ice crystals gain and lose water molecules simultaneously.

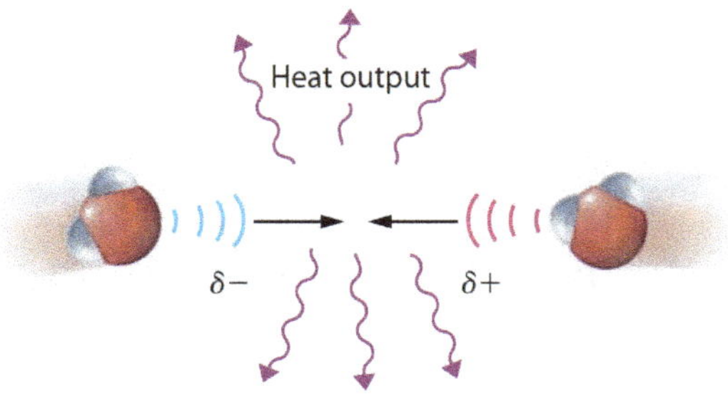

< Figure 8.7

As two water molecules come together to form a hydrogen bond, attractive electric forces cause them to accelerate toward each other. This results in an increase in their kinetic energies (the energy of motion), which is perceived on the macroscopic scale as heat energy.

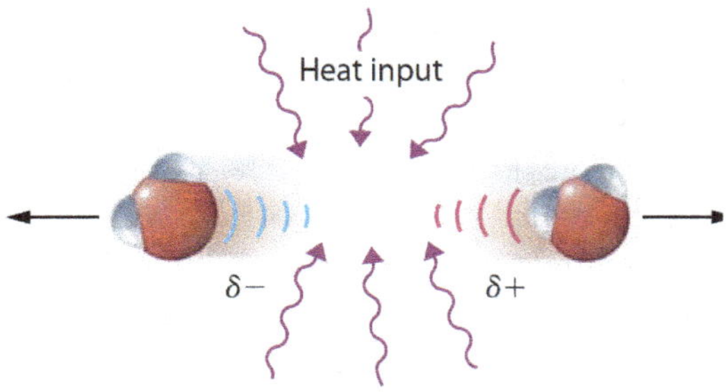

< Figure 8.8

Heat energy must be added in order to separate two water molecules held together by a hydrogen bond. This heat energy causes the molecules to vibrate so rapidly that the hydrogen bond breaks.

molecules leave the solid phase is greater than the rate at which they enter the solid phase. This imbalance can be compensated for by decreasing the temperature to below 0°C. At lower temperatures, water molecules in the liquid phase move more slowly and coalesce more easily. Thus, the rate of crystal formation is increased.

In general, adding anything to water lowers the freezing point. Antifreeze is a practical application of this process. The salting of icy roads is another.

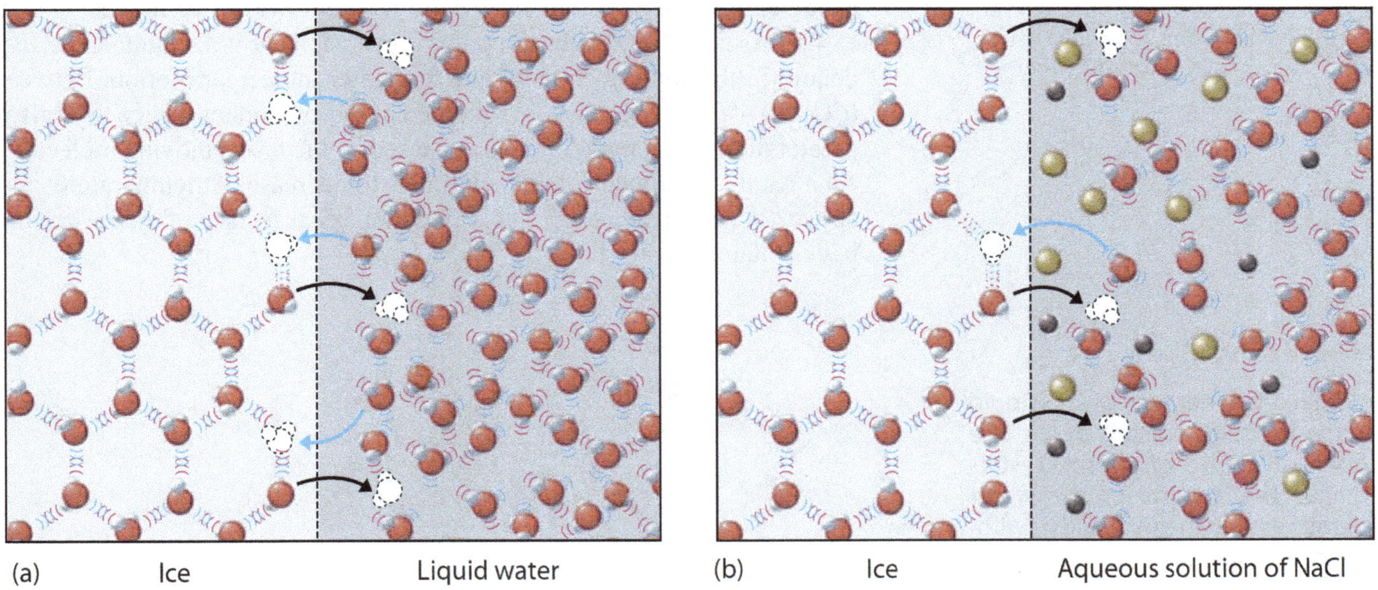

Λ Figure 8.9

(a) In a mixture of ice and liquid water at 0°C, the number of molecules entering the solid phase is equal to the number of molecules entering the liquid phase. (b) Adding a solute, such as sodium chloride, decreases the number of molecules entering the solid phase, because now there are fewer liquid molecules at the interface.

CONCEPT **CHECK**

How does a solute decrease the rate of freezing?

CHECK YOUR ANSWER

When a solute is present, there are fewer water molecules in contact with the surface of any ice crystal. This makes it more difficult for the ice crystal to grow.

Water Is Densest at 4°C

When the temperature of a substance is increased, its molecules vibrate faster and move farther apart on average. The result is that the substance expands. With few exceptions, all phases of matter—solids, liquids, and gases—expand when heated and contract when cooled. In many cases, these changes in volume are not very noticeable, but with careful observation, you can usually detect them. Telephone wires, for instance, are longer and sag more on a hot summer day than on a cold winter day. Metal lids on glass jars can often be loosened by heating them under hot water. If one part of a piece of glass is heated or cooled more rapidly than adjacent parts, the resulting expansion or contraction may break the glass.

Within any given phase, water also expands with increasing temperature and contracts with decreasing temperature. This is true of all three phases—ice, liquid water, and water vapor. Liquid water at near-freezing temperatures, however, is an exception.

READING CHECK

What forms in 0°C liquid water?

Liquid water at 0°C can flow just like any other liquid, but at 0°C, the temperature is cold enough that nanoscopic crystals of ice are able to form. These crystals slightly "bloat" the liquid water's volume, as shown in **Figure 8.10.** As the temperature is increased to above 0°C, more and more of these crystals collapse, and as a result, the volume of the liquid water *decreases*.

Figure 8.11 shows that between 0°C and 4°C, liquid water *contracts* as its temperature is raised. This contraction, however, continues only up to 4°C. As near-freezing liquid water is heated, there is a simultaneous tendency for the water to expand due to greater molecular motion. Between 0°C and 4°C, the decrease in volume caused by collapsing ice crystals is greater than the increase in volume caused by the faster-moving molecules. As a result, the water volume continues to decrease. At temperatures just above 4°C, expansion overrides contraction because most of the ice crystals have collapsed.

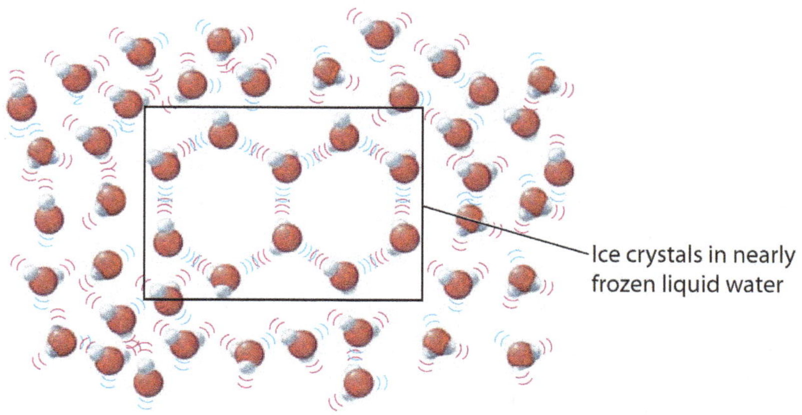

Figure 8.10 >

Within a few degrees of 0°C, liquid water contains crystals of ice. The open structure of these crystals makes the volume of the water slightly greater than it would be without the crystals.

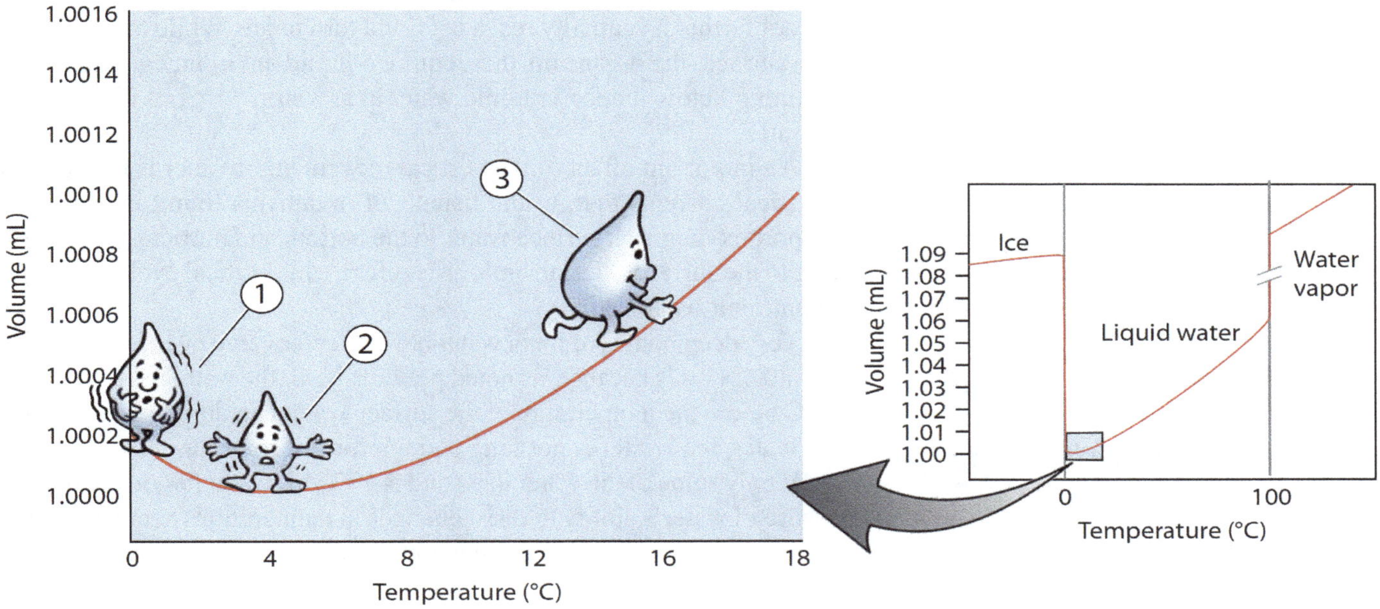

(1) Liquid water below 4°C is bloated with ice crystals.

(2) Upon warming, the crystals collapse, resulting in a smaller volume for the liquid water.

(3) Above 4°C, liquid water expands as it is heated because of greater molecular motion.

∧ Figure 8.11

Between 0°C and 4°C, the volume of liquid water decreases as the temperature increases. Above 4°C, water behaves the way all other substances do: its volume increases as its temperature increases. The volumes shown here are for a 1-gram sample.

Because of the effect of collapsing nanoscopic ice crystals, liquid water has its smallest volume and thus its greatest density at 4°C. (As discussed in Section 2.4, density is the amount of mass contained in a sample of anything divided by the volume of the sample.) By definition, 1 gram of pure water at this temperature has a volume of 1.0000 milliliter. As Figure 8.11 shows, 1 gram of liquid water at 0°C has an only slightly larger volume, 1.0002 milliliters. By comparison, 1 gram of ice at 0°C has a volume of 1.0870 milliliters. As can be seen in the small graph to the upper right in Figure 8.11, the volume of 1 gram of ice stays above 1.08 milliliters even below 0°C, meaning that even when ice is cooled to temperatures well below freezing, it is still less dense than liquid water.

Although liquid water at 4°C is only slightly denser than liquid water at 0°C, this small difference is of great importance in nature. Consider that if water were densest at its freezing point, as is true of most other liquids, the coldest water in a pond would settle to the bottom and the pond would freeze from the bottom up, destroying living organisms in winter months. Fortunately, this does not happen. As winter comes on and the temperature of the water drops, its density also drops. The entire volume of water in the pond does not cool all at once, however. Surface water cools first because it is in direct contact with the cold air. Being cooler than the underlying water, this surface water is denser and so sinks, with warmer water rising to replace it. That new batch of surface water then cools to the air temperature, gets denser as it does so, and sinks, only to be replaced by warmer water that cools. This process continues until the entire body of water has been cooled to 4°C. Then if the air temperature remains below 4°C, the surface water also cools to below 4°C. This surface water does not sink, however, because at this colder temperature, it is now less dense than the water below it. Thus, cooler, less dense water stays on the surface, where it can cool further, eventually reach 0°C, and turn to ice. While this ice forms at the surface, the organisms that require a liquid environment are happily swimming below the ice in liquid water at a "warm" 4°C, as **Figure 8.12** illustrates.

An important effect of the vertical movement of water is the creation of vertical currents that, to the benefit of organisms living in the water, transport oxygen-rich surface water to the bottom and nutrient-rich bottom water to the surface. Marine biologists refer to this vertical cycling of water and nutrients as *upwelling*.

CHEMICAL CONNECTIONS

How is a raindrop connected to a campfire?

Very deep bodies of fresh water are not ice-covered even in the coldest of winters. This is because, as noted previously, all the water must be cooled to 4°C before the temperature of the surface water can drop below 4°C. For deep water, the winter is not long enough for this to occur.

If only some of the water in a pond is 4°C, this water lies on the bottom. Because of water's ability to resist changes in temperature (Section 8.5) and its poor ability to conduct heat, the bottom of deep bodies of water in cold regions is a constant 4°C year-round.

CONCEPT **CHECK**

What was the precise temperature at the bottom of Lake Michigan on New Year's Eve in 1901?

CHECK YOUR ANSWER

If a body of water has 4°C water, then the temperature at the bottom of that body of water is 4°C, for the same reason that rocks are at the bottom. Both 4°C water and rocks are denser than water at any other temperature. If the body of water is deep and in a region of short summers, as is the case for Lake Michigan, the water at the bottom is 4°C year-round.

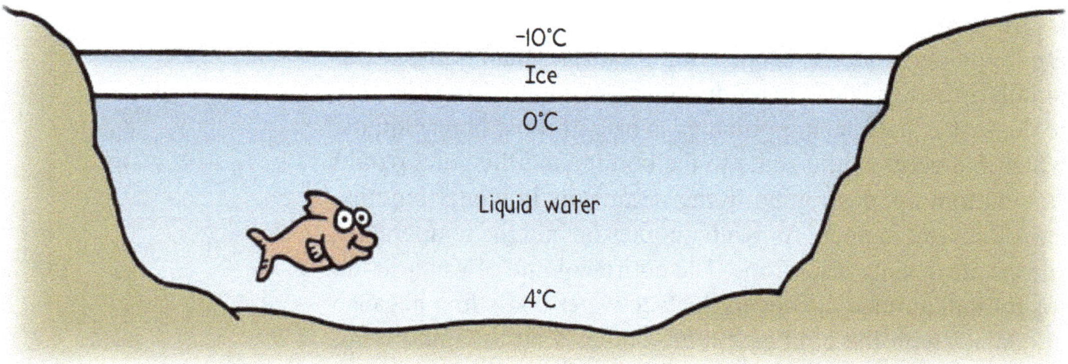

∧ Figure 8.12

As water cools to 4°C, it sinks. Then as water at the surface is cooled to below 4°C, it floats on top and can freeze. Only after surface ice forms can temperatures lower than 4°C extend down into the pond. This does not happen very readily, however, because surface ice insulates the liquid water from the cold air.

FOR YOUR INFORMATION

In most cases, the thermal expansion of liquids is greater than the thermal expansion of solids. This is why you should never top off your car's gas tank on a warm summer day. Gasoline is cooled by its underground storage at the gas station. After you place it in your car's tank, it warms and begins to expand. If the tank and the gasoline expanded at the same rate, they would expand together and no overflow would occur. Similarly, if the expansion of the glass of a thermometer were as great as the expansion of the mercury, the mercury would not rise with increasing temperature. The reason the mercury in the thermometer rises with increasing temperature is that the expansion of the liquid mercury is greater than the expansion of the glass.

8.3 The Stickiness of Water

In this section, we explore how water molecules in the liquid phase interact with one another via **cohesive forces**, which are forces of attraction between molecules of a single substance. For water, the cohesive forces are hydrogen bonds. We also explore how water molecules interact with other polar materials, such as glass, through **adhesive forces**, forces of attraction between molecules of two *different* substances.

Cohesive and adhesive forces involving water are dynamic. It is not one set of water molecules, for example, that holds a droplet of water to the side of a glass. Rather, the billions and billions of molecules in the droplet all take turns binding with the glass surface. Keep this in mind as you read this section and examine its illustrations, which, although informative, are merely freeze-frame depictions.

FOR YOUR INFORMATION

A related concept is viscosity, which is the resistance a fluid has to being deformed when flowing. In general, the stronger the intermolecular attractions within a fluid, the greater its viscosity. There is more hydrogen bonding in cold water than in hot water, which is why cold water is slightly more viscous than is hot water.

The Surface of Liquid Water Behaves Like an Elastic Film

Gently lay a dry paper clip on the surface of still water. If you're careful enough, the clip will rest on the surface, as shown in **Figure 8.13**. How can this be? Don't paper clips normally sink in water?

First, you should be aware that the paper clip is not floating *in* the water the way a boat floats. Rather, the clip is resting on the water surface. A slight depression in the surface is caused by the weight of the clip, which pushes down on the water much like the weight of a child pushes down on a trampoline. This elastic tendency found at the surface of a liquid is known as **surface tension**.

Surface tension in water is caused by hydrogen bonds. As shown in **Figure 8.14**, beneath the surface, each water molecule is attracted in every direction by neighboring molecules, with the result that there is no tendency to be pulled in any preferred direction. A water molecule on the surface, however, is pulled only by neighbors to each side and those below; there is no pull upward. The combined effect of these molecular attractions is thus to pull the molecule from the surface into the liquid. This tendency to pull surface molecules into the liquid causes the surface to become as small as possible, which means the surface behaves as if it were tightened into an elastic film. Lightweight objects that don't pierce the surface, such as a paper clip, are thus able to rest on the surface.

READING CHECK
What causes surface tension?

Surface tension accounts for the spherical shape of liquid drops. Raindrops, drops of oil, and falling drops of molten metal are all spherical

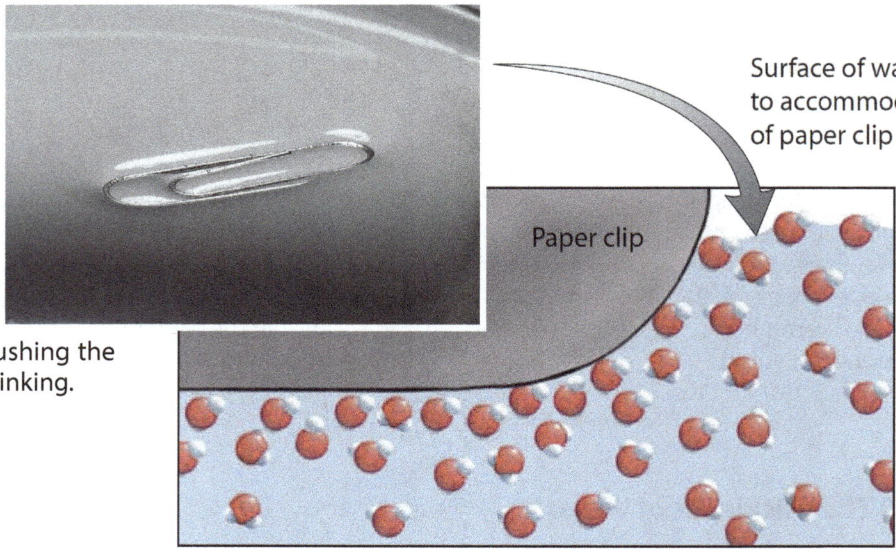

Figure 8.13 >

A paper clip rests on water, pushing the surface down slightly but not sinking.

because their surfaces tend to contract and this contraction forces each drop into the shape having the least surface area. This is a sphere, the geometric figure that has the least surface area for a given volume. In the weightless environment of an orbiting space station, a blob of water takes on a spherical shape naturally, as is shown in **Figure 8.15**. Back on the Earth, dewdrops on spider webs or drops on the downy leaves of plants are also spherical, except for the distortions caused by their weight.

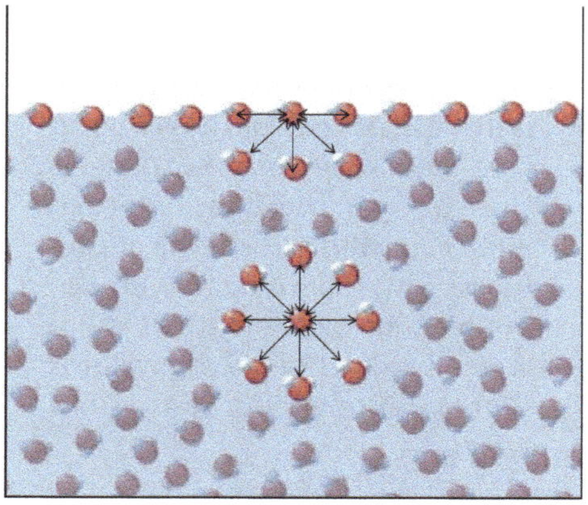

^ Figure 8.14

A molecule at the surface is pulled only sideways and downward by neighboring molecules. A molecule beneath the surface is pulled equally in all directions.

Surface tension is greater in water than in other common liquids because the hydrogen bonds in water are relatively strong. The surface tension in water is dramatically reduced, however, by the addition of soap or detergent. **Figure 8.16** shows that soap or detergent molecules tend to aggregate at the surface of water, with their nonpolar tails sticking out away from the water. At the surface, these molecules interfere with the hydrogen bonds between neighboring water molecules, thereby reducing the surface tension. Get a metal paper clip floating on the surface of some water and then carefully touch the water a few centimeters away with the corner of a bar of wet soap or a dab of liquid detergent. You will be amazed at how quickly the surface tension is reduced.

The strong surface tension of water is what prevents it from wetting materials that have nonpolar surfaces, such as waxy leaves, umbrellas, and freshly polished automobiles. Rather than wetting (spreading out evenly), the water beads. This is good if the idea is to keep water away. If we want to clean an object, however, the idea is to get it as wet as possible. This is another way in which soaps and detergents assist in cleaning, as **Figure 8.17** illustrates. By destroying water's surface tension, soaps and detergents

Figure 8.15 >

(a) The surface tension in a freely falling weightless blob of water causes the water to take on a spherical shape. (b) Small blobs of water resting on a surface would also be spheres if it weren't for the force of gravity, which squashes them into beads.

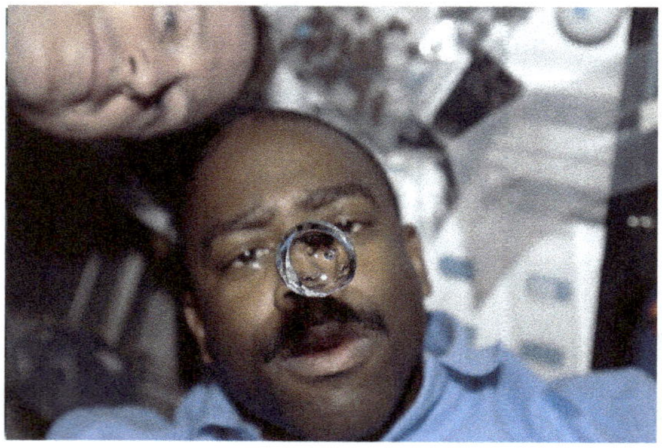

(a)

(b)

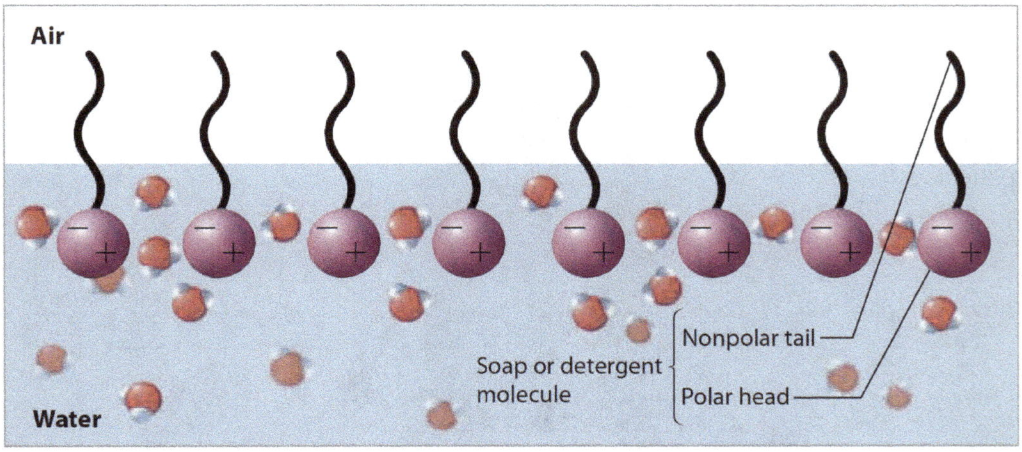

< **Figure 8.16**

Soap or detergent molecules align themselves at the surface of liquid water so that their nonpolar tails can escape the polarity of the water. This arrangement disrupts the water's surface tension.

enhance water's ability to wet. The nonpolar grime on dirty fabrics and dishes, for example, is then penetrated by the water more rapidly, making cleaning more efficient.

(a) (b)

∧ **Figure 8.17**

(a) Water beads on a surface that is clean and dry. (b) On a surface smeared with a thin film of detergent, water spreads evenly because the detergent has disrupted the water's surface tension.

Capillary Action—An Interplay of Adhesive and Cohesive Forces

Because glass is a polar substance, there are adhesive forces between glass and water. These adhesive forces are relatively strong, and the many water molecules adjacent to the inner surface of a glass container compete to interact with the glass. They do so to the point of climbing up the inner surface of the glass above the water surface. Take a close look at the tube of colored water in **Figure 8.18,** and you'll see that the water is curved up the sides of the glass. We call the curving of the water's surface (or the surface of any other liquid) at the interface between it and its container a *meniscus*.

Figure 8.19 illustrates what happens when a small-diameter glass tube is placed in water. (1) Adhesive forces initially cause a relatively steep meniscus. (2) As soon as the meniscus forms, the attractive cohesive forces among water molecules respond to the steepness by acting to minimize the surface area of the meniscus. The result is that the water level in the tube rises. (3) Adhesive forces will then cause the formation of another steep meniscus. (4) This is followed by the action of cohesive forces, which cause the steep meniscus to be "filled in." This cycle is repeated until the upward adhesive force equals the weight of the raised water in the tube. This rise of the liquid due to the interplay of adhesive and cohesive forces is called **capillary action**.

In a tube that has a smaller diameter, there is a smaller volume and less weight for a given height, and the water rises much higher, as **Figure 8.20** illustrates.

We see capillary action at work in many phenomena. If a paintbrush is dipped into water, the water rises into the narrow spaces between the bristles by capillary action. Hang your hair in the bathtub and water seeps up to your scalp the same way. This is how oil moves up a lamp wick and how water moves up a bath towel when one end hangs in water. Dip one end of a lump of sugar in coffee and the entire lump is quickly wet. The capillary action occurring between soil particles is important in bringing water to the roots of plants.

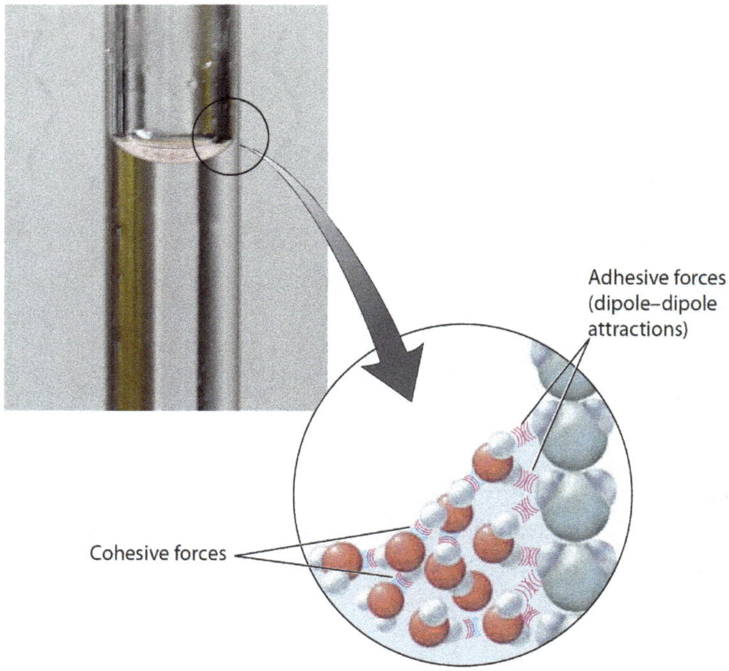

Figure 8.18 >

Adhesive forces between water and glass cause water molecules to creep up the sides of the glass, forming a meniscus.

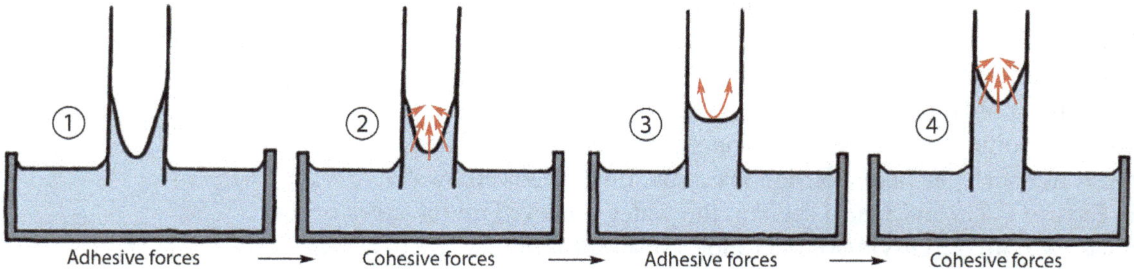

Figure 8.19 >

Water is drawn up a narrow glass tube by an interplay of adhesive and cohesive forces.

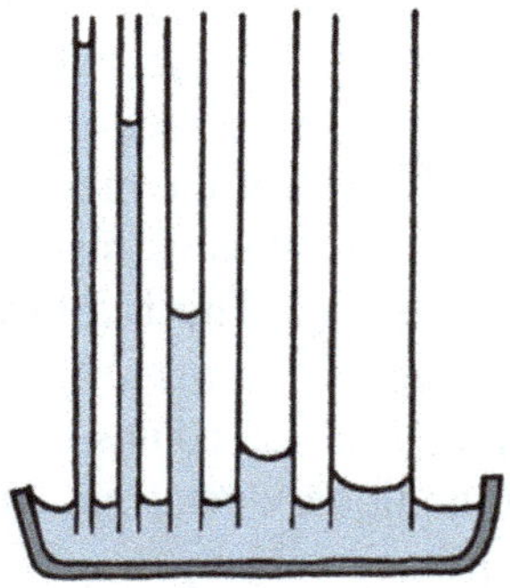

Figure 8.20 >

Capillary tubes. The smaller the diameter of the tube, the higher the liquid rises.

CONCEPT **CHECK**

An astronaut sticks a narrow glass tube into a blob of floating water while in orbit, and the tube fills with water. Why?

CHECK YOUR ANSWER

Capillary action causes the water to be drawn into the tube. In the freefall environment of an orbiting spacecraft, however, there is no downward force to stop this capillary action. As a result, the water continues to creep along the inner surface of the tube until the tube is filled.

8.4 Liquid and Gaseous Phases

Molecules of water in the liquid phase move about in all directions at different speeds. Some of these molecules may reach the liquid's surface moving fast enough to overcome the hydrogen bonds and escape into the gaseous phase. As presented in Section 2.6, this process of molecules converting from the liquid phase to the gaseous phase is called *evaporation* (also sometimes called *vaporization*). The opposite of evaporation is *condensation*—the changing of a gas to a liquid. At the surface of any body of water, there is a constant exchange of molecules from one phase to the other, as illustrated in **Figure 8.21**.

As evaporating molecules leave the liquid phase, they take their kinetic energy with them. This has the effect of lowering the average kinetic energy of all of the molecules remaining in the liquid, and the liquid is cooled, as **Figure 8.22** shows. Evaporation also has a cooling effect on the surrounding air, because liquid molecules that escape into the gaseous phase are moving relatively slowly compared to other molecules in the air. This makes sense when you consider that these newly arrived molecules lost a fair amount of their kinetic energy in overcoming the hydrogen bonds of the liquid phase. Adding these slower molecules to the surrounding air effectively decreases the average kinetic energy of all the molecules making up the air, and the air is cooled. So, no matter how you look at it, evaporation is a cooling process. **Figure 8.23** shows a useful application of this cooling effect.

As water cools, the rate of evaporation slows down, because fewer molecules have sufficient energy to escape the hydrogen bonds of the liquid phase. A higher rate of evaporation can be maintained if the water is in contact with a relatively warm surface, such as your skin. Body heat then flows from you into the water. In this way the water maintains a higher temperature, and evaporation continues at a relatively high rate. This is why you feel cool as you dry off after getting wet—you are losing body heat to the energy-requiring process of evaporation.

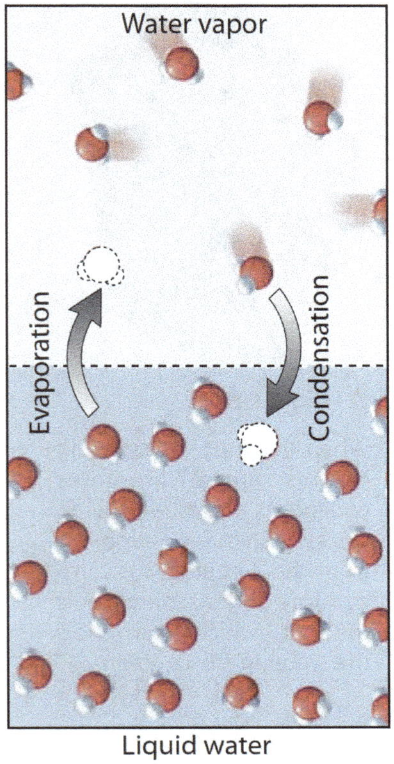

Figure 8.21

The exchange of molecules at the interface between liquid and gaseous water.

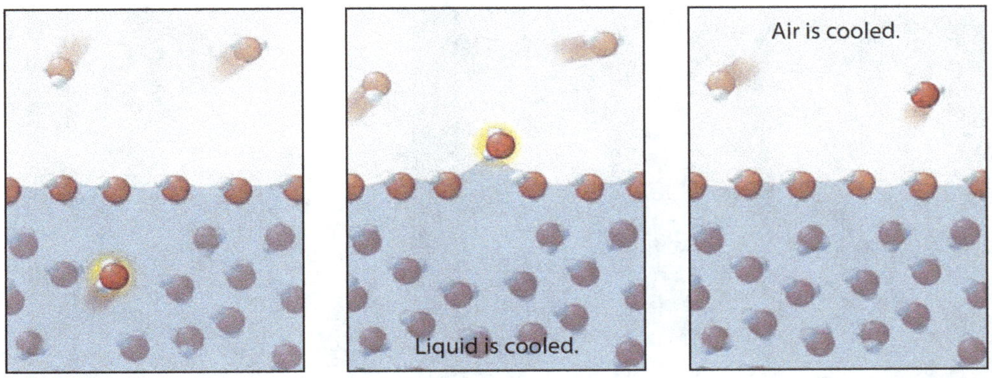

① Liquid water molecule having sufficient kinetic energy to overcome surface hydrogen bonding approaches liquid surface.

② Liquid water is cooled as it loses this high-speed water molecule.

③ Molecule enters gaseous phase, having lost kinetic energy in overcoming hydrogen bonding at the liquid surface. Air is cooled as it collects these slow-moving gaseous particles.

Figure 8.22

Evaporation is a cooling process.

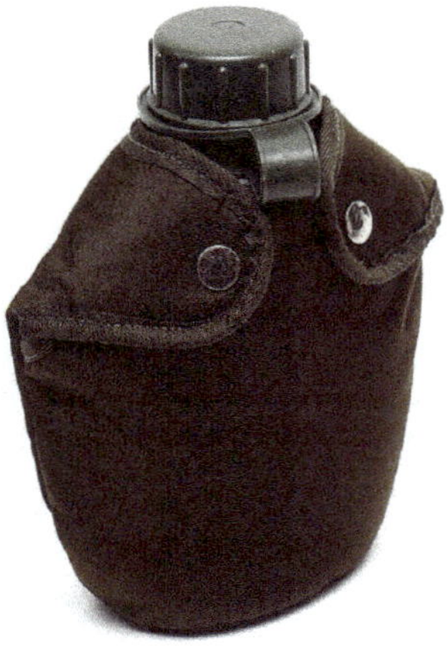

Figure 8.23

When wet, the cloth covering on this canteen promotes cooling. As the faster-moving water molecules evaporate from the wet cloth, its temperature decreases and cools the metal, which in turn cools the water in the canteen.

When your body overheats, your sweat glands produce perspiration. The evaporation of perspiration cools you and helps you maintain a stable body temperature. Many animals, such as those shown in **Figure 8.24**, do not have sweat glands and must cool themselves by other means.

Warm water evaporates, but so does cool water. The only difference is that cool water evaporates at a slower rate. Even frozen water "evaporates." This form of evaporation, in which molecules jump directly from the solid phase to the gaseous phase, is called *sublimation*, as introduced in Section 2.7. Because water molecules are so firmly held in the solid phase, frozen water does not release molecules into the gaseous phase as readily as liquid water does. Sublimation, however, does account for the loss of significant portions of snow and ice, especially on sunny, dry mountaintops. It's also why ice cubes left in the freezer for a long time tend to get smaller.

Condensation occurs as slow-moving water vapor molecules collide with and stick to the surface of a body of liquid water. Fast-moving water vapor molecules tend to bounce off each other or off the liquid surface, losing little of their kinetic energy. Only the slowest gas molecules condense into the liquid phase, as **Figure 8.25** illustrates. As this happens, energy is released as hydrogen bonds are formed. This energy is absorbed by the liquid and increases its temperature. Condensation involves the removal of slower-moving water vapor molecules from the gaseous phase. The average kinetic energy of the remaining water vapor molecules is therefore increased, which means that the water vapor is warmer. So, no matter how you look at it, condensation is a warming process.

A burn to your skin from 100°C water vapor is much more damaging than a burn from 100°C liquid water. Why? Remember that temperature and energy are not the same thing. When it comes to being burned, it's the energy that causes harm. The water vapor gives up considerable energy when it condenses to a liquid and wets the skin. This energy released by condensation is utilized in heating systems, such as the household radiator shown in **Figure 8.26**.

(a) (b)

Figure 8.24

(a) Dogs have no sweat glands (except between their toes). They cool themselves by panting. In this way, evaporation occurs in the mouth and the bronchial tract. (b) Pigs have no sweat glands and, therefore, cannot cool off by the evaporation of perspiration. Instead, they wallow in the mud to cool themselves.

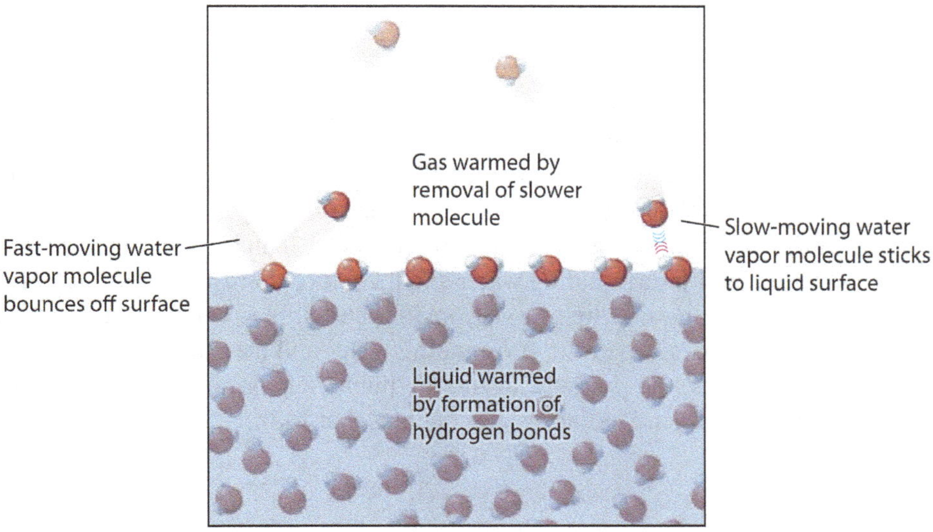

^ Figure 8.25

Condensation is a warming process.

^ Figure 8.26

Heat energy is given up by water vapor when the vapor condenses inside the radiator.

CONCEPT CHECK

If water were less "sticky," would you be cooled more or less by its evaporation?

CHECK YOUR ANSWER

Water molecules leave the liquid phase only when they have enough kinetic energy to overcome hydrogen bonding. It is hydrogen bonding that makes water sticky, so to say that water is less sticky is to say that hydrogen bonds are weaker than they actually are. Then at a given temperature, more molecules in the liquid phase would have sufficient kinetic energy to overcome the weaker hydrogen bonds and escape into the gaseous phase, carrying heat away from the liquid. The cooling power of evaporating water would therefore be greater. This is why less "sticky" substances, such as rubbing alcohol, have a noticeably greater cooling effect as they evaporate.

The water vapor in our atmosphere also gives up energy as it condenses. This is the energy source for many weather systems, such as hurricanes, which derive much of their energy from the condensation of water vapor contained in humid tropical air, as **Figure 8.27** illustrates. The formation of 1 inch of rain over an area of 1 square mile yields the energy equivalent of about 32,000 tons of exploded dynamite. We explore this further in the Chapter Essay on "Winds from Water".

^ Figure 8.27

As it condenses, the water vapor in humid tropical air releases ample quantities of heat. Continued condensation can sometimes lead to powerful storm systems, such as a hurricane, shown here from space.

After you take a shower, even a cold one, you are warmed by the heat energy released as the water vapor in the shower stall condenses. You quickly sense the difference if you step out of the stall, as the chilly guy in **Figure 8.28** is finding out. Away from the moisture, the rate of evaporation is much higher than the rate of condensation, and as a result, you feel chilly. When you remain in the shower stall, where the humidity is higher, the rate of condensation is increased, so you feel that much warmer. Now you know why you can dry yourself with a towel much more comfortably if you remain in the shower stall.

Spend a July afternoon in dry Tucson or Las Vegas and you'll soon notice that the evaporation rate is appreciably greater than the condensation rate. The result of this pronounced evaporation is a much cooler feeling than you would experience on a same-temperature July afternoon in New York City or New Orleans. In these humid locations, condensation outpaces evaporation, and you feel the warming effect as water vapor in the air condenses on your skin.

∧ Figure 8.28

If you're chilly outside the shower stall, step back inside and be warmed by the condensation of the excess water vapor there.

Boiling Is Evaporation beneath the Surface

When liquid water is heated to a sufficiently high temperature, bubbles of water vapor form beneath the surface, as we saw in Section 2.7. These bubbles are buoyed to the surface, where they escape, and we say the liquid is *boiling*. As shown in **Figure 8.29**, bubbles can form only when the pressure of the vapor inside them is equal to or greater than the combined pressure exerted by the surrounding water and the atmosphere above. At the boiling point of the liquid, the pressure inside the bubbles equals or exceeds the combined pressure of the surrounding water and the atmosphere. At lower temperatures, the pressure inside the bubbles is not enough, and the surrounding pressure collapses any bubbles that form.

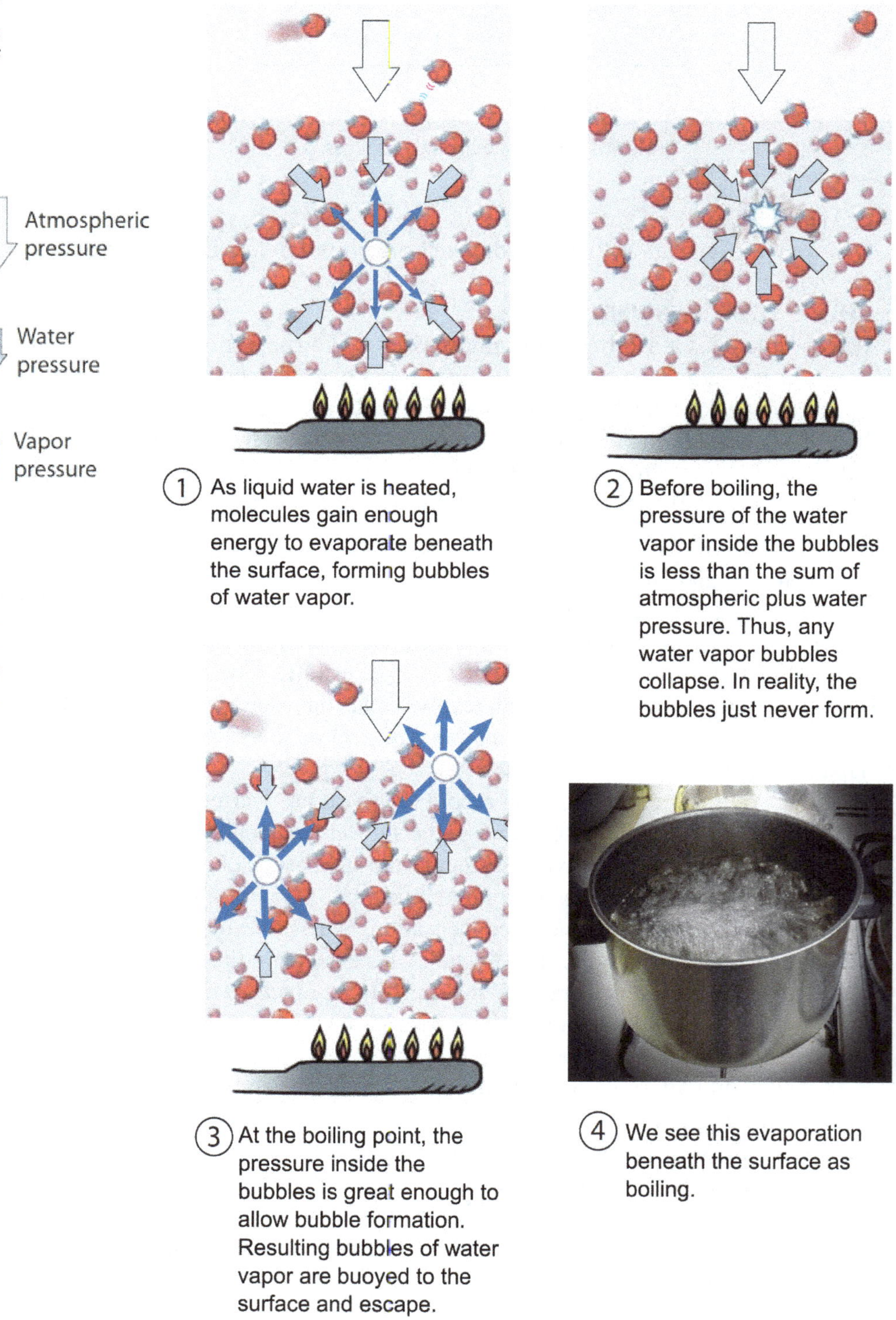

Figure 8.29 >

Boiling occurs when water molecules in the liquid move fast enough to generate bubbles of water vapor beneath the surface of the liquid.

READING CHECK

What does the pressure inside a bubble of boiling water equal or exceed?

At what point boiling begins depends not only on temperature but also on pressure. This includes the pressure from the weight of the surrounding liquid as well as the pressure from the weight of the atmosphere at the surface. As atmospheric pressure increases, the vapor molecules inside any bubbles that form must move faster in order to exert enough pressure from inside the bubble to counteract the additional atmospheric pressure. So increasing the pressure exerted on the surface of a liquid raises its boiling point. A cooking application of this effect of increased pressure is shown in **Figure 8.30**.

Conversely, lower atmospheric pressure (as at high altitudes) decreases the boiling point of the liquid, as **Figure 8.31** illustrates. In Denver, Colorado, the Mile-High City, for example, water boils at 95°C instead of the 100°C boiling temperature at sea level. If you try to cook food in boiling water that is cooler than 100°C, you must wait a longer time for proper cooking. A three-minute boiled egg in Denver is runny and undercooked. If the temperature of the boiling water were very low, food would not cook at all. As the German mountaineer Heinrich Harrer noted in his book *Seven Years in Tibet*, at an elevation of 4500 meters (15,000 feet) or higher, you can sip a cup of boiling tea without any danger of burning your mouth.

Figure 8.30

The tight lid of a pressure cooker holds pressurized vapor above the water surface, and this inhibits boiling. In this way, the boiling temperature of the water is increased. Any food placed in this hotter water cooks more quickly than food placed in water boiling at 100°C.

Boiling, like evaporation, is a cooling process. At first thought, this may seem surprising—perhaps because we usually associate boiling with heating. But heating water is one thing; boiling is another. As shown in **Figure 8.32**, boiling water is cooled by boiling as fast as it is heated by the energy from the heat source. So boiling water remains at a constant temperature. If cooling did not take place, continued application of heat to a pot of boiling water would raise the temperature of the water. The reason the pressure cooker in Figure 8.30 reaches higher temperatures is that boiling is forestalled by increased pressure, which in effect prevents cooling.

CHEMICAL CONNECTIONS

How is a hurricane connected to the hydrogen bond?

A simple experiment that dramatically shows the cooling effect of evaporation and boiling consists of a shallow dish of room-temperature water in a vacuum jar. When the pressure in the jar is slowly reduced by a vacuum pump, the water starts to boil. The boiling process takes heat away from the water, which consequently cools. As the pressure is further reduced, more and more of the slower-moving liquid molecules boil away. Continued

(a) (b)

Figure 8.31

The boiling point of water (as well as other liquids) decreases with increasing altitude. (a) At about 5000 feet elevation, full boiling water registers a temperature of about 95°C. (b) At even higher elevations, the boiling temperature of water can be much less.

CONCEPT CHECK

If the water level in a bowl of water remains unchanged from one day to the next, can you conclude that no evaporation or condensation is taking place?

CHECK YOUR ANSWER

Not at all, for there is much activity taking place at the molecular level. Both evaporation and condensation occur continuously and simultaneously. The fact that the water level remains constant indicates equal rates of evaporation and condensation—the number of H_2O molecules leaving the liquid surface by evaporation is equal to the number entering the liquid by condensation.

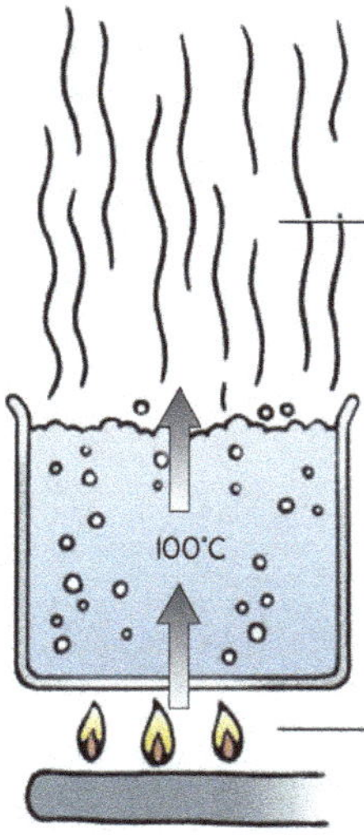

Figure 8.32

Heating warms the water from below, and boiling cools it from above. The net result is a constant temperature for the water.

boiling results in a lowering of the temperature until the freezing point of approximately 0°C is reached. Continued cooling by boiling causes ice to form over the surface of the bubbling water. Boiling and freezing take place at the same time! The frozen bubbles of boiling water in **Figure 8.33** are a remarkable sight.

Spray some drops of coffee into a vacuum chamber and they, too, boil until they freeze. Even after they are frozen, the water molecules continue to evaporate into the vacuum until all that is left to be seen are little crystals of coffee solids. This is how freeze-dried coffee is made. The low temperature of this process tends to keep the chemical structure of the coffee solids from changing. When hot water is added, much of the original flavor of the coffee is retained.

The refrigerator also employs the cooling effect of boiling. A liquid coolant that has a low boiling point is pumped into the coils inside the refrigerator, where the liquid boils (evaporates) and draws heat from the food stored in the refrigerator. Then the coolant in its gas phase, along with its added energy, is directed outside the refrigerator to coils located in the back, appropriately called condensation coils, where heat is given off to the air as the coolant condenses back to a liquid. A motor pumps the coolant through the system as it undergoes the cyclic process of vaporization and condensation. The next time you're near a refrigerator, place your hand near the condensation coils in the back; you'll feel the heat that has been extracted from inside.

An air conditioner employs the same principle, pumping heat energy from inside a building to outside the building. Turn the air conditioner around so that inside cold is pumped outside and the air conditioner becomes a type of heater known as a heat pump.

Figure 8.33

In a vacuum, water can freeze and boil at the same time. This is known as water's *triple point*, which is where it can exist as a solid, liquid, and vapor simultaneously.

8.5 Water's Specific Heat

Have you ever noticed that some foods stay hot much longer than others? The filling of a hot apple pie can burn your tongue while the crust does not, even when the pie was just taken out of the oven. A piece of toast may be comfortably eaten a few seconds after coming from a hot toaster, whereas you must wait several minutes before eating hot soup.

Different substances have different capacities for storing energy. This is because different materials absorb energy in different ways. The added energy may increase the jiggling motion of molecules, which raises the temperature, or it may pull apart the attractions among molecules and therefore go into potential energy, which does not raise the temperature. Generally, there is a combination of the two ways.

It takes 4.184 joules of energy to raise the temperature of 1 gram of liquid water by 1°C. As you can see in **Figure 8.34**, it takes only about one-ninth as much energy to raise the temperature of 1 gram of iron by the same amount. In other words, compared to iron, water absorbs more heat for the same change in temperature. We say water has a higher **specific heat**, defined as the quantity of heat required to change the temperature of 1 gram of a substance by 1°C.

READING CHECK

What happens to the energy added to a material?

We can think of specific heat as thermal inertia. As you learned in Section 2.3, *inertia* is a term used in physics to signify the resistance of an object to a change in its state of motion. Specific heat is like a thermal version of inertia because it signifies the resistance of a substance to a change in temperature. With a higher specific heat, more energy is required to change the temperature. Each substance has its own characteristic specific heat, which may be used to assist in its identification. Some typical values are given in **Table 8.1**.

FOR YOUR INFORMATION

A material with a high specific heat has a great capacity to hold onto heat, which is why specific heat is sometimes referred to as heat capacity. Water has a great capacity for holding onto heat. You can think of water as a very effective "heat sponge."

Guess why water has such a high specific heat. Once again, the answer is hydrogen bonds. When heat is applied to water, much of the heat is consumed in breaking hydrogen bonds. Broken hydrogen bonds are a form of potential energy (just as two magnets pulled apart are a form of potential energy). Much of the heat added to water, therefore, is stored as this potential energy. Consequently, less heat is available to increase the kinetic energy of the water molecules. Because temperature is a measure of kinetic energy, we find that as water is heated, its temperature rises slowly. By the same token, when water is cooled, its temperature drops slowly. Why? Because as the kinetic energy decreases, molecules slow down and more hydrogen

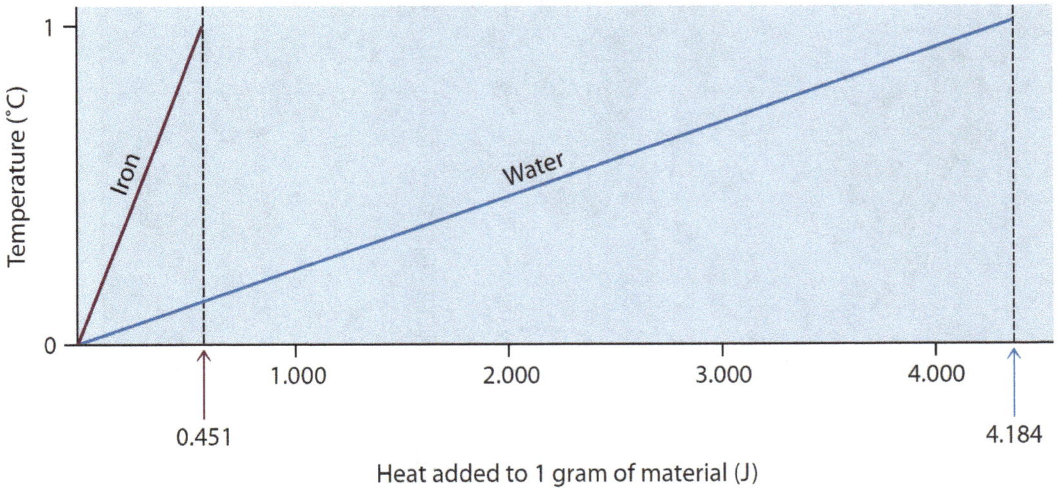

< Figure 8.34

It takes only 0.451 joule of heat to raise the temperature of 1 gram of iron by 1°C. A 1-gram sample of water, by contrast, requires a whopping 4.184 joules for the same temperature change.

> **FOR YOUR INFORMATION**
>
> Recall from Section 8.2 that adding salt lowers the freezing point of water. The reason is because the presence of a solute decreases the number of water molecules in contact with the surface of an ice crystal. Interestingly, salt water has a higher boiling point for similar reasons. In salt water, fewer water molecules at the liquid's surface are able to escape into the gaseous phase. This makes evaporation more difficult. Higher temperatures, therefore, are needed to get the salt water boiling.

TABLE 8.1 Specific Heat for Some Common Materials*

MATERIAL	SPECIFIC HEAT (J/g · °C)
Liquid ammonia, NH_3	4.70
Liquid water, H_2O	4.184
Ethylene glycol, $C_2H_6O_2$ (antifreeze)	2.42
Ice, H_2O	2.01
Water vapor, H_2O	2.0
Aluminum, Al	0.90
Iron, Fe	0.451
Silver, Ag	0.24
Gold, Au	0.13

*Specific heat assumes no change in the phase of a material.

bonds are able to re-form. This in turn releases heat that helps to maintain the temperature. So, as water cools down it actually heats itself back up by the formation of hydrogen bonds. That's why it takes so long for water to cool down.

Water's High Specific Heat Influences Global Climate

The tendency of liquid water to resist changes in temperature improves the climate in many places. For example, notice the high latitude of Europe in **Figure 8.35**. If water did not have a high specific heat, the countries of Europe would be as cold as the northeastern regions of Canada, because both Europe and Canada get about the same amount of sunlight per square kilometer of surface area. An ocean current carries warm water northeast from the Caribbean. The water holds much of its thermal energy long enough to reach the North Atlantic off the coast of Europe, where the water then cools. The energy released, 4.184 joules per degree Celsius for each gram of water that cools, is carried by the westerly winds (winds that blow west to east) over the European continent.

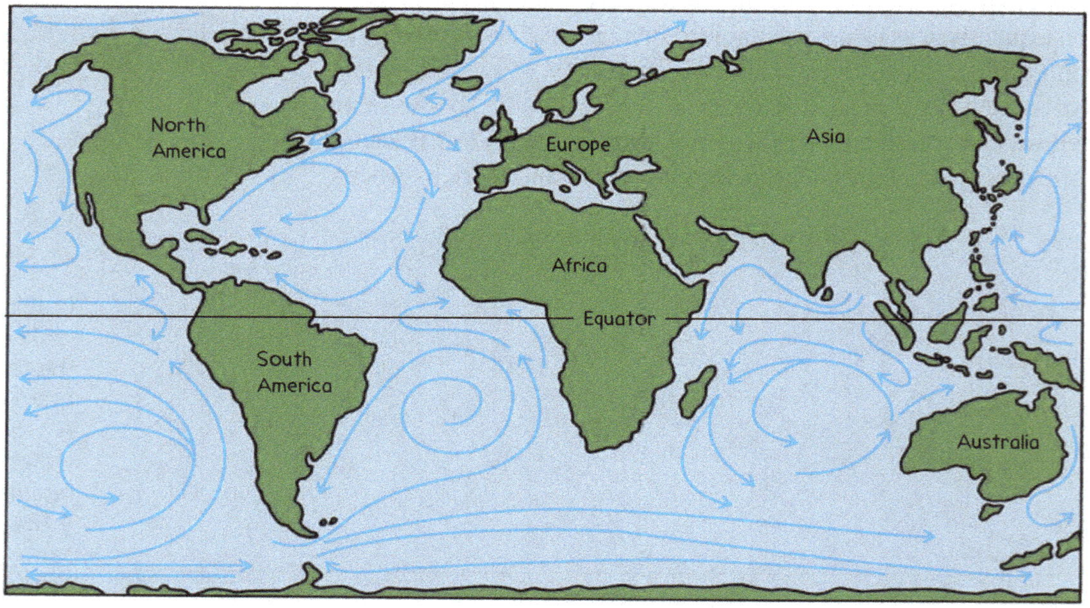

Λ Figure 8.35
Many ocean currents, shown in blue, distribute heat from the warmer equatorial regions to the colder polar regions

CONCEPT **CHECK**

Hydrogen bonds are not broken as heat is applied to ice (provided the ice doesn't melt) or water vapor. Would you therefore expect ice and water vapor to have specific heats that are greater or less than that of liquid water?

CHECK YOUR ANSWER

As Table 8.1 shows, the specific heats of ice and water vapor are about half that of liquid water. Only liquid water has a remarkable specific heat. This is because the liquid phase is the only phase in which hydrogen bonds are continually breaking and re-forming.

The winds in the latitudes of North America are westerly. On the western coast of the continent, therefore, air moves from the Pacific Ocean to the land. Because of water's high specific heat, ocean temperatures do not vary much from summer to winter. In winter, the water warms the air, which is then blown eastward over the coastal regions. In summer, the water cools the air and the coastal regions are cooled. On the eastern coast of the continent, the temperature moderating effects of the Atlantic Ocean are significant, but because the winds blow from the west—over land—temperature ranges in the east are much greater than in the west. San Francisco, for example, is warmer in winter and cooler in summer than is Washington, D.C., which is at about the same latitude.

Islands and peninsulas, because they are more or less surrounded by water, do not have the extremes of temperatures observed in the interior of a continent. The high summer temperatures and low winter temperatures common in Manitoba and the Dakotas, for example, are largely due to the absence of large bodies of water. Europeans, islanders, and people living near ocean air currents can be glad that water has such a high specific heat.

Public discussions of global climate change tend to focus on the average temperature of the atmosphere, which is increasing in concert with increasing levels of carbon dioxide, a potent greenhouse gas. An average rise of 3°C is understood to be problematic. This is discussed in more detail in Chapter 16. But what of the oceans, which are not only much more massive, but also have a significantly higher specific heat? The amount of energy required to raise the temperature of the oceans by 3°C is many orders of magnitude greater than that required to raise the temperature of the atmosphere.

Knowing about water's high specific heat, you'll understand that the bulk of solar radiation trapped by the atmosphere simply makes its way to the oceans. So, all of us, not just islanders, benefit from water's ability to absorb energy and thus moderate the climate. That's the good news.

The bad news is that direct measurements show the average surface ocean temperature has increased by about 0.7 °C over the past century as shown in **Figure 8.36**. This represents a colossal amount of energy. Because water expands upon warming, this has contributed to a rise in sea level as shown in **Figure 8.37**. Significantly, the northern polar ice cap is gradually disappearing and the warmer oceans are breeding grounds for more violent weather systems. Furthermore, warmer oceans are able to hold less carbon dioxide, which accelerates the increase in atmospheric carbon dioxide levels. Clearly, public discussions should also focus on the changing state of our oceans.

In summary, the oceans moderate our climate in great part because of water's high specific heat. Any temperature changes to the oceans will necessarily be slow so as not to be noticed over a single decade. But the changes that occur will also necessarily be enduring.

fyi FOR YOUR INFORMATION

Hot water has less surface tension than cold water does because its faster-moving molecules are not bonded as tightly. With its lower surface tension, hot water is more effective at penetrating the fabrics of our clothing, which is why so many people choose to wash their clothes with hot water. Hot water's low surface tension also permits the grease of soups to collect in liquidy blobs on the surface of soup. When the soup cools, the surface tension of the water increases and the grease spreads over the surface of the soup, making the soup taste more "greasy." Hot soup tastes different from cold soup primarily because the surface tension of the water in the soup changes with temperature.

Figure 8.36 >

This graph shows how the average surface temperature of the world's oceans has changed since 1880. Source: National Oceanic and Atmospheric Administration (NOAA) 2021.

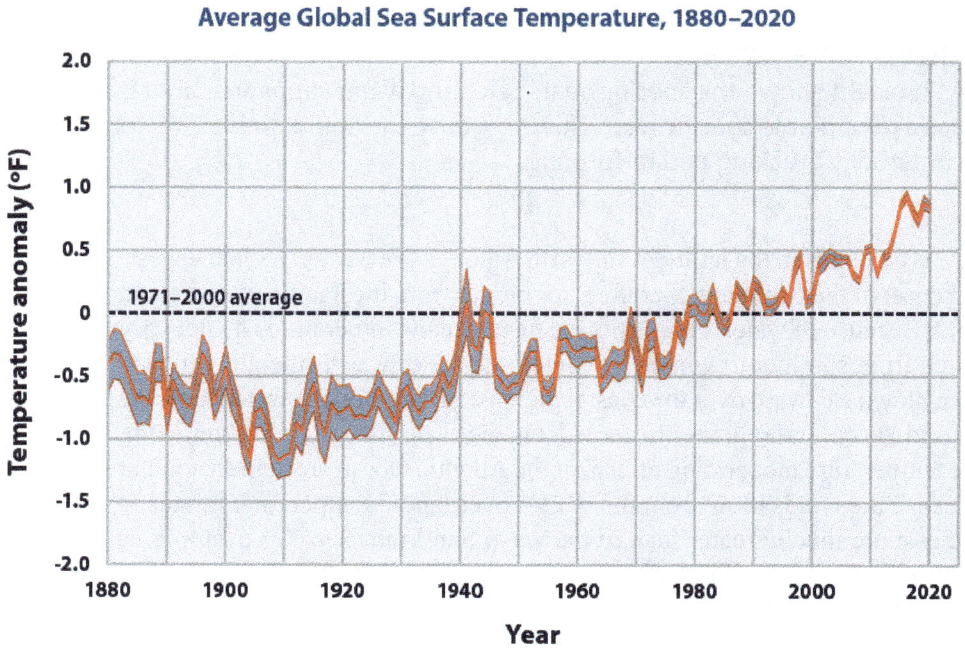

Data source: NOAA (National Oceanic and Atmospheric Administration). 2021. Extended reconstructed sea surface temperature (ERSST.v5). Accessed February 2021. www.ncdc.noaa.gov/data-access/marineocean-data/extended-reconstructed-sea-surface-temperature-ersst.

For more information, visit U.S. EPA's "Climate Change Indicators in the United States" at www.epa.gov/climate-indicators.

Figure 8.37 >

This graph shows how the average sea level of the oceans has changed since 1880. Measurements are from tidal records and, more recently, from satellite data. Source: National Oceanic and Atmospheric Administration (NOAA) 2021.

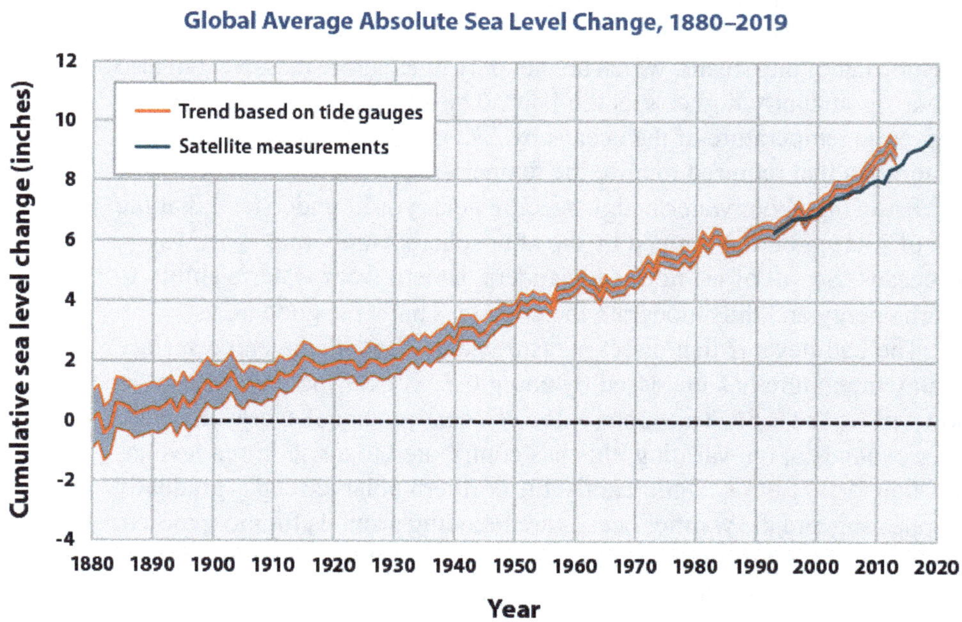

Data sources:

- CSIRO (Commonwealth Scientific and Industrial Research Organisation). 2017 update to data originally published in: Church, J.A., and N.J. White. 2011. Sea-level rise from the late 19th to the early 21st century. Surv. Geophys. 32:585–602. Accessed September 2017. www.cmar.csiro.au/sealevel/sl_data_cmar.html.
- NOAA (National Oceanic and Atmospheric Administration). 2021. Laboratory for Satellite Altimetry: Sea level rise. Accessed March 2021. www.star.nesdis.noaa.gov/sod/lsa/SeaLevelRise/LSA_SLR_timeseries_global.php.

For more information, visit U.S. EPA's "Climate Change Indicators in the United States" at www.epa.gov/climate-indicators.

FOR YOUR INFORMATION

The outer edge of a carousel platform always provides the fastest ride, because it travels a greater distance for each rotation. Similarly, because the Earth spins, those living closer to the equator travel farther, and hence faster, in a single day. When objects not fixed to the ground—such as air, water, airplanes, and ballistic missiles—move toward the equator, they are unable to keep up with the ever-faster moving ground beneath them. Because the Earth rotates to the east, these free-moving objects lag behind, which gives them an apparent motion to the west. Conversely, a free-moving object heading away from the equator deviates to the east, because it is maintaining the greater eastward- directed speed it had by being closer to the equator. This is called the *Coriolis effect*. Ocean currents heading toward the equator, for example, tend to deviate to the west, while currents heading away from the equator tend to deviate to the east. This explains why ocean currents in the northern hemisphere tend to circulate clockwise, while those in the southern hemisphere tend to circulate counterclockwise, as is evident in Figure 8.35. Why do you suppose airplane pilots need to be well acquainted with the Coriolis effect?

CONCEPT CHECK

Which has a higher specific heat, water or sand?

CHECK YOUR ANSWER

As suggested by the illustration, the temperature of water increases less compared to the temperature of sand in the same sunlight. Water therefore has the higher specific heat. Those who visit the beach frequently know that beach sand quickly turns hot on a sunny day, while the water remains relatively cool. At night, however, the sand feels quite cool, while the water's temperature feels about the same as it did during the day.

Calculation Corner: How Heat Changes Temperature

Heat must be applied to increase the temperature of a material. Conversely, heat must be withdrawn from a material in order to decrease its temperature. The amount of heat required for a given temperature change is calculated from the equation

$$\text{heat} = \text{specific heat} \times \text{mass} \times \text{temperature change}$$

We can use this formula for any material, provided there is no change of phase over the course of the temperature change. The value of the temperature change is obtained by subtracting the initial temperature T_i from the final temperature T_f:

$$\text{temperature change} = T_f - T_i$$

EXAMPLE 1

How much heat is required to increase the temperature of 1.00 gram of liquid water from an initial temperature of 30.0°C to a final temperature of 40.0°C?

ANSWER 1

The temperature change is

$$T_f - T_i = 40.0°\text{C} - 30.0°\text{C} = +10.0°\text{C}.$$

To find the amount of heat needed for this temperature change, multiply this positive change by the water's specific heat and mass:

$$\text{heat} = (4.184\ \text{J/g°C})(1.00\ \text{g})(10.0°\text{C}) = 41.8\ \text{J}$$

The temperature decrease that occurs when heat is removed from a material is indicated by a negative sign, as is shown in the next example.

EXAMPLE 2

A glass containing 10.0 grams of water at an initial temperature of 25.0°C is placed in a refrigerator. How much heat does the refrigerator remove from the water as the water is brought to a final temperature of 10.0°C?

ANSWER 2

The temperature change is

$$T_f - T_i = 10.0°\text{C} - 25.0°\text{C} = -15.0°\text{C}.$$

To find the heat removed, multiply this negative temperature change by the water's specific heat and mass:

$$\text{heat} = (4.184\ \text{J/g°C})(10.0\ \text{g})(-15.0°\text{C}) = -628\ \text{J}$$

YOUR TURN

1. A residential water heater raises the temperature of 100,000 grams of liquid water (about 26 gallons) from 25.0°C to 55.0°C. Show that 12,552,000 J of heat was applied.

2. Show that 402 J of heat must be extracted from a 10.0 gram ice cube (specific heat = 2.01 J/g°C) in order to bring its temperature from a chilly -10.0°C to an even chillier -30.0°C?

8.6 Phase Changes and Energy

Any phase change involves the breaking or forming of molecular attractions. The changing of a substance from a solid to a liquid to a gas, for example, involves the breaking of molecular attractions. Phase changes in this direction therefore require the input of energy. Conversely, the changing of a substance from a gas to a liquid to a solid involves the forming of molecular attractions. Phase changes in this direction therefore result in the release of energy. Both these directions are summarized in **Figure 8.38**.

Consider a 1-gram piece of ice at –50°C put on a stove to heat. A thermometer in the container reveals a slow increase in temperature up to 0°C, as shown in **Figure 8.39**. At 0°C, the temperature stops rising, even though heat is still being added, and for now, all the added heat goes into melting the 0°C ice, as indicated in **Figure 8.40**. This process of melting 1 gram of ice requires 335 joules. Only when all of the ice has melted does the temperature begin to rise again. Then for every 4.184 joules absorbed, the water increases its temperature by 1°C until the boiling temperature, 100°C, is reached. At 100°C, the temperature again stops rising even though heat is still being added, for now all the added heat is going into evaporating the liquid water to water vapor. The water must absorb a stunning 2259 joules of heat to evaporate all the liquid water. Finally, when all the liquid water has become vapor at 100°C, the temperature begins to rise once again and continues to rise as long as heat is added.

When the phase changes are in the opposite direction, the amounts of heat energy shown in Figure 8.39 are the amounts *released*—2259 joules per gram when water vapor condenses to liquid water and 335 joules per gram when liquid water turns to ice. The processes are reversible.

READING CHECK
What does the change of a substance's phase involve?

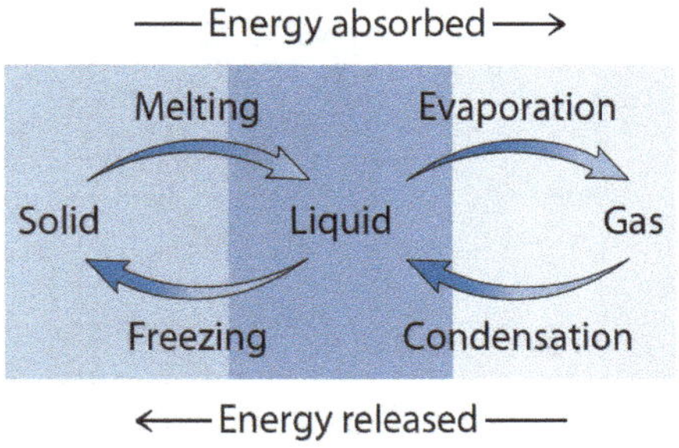

∧ Figure 8.38

Energy changes with change of phase.

CONCEPT CHECK

From Figure 8.39, deduce how much energy (in joules) is transferred when 1 gram of

a. water vapor at 100°C condenses to liquid water at 100°C.
b. liquid water at 100°C cools to liquid water at 0°C.
c. liquid water at 0°C freezes to ice at 0°C.
d. water vapor at 100°C turns to ice at 0°C.

CHECK YOUR ANSWER

a. 2259 joules
b. 418 joules
c. 335 joules
d. 3012 joules (2259 joules + 418 joules + 335 joules)

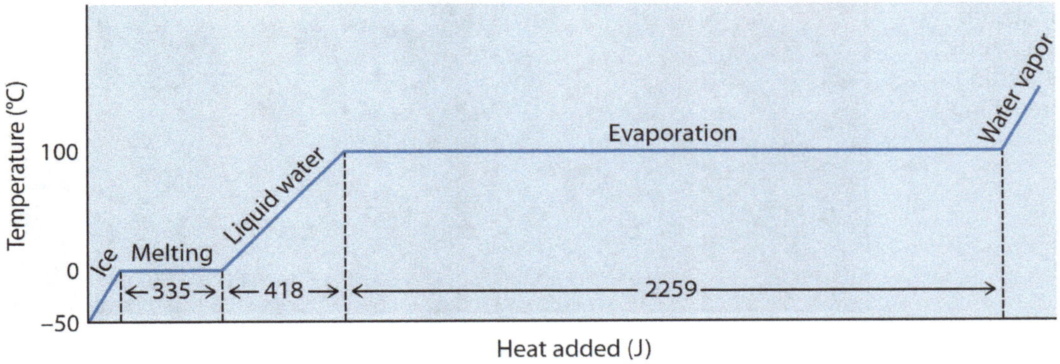

< Figure 8.39

A graph showing the heat energy involved in converting 1 gram of ice initially at -50°C to water vapor. The horizontal portions of the graph represent regions of constant temperature.

∧ Figure 8.40

Add heat to melting ice and there is no change in temperature. The heat is consumed in breaking hydrogen bonds.

The amount of heat energy required to change a solid to a liquid is called the **heat of melting**, and the amount of heat energy released when a liquid freezes is called the **heat of freezing***. Water has a heat of melting of +335 joules per gram. The positive sign indicates that this is the amount of heat energy that must be *added* to ice to melt it. Water's heat of freezing is –335 joules per gram. The negative sign indicates that this is the amount of heat energy that is released from liquid water as it freezes—the same amount that was required to melt it.

The amount of heat energy required to change a liquid to a gas is called the **heat of vaporization**. For water, this is +2259 joules per gram. The amount of heat energy released when a gas condenses is called the **heat of condensation**. For water, this is –2259 joules per gram. These values are high relative to the heats of vaporization and condensation for most other substances. This is due to the large amount of energy required to break the hydrogen bonds in the liquid to form the water vapor (heat of vaporization). Likewise, there is a large amount of energy released when the hydrogen bonds form as the water vapor condenses to the liquid (heat of condensation).

Although water vapor at 100°C and liquid water at 100°C have the same temperature, each gram of the vapor contains an additional 2259 joules of potential energy. As the molecules bind together to a liquid phase, this 2259 joules per gram is released to the surroundings in the form of heat. In other words, the potential energy of the far-apart water vapor molecules transforms to heat as the molecules get closer together. This is like the potential energy of two attracting magnets separated from each other; when released, their potential energy is converted first to kinetic energy and then to heat as the magnets strike each other.

CHEMICAL CONNECTIONS

How is the time it takes to warm a pot of water connected to the consistently warm climate of Hawaii?

Water's high heat of vaporization allows you to *briefly* touch your wetted finger to a hot skillet or hot stove without harm. This is because energy that ordinarily would go into burning your finger goes instead into changing the phase of the moisture on your finger from liquid to vapor. You can judge the hotness of a clothes iron in the same way—with a *wet* finger.

The firefighters in **Figure 8.41** know that certain types of flames are best extinguished with a fine mist of water rather than a steady stream. The fine mist readily turns to water vapor and in doing so quickly absorbs heat energy and cools the burning material.

Figure 8.41 >

Water extinguishes a flame by absorbing much of the heat the fire needs to sustain itself, as well as by wetting, which blocks oxygen from reaching the burning material.

*The term *heat of fusion* is commonly used to indicate either heat of melting or heat of freezing.

Water's high heat of vaporization makes walking barefooted on red-hot coals more comfortable, as shown in **Figure 8.42**. When your feet are wet, either because they are perspiring or because you are stepping off wet grass, much of the heat from the coals is absorbed by the water and not by your skin. (Firewalking also relies on the fact that wood is a poor conductor of heat, even when it is in the form of red-hot coals.)

Many of the properties of water we have explored in this chapter at the molecular level are nicely summarized in the scene shown in **Figure 8.43**. First, notice that the massive bear is jumping from floating ice. The ice floats because hydrogen bonds hold together the H_2O molecules in the ice in an open crystalline structure that makes the ice less dense than the liquid water. Because so much energy is required both to melt ice and to evaporate liquid water, most of the Arctic ice remains in the solid phase throughout the year. Where the seawater is in contact with the ice, the temperature is lower than 0°C because salts dissolved in the seawater inhibit the formation of ice crystals and thereby lower the freezing point of the seawater. The high specific heat of the Arctic Ocean beneath the bear moderates the Arctic climate. It gets cold in the Arctic in winter, yes, but not as cold as it gets in Antarctica, where the specific heat of the mile-thick ice is only half that of the liquid water predominating in the Arctic. Covered by ice with its much lower specific heat, Antarctica experiences much greater extremes in temperature than does the Arctic.

∧ **Figure 8.42**

Tracy Suchocki walks with wetted bare feet across red-hot wood coals without harm.

CONCEPT **CHECK**

Can you add heat to ice without melting it?

CHECK YOUR ANSWER

A common misconception is that ice cannot have a temperature lower than 0°C. In fact, ice can have any temperature below 0°C, down to absolute zero, −273°C. Adding heat to ice below 0°C raises its temperature, say, from −200°C to −100°C. So long as its temperature stays below 0°C, the ice does not melt.

In the previous chapter, we focused on the physical behavior of molecules and ions in general. In this chapter, the main focus has been on the extraordinary physical behavior of water molecules in particular. For the next several chapters, we shift our attention to the *chemical* behavior of molecules and ions, which change their fundamental identities as they chemically react with one another.

< **Figure 8.43**

Besides color, what remarkable properties of water can you find in this photograph?

Concept Review

Chapter 8

Summary of Terms

Adhesive Force An attractive force between molecules of two different substances.

Capillary Action The rising of liquid into a small vertical space due to the interplay of cohesive and adhesive forces.

Cohesive Force An attractive force between molecules of the same substance.

Heat of Condensation The energy released by a substance as it transforms from gas to liquid.

Heat of Freezing The heat energy released by a substance as it transforms from liquid to solid.

Heat of Melting The heat energy absorbed by a substance as it transforms from solid to liquid.

Heat of Vaporization The heat energy absorbed by a substance as it transforms from liquid to gas.

Meniscus The curving of the surface of a liquid at the interface between the liquid surface and its container.

Specific Heat The quantity of heat required to change the temperature of 1 gram of a substance by 1 Celsius degree.

Surface Tension The elastic tendency found at the surface of a liquid.

Review Questions

8.1 Open Structured Crystals

1. What accounts for the fact that ice is less dense than water?
2. What is inside one of the open spaces of an ice crystal?
3. What happens to ice when great pressure is applied to it?

8.2 Melting and Freezing

4. What is released when a hydrogen bond forms between two water molecules?
5. When the temperature of 0°C liquid water is increased slightly, does the water undergo a net expansion or a net contraction?
6. At what temperature do the competing effects of contraction and expansion produce the smallest volume for liquid water?

8.3 The Stickiness of Water

7. What is the difference between cohesive forces and adhesive forces?
8. In what direction is a water molecule on the surface pulled?
9. Does liquid water rise higher in a narrow tube or a wide tube?

8.4 Liquid and Gaseous Phases

10. Do all the molecules in a liquid have about the same speed?
11. Why do we feel uncomfortably warm on a hot, humid day?
12. Is it the pressure or the higher temperature that cooks food faster in a pressure cooker?
13. What condition permits liquid water to boil at a temperature below 100°C?

8.5 Water's Specific Heat

14. Is it easy or difficult to change the temperature of a substance that has a low specific heat?
15. Does a substance that heats up quickly have a high or a low specific heat?

16. How does the specific heat of liquid water compare with the specific heats of other common materials?

17. What within liquid water is most responsible for its unusual specific heat?

8.6 Phase Changes and Energy

18. When liquid water freezes, is heat released to the surroundings or absorbed from the surroundings?

19. Why doesn't the temperature of melting ice rise as the ice is heated?

20. How much heat is needed to melt 1 gram of ice? Give your answer in joules.

21. Why does it take so much more energy to boil 10 grams of liquid water than to melt 10 grams of ice?

Quantitative Questions

22. A walnut stuck to a pin is burned beneath a can containing 100.0 grams of water at 21°C. After the walnut has completely burned, the water's final temperature is 28°C. How much heat energy came from the burning walnut?

23. How much heat is required to raise the temperature of 100,000 grams of iron by 30 C°?

24. By how much will the temperature of 5.0 grams of liquid water increase upon the addition of 230 joules of heat?

25. How much heat is required to raise the temperature of 1.00 gram of water from –5.00°C to +5.00°C?

26. How much heat is required to raise the temperature of 1.00 gram of water from absolute zero, -273°C, to 100. °C?

27. How much energy is required to transform 1.00 gram of water from 100°C liquid to 100°C water vapor?

28. Why does transforming water from 100°C liquid to 100°C gas require much more energy than to raise the temperature of the same amount of water from absolute zero all the way to 100°C liquid?

29. To increase the temperature of 575 grams of ocean water by 5.0°C requires 2674 J. Use this information to calculate ocean water's specific heat.

30. Rank the following in order of decreasing melting point. A 0.1 M aqueous solution of

    a) sodium bromide, NaBr.

    b) magnesium chloride, $MgCl_2$.

    c) scandium iodide, ScI_3.

Solutions (Odd-Numbered)

1. Ice is less dense than water because water expands as it freezes. Each water molecule in the solid phase occupies more space than it does in the liquid phase.

3. As great pressure is applied, the open-pockets in the crystalline structure of ice may collapse, which results is a melting of the ice. This occurs primarily where the pressure is being applied, such as at the bottom of a heavy glacier.

5. When the temperature of 0°C liquid water is increased slightly, it undergoes a net contraction. Water at 4°C is more dense (less volume = more contracted) than 0°C.

7. Cohesive forces result from the molecular attractions within a material while adhesive forces result from the molecular attractions occurring between two different materials.

9. Water will rise higher in a narrow tube because there is more surface area of contact between the water and the wall of the tube relative to the weight of the water contained within the tube.

11. Condensation counteracts evaporation on a hot and humid day. The vapor in the air from the humidity condenses on your skin to make you feel uncomfortably warm.

13. Water can boil at temperatures less than 100°C at higher altitudes where the atmospheric pressure is less.

15. A substance that heats up quickly will have a low specific heat.

17. Hydrogen bonds.

19. The temperature of melting ice does not rise as it is heated because the energy being supplied by the added heat is working to break the hydrogen bonds.

21. It takes so much more energy to boil water than to melt it because in a sample of steam, the molecules are relatively free of one another and are not bound together as they are in the liquid phase.

23. From Table 8.1, the specific heat of iron is 0.451 J/g°C. So, the total amount of heat required is:

    Heat = 100,000 g × 0.451 J/g°C × 30°C = +1,343,000 joules

    which is about 10 million joules less than that required for water.

25. This is a three part calculation. First you need to calculate the amount of heat required to raise the water's temperature from –5.00° to 0.00°C. Then you need to calculate the amount of heat required to transform the 1.00 gram of ice into liquid water. Third, you need to calculate the amount of heat required to raise the water's temperature from 0.00°C to +5.00°C. From Table 8.1 you have that the specific heat of ice is 2.01 J/g°C. Note this calculation provides three significant figures.

    1) Heat = (2.01 J/g°C)(1.0 g)(+5.00°C) = 10.1 J

    2) Heat = (1.00 g)(+335 J/g) = +335 J

    3) Heat = (4.18 J/g°C)(1.00 g)(+5.00°C) = +20.9 J

    Total heat = 10.1 J + 335 J + 20.9 J = 366 J

27. Calculate this answer using water's heat of vaporization, which is +2259 J/g

    Heat = (1.00 g)(+2259 J/g) = 2260 J (to three sig figs)

29. Heat = (specific heat)(mass)(temperature change)

    With a little algebra this transforms into:

    Specific heat = Heat/(mass)(temperature change)

    Specific heat = (2674 J)/(575 g)(5.0°C)

    = 0.93 J/g°C (to two significant figures)

Solutions to Chapter 8 Calculation Corner

How Heat Changes Temperature

1. The temperature change is final temperature minus initial temperature: 55.0°C - 25.0°C = +30.0 C°. Multiply this positive temperature change by the water's specific heat and mass:

   heat = (4.184 J/g °C)(100,000 g)(+30.0 C°)

   = 12,552,000 J

   This large number helps to explain how an electric water heater consumes about 25 percent of all household electricity. With the proper number of significant figures (see Chapter 1), this answer should be expressed as 10,000,000 joules.

2. The temperature change is -30.0°C - (-10.0°C) = -20.0 C°. Multiply this negative temperature change by the ice's specific heat and mass:

   heat = (2.01 J/g °C)(10.0 g)(-20.0 C°) = -402 J

   The next time you're near a refrigerator/freezer, place your hand near its back, and you'll feel the heat that has been extracted from the food inside. Notice that this answer has the opposite sign (-) as the previous answer (+). The reason for this is because the direction of heat flow is reversed. .

Chapter 8: Contextual Chemistry Essay

Winds From Water

Earth's atmosphere would be a boring place if it weren't for the moisture it contains. The vastly different specific heats of the oceans and continents also adds for excitement. But first imagine an Earth with no oceans, very dry air, and just a smooth surface of land. What would happen? Air close to the surface would be warmed by the solar heated surface. Air at higher elevations is farther from that warm surface and would therefore be much cooler.

On this pretend Earth, consider a large parcel of cool air way up high in the sky. If this parcel sinks downward, the greater atmospheric pressure closer to the surface causes it to compress, as shown in **Figure A**. This compression causes the parcel to heat up, like compressed air in a bicycle pump. As the parcel heats up, it expands thus becoming less dense than the surrounding lower air, which is why it rises back upward like a hot air balloon. The end result is that parcels of cold air tend to stay aloft. They can't sink any more than a cork floating on water.

Conversely, a parcel of warm air at the surface expands as it rises upward. As it expands, it cools down. As it cools down it becomes denser than the surround upper air and sinks back downward. So, on our idealized extremely dry planet, the atmosphere defaults to a natural gradient of cooler air aloft and warm air close to the surface.

If a parcel of rising surface air contains water vapor, as within our real Earth's atmosphere, then the situation is quite different. As the moist rising air cools, much of the water vapor condenses into tiny droplets, which form a cloud, as shown in **Figure B**. The condensation of the water, in turn, releases large amounts of heat, as was described in Section 8.6. This heat of condensation compensates for the cooling that occurs as the rising air expands. As a result, a rising parcel of moist air, due to the heat of condensation, is able to remain aloft.

This rising parcel of air leaves a partial vacuum in its wake, or, rather, a region of low pressure where there are fewer air molecules. Surrounding surface air flows inward toward this low-pressure region creating what we feel as wind. If the incoming air is also moist, it too will be able to rise resulting in an updraft. This process can continue until a huge cloud is formed creating what we recognize as a thunderstorm. Eventually, downdrafts complement the updrafts resulting in strong winds, heavy rain, and lightning. So, you see, the presence of water makes the weather rather exciting.

As wind rushes into the low-pressure region, you might think all these winds meet up at a single central

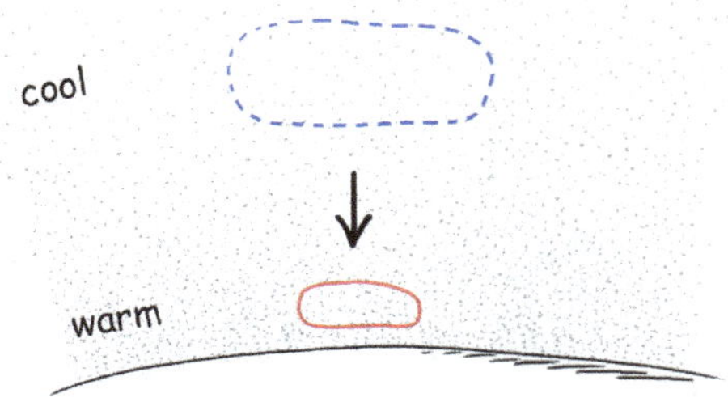

Λ Figure A

An upper parcel of air is compressed if drawn downward. The compression causes it to warm up and expand (not shown) to become less dense than the surrounding air causing it to rise back upward.

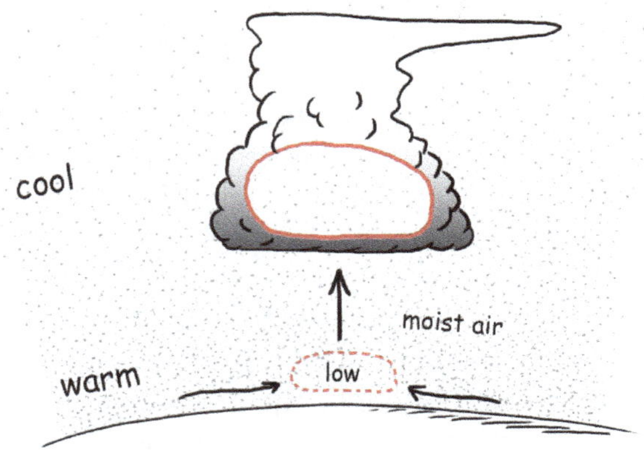

Λ Figure B

Water condensing out of rising moist air forms clouds, releasing heat, which allows the parcel of air to remain aloft.

spot. They would, except for that the Earth rotates. The regions of the storm system closer to the equator are spinning around Earth's axis faster than those farther from the equator. Winds moving toward the north or south pole are deflected a bit to the east, while winds moving toward the equator are deflected a bit to the west. Therefore, the winds don't meet exactly at the center of low-pressure. Rather their directions are skewed into counterclockwise rotation when in the northern hemisphere and clockwise when in the southern hemisphere.

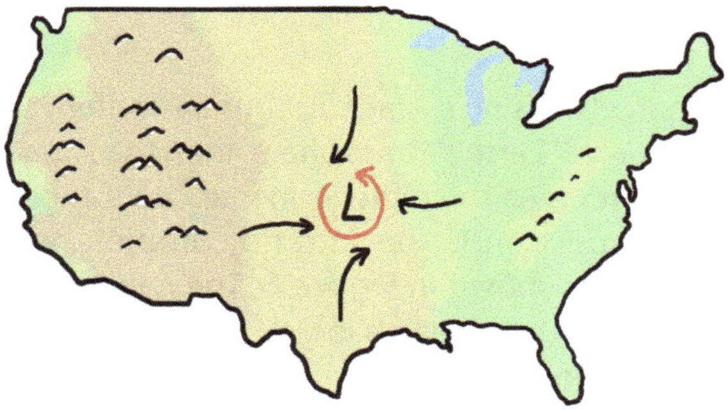

Λ Figure C

Within North America, southward moving air veers westward while northward moving air veers eastward causing a counterclockwise rotation around a low-pressure zone. This happens because Earth rotates on its axis.

For a large thunderstorm, this spin plays a role in the formation of a tornado, which is a region of intense low pressure that sucks everything upward in a spiral fashion. The wind speeds are quite high because of the conservation of angular momentum. When an ice skater pulls her arms inward during a spin—she spins faster. Similarly, the tornado is a region where the storm's angular momentum is focused into a small diameter—the smaller the diameter, the faster the wind speeds and the greater the tornado's destructive power.

Thunderstorms are fed by warm moist air. So, they rarely happen in the winter when the air is not only cold but dry. Thunderstorms are quite common over maritime tropical regions, usually in the afternoon when the air is warmest. Under the right conditions, multiple tropical thunderstorms can join into a single much larger system. A region of very low pressure develops about which these merging multiple thunderstorms rotate. Moist air rushes inward and then spirals upward driven by the warmth released upon the condensation of that moist air. The system starts out as a tropical storm, which can potentially grow into a hurricane that turns like a pinwheel due to Earth's rotation—always counterclockwise in the northern hemisphere and clockwise in the southern hemisphere.

Λ Figure D

Thunderheads and tornados gain their energy from the condensation of water.

A hurricane is a moisture-driven positive feedback loop, which means it reinforces itself so long as there is a source of warm ocean water. A parcel of warm moist air rises and continues to rise because of the released heat during condensation. This draws in more warm moist air that also rises, thus pulling in even more warm moist air. This cycle—essentially a natural heat engine—continues to build on itself creating a storm of planetary proportion. The upward lift of a hurricane is so powerful that it can raise the sea level beneath it by several meters. The heightened sea level is called a *storm surge*. An active hurricane over warm waters can be stopped if strong upper-level winds from outside the system are able to

Λ Figure E

Hurricanes arise from the evaporation followed by condensation of water found within warm moist tropical air.

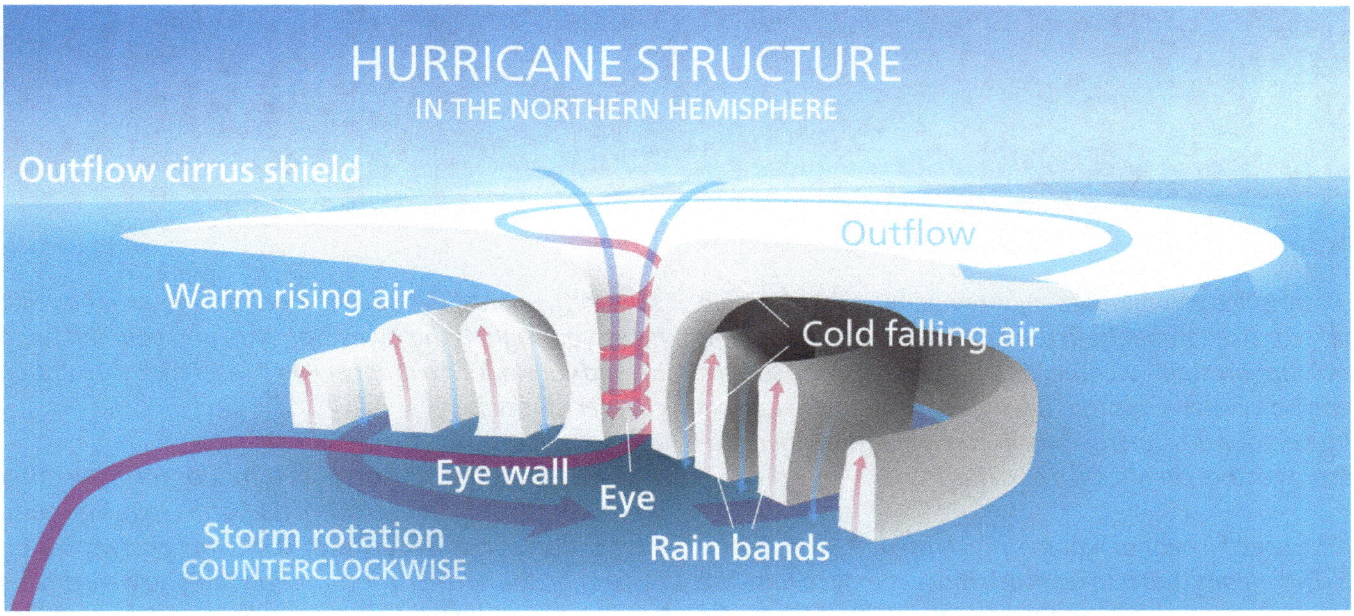

Λ Figure F
The anatomy of a hurricane. Moist warm air is drawn to the central eye, which is a region of intense low atmospheric pressure.

disrupt the upward flow pattern. Once over land, the hurricane is deprived of its warm water energy source and eventually dies out, but not without first causing massive destruction from winds, heavy rains, and storm surge flooding to the coastal communities.

CONCEPT **CHECK**
Why are there practically no winds on Venus?

CHECK YOUR ANSWER
One of the main reasons there are practically no winds on Venus is because there are no oceans on this planet and very little water within the atmosphere.

Think and Discuss

1. As described in Section 8.5, water has a high specific heat. Does water's high specific heat cause more dramatic weather or less dramatic weather?

2. Which would be of greater concern: the average temperature of the atmosphere going up by 3°C or the average temperature of the oceans going up by the same amount? Why?

3. How can you simulate the spin of a hurricane north of the equator using a pencil held between the palms of your hands? Should your left hand come toward you or away from you?

4. Why do tornados in the northern hemisphere tend to track toward the northeast?

Author Responses to Think and Discuss

1. Both! Island communities enjoy temperatures that remain fairly constant throughout the year. Stable temperature, in turn, support stable weather patterns. Water's high specific heat, however, means that it is able to hold onto relatively large amounts of energy, which can be used to drive dramatic weather events, such as a hurricane.

2. The specific heat of water is much greater than that of air. This means that it takes much more energy to raise the temperature of water than the temperature of air by the same number of degrees. Mostly, however, the combined mass of the oceans is billions of times greater than the mass of the atmosphere. So, if our oceans went up by 3°C, then that would mean our planet has absorbed that much more thermal energy. The oceans would expand, which would raise sea levels significantly. Furthermore, the warm ocean would warm up the air as well. In short, our oceans going up by 3°C would be much, much more problematic than the atmosphere going up by 3°C. Though there is evidence indicating that a 3°C increase in average atmospheric temperatures is also fraught with threats to living systems on this planet, including ourselves.

3. Hold the pencil between the palm of your hands and move your right hand away from you and your left hand toward you. In such a manner you'll see the pencil rotating counter-clockwise.

4. At mid-latitudes, where most tornados form, weather prevails eastward. But because of the Earth's spin, the southern portion of a thunderstorm is moving eastward a bit faster than the northern portion. Imagine you were driving a bulldozer eastward in a parking lot and your southern treads started moving faster than your northern treads. What would happen? You would find yourself veering to the north (to your left). Similarly, north of the equator, thunderstorms at mid-latitudes, and the tornados they may contain, tend to track toward the northeast. If a tornado is coming your way directly from the southwest, and you don't have safe shelter, you would be best to head away from it as fast as you can in a southeast or northwest direction.

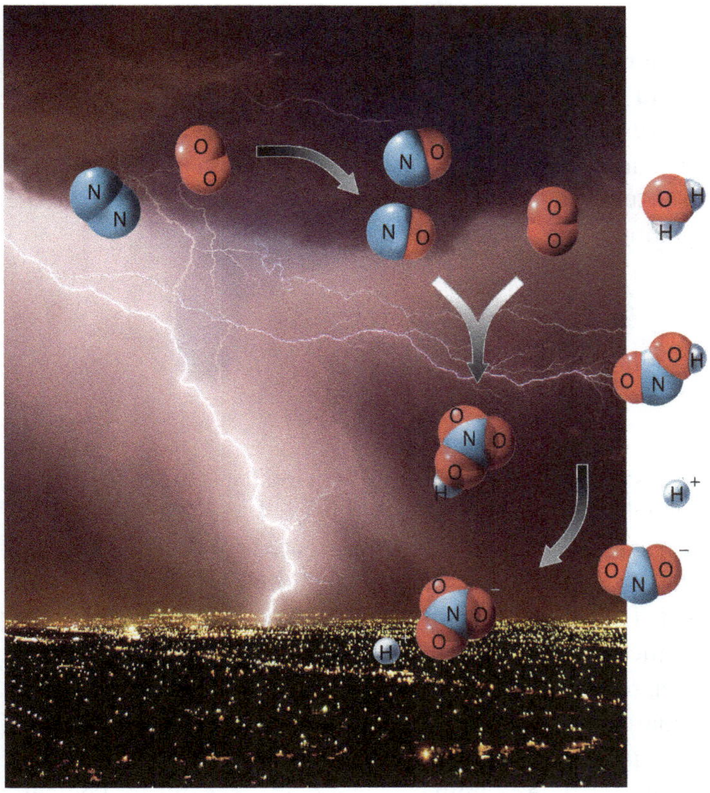

Chapter 9

How Chemicals React

THE MAIN IDEA

Atoms change partners during a chemical reaction.

^ The atmosphere is awash with chemicals such as water, nitrogen, and oxygen, that are forced to undergo chemical changes during a high-energy lightning bolt.

Chemists have learned how to control chemical reactions to produce many useful materials— nitrogen-based fertilizers from the atmosphere, metals from rocks, plastics and pharmaceuticals from petroleum. These materials and the thousands of others produced by chemical reactions have dramatically improved our standard of living, as has the abundant energy released when fossil fuels take part in the chemical reaction called combustion. What happens in a chemical reaction? Why are new materials produced? Why must chemical equations always be balanced? If two molecules meet, will they always react to form new molecules? What is a catalyst, and what role do catalysts play in our environment? Why do some chemical reactions, such as the burning of wood, produce energy, while other reactions, such as those occurring in the cooking of food, require the input of energy? What is the ultimate driving force for all chemical reactions? The goal of this chapter is to provide you with a stronger handle on the basics of chemical reactions introduced in Chapter 3.

9.1 Chemical Equations

As was discussed in Chapter 3, during a chemical reaction, atoms rearrange to form one or more new compounds. This process is neatly summed up in written form as a **chemical equation**. A chemical equation shows the reacting substances, called **reactants**, before an arrow that points to the newly formed substances, called **products**:

$$\text{reactants} \longrightarrow \text{products}$$

Typically, reactants and products are represented by their elemental or chemical formulas. Sometimes molecular models or, simply, names may be used instead. Phases are also often shown: (*s*) for solid, (*l*) for liquid, and (*g*) for gas. Compounds dissolved in water are designated (*aq*) for aqueous solution. Finally, numbers are placed in front of the reactants or products to show the ratio in which they either combine or form. These numbers are called *coefficients*, and they represent numbers of individual atoms and molecules. For instance, to represent the chemical reaction in which carbon burns in the presence of oxygen to form gaseous carbon dioxide, we write the chemical equation using coefficients of 1:

$$1\,C\,(s) + 1\,O_2\,(g) \longrightarrow 1\,CO_2\,(g)$$

Reactants Products

One of the most important principles of chemistry is the **law of mass conservation**, which we discussed in Sections 1.4 and 2.2. The law of mass conservation states that matter is neither created nor destroyed during a chemical reaction. The atoms present at the beginning of a reaction are merely rearranged. This means that no atoms are lost or gained during any reaction. The chemical equation must therefore be *balanced*. In a balanced equation, each atom must appear on both sides of the arrow the same number of times. The equation for the formation of carbon dioxide is balanced because each side shows one carbon atom and two oxygen atoms. You can count the number of each type of atom in the models to see this for yourself.

In another chemical reaction, two hydrogen gas molecules, H_2, react with one oxygen gas molecule, O_2, to produce two molecules of water, H_2O, in the gaseous phase:

$$2\,H_2\,(g) + 1\,O_2\,(g) \longrightarrow 2\,H_2O\,(g)$$

This equation for the formation of water is also balanced—there are four hydrogen and two oxygen atoms before and after the arrow.

A coefficient in front of a chemical formula tells us the number of times that element or compound must be counted. For example, 2 H_2O indicates two water molecules, which contain a total of four hydrogen atoms and two oxygen atoms.

By convention, the coefficient 1 is omitted; so, the preceding chemical equations are typically written

$$C\,(s) + O_2\,(g) \rightarrow CO_2\,(g) \qquad \text{(balanced)}$$

$$2\,H_2\,(g) + O_2\,(g) \rightarrow 2\,H_2O\,(g) \qquad \text{(balanced)}$$

CONCEPT **CHECK**

How many oxygen atoms are indicated by the following balanced equation?

$3\,O_2\,(g) \rightarrow 2\,O_3\,(g)$

CHECK YOUR ANSWER

There are six oxygen atoms. Before the reaction, these six oxygen atoms are found in three O_2 molecules. After the reaction, these same six atoms are found in two O_3 molecules.

An unbalanced chemical equation shows the reactants and products without the correct coefficients. For example, the equation →

$$NO\ (g) \rightarrow N_2O\ (g) + NO_2\ (g) \quad \text{(unbalanced)}$$

is not balanced because there are one nitrogen atom and one oxygen atom before the arrow but three nitrogen atoms and three oxygen atoms after the arrow.

You can balance unbalanced equations by adding or changing coefficients to produce correct ratios. It's important not to change subscripts, however, because to do so changes the compound's identity—H_2O is water, but H_2O_2 is hydrogen peroxide. For example, to balance the preceding equation, place a 3 before the NO:

$$3\ NO\ (g) \rightarrow N_2O\ (g) + NO_2\ (g) \quad \text{(balanced)}$$

Now there are three nitrogen atoms and three oxygen atoms on each side of the arrow, and the law of mass conservation is not violated.

Practicing chemists develop a skill for balancing equations. This skill involves creative energy and, like other skills, improves with experience. More important than being an expert at balancing equations is knowing why they need to be balanced. And the reason is the law of mass conservation, which tells us that atoms are neither created nor destroyed in a chemical reaction—they are simply rearranged. So, every atom present before the reaction must be present after the reaction, even though the groupings of atoms are different.

READING CHECK
Why must the chemical equation be balanced?

Many students enjoy learning how to balance chemical equations. Perhaps this is because balancing an equation is akin to solving a crossword or Sudoku puzzle. There are many different approaches one can take to balancing a chemical equation. There are also some crafty tricks one can perform to solve some difficult equations. Your instructor may share with you some of his or her favorite approaches and/or tricks to balancing equations. To get you started, we offer you the following guidelines, which you can follow as you work to balance the chemical equations appearing in the chapter review and elsewhere.

A Quick Guide to Balancing Chemical Equations

Unbalanced equations are balanced by changing the *coefficients*. Subscripts should **never** be changed, because this changes the chemical's identity—H_2O is water, but H_2O_2 is hydrogen peroxide. Also, a coefficient is a multiplier for all the atoms within a molecule. Writing 2 Fe_2O_3, for example, means four Fe atoms and six O atoms.

1. Focus on balancing only one element at a time. Start with the leftmost element and modify the coefficients so that this element appears on both sides of the arrow the same number of times.

2. Move to the next element and modify the coefficients so as to balance this element. Do not worry if you incidentally unbalance the previous element. You will come back to it in subsequent steps.

3. Continue from left to right focusing on only one element at a time. Change coefficients as needed for that one element and don't worry about messing up the other elements by doing so.

4. Repeat steps 1–3 until all the elements are balanced.

CONCEPT **CHECK**

Write a balanced equation for the reaction showing hydrogen gas, H_2, and nitrogen gas, N_2, forming ammonia gas, NH_3:

$$\_\_\_H_2\,(g) + \_\_\_N_2\,(g) \rightarrow \_\_\_NH_3\,(g)$$

CHECK YOUR ANSWER

Initially, we see two hydrogen atoms before the reaction arrow and three on the right. This can be remedied by placing a coefficient of 3 by the hydrogen, H_2, and a coefficient of 2 by the ammonia, NH_3. This makes for six hydrogen atoms both before and after the reaction arrow. Meanwhile, the coefficient of 2 by the ammonia also makes for two nitrogen atoms after the arrow, which balances out the two nitrogen atoms appearing before the arrow. The full balanced equation, therefore, is

$$3\,H_2\,(g) + N_2\,(g) \longrightarrow 2\,NH_3\,(g)$$

Types of Chemical Equations

Chemical equations can be classified based upon how the atoms of a reaction rearrange to become products. Five categories include: synthesis, decomposition, single replacement, double replacement, and combustion.

Synthesis

Within a synthesis reaction two or more reactants combine to make a product containing the atoms of the reactants in some combination. It follows the general formula A + B $\rightarrow$ AB. Some examples include the formation of ammonia, NH_3, and water, H_2O:

$$3\,H_2 + N_2 \rightarrow 2\,NH_3$$

$$2\,H_2 + O_2 \rightarrow 2\,H_2O$$

Decomposition

Within a decomposition reaction, a reactant consisting of a grouping of multiple atoms splits apart into multiple products. A decomposition reaction follows the general formula AB $\rightarrow$ A + B. Some examples include the decomposition of calcium carbonate, $CaCO_3$, and hydrogen peroxide, H_2O_2:

$$CaCO_3 \rightarrow CaO + CO_2$$

$$2\,H_2O_2 \rightarrow 2\,H_2O + O_2$$

Single Replacement

Within a single replacement reaction, an atom (or polyatomic ion) jumps from one reactant to another. A single replacement reaction follows the general formula AX + B $\rightarrow$ A + BX. Some examples include the zinc, Zn, plus copper chloride, $CuCl_2$, reaction and the potassium, K, plus water, H_2O, reaction:

$$Zn + CuCl_2 \rightarrow ZnCl_2 + Cu$$

$$2\ K + 2\ H_2O \rightarrow 2\ KOH + H_2$$

Double Replacement

Within the double replacement reaction, an atom (or polyatomic ion) jumps from each reactant—there is a double switch. A double replacement reaction follows the general formula $AY + BX \rightarrow BY + AX$. Some examples include the silver nitrate, $AgNO_3$, plus sodium chloride, NaCl reaction and the hydrogen chloride, HCl, plus sodium hydroxide, NaOH, reaction:

$$AgNO_3 + NaCl \rightarrow AgCl + NaNO_3$$

$$HCl + NaOH \rightarrow HOH + NaCl$$

Combustion Reaction

Lastly, there's the combustion reaction, which we'll explore in more detail in Chapter 11. A combustion reaction is the burning that occurs when a carbon compound, such as is found within wood, is ignited to react with oxygen, O_2. The products of a combustion reaction include carbon dioxide, CO_2, and water, H_2O plus lots of energy. Some examples include the combustion of methane, CH_4, and glucose, $C_6H_{12}O_6$:

$$CH_4 + 2\ O_2 \rightarrow CO_2 + 2\ H_2O$$

$$C_6H_{12}O_6 + 6\ O_2 \rightarrow 6\ CO_2 + 6\ H_2O$$

CONCEPT **CHECK**

Balance and then interpret the following chemical equations as representing *synthesis, decomposition, single replacement, double replacement,* or a *combustion reaction*:

a) ___ $KClO_3$ (*s*) → ___ KCl (*s*) + ___ O_2 (*g*)

b) ___ $BaCl_2$ (*aq*) + ___ Na_2SO_4 (*aq*) → ___ $BaSO_4$ (*s*) + ___ NaCl (*aq*)

c) ___ C_3H_8O (*l*) → ___ C_3H_6 (*l*) + ___ H_2O (*l*)

d) ___ C_2H_2 (*g*) + ___ O_2 (*g*) → ___ CO_2 (*g*) + ___ H_2O (*g*)

e) ___ H_2 (*g*) + ___ Cl_2 (*g*) → ___ HCl (*g*)

f) ___ NH_3 (*g*) → ___ N_2 (*g*) + ___ H_2 (*g*)

g) ___ Cl_2 (*g*) + ___ NaBr (*aq*) → ___ NaCl (*aq*) + ___ Br_2 (*g*)

h) ___ N_2 (*g*) + ___ H_2 (*g*) → ___ NH_3 (*g*)

i) ___ Na (*l*) + ___ KCl (*l*) → ___ NaCl (*l*) + ___ K (*l*)

j) ___ $Pb(NO_3)_2$ (*aq*) + ___ KI (*aq*) → ___ KNO_3 (*aq*) + ___ PbI_2 (*s*)

CHECK YOUR ANSWERS

a) $2\ KClO_3\ (s) \rightarrow 2\ KCl\ (s) + 3\ O_2\ (g)$	*decomposition*
b) $BaCl_2\ (aq) + Na_2SO_4\ (aq) \rightarrow BaSO_4\ (s) + 2\ NaCl\ (aq)$	*double replacement*
c) $C_3H_8O\ (l) \rightarrow C_3H_6\ (l) + H_2O\ (l)$	*decomposition*
d) $2\ C_2H_2\ (g) + 5\ O_2\ (g) \rightarrow 4\ CO_2\ (g) + 2\ H_2O\ (g)$	*combustion*
e) $H_2\ (g) + Cl_2\ (g) \rightarrow 2\ HCl\ (g)$	*synthesis*
f) $2\ NH_3\ (g) \rightarrow N_2\ (g) + 3\ H_2\ (g)$	*decomposition*
g) $Cl_2\ (g) + 2\ NaBr\ (aq) \rightarrow 2\ NaCl\ (aq) + Br_2\ (g)$	*single replacement*
h) $N_2\ (g) + 3\ H_2\ (g) \rightarrow 2\ NH_3\ (g)$	*synthesis*
i) $Na\ (l) + KCl\ (l) \rightarrow NaCl\ (l) + K\ (l)$	*single replacement*
j) $Pb(NO_3)_2\ (aq) + 2\ KI\ (aq) \rightarrow 2\ KNO_3\ (aq) + PbI_2\ (s)$	*double replacement*

9.2 Measuring Molecules

In any chemical reaction, a specific number of reactants react to form a specific number of products. For example, when carbon and oxygen combine to form carbon dioxide, they always combine in the ratio of one carbon atom to one oxygen molecule. A chemist who wants to carry out this reaction in the laboratory would be wasting chemicals and money if she were to try to combine, say, four carbon atoms for every one oxygen molecule. The excess carbon atoms would have no oxygen molecules to react with and would remain unchanged.

But how is it possible to measure out a specific number of atoms or molecules? Rather than counting these particles individually, chemists can use a scale that measures the mass of bulk quantities. Because different atoms and molecules have different masses, however, a chemist can't simply measure out equal masses of each. Say, for example, he needs the same number of carbon atoms as oxygen molecules. Measuring equal masses of the two materials would not provide equal numbers.

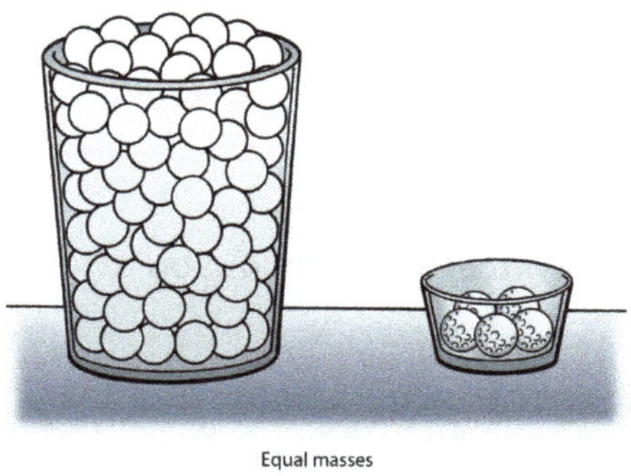

∧ Figure 9.1

For the same mass, the number of Ping-Pong balls is greater than the number of golf balls because a single Ping-Pong ball is lighter than a single golf ball.

Consider this analogy: You know that Ping-Pong balls are much lighter than golf balls. So, if you were to measure out 1 kilogram of each, you understand that you would need many more Ping-Pong balls to create a full kilogram, as shown in **Figure 9.1**. You understand that the *number* of something and the mass of something are two different things and that they should never be confused with each other.

CONCEPT **CHECK**

A customer wants to buy a 1:1 mixture of blue and red jelly beans. Each blue bean is twice as massive as each red bean. If the clerk measures out 5 pounds of red beans, how many pounds of blue beans must she measure out?

CHECK YOUR ANSWER

Because each blue jelly bean has twice the mass of each red one, the clerk needs to measure out twice as much mass of blues in order to have the same count, which means 10 pounds of the blues. If the clerk did not know that the blue beans were twice as massive as the red ones, she would not know what mass of blues was needed for the 1:1 ratio.

Because different atoms and molecules have different masses, there are different numbers of them in a sample of the same mass. Carbon atoms are less massive than oxygen molecules. That's why there are more carbon atoms in a 1-gram sample of carbon than there are oxygen molecules in a 1-gram sample of oxygen. So clearly, equal masses of these two substances do not yield equal numbers of carbon atoms and oxygen molecules.

If we know the *relative masses* of different materials, we can measure equal numbers. Golf balls, for example, are about 20 times more massive than Ping-Pong balls, which is to say the relative mass of Ping-Pong balls to golf balls is 20 to 1. Measuring out 20 times as much mass of golf balls as of Ping-Pong balls, therefore, gives equal numbers of each, as is shown in **Figure 9.2.**

Where do we turn to find the relative masses of atoms and molecules? You guessed it! The masses of elements shown in the periodic table are relative masses. Using these masses, we can measure out equal numbers of atoms or molecules.

For example, as illustrated in **Figure 9.3**, the mass of carbon is 12.011 amu. (As discussed in Section 4.4, one *atomic mass unit* (amu) equals 1.661×10^{-24} gram.) The **formula mass** of a substance is the sum of the atomic masses of the elements in its chemical formula. Consider, the formula mass of an oxygen molecule, O_2, is about 32 amu (15.999 amu + 15.999 amu). A carbon atom, therefore, is about 12/32 as massive as an oxygen molecule. To measure out equal numbers of carbon atoms and oxygen molecules, we could measure out 12 grams of carbon and 32 grams of molecular oxygen. Any proportion equal to 12/32, such as 6/16 or 3/8, would also do. For example, 3 grams of carbon would have the same number of particles as 8 grams of molecular oxygen.

READING CHECK

Samples of two different materials have the same mass. Do they necessarily have the same number of atoms?

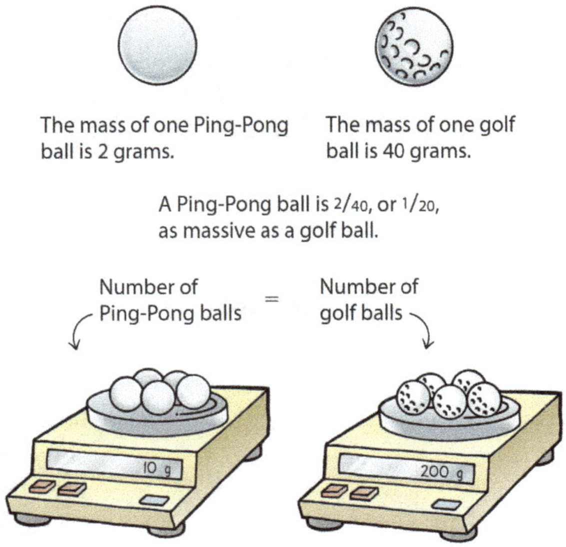

Figure 9.2

The number of golf balls in 200 grams of golf balls equals the number of Ping-Pong balls in 10 grams of Ping-Pong balls.

FOR YOUR INFORMATION

The number of atoms in your breath is about 100 times the number of breaths in the atmosphere.

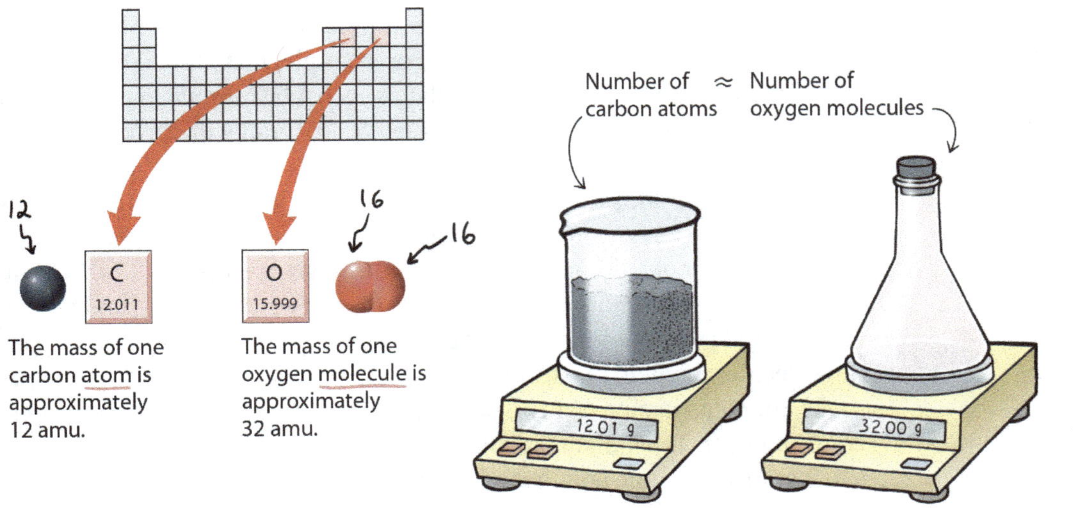

Figure 9.3

Measuring carbon, C, and oxygen, O_2, in a 12:32 ratio provides the same number of particles. The same would be true if they were measured in a 6 to 16 ratio, or even a 3 to 8 ratio. (Note: the volumes shown are exaggerated. Twelve grams of solid carbon would occupy only about 5 mL, while 32 grams of gaseous oxygen, O_2, would occupy about 22.4 L.)

CONCEPT **CHECK**

1. Reacting 3 grams of carbon, C, with 8 grams of molecular oxygen, O_2, results in 11 grams of carbon dioxide, CO_2. Does it follow that 1.5 grams of carbon will react with 4 grams of oxygen to form 5.5 grams of carbon dioxide?

2. Would reacting 5 grams of carbon with 8 grams of oxygen also result in 11 grams of carbon dioxide?

CHECK YOUR ANSWER

1. Yes, it does. The quantities are only half as much, but their ratio is the same as when 11 grams of carbon dioxide is formed: 1.5/4/5.5 = 3/8/11.

2. A common error of many students is to think that no reaction will occur if the proper ratios of reactants are not provided. You should realize, however, that in this reaction, only 3 of the 5 grams of carbon can react with 8 grams of oxygen to form 11 grams of carbon dioxide. Two grams of carbon will be unreacted after the reaction. Reacting this remaining 2 grams of carbon would require more oxygen.

9.3 Grams to Moles

Atoms and molecules react in specific ratios by number. In the laboratory, however, chemists work with bulk quantities of materials, which are measured by grams. Chemists therefore need to know the relationship between the grams of a given sample and the number of atoms or molecules contained in that mass. The key to this relationship is the *mole*. Recall from Section 7.3 that we use the word *mole* to indicate a number, much as we use the word *dozen* to indicate a number. While a dozen is equal to 12, a mole is equal to a number that is incredibly large: 6.02×10^{23}. This number is known as **Avogadro's number**, in honor of the 18th-century scientist Amedeo Avogadro. Again, for emphasis, one mole is equal to 6.02×10^{23}.

As **Figure 9.4** illustrates, if you express the numeric value of the atomic mass of any element in *grams*, the number of atoms in a sample of the element having this mass is always 6.02×10^{23}, which is 1 mole. For example, a 22.990-gram sample of sodium metal, Na (atomic mass = 22.990 amu), contains 6.02×10^{23} sodium atoms, and a 207.2-gram sample of lead, Pb (atomic mass = 207.2 amu), contains 6.02×10^{23} lead atoms.

The same concept holds for molecules (or any compound). Express the numeric value of the formula mass in *grams* and a sample having that mass contains 6.02×10^{23} molecules. For example, there are 6.02×10^{23} O_2 molecules in 31.998 grams of molecular oxygen, O_2 (formula mass = 31.998 amu), and 6.02×10^{23} CO_2 molecules in 44.009 grams of carbon dioxide, CO_2 (formula mass = 44.009 amu).

The **molar mass** of any substance, be it element or compound, is defined as the mass of 1 mole of the substance. Thus, the units of molar mass are *grams per mole*. For instance, the atomic mass of carbon is 12.011 amu, which means that 1 mole of carbon has a mass of 12.011 grams, and we say that the molar mass of carbon is 12.011 grams per mole. The molar mass of molecular oxygen, O_2 (formula mass 31.998 amu), is 31.998 grams per mole. For convenience, values such as these are often rounded to the nearest whole number. The molar mass of carbon, therefore, might also be presented as 12 grams per mole and that of molecular oxygen as 32 grams per mole.

fyi FOR YOUR INFORMATION

An Avogadro's number of grains of sand would cover the United States to a depth of about 2 meters. A stack of an Avogadro's number of pennies would be about 800,000 trillion kilometers tall, which is about the diameter of our galaxy. Placed side by side, these pennies would reach to the Andromeda galaxy, which is about 2.5 million light-years away. Avogadro's number is beyond huge, yet that's the number of water molecules found in a mere 18 grams of water. Water molecules are that small!

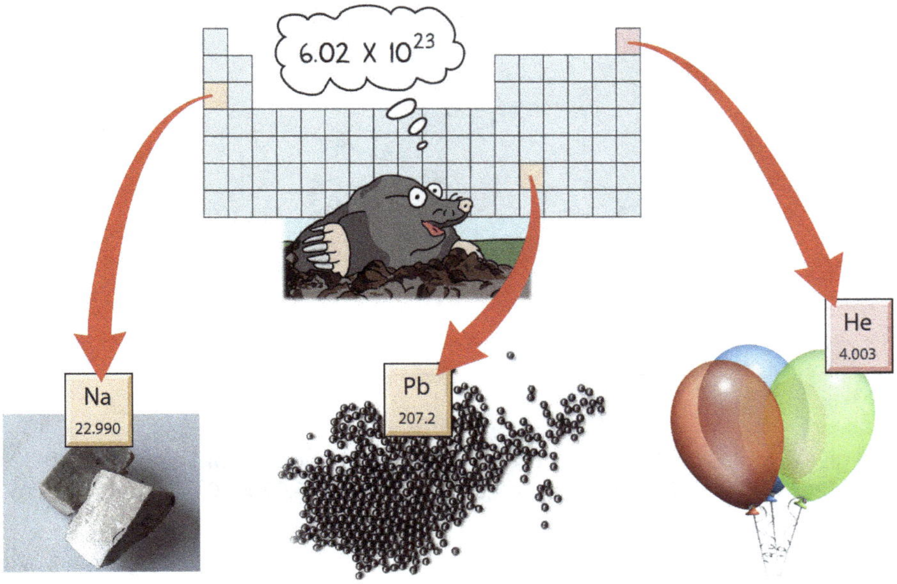

22.990 g
6.02×10^{23} atoms,
which is 1 mole

207.2 g
6.02×10^{23} atoms,
which is 1 mole

4.003 g
6.02×10^{23} atoms,
which is 1 mole

< Figure 9.4

Express the numeric value of the atomic mass of any element in grams, and that many grams contain 6.02×10^{23} atoms.

Because 1 mole of any substance always contains 6.02×10^{23} particles, the mole is an ideal unit for chemical reactions. For example, 1 mole of carbon (12 grams) reacts with 1 mole of molecular oxygen (32 grams) to give 1 mole of carbon dioxide (44 grams).

In many instances, the ratio in which chemicals react is not 1:1. As shown in **Figure 9.5**, for example, 2 moles (4 grams) of molecular hydrogen react with 1 mole (32 grams) of molecular oxygen to give 2 moles (36 grams) of water. Note how the coefficients of the balanced chemical equation can be conveniently interpreted as the number of moles of reactants or products. A chemist therefore need only convert these numbers of moles to grams in order to know how much mass of each reactant he should measure out to have the proper proportions.

Cooking and chemistry are similar in that both require measuring ingredients. Just as a cook looks to a recipe to find the necessary quantities measured by the cup or the tablespoon, a chemist looks to the periodic table to find the necessary quantities measured by the number of grams per mole for each element or compound.

CONCEPT CHECK

1. How many atoms are in a 6.941-gram sample of lithium, Li (atomic mass = 6.941 amu)?

2. How many molecules are in an 18.015-gram sample of water, H_2O (formula mass = 18.015 amu)?

CHECK YOUR ANSWER

1. Because this number of grams of lithium is numerically equal to the atomic mass, there are 6.02×10^{23} atoms in the sample, which is 1 mole of lithium atoms.

2. Because this number of grams of water is numerically equal to the formula mass, there are 6.02×10^{23} water molecules in the sample, which is 1 mole of water molecules.

$2\,H_2$	$+$	$1\,O_2$	$\longrightarrow$	$2\,H_2O$
2 moles		1 mole		2 moles
which is		which is		which is
4 grams		32 grams		36 grams
which is		which is		which is
12.04×10^{23} molecules		6.02×10^{23} molecules		12.04×10^{23} molecules

Figure 9.5

Two moles of H_2 reacts with 1 mole of O_2 to give 2 moles of H_2O. This is the same as saying 4 grams of H_2 reacts with 32 grams of O_2 to give 36 grams of H_2O or, equivalently, that 12.04×10^{23} H_2 molecules react with 6.02×10^{23} O_2 molecules to give 12.04×10^{23} H_2O molecules.

CONCEPT **CHECK**

What is the molar mass of water (formula mass = 18 amu)?

CHECK YOUR ANSWER

From the formula mass, you know that 1 mole of water has a mass of 18 grams. Therefore, the molar mass is 18 grams per mole.

CHEMICAL CONNECTIONS

How are the atoms in your lungs connected to the atoms that were once in the lungs of Leonardo da Vinci?

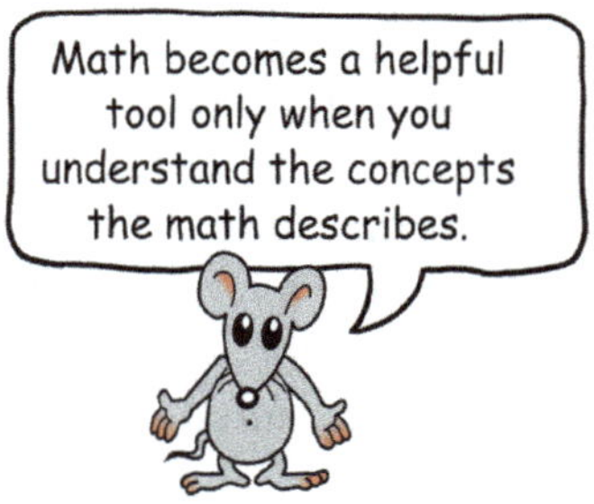

fyi FOR YOUR INFORMATION

Pi is the ratio of a circle's circumference to diameter. As the circumference gets bigger, so does the diameter. The ratio of the two remains the same, which is why the value of pi is constant. Similarly, Avogadro's number is the ratio of relative mass expressed in amu to mass expressed in grams. For example, a carbon atom has a relative mass of 12.011 amu. It's mass in grams is 1.99×10^{-23}. The ratio of these two numbers is 6.02×10^{23}. This is Avogadro's number, which stays constant because it is simply a ratio. An atom with a larger mass in amu will necessarily have a larger mass in grams. The value of Avogadro's number was not known until scientists were able to estimate the mass of single atoms and molecules.

Calculation Corner: Masses of Reactants and Products

The coefficients of a chemical equation tell us the ratio by which reactants react and products form. The following equation, for example, tells us that every 1 mole of methane, CH_4, reacts to produce 2 moles of water, H_2O.

$$CH_4 + 2\,O_2 \rightarrow CO_2 + 2\,H_2O$$

If you were given 16 g of methane, CH_4, how many grams of water, H_2O, would form? From the text, you should know that 16 g of methane, CH_4, is 1 mole (formula mass 16 amu). So, 16 g of methane would yield 2 moles of water. But how many grams are 2 moles of water? Well, if 1 mole of water, H_2O, equals 18 g (formula mass 18 amu), then 2 moles equals 36 g. Thus, 16 g of methane, CH_4, would yield 36 g of water, H_2O. Let's look at this process from a step-by-step mathematical point of view.

SAMPLE PROBLEM 1

What mass of water is produced when 16 g of methane, CH_4 (formula mass 16 amu), reacts with an unlimted supply of oxygen, O_2, in the following reaction? $CH_4 + 2\,O_2 \rightarrow CO_2 + 2\,H_2O$

SOLUTION:

Step 1. Convert the given mass to moles:

$$\text{Conversion factor}$$
$$(16\ \text{g}\ \cancel{CH_4})\left(\frac{1\ \text{mole}\ CH_4}{16\ \text{g}\ \cancel{CH_4}}\right) = 1\ \text{mole}\ CH_4$$

Step 2. Use the coefficients of the balanced equation to find out how many moles of H_2O are produced from this many moles of CH_4:

$$\text{Conversion factor}$$
$$(\cancel{1\ \text{mole}\ CH_4})\left(\frac{2\ \text{moles}\ H_2O}{\cancel{1\ \text{mole}\ CH_4}}\right) = 2\ \text{moles}\ H_2O$$

Step 3. Now that you know how many moles of H_2O are produced, convert this value to grams of H_2O:

$$\text{Conversion factor}$$
$$(\cancel{2\ \text{moles}\ H_2O})\left(\frac{18\ \text{g}\ H_2O}{\cancel{1\ \text{mole}\ H_2O}}\right) = 36\ \text{g}\ H_2O$$

Steps 1 through 3 can be combined into a single equation that starts with what you are given (16 g of CH_4) and calculates that which is asked for:

$$(16\ \text{g}\ \cancel{CH_4})\underbrace{\left(\frac{1\ \text{mole}\ \cancel{CH_4}}{16\ \text{g}\ \cancel{CH_4}}\right)}_{\text{step 1}}\underbrace{\left(\frac{2\ \text{moles}\ \cancel{H_2O}}{1\ \text{mole}\ \cancel{CH_4}}\right)}_{\text{step 2}}\underbrace{\left(\frac{18\ \text{g}H_2O}{1\ \text{mole}\ \cancel{H_2O}}\right)}_{\text{step 3}} = 36\ \text{g}\ H_2O$$

This technique of combining a series of conversion factors is an application of what you learned in the unit conversion Calculation Corner of Chapter 1. Also, within each conversion factor, the unit that you want to change goes in the denominator and that which you want to obtain goes in the numerator.

The method of converting grams of a substance to moles (step 1), then from moles of this substance to moles of that substance (step 2), followed by moles of that substance to grams (step 3) is called *stoichiometry*. Using stoichiometry, we can calculate the amounts of reactants or products in any chemical reaction.

The methods of stoichiometry are developed much further in upper level chemistry. For our current purposes, we're simply introducing the main idea, which how we keep tabs on atoms and molecules as they react to form products.

Nonetheless, for a special assignment, you might try your analytical thinking skills on the following problems. First, try to deduce the answer based on what you know about the law of mass conservation, and then follow the mathematical steps to check your answers.

SAMPLE PROBLEM 2

Show that 44 g of carbon dioxide, CO_2, is produced when 16 g of methane, CH_4, reacts with an unlimited supply of oxygen, O_2. How many grams of oxygen, O_2, are actually needed for this reaction?

SOLUTION:

The 16 g of methane, CH_4, is one mole, which reacts with oxygen to produce one mole of carbon dioxide, CO_2, which you know because of the coefficients in the balanced chemical equation shown previously. One mole of carbon dioxide (formula mass 44 amu) is 44 g. So, the 16 g of methane reacts with oxygen to produce 44 g of carbon dioxide plus 36 g of water. The mass of the products (44 g + 36 g = 80 g) must be equal to the mass of the reactants (16 g + ? = 80 g). We can calculate that the methane reacted with 64 g of oxygen, which, interestingly enough, is 2 moles, as we see from the balanced equation.

The mass of carbon dioxide produced can also be calculated as follows:

$$(16\ \text{g}\ \cancel{CH_4})\left(\frac{1\ \text{mole}\ \cancel{CH_4}}{16\ \text{g}\ \cancel{CH_4}}\right)\left(\frac{1\ \text{mole}\ \cancel{CO_2}}{1\ \text{mole}\ \cancel{CH_4}}\right)\left(\frac{44\ \text{g}\ CO_2}{1\ \text{mole}\ \cancel{CO_2}}\right) = 44\ \text{g}\ CO_2$$

Similarly, the mass of oxygen can be calculated as follows:

$$(16\ \text{g}\ \cancel{CH_4})\left(\frac{1\ \text{mole}\ \cancel{CH_4}}{16\ \text{g}\ \cancel{CH_4}}\right)\left(\frac{2\ \text{moles}\ \cancel{O_2}}{1\ \text{mole}\ \cancel{CH_4}}\right)\left(\frac{32\ \text{g}\ O_2}{1\ \text{mole}\ \cancel{O_2}}\right) = 64\ \text{g}\ O_2$$

SAMPLE PROBLEM 3

How many grams of ozone (O_3, 48 amu) can be produced from 64 g of oxygen (O_2, 32 amu) in the reaction shown?

$$3\,O_2 \rightarrow 2\,O_3$$

SOLUTION:

According to the law of mass conservation, the amount of mass in the products must equal the amount of mass in the reactants. Given that this reaction involves only one reactant and one product, you should not be surprised to learn that 64 g of reactant produces 64 g of product. Here are the stepwise calculations, which start with *grams* of oxygen and then finds the *moles* of oxygen. The moles of oxygen are then converted to moles of ozone using the coefficients of the equation. This is followed by a conversion of *moles* of ozone to *grams* of ozone:

$$(64\ \text{g}\ \cancel{O_2})\left(\frac{1\ \text{mole}\ \cancel{O_2}}{32\ \text{g}\ \cancel{O_2}}\right)\left(\frac{2\ \text{moles}\ \cancel{O_3}}{3\ \text{moles}\ \cancel{O_2}}\right)\left(\frac{48\ \text{g}\ O_3}{1\ \text{mole}\ \cancel{O_3}}\right) = 64\ \text{g}\ O_3$$

SAMPLE PROBLEM 4

What mass of nitrogen monoxide (NO, 30 amu) is formed when 28 g of nitrogen (N_2, 28 amu) reacts with 32 g of oxygen (O_2, 32 amu) in the reaction shown?

$$N_2 + O_2 \rightarrow 2\,NO$$

SOLUTION:

There are several ways to answer this question. One way would be to recognize that 28 g of N_2 is 1 mole of N_2 and 32 g of O_2 is 1 mole of O_2. According to the balanced equation, combining 1 mole of N_2 with 1 mole of O_2 yields 2 moles of NO. Because the mass of 1 mole of NO is 30 g, the mass of 2 moles of NO must be 30 g × 2 = 60 g, which is simply the sum of all the reactants (28 g + 32 g = 60 g). To calculate step-wise, you can start with either the 28 g of N_2 or the 32 g of O_2.

$$(28 \text{ g } N_2)\left(\frac{1 \text{ mole } N_2}{28 \text{ g } N_2}\right)\left(\frac{2 \text{ moles NO}}{1 \text{ mole } N_2}\right)\left(\frac{30 \text{ g NO}}{1 \text{ mole NO}}\right) = 60 \text{ g NO}$$

In this reaction, N_2 and O_2 were combined in exactly the right amounts. Do you recognize that there would be too much oxygen if 28 g of N_2 were reacted with 40 g of O_2? If that were the case, then only 32 g of the 40 g of O_2 would react. There would be 8 g of unreacted O_2 left over. Stoichiometry is an area of chemistry rich in opportunities for analytical thinking.

YOUR TURN

1. How many grams of iron oxide, Fe_2O_3, can be produced from the reaction of 25.0 g of iron, Fe, with an unlimited amount of oxygen, O_2? (Not sure where to start? Here's a helpful rule: always start your calculation by writing down that which you are given, which in this case is 25.0 g of Fe. Then proceed with the conversions.)

$$4\text{ Fe} + 3\text{ O}_2 \rightarrow 2\text{ Fe}_2\text{O}_3$$

2. How many grams of iron oxide, Fe_2O_3, can be produced from the reaction of 25.0 g of oxygen, O_2, with an unlimited amount of iron, Fe?

3. How many grams of oxygen, O_2, are required to react with 25.0 g of iron, Fe, to form iron oxide, Fe_2O_3?

4. A 25.0-g sample of iron, Fe, powder is mixed with 25.0 g of oxygen, O_2, gas and heated until all of the Fe has reacted. How many grams of O_2 remain after the reaction is complete?

Empirical Formulas

Consider the glucose molecule, which has the following structure:

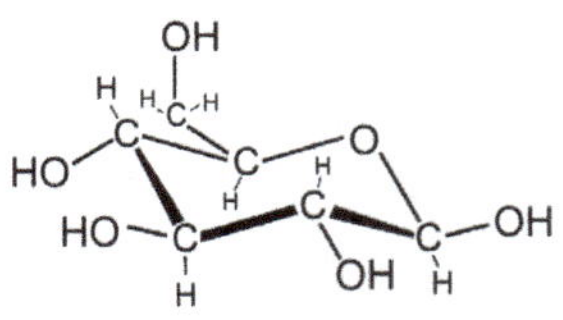

Glucose, $C_6H_{12}O_6$

The molecular formula for this molecule is obtained by counting the number of different types of atoms within that molecule. With six carbons, twelve hydrogens, and six oxygens the molecular formula is $C_6H_{12}O_6$.

Back in the day—over a century ago, before the advent of modern instruments—how exactly did chemists find the formula for a compound? One method was *combustion analysis*. They would burn the compound under carefully controlled conditions such that they could analyze relative amounts of the remaining elements, which is the percent composition. Combusting glucose, for example, would reveal experimental data looking something like this:

Percent Composition (Carbon): 40.123

Percent Composition (Hydrogen): 6.670

Percent Composition (Oxygen): 53.332

From this raw data, the chemist could calculate an *empirical formula*, which shows the lowest ratio of atoms. This calculation requires dividing the percent composition by the atomic mass of each atom. This provides the relative ratio of elements, looking something like this:

Relative Ratio (Carbon): 40.123/12.011 = 3.341

Relative Ratio (Hydrogen): 6.670/1.008 = 6.617

Relative Ratio (Oxygen): 53.332/15.999 = 3.333

The chemist would then divide each by the lowest number and then round off to the closest whole number. This rounding-off accommodated for experimental errors from the combustion analysis.

Relative Ratio (Carbon): 3.341/3.333 = 1

Relative Ratio (Hydrogen): 6.617/3.333 = 2

Relative Ratio (Oxygen): 3.333/3.333 = 1

The chemist would then know the empirical formula, which for glucose is $C_1H_2O_1$. The empirical formula shows the elements of a compound in their lowest whole-number ratio. Back in the day, it was a chemist's first step for discovering a compound's molecular formula, which would be some multiple of the empirical formula. To discern the molecular formula required other innovative techniques such as crystallography.

Percent Yield

Methyl salicylate is the compound responsible for the smell and taste of wintergreen, a common ingredient of many candies. It can be synthesized from the starting materials salicylic acid and methanol:

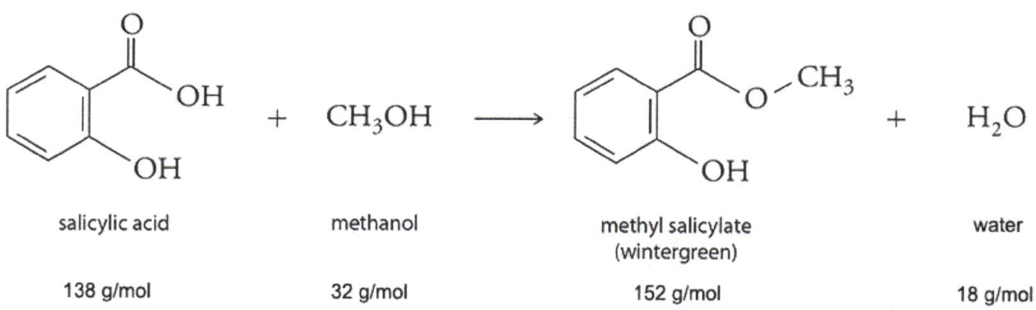

According to this equation, methyl salicylate forms from salicylic acid in a 1:1 mole ratio. Thus, if you were to start with 138 grams of salicylic acid (1 mole), then, ideally, a total of 152 grams of methyl salicylate (1 mole) can be obtained. The reality is that not all of the salicylic acid will be converted to methyl salicylate. Why not? There can be other reactions that occur instead. Chemists will usually have to control the conditions carefully to assure the reaction they want to occur is optimized. Also, some of the product might be lost as the chemist tries to isolate and purify it.

This is why chemists use the concept of **percent yield**, which is a measure of how much product was obtained relative to the amount of product that could be obtained under ideal conditions. For example, we know that from 138 grams of salicylic acid, the theoretical yield of methyl salicylate is 152 grams. Let's say the actual yield was 142 grams. The percent yield is then calculated by this formula:

$$\text{Percent Yield} = \frac{\text{Actual Yield}}{\text{Theoretical Yield}} \times 100$$

$$\text{Percent Yield} = (142/152) \times 100 = 93.4\%$$

Because of the reality of conditions in the laboratory, a 100% yield is not to be expected. That said, chemists will spend much time, energy, and thought to optimize their actual yields to be as close to 100% as is possible and practical.

fyi FOR YOUR INFORMATION

If the molecular formula is already known, then the percent composition of each element can be calculated with no need for any experimental methods. Glucose has the molecular formula $C_6H_{12}O_6$ with a formula mass of 180.156 amu. Six carbons is 72.066 amu. Twelve hydrogens is 12.096 amu. Six oxygens is 95.994 amu. Thus:

Percent Composition (Carbon): (72.066/180.156) × 100 = 40.002
Percent Composition (Hydrogen): (12.096/180.156) × 100 = 6.714
Percent Composition (Oxygen): (95.994/180.156) × 100 = 53.284

9.4 Exothermic or Endothermic

Once a reaction is complete, there may be either a net release or a net absorption of energy. Reactions in which there is a net release of energy are called **exothermic**. Rocket ships lift into space and campfires glow red hot as a result of exothermic reactions. Reactions in which there is a net absorption of energy are called **endothermic**. Photosynthesis, for example, involves a series of endothermic reactions that are driven by the energy of sunlight. Both exothermic and endothermic reactions, illustrated in **Figure 9.6**, can be understood through the concept of bond energy.

< Figure 9.6

For the chemical reactions that occur in a photosynthetic plant, there is a net absorption of energy. For the chemical reactions that occur when wood is burning, there is a net release of energy.

READING CHECK

What do the breaking and forming of chemical bonds involve?

During a chemical reaction, chemical bonds are broken and atoms rearranged to form new chemical bonds. Such breaking and forming of chemical bonds involves changes in energy. As an analogy, consider a pair of magnets. To separate them requires an input of "muscle energy." Conversely, when the two separated magnets collide, they become slightly warmer than they had been, and this warmth is evidence of energy released. Energy must be absorbed by the magnets if they are to move apart, and energy is released as they come together. The same principle applies to atoms. To pull bonded atoms apart requires an energy input. When atoms combine, there is an energy output, usually in the form of faster-moving atoms and molecules, electromagnetic radiation, or both.

The amount of energy required to pull two bonded atoms apart is the same as the amount released when they are brought together. This energy is called **bond energy**. Each chemical bond has its own characteristic bond energy. The hydrogen–hydrogen bond energy, for example, is 436 kilojoules per mole. This means that 436 kilojoules of energy is absorbed as 1 mole of hydrogen–hydrogen bonds break apart and 436 kilojoules of energy is released upon the formation of 1 mole of hydrogen–hydrogen bonds. Different bonds involving different elements have different bond energies, as **Table 9.1** shows. You can refer to the table as you study this section, but please do not memorize these bond energies. Instead, focus on understanding what they mean.

By convention, a positive bond energy represents the amount of energy absorbed as a bond breaks and a negative bond energy represents the amount of energy released as a bond forms. Thus, when you are calculating the net energy released or absorbed during a reaction, you need to be careful about plus and minus signs. It is standard practice when doing such calculations to assign a plus sign to energy absorbed and a minus sign to energy released. For instance, when dealing with a reaction in which 1 mole of H–H bonds are broken, you write +436 kilojoules to indicate energy absorbed. The positive sign indicates that the molecules gained energy, which was used to break the bonds. When dealing with the formation of 1 mole of H–H bonds, you write –436 kilojoules to indicate energy released. The negative sign indicates that the molecules lost energy, which was released to the environment. We'll do some sample calculations in a moment.

TABLE 9.1 Selected Bond Energies

BOND	BOND ENERGY (KJ/MOLE)	BOND	BOND ENERGY (KJ/MOLE)
H–H	436	N–N	159
H–C	414	O–O	138
H–N	389	Cl–Cl	243
H–O	464	C=O	803
H–F	569	N=O	631
H–Cl	431	O=O	498
C–O	351	C≡C	837
C–C	347	N≡N	946

An Exothermic Reaction: A Net Release of Energy

For most chemical reactions, the total amount of energy absorbed in breaking bonds in reactants is different from the total amount of the energy released as bonds form in the products. Consider the reaction in which hydrogen and oxygen react to form water:

In the reactants, hydrogen atoms are bonded to hydrogen atoms and oxygen atoms are double-bonded to oxygen atoms. The total amount of energy absorbed as these bonds break is +1370 kilojoules. Note that we use the plus sign to indicate the amount of energy *absorbed* to break bonds.

Type of bond	Number of bonds	Bond energy	Total energy
H—H	2 moles	+436 kJ/mole	+872 kJ
O=O	1 mole	+498 kJ/mole	+498 kJ
		Total energy absorbed:	+1370 kJ

In the products, there are four moles of hydrogen–oxygen bonds. The total amount of energy released as these bonds form is –1856 kilojoules. Note that we use the minus sign to indicate the amount of energy *released* as bonds are formed.

Type of bond	Number of bonds	Bond energy	Total energy
H—O	4 moles	–464 kJ/mole	–1856 kJ
		Total energy released:	–1856 kJ

The amount of energy released in this reaction exceeds the amount of energy absorbed. The net energy of the reaction is found by adding the two quantities:

$$\begin{aligned}\text{net energy of reaction} &= \text{energy absorbed} + \text{energy released}\\ &= +1370\text{ kJ} + (-1856\text{ kJ})\\ &= -486\text{ kJ}\end{aligned}$$

The negative sign on the net energy indicates that there is a net release of energy, so the reaction is exothermic. For any exothermic reaction, energy can be considered a product and is thus sometimes included after the arrow of the chemical equation:

$$2\,H_2 + O_2 \rightarrow 2\,H_2O + \text{energy}$$

In an exothermic reaction, the potential energy of atoms in the product molecules is lower than their potential energy in the reactant molecules. This is illustrated in the reaction profile shown in Figure 9.7. The lowered potential energy of the atoms in the product molecules is due to their being more tightly held together. This is analogous to two attracting magnets, whose potential energy decreases as they come closer together. The loss of potential energy is balanced by a gain in kinetic energy. Like two free-floating magnets coming together and accelerating to higher speeds, the potential energy of the reactants is converted to faster-moving atoms and molecules, electromagnetic radiation, or both. This kinetic energy released by the reaction is equal to the difference between the potential energy of the reactants and the potential energy of the products, as is indicated in **Figure 9.7**.

FOR YOUR INFORMATION

The bonds that form in the product molecules are not the same bonds that broke in the reactant molecules. The amount of energy released upon product formation, therefore, will not be the same as the amount of energy consumed in reactant deformation.

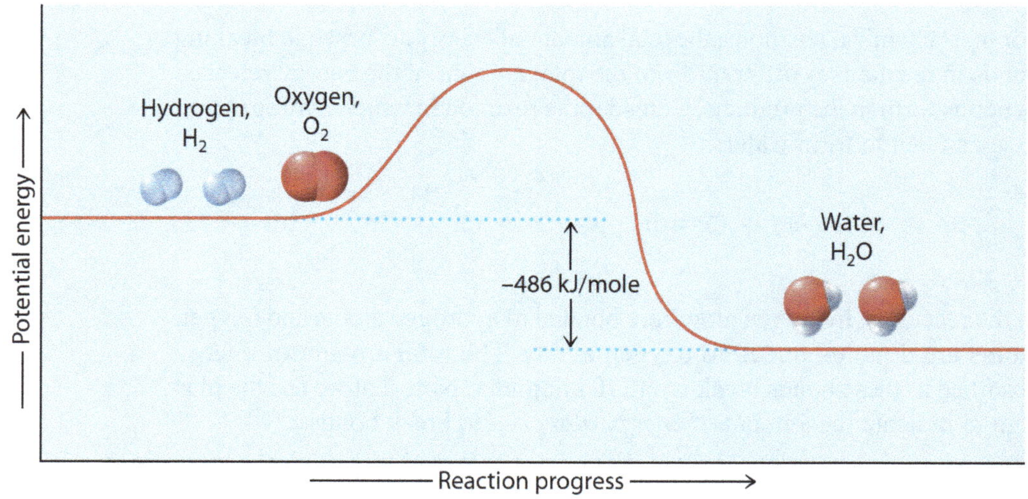

Figure 9.7 >

In an exothermic reaction, the product molecules are at a lower potential energy than the reactant molecules. The net amount of energy released by the reaction is equal to the difference in the potential energies of the reactants and the products. In Section 9.6, we explore why graphs such as this have a bump in the middle.

The total amount of energy released in an exothermic reaction depends on the amounts of the reactants. The reaction of large amounts of hydrogen and oxygen, for example, provided the energy to lift the Space Shuttle shown in **Figure 9.8** into orbit. There were two compartments in the large central tank, to which the orbiter was attached—one filled with liquid hydrogen and the other filled with liquid oxygen. Upon ignition, these two liquids mixed and reacted chemically to form water vapor, which produced the needed thrust as it is expelled the rocket cones. Additional thrust was obtained from a pair of solid-fuel rocket boosters containing a mixture of ammonium perchlorate, NH_4ClO_4, and powdered aluminum, Al. Upon ignition, these chemicals reacted to form products that were expelled at the rear of the rocket. The balanced equation representing this aluminum reaction is

$$3\,NH_4ClO_4 + 3\,Al \rightarrow Al_2O_3 + AlCl_3 + 3\,NO + 6\,H_2O + \text{energy}$$

< Figure 9.8

Space rockets use exothermic chemical reactions to lift off from the Earth's surface.

CONCEPT **CHECK**

Where does the net energy released in an exothermic reaction go?

CHECK YOUR ANSWER

This energy goes into increasing the speeds of reactant atoms and molecules and often into electromagnetic radiation such as light.

An Endothermic Reaction: A Net Absorption of Energy

When the amount of energy released in product formation is less than the amount of energy absorbed when reactant bonds break, the reaction is endothermic. An example is the reaction of atmospheric nitrogen and oxygen to form nitrogen monoxide:

$$N \equiv N + O = O \rightarrow N = O + N = O$$

The amount of energy absorbed as the chemical bonds in the reactants break is

Type of bond	Number of bonds	Bond energy	Total energy
$N \equiv N$	1 mole	+946 kJ/mole	+946 kJ
$O = O$	1 mole	+498 kJ/mole	+498 kJ
		Total energy absorbed:	+1444 kJ

The amount of energy released upon the formation of bonds in the products is

Type of bond	Number of bonds	Bond energy	Total energy
$N = O$	2 moles	−631 kJ/mole	−1262 kJ
		Total energy released:	−1262 kJ

As before, the net energy of the reaction is found by adding the two quantities:

$$\begin{aligned}\text{net energy of reaction} &= \text{energy absorbed} + \text{energy released}\\ &= +1444\text{ kJ} + (-1262\text{ kJ})\\ &= +182\text{ kJ}\end{aligned}$$

The positive sign indicates that there is a net absorption of energy, meaning the reaction is endothermic. For any endothermic reaction, energy can be considered a reactant and is thus sometimes included before the arrow of the chemical equation:

$$\text{energy} + N_2 + O_2 \rightarrow 2\,NO$$

In an endothermic reaction, the potential energy of atoms in the product molecules is higher than their potential energy in the reactant molecules. This is illustrated in the reaction profile shown in **Figure 9.9**. Raising the potential energy of the atoms in the product molecules requires a net input of energy, which must come from some external source, such as electromagnetic radiation, electricity, or heat. Nitrogen and oxygen react to form nitrogen monoxide with the application of large amounts of heat, as occurs adjacent to a lightning bolt or in an internal-combustion engine.

fyi FOR YOUR INFORMATION

NASA scientists routinely test various materials for their durability against atomic oxygen, O, which is abundant in low orbit. They discovered that atomic oxygen effectively transforms surface organic materials into gaseous carbon dioxide. The scientists realized atomic oxygen's usefulness for restoring fine art paintings damaged by smoke or other organic contaminants. Together with art conservationists, they used atomic oxygen to restore certain damaged paintings, and it worked spectacularly.

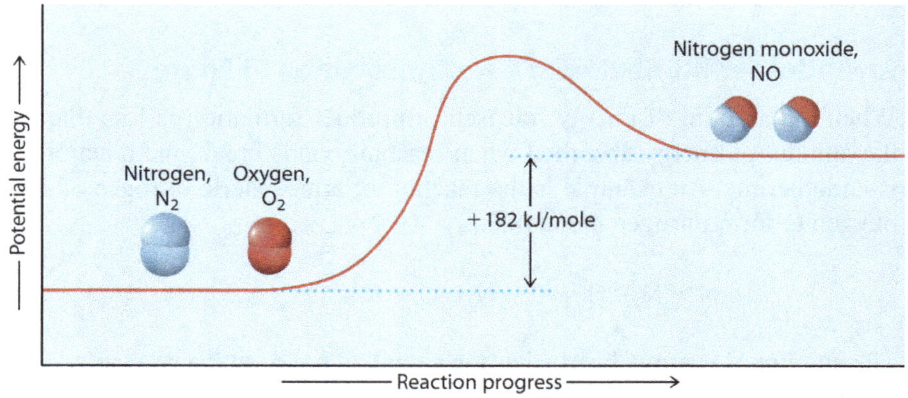

Figure 9.9 >

In an endothermic reaction, the product molecules are at a higher potential energy than the reactant molecules. The net amount of energy absorbed by the reaction is equal to the difference in the potential energies of the reactants and the products.

CONCEPT **CHECK**

Should the following reaction be endothermic or exothermic?

$O_2NNO_2 \rightarrow O_2N + NO_2$

CHECK YOUR ANSWER

No calculations are necessary. This reaction is endothermic because it involves only the breaking of a chemical bond. For practice with examples requiring calculations, see the questions in the Chapter Review.

Energy Is Conserved in a Chemical Reaction

We have been focusing on the key role energy plays in chemical reactions. This is an area of science known as **thermodynamics**, which stems from Greek words meaning "movement of heat." The concepts of exothermic and endothermic reactions fit neatly within what is known as the *first law of thermodynamics*, which can be paraphrased as follows:

First Law of Thermodynamics

Energy is conserved. It may be converted from one form to another, say, from potential to kinetic energy, but the total amount of energy remains the same. The energy that an exothermic reaction releases always goes somewhere into the environment, often in the form of thermal energy (heat), but other forms, such as light energy or electric energy, are also possible. Likewise, the energy that an endothermic reaction absorbs always comes from somewhere in the environment, again, in various forms.

This is a common sense type of law, from which we derive the expression "You can't get something for nothing." Energy doesn't just appear or disappear. It either comes from somewhere or goes to somewhere. In the case of an exothermic reaction, energy comes from the formation of bonds in which atoms are held more tightly but vibrate faster. This is a downhill transformation of potential energy into kinetic energy. In the case of an endothermic reaction, energy is absorbed in the formation of bonds in which atoms are held more loosely and vibrate slower. This is an uphill transformation of kinetic energy into potential energy.

The first law of thermodynamics tells us that energy can be neither created nor destroyed. This is an important law to understand, but it does not tell us the whole picture. Why, for example, do reactants react to form products? By doing nothing instead, they are following the first law of thermodynamics quite perfectly. What then compels them to move forward in the formation of products? The answer to this question is aptly provided by the *second law of thermodynamics*, which we explore next.

9.5 Entropy and Chemical Reactions

Energy tends to disperse. It flows from where it is concentrated to where it is spread out. The energy of a hot pan, for example, does not stay concentrated in the pan once the pan is taken off the stove. Instead, the energy spreads away from the pan into the cooler surroundings. Similarly, the concentrated chemical energy in gasoline disperses, when ignited, in the formation of many very hot smaller molecules that scatter explosively. Some of this released thermal energy is used by the engine to get the car moving. The rest spreads into the engine block and radiator fluid and out the exhaust pipe.

Scientists consider this tendency of energy to disperse to be one of the central reasons for both physical and chemical changes. In other words, changes that result in energy spreading out tend to occur on their own—they are favored. This includes the cooling down of a hot pan or the burning of ignited gasoline. In both cases, there is a dispersal of energy to the environment.

READING CHECK

What is one of the central reasons for the occurrence of any physical or chemical change?

The opposite holds true, too. Changes that result in the concentration of energy do *not* tend to occur—they are not favored. Heat from the room, for example, will never spontaneously move into a pan to heat it up. Likewise, low-energy exhaust molecules coming out of a car's tailpipe will not spontaneously come back together to form higher-energy gasoline molecules. The natural flow of energy is always a one-way trip from where it is concentrated to where it is less concentrated, or "spread out."

That energy tends to disperse is spelled out by the *second law of thermodynamics*, which can be paraphrased as follows:

> *Any process that happens by itself results in the net dispersal of energy. For example, heat naturally flows from a higher-temperature object to a lower-temperature object because in doing so, energy is dispersed from where it is concentrated (a hot pan) to where it is spread out (the cooler kitchen).*

Second Law of Thermodynamics

Entropy is a measure of this natural spreading of energy. Wherever there is a spreading of energy, there is a corresponding *increase* in entropy. Applied to chemistry, entropy helps us to answer a most fundamental question: if you take two materials and put them together, will they react to form new materials? If the reaction results in the dispersal of energy (an overall increase in entropy), then the answer is yes. Conversely, if the reaction results in the concentration of energy (an overall decrease in entropy), then the reaction will not occur by itself. Instead, such a reaction will occur only if an external source of energy is supplied to it.

CHEMICAL CONNECTIONS

How is the dispersal of energy related to the concept of time?

Using this concept of entropy, you are now in a position to understand why exothermic reactions are self-sustaining—occurring on their own without need of external help—while most endothermic reactions need a continual prodding. Exothermic reactions spread energy out to the surroundings, much like a cooling hot pan. This is an increase in entropy; hence, exothermic reactions are favored to occur. An endothermic reaction, by contrast, requires that energy from the surroundings be absorbed by the reactants. This is a concentration of energy, which is counter to energy's natural tendency to disperse. Endothermic reactions, therefore, can progress from reactants to products only with the continual input of energy. But from

Figure 9.10

The Sun is truly a "hothouse"—dispersing enormous amounts of energy from exothermic nuclear reactions. A tiny fraction of the Sun's energy is used to drive photosynthesis, which is vital for plants and plant-eating creatures like us.

where might this energy come? The answer is "from some self-sustaining exothermic reaction occurring elsewhere."

The classic example is photosynthesis, which is an endothermic reaction by which plants use solar energy to create carbohydrates and oxygen from carbon dioxide and water, as represented by the following equation:

$$\underset{\text{}}{\text{sunlight}} + \underset{\text{carbon dioxide}}{6\,CO_2} + \underset{\text{water}}{6\,H_2O} \rightarrow \underset{\text{carbohydrate}}{C_6H_{12}O_6} + \underset{\text{oxygen}}{6\,O_2}$$

In photosynthesis, energy dispersed from the Sun becomes contained within the carbohydrate and oxygen products, which, of course, are the primary fuels of living organisms (**Figure 9.10**). Likewise, most modern materials, such as plastics, synthetic fibers, pharmaceuticals, fertilizers, and metals such as iron and aluminum, are made or purified using endothermic reactions. Our ability to produce these new and useful materials has been the hallmark of modern chemistry. Creating these products, however, necessarily requires the input of energy, which we must obtain from some external source, such as electricity from a power plant that burns fossil or nuclear fuels.

While on the subject of the second law of thermodynamics, we would be remiss not to discuss the close relationship between entropy and our psychological sense of time. That energy tends to spread out is part of our human experience. We *expect* a hot pan to cool, just as we *expect* hot gases to come out of an exhaust pipe. But what if we saw the reverse? For example, what would we think if we saw smoke moving *into* a smokestack? Or what

CONCEPT **CHECK**

Sugar crystals form naturally within a supersaturated solution of sugar water. Does the formation of these crystals result in an increase or a decrease in entropy?

CHECK YOUR ANSWER

The formation of these sugar crystals results in an increase in entropy. Your clue to an increase in entropy here is that the crystals form "on their own," a spontaneous process and thus one that must result in an entropy increase. Recall from Section 8.6 that energy is released when molecules come together to form a solid (the heat of freezing). This release of heat involves the spreading out of energy, which is, by definition, an increase in entropy.

if we saw a diver fly out of the water and rise upward to the diving platform? If we were watching these energy-concentrating events on video, we would quickly conclude that the video was running backward. However, if in real life we actually saw such things—if we could survive the shock—we would immediately sense that time itself was running backward. Thus, when we watch smoke shoot out of a smokestack and a diver diving into water, as shown in **Figure 9.11,** we have the sense that time is moving forward. The second law of thermodynamics, therefore, gives us our psychological sense of time; it is the "arrow of time."

Students often state the laws of thermodynamics this way: you can't win because you can't get any more energy out of a system than you put into it. You can't break even because no matter what you do, some of your energy will be dispersed as ambient heat. Finally, you can't get out of the game because you depend on entropy-increasing processes, such as solar thermonuclear fusion, to remain alive.

fyi FOR YOUR INFORMATION

There are examples of endothermic reactions that proceed spontaneously absorbing heat from the environment. A classic example is the mixing of ions (salt) in water, as presented as one of the Quick Activities for this chapter. In such cases, the environment disperses energy to the dissolving ions, which then spread this energy more widely into the volume of the solvent.

CONCEPT CHECK

The energy of a diver diving into a pool is dispersed as lots of moving water and a little heat after the diver hits the water. How, then, can the diver get back up to the platform?

CHECK YOUR ANSWER

As a living organism, the diver has a supply of biochemical energy, obtained ultimately from photosynthesis, that she can tap to climb upward against gravity to get back to the diving platform.

< Figure 9.11

As the diver dives, her potential energy is converted into kinetic energy. As she splashes into the water, this kinetic energy is spread to make the water molecules move faster and to heat up just a little. Will this dissipated energy reconcentrate itself to push her back through the air up to the cliff? Our common experience with the second law of thermodynamics says no.

9.6 Chemical Kinetics

How exactly do reactants react to form products? This is the focus of the remaining sections of this chapter. We begin by showing how the speed of a reaction depends on the concentration and temperature of reacting molecules.

Some chemical reactions, such as the rusting of iron, are slow, while others, such as the burning of gasoline, are fast. The speed of any reaction is indicated by its *reaction rate*, which is an indicator of how quickly the reactants transform to products. As shown in **Figure 9.12**, initially, a flask may contain only reactant molecules. Over time, these reactants form product molecules, and as a result, the concentration of product molecules increases. The **reaction rate**, therefore, can be defined either as how quickly the concentration of products increases or as how quickly the concentration of reactants decreases.

READING CHECK

What must two molecules do in order to react with each other?

What determines the rate of a chemical reaction? Once again, atomic theory guides us to an explanation. The answer is complex, but one important factor is that reactant molecules must physically come together. Because molecules move rapidly, this physical contact is appropriately described as a collision. We can illustrate the relationship between molecular collisions and reaction rate by considering the reaction of gaseous nitrogen and gaseous oxygen to form gaseous nitrogen monoxide, as shown in **Figure 9.13.**

Because reactant molecules must collide in order for a reaction to occur, the rate of a reaction can be increased by increasing the number of collisions. An effective way to increase the number of collisions is to increase the concentration of the reactants. **Figure 9.14** shows that with higher concentrations, there are more molecules in a given volume, which makes collisions between molecules more probable. As an analogy, consider a group of people on a dance floor—as the number of people increases, so does the rate at which they bump into one another. An increase in the concentration of nitrogen and oxygen molecules, therefore, leads to a greater number of

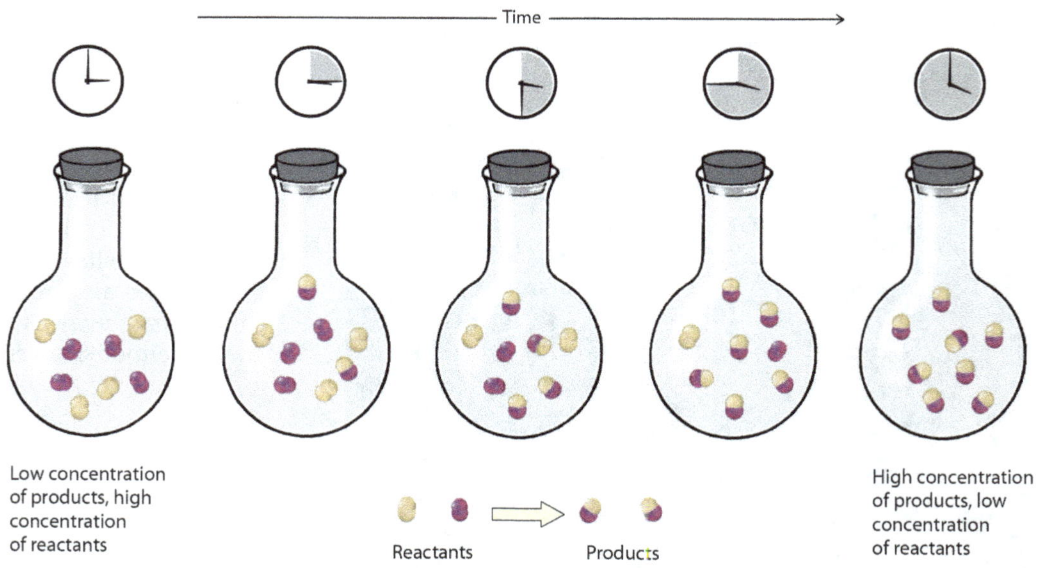

Figure 9.12

Over time, the reactants in this reaction flask may transform to products. If this happens quickly, the reaction rate is fast. If it happens slowly, the reaction rate is slow.

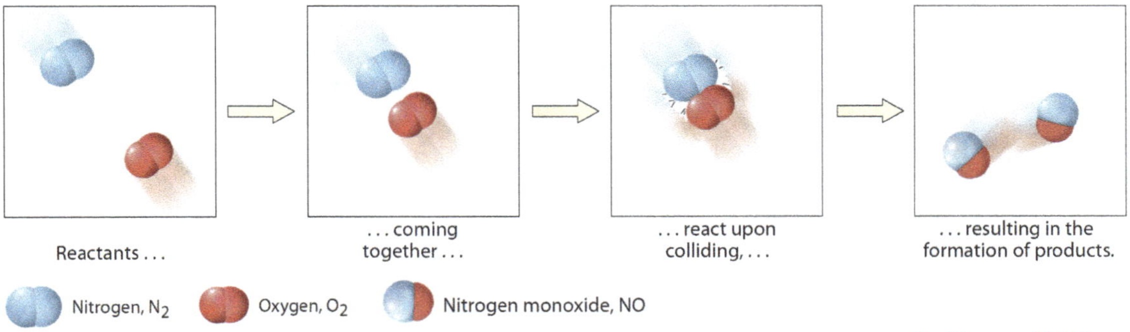

Figure 9.13

During a reaction, reactant molecules collide with one another.

collisions between these molecules—hence, a greater number of nitrogen monoxide molecules will form in a given period of time.

Not all collisions between reactant molecules lead to products, however, because the molecules must collide in a certain orientation in order to react. Nitrogen and oxygen, for example, are much more likely to form nitrogen monoxide when the molecules collide in the parallel orientation shown in Figure 9.13. When they collide in the perpendicular orientation shown in **Figure 9.15,** nitrogen monoxide does not form. For larger molecules, which can have numerous orientations, this orientation requirement is even more restrictive.

A second reason that not all collisions lead to product formation is that the reactant molecules must also collide with enough kinetic energy to break their bonds. Only then is it possible for the atoms in the reactant molecules to change bonding partners and form product molecules. The bonds in N_2 and O_2 molecules, for example, are quite strong. In order for these bonds to be broken, collisions between the molecules must contain enough energy to break the bonds. As a result, collisions between slow-moving N_2 and O_2 molecules, even those that collide in the proper orientation, may not form NO, as shown in **Figure 9.16**.

The higher the temperature of a material, the faster its molecules move and the more forceful and frequent the collisions between them. Higher temperatures, therefore, increase reaction rates. The nitrogen and oxygen molecules that make up our atmosphere, for example, are continually colliding with one another. At the normal temperatures of our atmosphere, however, these molecules do not generally have sufficient kinetic energy for the formation of nitrogen monoxide. The heat of a lightning bolt, as

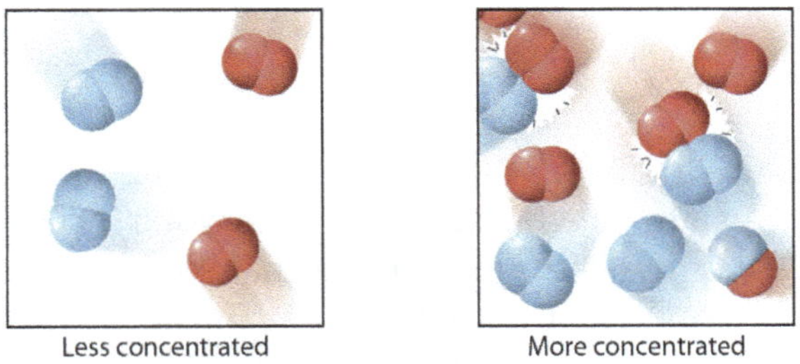

Figure 9.14

The more concentrated a sample of nitrogen and oxygen, the greater the probability that N_2 and O_2 molecules will collide and form nitrogen monoxide.

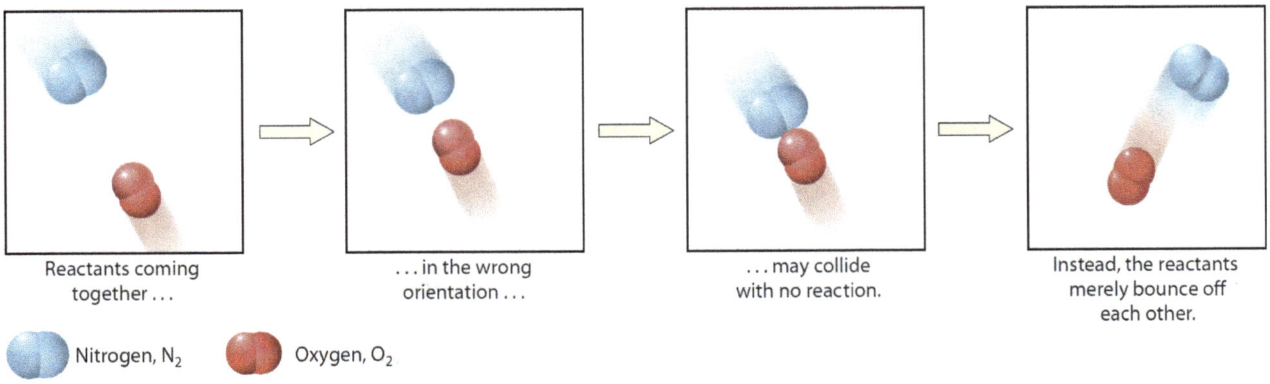

Figure 9.15

The orientation of reactant molecules in a collision can determine whether a reaction occurs. A perpendicular collision between N_2 and O_2 tends not to result in formation of a product molecule.

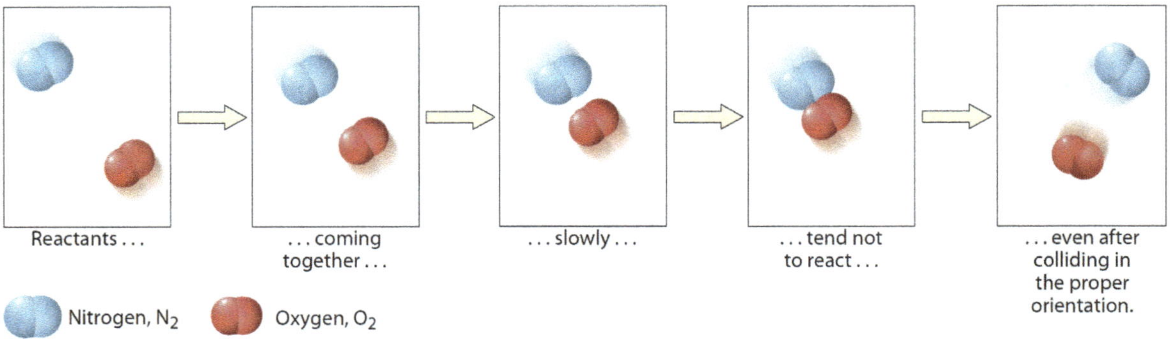

∧ Figure 9.16

Slow-moving molecules may collide with insufficient force to break their bonds. As a result, they cannot react to form product molecules.

shown in the opening photo of this chapter, dramatically increases the kinetic energy of these molecules, to the point that a large portion of the collisions in the vicinity of the bolt result in the formation of nitrogen monoxide. The nitrogen monoxide formed in this manner undergoes further atmospheric reactions to form chemicals known as nitrates, which plants depend on for survival. This is an example of *nitrogen fixation*, which you may have learned about in a biology class.

CONCEPT **CHECK**

An internal-combustion engine works by drawing a mixture of air and gasoline vapors into a chamber. The action of a piston then compresses these gases into a smaller volume prior to ignition by the spark of a spark plug. What is the advantage of squeezing the vapors into a smaller volume?

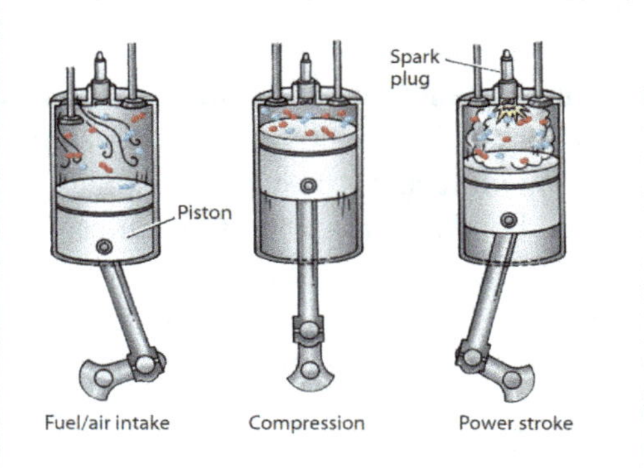

CHECK YOUR ANSWER

Squeezing the vapors into a smaller volume effectively increases their concentration and hence the number of collisions between molecules. This, in turn, promotes the chemical reaction. As per the ideal gas law discussed in Section 2.8, compression also increases the temperature, which further favors the chemical reaction.

The energy required to break bonds can also come from the absorption of electromagnetic radiation. As the radiation is absorbed by reactant molecules, the atoms in the molecules may start to vibrate so rapidly that the bonds between them are easily broken. In many instances, the direct absorption of electromagnetic radiation is sufficient to break chemical bonds and to initiate a chemical reaction. The common atmospheric pollutant nitrogen dioxide, NO_2, for example, may transform to nitrogen monoxide and atomic oxygen merely upon exposure to sunlight:

$$NO_2 + \text{sunlight} \rightarrow NO + O$$

Activation Energy Is Needed for Reactants to React

Whether they result from collisions, from the absorption of electromagnetic radiation, or both, broken bonds are a necessary first step in most chemical reactions. The energy required for this initial breaking of bonds can be viewed as an energy barrier. The minimum energy required to overcome this *energy barrier* is known as the **activation energy** (E_a).

In the reaction between nitrogen and oxygen to form nitrogen monoxide, the activation energy is so high (because the bonds in N_2 and O_2 are strong) that only the fastest-moving nitrogen and oxygen molecules possess

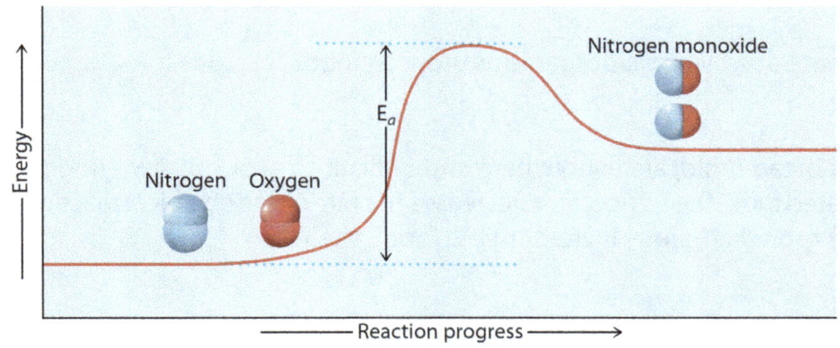

< **Figure 9.17**

Reactant molecules must gain a minimum amount of energy, called the activation energy, E_a, before they can transform to product molecules.

sufficient energy to react. **Figure 9.17** shows the activation energy in this chemical reaction as a vertical hump.

The activation energy of a chemical reaction is analogous to the energy a car needs to drive over the top of a hill. Without sufficient energy to climb to the top of the hill, it isn't possible for the car to get to the other side. Likewise, reactant molecules can transform to product molecules only if the reactant molecules possess an amount of energy equal to or greater than the activation energy.

At any given temperature, there is a wide distribution of kinetic energies in reactant molecules. Some are moving slowly, and others are moving quickly. As we discussed in Section 2.6, the temperature of a material is related to the *average* of all these kinetic energies. The few fast-moving reactant molecules in **Figure 9.18** have enough energy to pass over the energy barrier and are the ones to transform to product molecules.

When the temperature of the reactants is increased, the number of reactant molecules possessing sufficient energy to pass over the barrier also increases, which is why reactions are generally faster at higher temperatures. Conversely, at lower temperatures, fewer molecules have sufficient energy to pass over the barrier. Hence, reactions are generally slower at lower temperatures.

Most chemical reactions are influenced by temperature in this manner, including those reactions occurring in living bodies. The body temperature of animals such as humans, that regulate their internal temperature, is fairly constant. However, the body temperature of some animals, such as the alligator shown in **Figure 9.19**, rises and falls with the temperature of the environment. On a warm day, the chemical reactions occurring in an alligator are "up to speed" and the animal is more active. On a chilly day, however, the chemical reactions proceed at a lower rate; as a consequence, the alligator's movements are unavoidably sluggish.

fyi FOR YOUR INFORMATION

The life sciences involve fantastic applications of chemistry, nitrogen fixation being just one example. Others include photosynthesis, cellular respiration, and molecular genetics. So, there are distinct advantages to learning about chemistry before advancing to the life sciences.

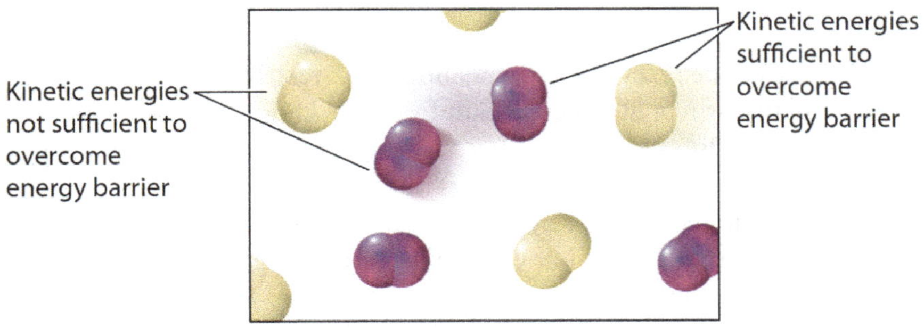

< **Figure 9.18**

Because fast-moving reactant molecules possess sufficient energy to pass over the energy barrier, they are the ones to transform to product molecules.

CONCEPT **CHECK**

What kitchen device is used to lower the rate at which microorganisms grow on food?

CHECK YOUR ANSWER

The refrigerator. Microorganisms, such as bread mold, are everywhere and difficult to avoid. By lowering the temperature of microorganism-contaminated food, the refrigerator decreases the rate of the chemical reactions that these microorganisms depend on for growth, thereby increasing the food's shelf life.

Figure 9.19 >

Drivers beware! Alligators can become immobilized on pavement after being caught in the cold night air. By mid-morning, shown here, the temperature becomes warm enough to allow the alligator to get up and walk away.

9.7 Chemical Catalysts

READING CHECK

A catalyst speeds up a chemical reaction by lowering what?

As discussed in the previous section, a chemical reaction can be made to go faster by increasing the concentration of the reactants or by increasing the temperature. A third way to increase the speed of a reaction is to add a **catalyst**, which is any substance that increases the rate of a chemical reaction without itself being consumed.

To learn how catalysts work, let's consider some chemical reactions that take place in our atmosphere. One of the most important reactions occurs at altitudes higher than 20 km, within a region known as the *stratosphere*. Here the Sun's ultraviolet (UV) rays cause oxygen molecules to transform into ozone molecules, as follows:

$$\underset{\text{molecular oxygen}}{3\,O_2(g)} + UV \rightarrow \underset{\text{ozone}}{2\,O_3(g)}$$

Furthermore, the ozone itself, once formed, is able to transform ultraviolet light into heat:

$$\underset{\text{ozone}}{O_3(g)} + UV \rightarrow \underset{\text{ozone}}{O_3(g)} + \text{heat}$$

In all, these ozone reactions prevent about 95 percent of the ultraviolet rays that come to our planet from the Sun from reaching its surface. This is of great benefit to life on the Earth because ultraviolet rays are most harmful to living tissue. Stratospheric ozone is our planet's safety shield.

Starting in the 1970s, scientists began to recognize that human-made chlorofluorocarbons, also known as CFCs, posed a significant threat to stratospheric ozone. Because CFCs are inert gases, they were once commonly used in air conditioners and aerosol propellants. Two of the most frequently used CFCs are shown in **Figure 9.20.**

FOR YOUR INFORMATION

The stratospheric ozone layer is about 10 kilometers thick, but the concentration of ozone, O_3, within the layer is quite low. If this ozone layer were brought down to Earth's surface, atmospheric pressure would squeeze it to a thickness of only 3 millimeters. Not much ozone is up there protecting us. But thankfully, even small amounts of ozone are very good at shading us from solar ultraviolet light.

The CFCs, it was discovered, drift their way to the stratosphere, where the intense ultraviolet rays break apart these molecules, producing chlorine atoms. The chlorine atoms, in turn, speed up the rate at which ozone molecules are destroyed. They do this by offering a reaction pathway that has a lower activation energy, as shown in **Figure 9.21**.

Chlorine atoms provide an alternate pathway involving intermediate reactions, each having a lower activation energy than the uncatalyzed reaction. This alternate pathway involves two steps. Initially, the chlorine reacts with the ozone to form chlorine monoxide and oxygen:

Cl	+	O_3	$\longrightarrow$	ClO	+	O_2
Chlorine		Ozone		Chlorine monoxide		Oxygen

The chlorine monoxide then reacts with another ozone molecule to re-form the chlorine atom, as well as to produce two additional oxygen molecules:

ClO	+	O_3	$\longrightarrow$	Cl	+	$2\,O_2$
Chlorine monoxide		Ozone		Chlorine		Oxygen

A catalyst speeds up a chemical reaction, but interestingly, it is not destroyed by the reaction it catalyzes. Notice that although chlorine is depleted in the first reaction, it is regenerated in the second reaction. As a result, there is no net consumption of chlorine. Instead, it survives; so, it can speed up the reaction repeatedly. One chlorine atom in the ozone layer, for example, can catalyze the transformation of 100,000 ozone molecules to oxygen molecules in the one or two years before the chlorine atom is removed by natural processes. In short, chlorine is bad news for stratospheric ozone, which is bad news for life on the Earth.

The fragility of stratospheric ozone came to the world's attention in 1985, with the discovery of a seasonal depletion of stratospheric ozone over the Antarctic continent, a phenomenon known as the *ozone hole*. The fact that chlorine atoms play an active role in the destruction of Antarctic ozone is revealed by measuring chlorine monoxide concentrations over this region, as shown in **Figure 9.22**. Furthermore, the satellite images show that the shape of the ozone hole typically matches the shape of a map showing chlorine monoxide distribution (**Figure 9.23**).

There has been an unprecedented level of international cooperation toward banning ozone-destroying substances such as CFCs. As a result, stratospheric concentrations of chlorine have been declining. Even with these treaties in place, however, the ozone-destroying actions of CFCs will be with us for some time. Atmospheric CFC levels are not expected to drop back to the levels found before the ozone hole was formed until sometime in the 22nd century.

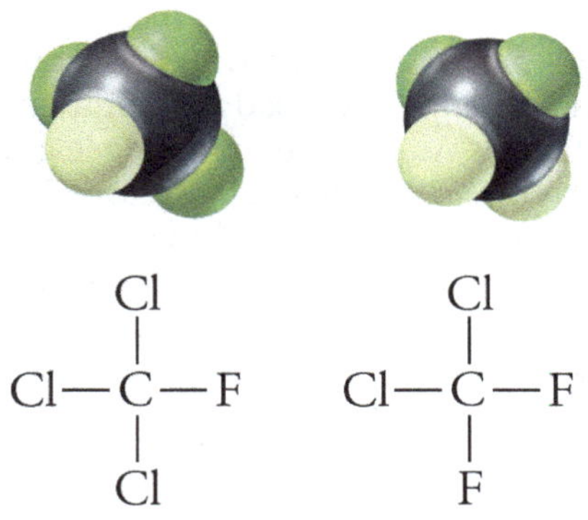

CFC-11 CFC-12

Λ Figure 9.20

Two of the most common CFCs, also known as freons, were CFC-11, trichlorofluoromethane, and CFC-12, dichlorodifluoromethane. At the height of CFC production in 1988, some 1.13 million tons was produced worldwide. Because of their inertness, CFCs were once thought to pose little threat to the environment.

V Figure 9.21

(a) The high activation energy indicates that only the most energetic ozone molecules can react to form oxygen molecules. (b) The presence of chlorine atoms allows conversion of ozone to oxygen with a lower activation energy, which means more ozone molecules have sufficient energy to form the product. (Note that the convention is to write the catalyst above the reaction arrow.)

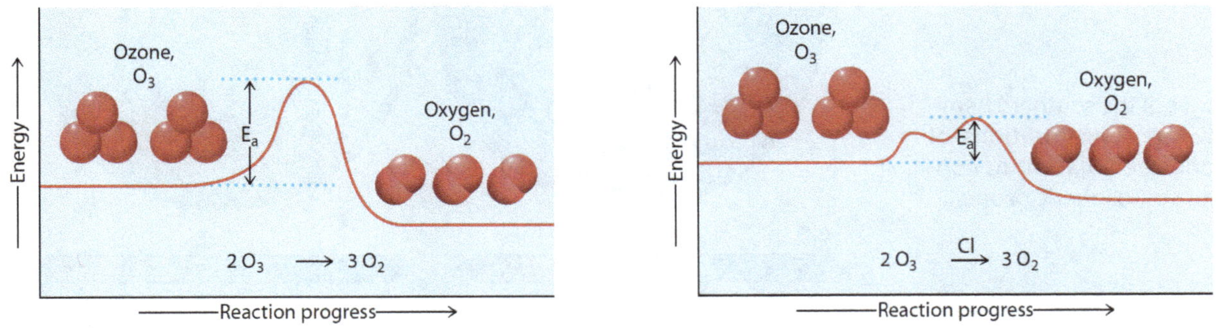

(a) Without catalyst (b) With chlorine catalyst

Figure 9.22 >

Concentrations of stratospheric ozone and chlorine monoxide in southern latitudes. As chlorine monoxide levels increase, ozone levels decrease. The yellow highlighting shows where small fluctuations in ClO concentrations result in large fluctuations in O_3 concentrations. This is consistent with catalytic behavior.

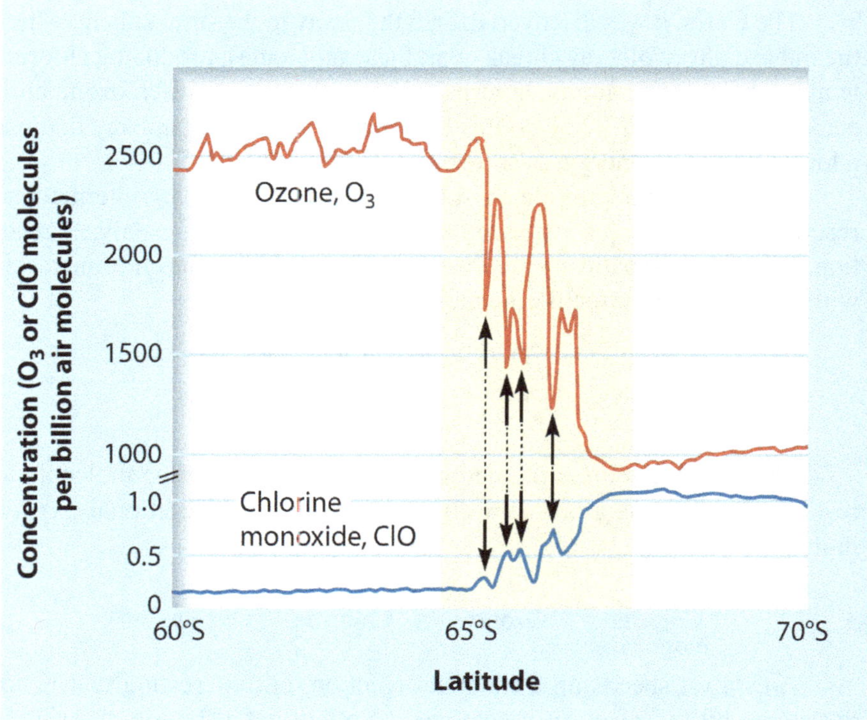

FOR YOUR INFORMATION

Numerous oil-drilling sites in Siberia were once allowed to vent natural gas freely into the atmosphere. After the fall of the Soviet Union, these wells were capped to prevent this venting. Within weeks, instruments at the Mauna Loa weather observatory on the other side of the planet noted a significant drop in atmospheric levels of methane and its by-product, carbon dioxide. The effect that we humans have on global atmospheric conditions is very measurable.

Catalysts are also Most Beneficial

Chemists have been able to harness the power of catalysts for numerous beneficial purposes. The exhaust that comes from an automobile engine, for example, contains a wide assortment of pollutants, such as nitrogen monoxide, carbon monoxide, and uncombusted fuel vapors (hydrocarbons). To reduce the amount of these pollutants entering the atmosphere, most automobiles are equipped with *catalytic converters*, as shown in **Figure 9.24**. Metal catalysts in a converter speed up reactions that convert exhaust pollutants to less toxic substances. Nitrogen monoxide is transformed to nitrogen and oxygen, carbon monoxide is transformed to carbon dioxide, and unburned fuel is converted to carbon dioxide and water vapor. Because catalysts are not consumed by the reactions they speed up, a single catalytic converter may continue to operate effectively for the lifetime of the car.

Figure 9.23 >

Satellite images of the southern hemisphere showing concentrations of chlorine monoxide adjacent to concentrations of stratospheric ozone.

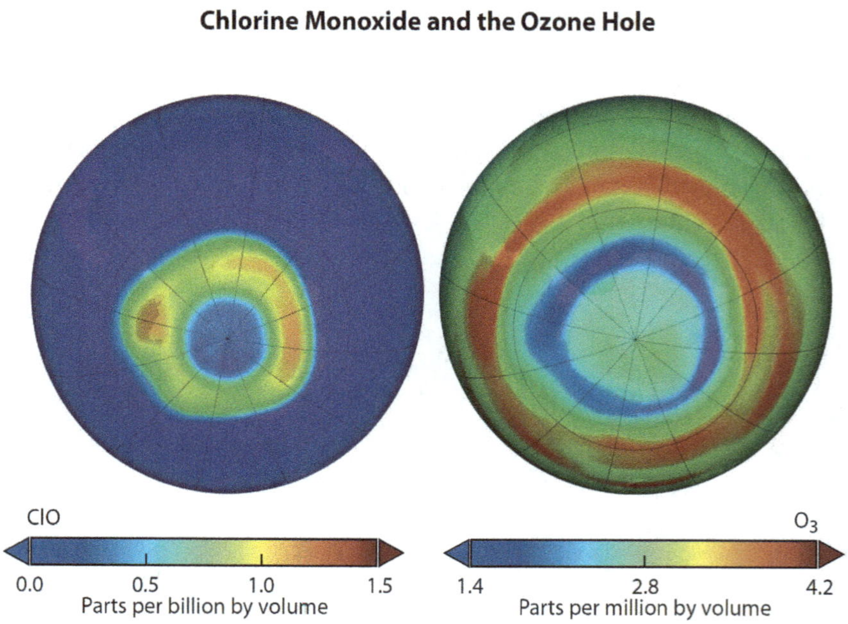

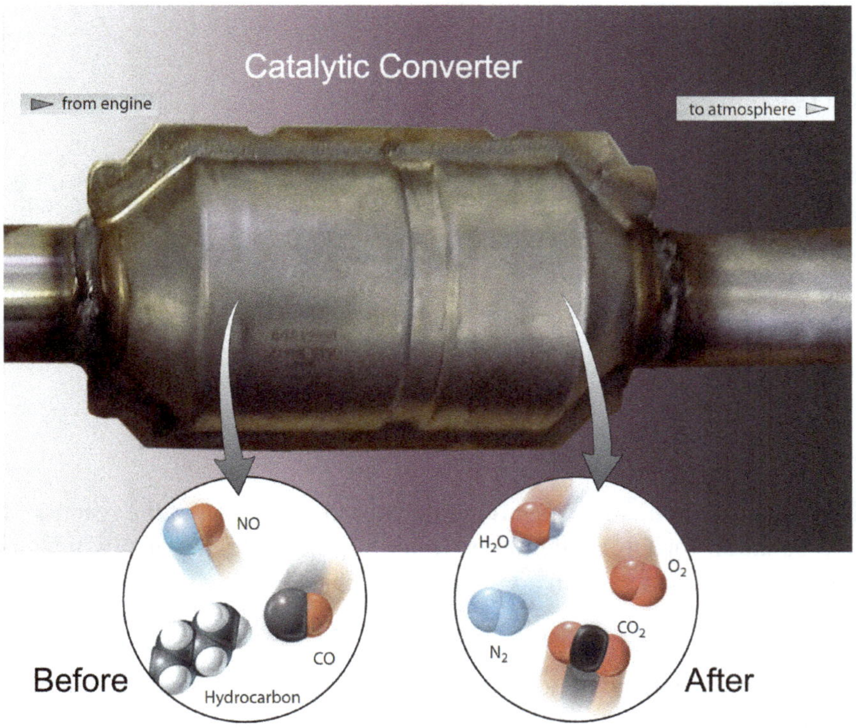

< Figure 9.24

A catalytic converter reduces the pollution caused by automobile exhaust by converting such harmful combustion products as NO, CO, and hydrocarbons to harmless N_2, O_2, CO_2, and H_2O. The catalyst is typically platinum, Pt; palladium, Pd; or rhodium, Rd.

Catalytic converters, along with microchip-controlled fuel–air ratios, have led to a significant drop in the per-vehicle emission of pollutants. This improvement, however, has been offset by an increase in the number of cars being driven, as exemplified by the traffic jam shown in **Figure 9.25**.

Catalysts lower manufacturing costs by lowering required temperatures and by providing greater product yields without being consumed. Indeed, more than 90 percent of all manufactured goods are produced with the assistance of catalysts. The production of polymer plastics, as described in Chapter 12, is particularly dependent upon catalysts. Also, without catalysts, the price of gasoline would be much higher, as would be the price of such consumer goods as rubber, pharmaceuticals, automobile parts, clothing, and food grown with chemical fertilizers. Living organisms rely on special types of catalysts known as *enzymes*, which allow exceedingly complex biochemical reactions to occur with ease.

fyi FOR YOUR INFORMATION

Chlorofluorocarbons are distributed fairly evenly across the Earth's stratosphere. Over the poles, however, stratospheric clouds are able to form during the dark and extremely cold winters. CFC molecules bind to ice crystals within these clouds. When hit by sunlight in the spring, the crystal-bound CFC molecules break apart, producing a large dose of ozone-destroying chlorine atoms. In Antarctica, this process reaches a peak in September, which is when the largest ozone holes are typically observed. The ozone hole over the South Pole is typically more intense than the one over the North Pole. The reason is because the geography of the Southern Ocean favors a stable vortex around the South Pole, which, in turn, favors the formation of the stratospheric clouds.

< Figure 9.25

The exhaust from automobiles today is much cleaner than before the advent of the catalytic converter, but there are many more cars on the road. In 1960, there were about 74 million registered motor vehicles in the United States. By 2020, there were more than 276 million.

CONCEPT **CHECK**

How is a catalyst different from a chemical reactant?

CHECK YOUR ANSWER

A catalyst is not used up during a chemical reaction. Instead, it is released as its original self. The catalyst can then serve to catalyze more reactions. A chemical reactant, by contrast, is consumed during a reaction as it is transformed into a chemical product.

9.8 Chemical Equilibrium

CHEMICAL CONNECTIONS

How is the invention of air conditioning connected to the quality of sunshine in Antarctica?

The arrow of a chemical equation indicates the direction of the reaction. In the equation for the formation of water, for example, the arrow indicates that hydrogen and oxygen combine to form water:

$$2\ H_2 + O_2 \rightarrow 2\ H_2O$$

The reverse reaction is also possible; that is, water can transform back into hydrogen and oxygen:

$$2\ H_2 + O_2 \leftarrow 2\ H_2O$$

Most chemical reactions are *reversible*, but the extent of the reverse reaction depends very much on conditions. Under usual conditions, water does not readily convert to oxygen and hydrogen. When electricity passes through water, however, the water readily transforms into O_2 and H_2 (**Figure 9.26**).

Consider nitrogen dioxide, NO_2, which is a brown gas and one of the toxic components of air pollution. At room temperature nitrogen dioxide molecules join to become colorless dinitrogen tetroxide, N_2O_4:

$$NO_2 + NO_2 \rightarrow N_2O_4$$

Once dinitrogen tetroxide molecules are formed, they can break apart to re-form nitrogen dioxide:

$$NO_2 + NO_2 \leftarrow N_2O_4$$

Figure 9.26 >

In principle, applying a strong electric current to the ocean for an incredibly long time would cause all the water to be converted into gaseous hydrogen and oxygen. The first strike of a match, however, would initiate a massive explosion as the two gases exothermically convert back into water. The point being that chemical reactions are reversible—some more readily than others.

Of course, once the nitrogen dioxide molecules are re-formed, they can get together to re-form the dinitrogen tetroxide. The net result is two competing reactions that keep occurring simultaneously. We can depict this using double arrows:

$$NO_2 + NO_2 \rightleftharpoons N_2O_4$$

Initially, we may start with pure nitrogen dioxide, NO_2. Soon, however, the percentage of nitrogen dioxide decreases and the percentage of dinitrogen tetroxide, N_2O_4, increases. This shift in amounts continues to a point where the two reactions balance each other and the percentages of NO_2 and N_2O_4 settle to constant values. At room temperature this is about 31% NO_2 and 69% N_2O_4. At this point, the forward and reverse reactions have achieved **chemical equilibrium**, where the rate at which products are formed is equal to the rate at which they are converted back into reactants (**Figure 9.27**).

Chemists do not usually express the balance of reactants and products at equilibrium in terms of percentages. Rather, they use *equilibrium constants*. An equilibrium constant is a number that relates the amounts of reactants to products at equilibrium. For example, consider the reaction in which *a* moles of reactant A and *b* moles of reactant B react to give *c* moles of product C and *d* moles of product D at equilibrium:

$$a\text{A} + b\text{B} \rightleftharpoons c\text{C} + d\text{D}$$

Because the reaction is at equilibrium, there is no net change in the amounts of chemical reactants and products—A, B, C, and D—at any time. Therefore, we can use a number—we call an equilibrium constant, K_{eq}—with the following mathematical form to express the relative concentrations of reactants and products at equilibrium:

$$K_{eq} = \frac{[\text{C}]^c \times [\text{D}]^d}{[\text{A}]^a \times [\text{B}]^b}$$

Equilibrium constants provide valuable information in a shorthand format. Can you see that, since an equilibrium constant is a ratio of products to reactants, a value of K_{eq} greater than 1 means products are favored over reactants? Can you also see that if K_{eq} is less than 1, reactants are favored at equilibrium? In other words, a large K_{eq} means products are favored while a small K_{eq} means the reactants are favored.

To help with understanding chemical equilibrium, let's step back and use an analogy. Chemical equilibrium is like a store in which there is a steady flow of customer traffic (**Figure 9.28**). Think of people inside the store as product molecules and those outside as reactant molecules. The steady customer flow means that the number of people entering the store in any given time interval equals the number of people leaving. As a result, the number of people inside the store never changes. The individual people in the store change as Ms. X and Mr. Y enter and Mr. A and Ms. B leave, but the total number inside remains constant.

In the same way, the number of N_2O_4 molecules in the flask of Figure 9.27 stays constant because the rate at which N_2O_4 molecules convert to NO_2 molecules equals the rate at which N_2O_4 molecules are formed from NO_2 molecules.

∧ Figure 9.27

The gases NO_2 and N_2O_4 in this bottle are in equilibrium at 16% NO_2 and 84% N_2O_4. Although the nonchanging brown color seems to indicate no activity in the flask, individual molecules are continuously converting back and forth with a balance between forward and reverse reactions.

Figure 9.28 >

If customers enter and leave the department store at the same rate, then the number of customers within the store remains constant. This is an example of customer equilibrium.

READING CHECK

Does chemical equilibrium mean the amounts of reactants and products are the same?

Chemical equilibrium does NOT mean that the amounts of reactants and products are equal. Just as the equilibrium point in our store analogy might be 10 people in the store and 100 on the sidewalk, the NO_2/N_2O_4 reaction reaches an equilibrium point when the $NO_2:N_2O_4$ ratio is 31:69. The important thing to remember is that chemical equilibrium is a dynamic state. Although the concentrations of products and reactants remain constant, individual molecules are continuously changing back and forth.

If the store holds a big sale, the equilibrium of customers is affected. At the beginning of the sale, the rate at which people enter the store will be greater than the rate at which they leave. Hence the number of people inside the store increases. There is a limit to the number of customers that the store can hold, however. Eventually, a doorman is hired to allow people to enter only at the rate people leave. Thus, a new equilibrium is established. In this case, however, there are now more people in the store than there were before the sale. A new "customer equilibrium" has been achieved, which was the objective of the sale.

Likewise, we find that chemical equilibria are affected by prevailing conditions. The formal statement of this is known as **Le Chatelier's principle**:

> **If a stress is applied to a system in chemical equilibrium, the system changes to relieve the stress.**

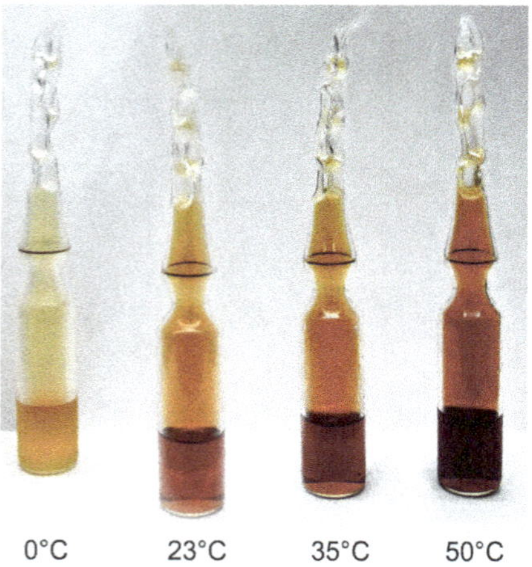

^ Figure 9.29

Lower temperatures favor the formation of colorless dinitrogen tetroxide, and the mixture is a much lighter brown as seen in the ampoule at 0°C. With higher temperatures, the formation of brown nitrogen dioxide is favored and the mixture turns darker brown as seen in the ampoule at 50°C.

For example, when dinitrogen tetroxide is exposed to heat, it tends to break down into nitrogen dioxide. Increasing the temperature of a flask of NO_2 and N_2O_4 increases the rate of nitrogen dioxide formation until a new equilibrium is reached as shown in **Figure 9.29**. At this higher-temperature equilibrium, nitrogen dioxide, NO_2, predominates and the overall color of the reaction mixture is darker brown. This is one of the reasons air pollution becomes particularly noticeable on warm days: the higher the temperature, the greater the concentration of brown nitrogen dioxide (**Figure 9.30**).

Pressure as well as temperature can affect the chemical equilibrium of reacting gases. As we saw in studying the gas laws, we can increase the pressure of a gas by decreasing the volume of its container. This brings the

Figure 9.30

(a) Much of the brown color of air pollution is from the nitrogen dioxide, NO_2, it contains. (b) Prior to air pollution measures, brown colored skies were a particular challenge during the warmer summer months as shown in this archived photo of downtown Cleveland, Ohio, in 1973.

gaseous molecules closer together, which has the effect of increasing the number of collisions among the molecules (**Figure 9.31**).

Take a container filled with an equilibrium mixture of NO_2 gas and N_2O_4 gas and squeeze it to a smaller volume. The rate at which the molecules collide with one another increases. This is advantageous for the *formation* of colorless N_2O_4, which requires the coming together (collision) of two NO_2 molecules. Under higher pressures, therefore, the rate of N_2O_4 formation becomes greater than the rate of NO_2 formation and the gaseous mixture becomes lighter in color, as shown in **Figure 9.32**.

Figure 9.31

Gas particles at higher pressures are closer together, which increases the rate at which they collide with one another.

CONCEPT **CHECK**

After a mixture of nitrogen dioxide and dinitrogen tetroxide initially at room temperature is brought to 100°C and kept at that temperature for a few minutes, which is greater—the rate of nitrogen dioxide formation or the rate of dinitrogen tetroxide formation?

CHECK YOUR ANSWER

Neither! After a few minutes a new equilibrium is achieved. It's like the store holding a sale. Initially, equilibrium is lost as more customers pile in. Eventually, however, the store becomes so packed that the rate at which people come in matches the rate at which they go out. Similarly, with the application of heat, the rate of NO_2 formation will initially be greater than the rate of N_2O_4 formation. As the N_2O_4 becomes depleted, however, the rate of NO_2 formation decreases to the point that the forward and reverse reactions become matched as they were before.

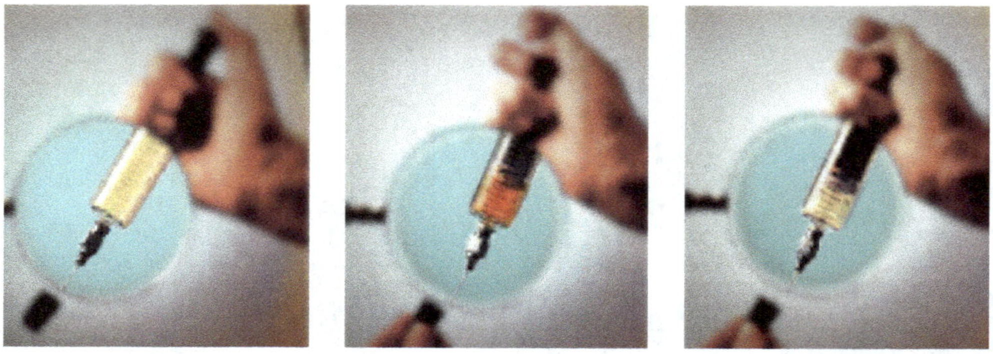

(a) (b) (c)

∧ Figure 9.32

(a) A syringe containing an equilibrium mixture of nitrogen dioxide and dinitrogen tetroxide. (b) Compressing the gas initially shows a deepening of the brown color as brown NO_2 molecules are bunched together. (c) The higher proportion of NO_2 molecules, however, favors a greater number of collisions among these molecules, which upon colliding convert to the colorless N_2O_4 molecules. This is evidenced by a rapid lightening of the brown color soon after the NO_2 molecules have been compressed.

CONCEPT CHECK

Would an increase in pressure favor the formation of gaseous sulfur trioxide, SO_3, from gaseous sulfur dioxide, SO_2, and oxygen, O_2?

$$2\ SO_2(g) + O_2(g) \rightleftharpoons 2\ SO_3$$

CHECK YOUR ANSWER

Yes. an increase in pressure favors a greater number of collisions, which has a greater benefit for the formation of SO_3. Incidentally, SO_2 is an industrial pollutant. In the atmosphere it transforms into SO_3, as shown above. The SO_3 then reacts with water to form sulfuric acid, H_2SO_4—a prime component of acid rain.

Concept Review

Chapter 9

Summary of Terms

Activation energy The minimum energy required in order for a chemical reaction to proceed.

Avogadro's number The number of particles—6.02×10^{23}—contained in 1 mole of anything.

Bond energy The amount of energy required to pull two bonded atoms apart, which is the same as the amount of energy released when the two atoms are brought together into a bond.

Catalyst Any substance that increases the rate of a chemical reaction without itself being consumed by the reaction.

Chemical equation A representation of a chemical reaction in which reactants are drawn before an arrow that points to the products.

Chemical equilibrium A state in which the forward and reverse rates of a reaction are equal.

Endothermic A term that describes a chemical reaction in which there is a net absorption of energy.

Entropy A measure of an amount of energy that has been dispersed. Wherever there is a spreading of energy, there is a corresponding increase in entropy.

Exothermic A term that describes a chemical reaction in which there is a net release of energy.

Formula mass The sum of the atomic masses of the elements in a chemical formula.

Law of mass conservation Matter is neither created nor destroyed during a chemical reaction—atoms merely rearrange, without any apparent loss or gain of mass, to form new molecules.

Le Chatelier's principle If a stress is applied to a system in chemical equilibrium, the system changes to relieve the stress.

Molar mass The mass of 1 mole of a substance.

Products The new materials formed in a chemical reaction.

Percent Yield A measure of how much product was obtained relative to the amount of product that could be obtained under ideal conditions.

Reactants The reacting substances in a chemical reaction.

Reaction rate A measure of how quickly the concentration of products in a chemical reaction increases or the concentration of reactants decreases.

Thermodynamics An area of science concerned with the role energy plays in chemical reactions and other energy dependent processes.

Review Questions

9.1 Chemical Equations

1. What is the purpose of coefficients in a chemical equation?

2. How many oxygen atoms are indicated on the right side of this balanced chemical equation:

$$4\ Cr(s) + 3\ O_2(g) \rightarrow 2\ Cr_2O_3(g)$$

3. Why is it important that a chemical equation be balanced?

4. Why is it important never to change a subscript in a chemical formula when balancing a chemical equation?

9.2 Measuring Molecules

5. Why don't equal masses of golf balls and Ping-Pong balls contain the same number of balls?

6. Why don't equal masses of carbon atoms and oxygen molecules contain the same number of particles?

7. What is the formula mass of nitrogen monoxide, NO, in atomic mass units?

9.3 Grams to Moles

8. If you had 1 mole of marbles, how many marbles would you have? How about 2 moles?

9. How many moles of water are there in 18 grams of water?

10. How many molecules of water are there in 18 grams of water?

11. Why is saying you have 1 mole of water molecules the same as saying you have 6.02×10^{23} water molecules?

12. What is the mass of an oxygen atom in atomic mass units?

9.4 Exothermic or Endothermic

13. If it takes 436 kilojoules to break a bond, how many kilojoules are released when the same bond is formed?

14. What is released by an exothermic reaction?

15. What is absorbed by an endothermic reaction?

9.5 Entropy and Chemical Reactions

16. As energy disperses, where does it go?

17. What is always increasing?

18. Why are exothermic reactions self-sustaining?

9.6 Chemical Kinetics

19. Why don't all collisions between reactant molecules lead to product formation?

20. What generally happens to the rate of a chemical reaction with increasing temperature?

21. Which reactant molecules are the first to pass over the energy barrier?

22. What term is used to describe the minimum amount of energy required in order for a reaction to proceed?

9.7 Chemical Catalysts

23. What catalyst is effective in the destruction of atmospheric ozone, O_3?

24. What does a catalyst do to the energy of activation for a reaction?

25. What net effect does a chemical reaction have on a catalyst?

26. Why are catalysts so important to our economy?

9.8 Chemical Equilibrium

27. What is true about a chemical reaction in equilibrium?

28. How can the chemical equilibrium of a reaction be changed?

29. Which molecule is deeper brown in color: nitrogen dioxide, NO_2, or dinitrogen tetroxide, N_2O_4?

Quantitative Questions

30. Show that there are 1.0×10^{22} carbon atoms in a one carat pure diamond, which has a mass of 0.20 grams.

31. How many gold atoms are there in a 5.00 gram sample of pure gold, Au (197 amu)?

32. Show that one mole of $KClO_3$ contains 122.55 grams.

33. Small samples of oxygen gas needed in the laboratory can be generated by a number of simple chemical reactions, such as

$$2\ KClO_3(s) \rightarrow 2\ KCl(s) + 3\ O_2(g)$$

According to this balanced chemical equation, how many moles of oxygen gas are produced from the reaction of two moles of $KClO_3$ solid?

34. Small samples of oxygen gas needed in the laboratory can be generated by a number of simple chemical reactions, such as

$$2\ KClO_3(s) \rightarrow 2\ KCl(s) + 3\ O_2(g)$$

What mass of oxygen (in grams) is produced when 122.55 grams of $KClO_3$ (formula mass 122.55 atomic mass units) takes part in this reaction?

35. Show that the formula mass of 2-propanol, C_3H_8O, is 60 amu, and that the formula mass of propene, C_3H_6, is 42 amu, and also that the formula mass of water, H_2O, is 18 amu.

36. How many grams of water, H_2O, and propene, C_3H_6, can be formed from the reaction of 6.0 grams of 2-propanol, C_3H_8O?

$$C_3H_8O \rightarrow C_3H_6 + H_2O$$

37. A 16 g sample of methane, CH_4, is combined with a 16 g sample of molecular oxygen, O_2, in a sealed container. Upon ignition, what is the maximum amount of carbon dioxide, CO_2, that can be formed?

$$CH_4 + 2\ O_2 \rightarrow CO_2 + 2\ H_2O$$

38. Use the bond energies in Table 9.1 and the accounting format shown in Section 9.4 to determine whether these reactions are exothermic or endothermic:

$H_2 + Cl_2 \rightarrow 2\ HCl$

$2\ HC{\equiv}CH + 5\ O_2 \rightarrow 4\ CO_2 + 2\ H_2O$

39. Use the bond energies in Table 9.1 to determine whether these reactions are exothermic or endothermic:

a. 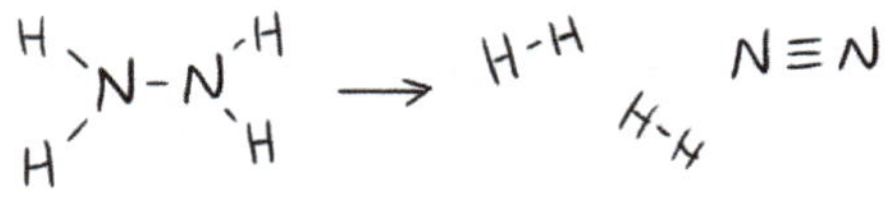

b. 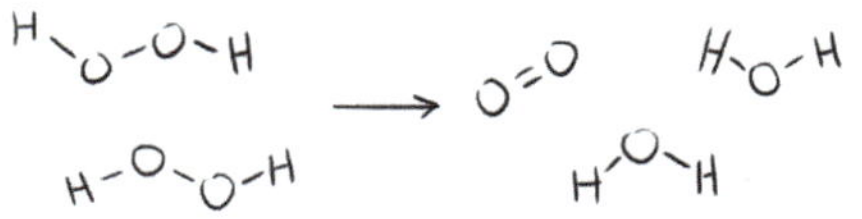

40. Show that the equilibrium constant for the following reaction is 56

$H_2\ (g)\ +\ I_2\ (g).\ \rightarrow\ HI\ (g)\ +\ HI\ (g)$

Equilibrium concentrations:

$H_2\ (g) = 0.0037$ moles per liter

$I_2\ (g) = 0.0037$ moles per liter

$HI\ (g) = 0.0276$ moles per liter

41. The equilibrium constant for the following equation is 3.5×10^8:

$N_2\ (g)\ +\ 3\ H_2\ (g)\ \rightarrow\ 2\ NH_3\ (g)$

Calculate the equilibrium constant for the reverse reaction:

$2\ NH_3\ (g)\ \rightarrow\ N_2\ (g)\ +\ 3\ H_2\ (g)$

Solutions (Odd-Numbered)

1. Coefficients are used to show the ration in which reactants combine and products for in a chemical reaction.

3. A chemical equation must be balanced because the law of conservation of mass says that mass can neither be created nor destroyed. There must be the same number of each atom on both sides of the equation.

5. The relative mass of golf balls is greater than that of ping pong ball balls therefore, it would take more ping pong balls to equal the same mass of golf balls.

7. The formula mass of NO is 30.006 amu

9. For water, 18 grams is one mole.

11. One mole of water has 6.02×10^{23} water molecules.

13. The amount of energy released when the bond is formed equals the amount of energy needed to break the bond, which is 436 kJ.

15. Energy is consumed by an endothermic reaction.

17. The net entropy of the universe is always increasing.

19. Reactants must collide in a certain orientation with enough energy in order to react.

21. Assuming the reactant molecules are already mixed together, the first to react are those with sufficient kinetic energy and the proper orientation.

23. Atomic chlorine is a catalyst for the destruction of ozone.

25. A catalyst is unchanged by a chemical reaction.

27. The rate of forward and reverse reactions is equal.

29. Nitrogen dioxide, NO_2, is deeper brown in color.

31. (5.00 g gold)(1 mole gold/197 g gold)(6.02×10^{23} atoms/1 mole) = 1.53×10^{22} gold atoms

33. The coefficients within this balanced equation tell us the ratio by which reactants react to form products. Accordingly, three moles of oxygen gas are produced for the reaction of every two moles of $KClO_3$ solid. Do you also see that only 1.5 moles of oxygen gas would be produced from the reaction of one mole of $KClO_3$ solid?

35. Use the periodic table to find the masses of all the atoms within each molecule. Add these masses together and you'll come up with the given formula masses. Use these masses to help answer the next question.

37. For this reaction, is there enough oxygen to react with all of the methane? Is there enough methane to react with all of the oxygen? The sure-fire way is to work out how much of one reactant is needed in order for all of the other reactant to be consumed. According to the following calculation, 16 g of CH_4, would require 64 g of O_2:

(16 g CH_4)(1 mole CH_4/16 g CH_4)(2 moles O_2/1 mole CH_4)(32 g O_2/1 mole O_2) = 64 g O_2

But there is only 16 g of O_2, which means that not all of the CH_4 is going to be able to react. How much of the CH_4 will react? That can be calculated as follows:

(16 g O_2)(1 mole O_2/32 g O_2)(1 mole CH_4/2 mole O_2)(16 g CH_4/1 mole CH_4) = 4 g CH_4

The maximum amount of CO_2 that can be formed is calculated as follows:

(16 g O_2)(1 mole O_2/32 g O_2)(1 mole CO_2/2 mole O_2)(44 g CO_2/1 mole CO_2) = 11 g CO_2

39a. Energy to break bonds:
Energy released from bond formation:

N-N = 159 kJ	
N-H = 389 kJ	
N-H = 389 kJ	H-H = 436 kJ
N-H = 389 kJ	H-H = 436 kJ
N-H = 389 kJ	N≡N = 946 kJ

Total = 1715 kJ absorbed
Total = 1818 kJ released

NET = 1715 kJ absorbed - 1818 released = -103 kJ released (exothermic)

39b. Energy to break bonds:
Energy released from bond formation:

O-O = 138 kJ

H-O = 464 kJ	O=O = 498 kJ
H-O = 464 kJ	H-O = 464 kJ
O-O = 138 kJ	H-O = 464 kJ
H-O = 464 kJ	O-H = 464 kJ
H-O = 464 kJ	O-H = 464 kJ

Total = 2132 kJ absorbed
Total = 2354 kJ released

NET = 2132 kJ absorbed - 2354 kJ released = -222 kJ released (exothermic)

41. For the reverse reaction, the products become the reactants and the reactants become the products. To calculate the equilibrium constant in such a situation, the products and reactant switch places in the equilibrium constant expression—products are placed in the denominator, while the reactants are placed in the numerator. The equilibrium constant of 3.5×10^8 therefore transforms into $1/(3.5 \times 10^8) = 2.9 \times 10^{-9}$.

Solutions to Chapter 9

Calculation Corner

Masses of Reactants and Products

1. With an unlimited amount of O_2 available, 35.8 g of Fe_2O_3 can be produced:

$$(25.0\ \text{g Fe})\left(\frac{1\ \text{mole Fe}}{55.8\ \text{g Fe}}\right)\left(\frac{2\ \text{moles } Fe_2O_3}{4\ \text{moles Fe}}\right)\left(\frac{159.7\ \text{g } Fe_2O_3}{1\ \text{mole } Fe_2O_3}\right) = 35.8\ \text{g } Fe_2O_3$$

2. With an unlimited amount of Fe available, 83.2 g of Fe_2O_3 can be produced:

$$(25.0\ \text{g } O_2)\left(\frac{1\ \text{mole } O_2}{32.0\ \text{g } O_2}\right)\left(\frac{2\ \text{moles } Fe_2O_3}{3\ \text{moles } O_2}\right)\left(\frac{159.7\ \text{g } Fe_2O_3}{1\ \text{mole } Fe_2O_3}\right) = 83.2\ \text{g } Fe_2O_3$$

3. The 25.0 g of Fe will require 10.8 g of O_2:

$$(25.0\ \text{g Fe})\left(\frac{1\ \text{mole Fe}}{55.8\ \text{g Fe}}\right)\left(\frac{3\ \text{moles } O_2}{4\ \text{moles Fe}}\right)\left(\frac{32\ \text{g } O_2}{1\ \text{mole } O_2}\right) = 10.8\ \text{g } O_2$$

4. From the previous question we know that 25.0 g of Fe requires 10.8 g of O_2. For this question, however, we were supplied with 25.0 g of O_2, which is much more than is needed. After the reaction is complete, there should be 25.0 g – 10.8 g = 14.2 g of O_2 remaining after the reaction is complete.

Chapter 9: Contextual Chemistry Essay

Mercury Emissions

Of all the metals in the periodic table, mercury, Hg (atomic number 80), is the only one to exist as a liquid at ambient temperatures. Mercury is also volatile, which means that uncontained mercury atoms evaporate into the atmosphere. Today, the atmosphere carries a load of about 5000 tons of mercury. Of this amount, about 2900 tons are from current human activities, such as the burning of coal, and the remaining 2100 tons appear to be from natural sources, such as out-gassing from the Earth's crust and oceans. Since the mid-19th century, however, humans have emitted an estimated 200,000 tons of mercury into the atmosphere, most of which has since subsided onto the land and into sea. It is probable, therefore, that a large portion of the mercury emitted from "natural" sources is actually the re-emission of mercury originally put there by humans over the last 150 years.

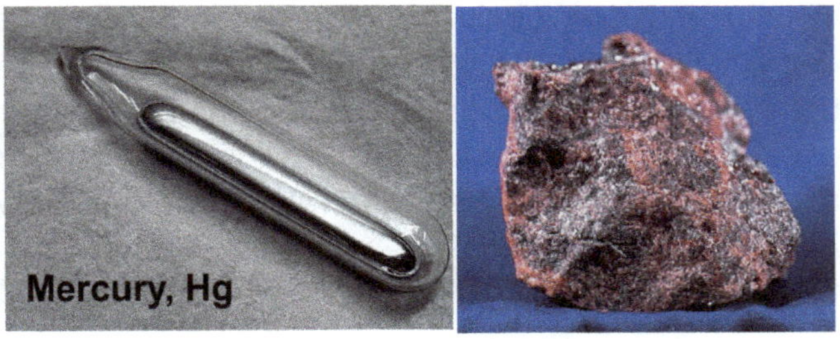

∧ Figure A

(a) Mercury is a dense liquid metal at room temperature. This, plus other unique properties, makes this material useful for many industrial purposes. The element, however, is also a potent toxin. (b) It can be produced from mercury containing minerals, such as cinnabar, which is a naturally occurring form of mercury sulfide, HgS.

Mercury is a poison to the nervous system. Its most dangerous form is that of the methyl mercury ion, CH_3Hg^+, which forms from elemental mercury within aquatic habitats. This form of mercury tends to bioaccumulate, so that organisms higher up in the marine food chain, such as pike, tuna, and swordfish, tend to have the highest levels. People who eat these fish regularly may be exposing themselves to high levels of mercury.

Typical consequences of mercury poisoning include a loss of mental focus and personality changes. In babies and children, the damaging effects are more severe, because the mercury disrupts brain development. Pregnant or nursing mothers are advised to avoid mercury-tainted fish, because methyl mercury passes through the placenta and into breast milk.

Because of a growing awareness of the dangers of mercury and because of imposed governmental regulations, over the past several decades, there has been a gradual phasing out of the use of mercury. This includes use in various commercial products, such as thermometers, and manufacturing processes, most notably in the manufacture of another element, chlorine. As shown in Figure B, the most significant source of human produced atmospheric mercury is the small-scale production of gold, which, worldwide, has remained largely unregulated.

The History of Mercury Regulations in the U.S.

The 1970 Clean Air Act exempted then existing coal-burning power plants. It wasn't until the Clean Air Act Amendments of 1990 that Congress gave the EPA the power to create regulations that would force coal-burning power plants to be fitted with mercury-reducing mechanisms. The EPA was to provide these regulations by 1994, but for a number of political reasons, this never happened. A committee of state regulators and pollution control experts was therefore convened to propose regulations. Their recommendations were put forth at the end of 2000 in a report that declared that U.S. mercury emissions could be reduced by up to 90 percent by 2007 using currently available technology. Coal-burning utilities disagreed, as did the newly elected Bush administration, which disbanded the committee.

Going back to the drawing board, the EPA took another two years to develop alternative regulations that called for a 34-ton cap (a 30 percent reduction) on the annual mercury output of coal utilities by 2007. This cap was set because it was achievable via regulations

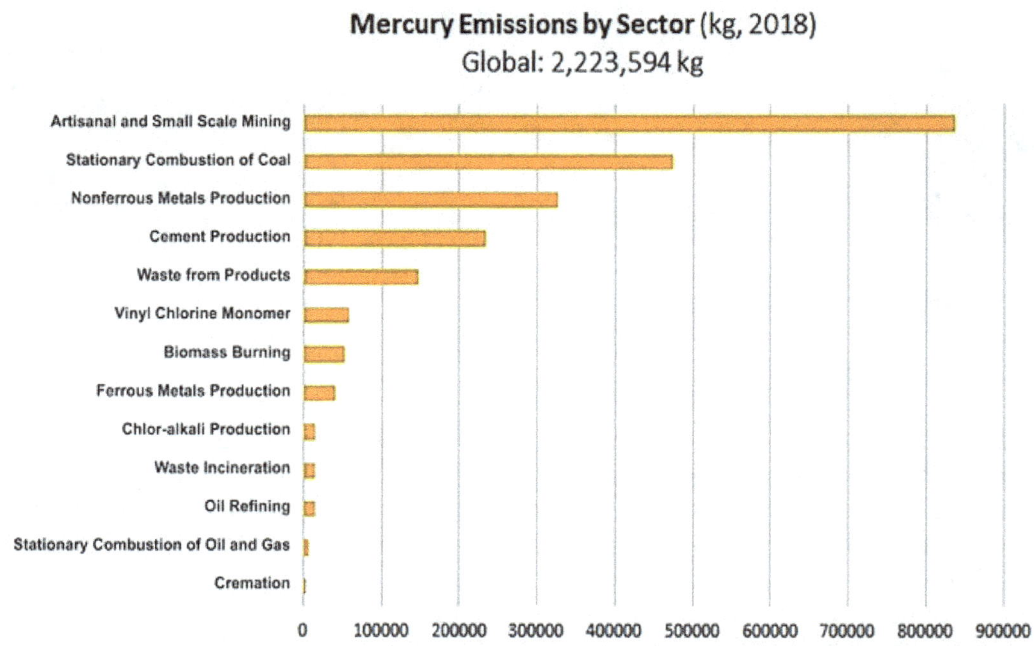

Λ Figure B

Global sources of mercury. Source: unep.org/globalmercurypartnership/

designed to curb other pollutants—no mercury-specific technologies would be needed. The cap would then be gradually lowered to 15 tons (a 70 percent reduction) by 2018. State regulators and environmental groups, as well as the EPA's oversight agency, the Office of the Inspector General, cited these regulations as too lenient and a violation of the 1990 Clean Air Act Amendment.

To override the Clean Air Act and to legitimize the EPA's proposed regulations, the Bush administration and like-minded members of Congress introduced the Clear Skies Act of 2003, which adopted the EPA's proposed mercury regulations. By March 2005, however, Congress remained deadlocked on approving this act. In response, the Bush administration took the essential provisions of the Clear Skies Act and ordered the EPA to implement them through regulations regardless of conflicts with the Clean Air Act. This resulted in the EPA-mandated Clean Air Rules, one of which was the Clean Air Mercury Rules (CAMR). Litigation culminated in the U.S. Supreme Court's decision in February 2009 upholding a lawsuit claiming these rules to be fundamentally flawed. The EPA was once again ordered back to the drawing board to come up with regulations more in line with the original Clean Air Act Amendment of 1990.

Λ Figure C

Gold and mercury are very different elements, yet they are very much related, which stems from the fact that they sit right next to each other in the periodic table. Most notable, gold dissolves into liquid mercury to form a mixture, known as an *amalgam*. This property makes it easy to isolate gold from gold ore. Worldwide, about 15 million small scale gold miners using this poisonous technique are responsible for the release of about 1,400 tons of mercury each year.

Λ Figure D

The three towers of a 2280 megawatt coal-fired power station.

This led to the Mercury and Air Toxics Standards, which the EPA released in December 2011 (www. epa.gov/mats), which continues to operate today. These regulations target primarily coal-fired power plants. Interestingly, by the time these regulations were released, about half of these power plants had already implemented the necessary air pollution control measures. But under these standards, less efficient older power plants are pressured to shut down and all the power plants are required to come into compliance. The EPA estimates that these regulations prevent about 11,000 premature deaths, 4,700 heart attacks, and 130,000 asthma attacks each year.

Mercury Emissions: A Global Issue

Meanwhile, a 2009 study by the U.S. Geological Survey showed that total mercury levels in North Pacific Ocean water had risen about 30 percent over the previous 20 years. The study forecasted another 50 percent increase by 2050. According to the study, this increase in mercury is attributable to increases in global mercury atmospheric emission rates, particularly from Asia.

Besides the United States, other major contributors to mercury pollution include China, India, and Russia, but all nations contribute to some extent. Atmospheric mercury remains airborne for about a year, which allows it to reach all regions of the planet. There are definitely "hot spots" that occur within 100 kilometers downwind of a coal-burning power plant, but on average, much of the mercury North Americans are exposed to originated in Asia. Similarly, much of the mercury Europeans are exposed to originated in North America. Asians get it from Europe. All of us are exposed to mercury arising from our past. Atmospheric mercury is a global problem, as is human inertia when it comes to finding solutions.

CONCEPT CHECK

There are about 4.8 billion mercury atoms in each breath of air that enters your lungs. What is the source of most of these mercury atoms?

CHECK YOUR ANSWER

Almost all of these mercury atoms were put into the environment by recent or past human activities. Note that this number of mercury atoms is actually quite small compared to the number of other atoms in the air. The greatest threat arises when mercury in aquatic environments is transformed into the methyl mercury ions that bioaccumulate in the food chain.

THINK AND DISCUSS

1. A key argument for the Clear Skies Act of 2003 was that it was practical and took into consideration economic factors. A key argument against this act was that it was less than ideal. Ultimately, because it was less than ideal, the act was defeated, which furthered the delay of

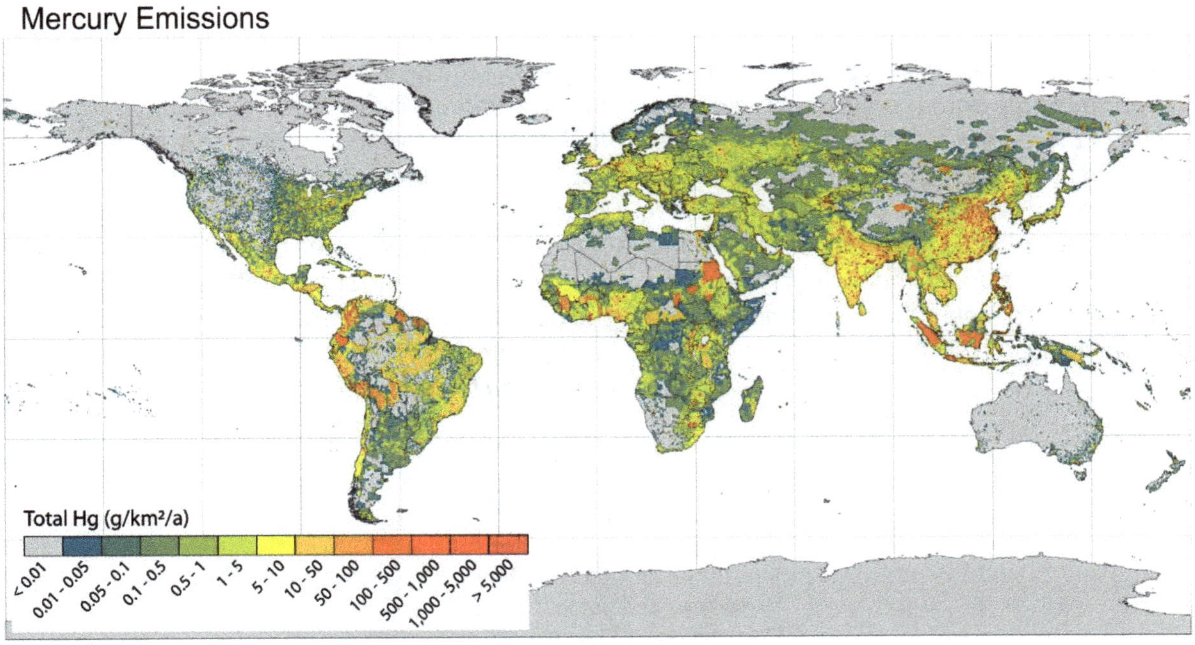

Λ Figure E

Geospatially distributed (total) mercury emissions to air from anthropogenic sources in 2015 ($g/km^2/a$) from all sectors. Source: epa.gov/international-cooperation/mercury-emissions-global-context

needed mercury emission regulations. Would you have voted for or against this act? Why?

2. Is it reasonable and practical to regulate the sale of mercury? Consider that small-scale gold miners around the world use mercury to help them extract gold.

3. The cost of equipping coal-burning utilities with mercury-specific anti-pollution technologies gets passed along to the consumer. The exemption for coal-burning utilities given by the 1970 Clean Air Act has therefore saved consumers billions of dollars on their utility bills. How might this be offset by medical bills? Discuss the value of long-term thinking over short-term thinking.

4. Even if the United States eliminated all of its mercury emissions, humans would still be exposed to about the same amount of mercury because atmospheric mercury is such a global problem. So, maybe it doesn't really matter that the U.S. has been so tardy in developing mercury regulations. What is really needed is a global effort. What do you think?

Author Responses to Think and Discuss

1. Responses will vary. As a politician, however, you would need to be able to defend how you voted to your constituents. Vote for the act and your constituents might see you as against the environment. You would then need to stress that you were actually for the environment because, although less than ideal, this act was practical and would significantly lower mercury levels. But then your constituents might then wonder why you settled for less than ideal. They might then start wishing for a candidate who better stuck to his ideals. Then again, your constituents might be widely employed by the coal industry. How would you defend your voting against the act?

2. If regulations lead to higher prices, it could be that these gold miners would simply obtain mercury from a black market. The root of the issue here is similar to the "Tragedy of the Commons" dilemma, which you should learn more about via the internet. Any effective means of minimizing the use of mercury should include a strong educational component so that people understand why its use should be minimized. It would also help if the miners had alternate ways of making of living.

3. It would depend on one's point of view. Billions of dollars spread out over millions of people over several decades amounts to not a whole lot from the point of view of the individual. But every bit helps. Think of the extra pizzas that were bought on account of lower utility bills. Turning to the environment, however, we find it necessary to think long term. The economy goes up and down, while mercury levels in the atmosphere do nothing but rise. Here in the early 21st century we are paying the environmental price for those few extra pizzas back in the mid-1970s. What sort of considerations should we have for people of the early 22nd century?

4. Why should the individual vote? What difference could a single vote make in an election? Consider that if everyone had this attitude, then no one would vote. Perhaps that's the logic that helps to bring dictators to power—a non-vote is usually a vote against one's interests. Is it futile for any one country to make an effort to curb its air pollution? If that's the common attitude, then the world is in big trouble. The global effort needs to start somewhere. Individuals speaking up to our elected officials is a good place to start. The mercury problem is not "somewhere else". It's right there in the air you breathe and the food you eat.

Chapter 10

Acids and Bases

THE MAIN IDEA

Acids donate protons and bases accept them.

∧ The carbon dioxide from ancient atmospheres is captured within limestone sediments, such as those of the White Cliffs of Dover along the southern coast of England.

As rainwater falls, it absorbs atmospheric carbon dioxide. Once in the rainwater, the carbon dioxide reacts to form carbonic acid, H_2CO_3, which, as we discuss in this chapter, makes rainwater naturally acidic. The acidic rain falls into the oceans, which happen to be basic because of the many alkaline minerals they contain. The carbonic acid of the rain reacts with the bases of the oceans to form salts, such as calcium carbonate, $CaCO_3$. The consequence is the removal of massive amounts of atmospheric carbon dioxide, which is a notable greenhouse gas. Furthermore, the calcium carbonate in the oceans is taken up by countless marine organisms that use this compound to form their protective shells. Upon death, the bulk of these creatures sink to the ocean floor, where over millions of years, their shells transform into a hardened rock known as limestone. Through geologic processes, ocean floors can be lifted to become parts of continents. The rock beneath the Great Plains of the United States, for example, is made of an ancient limestone sea floor, as are the impressive White Cliffs of Dover, shown on this page. These rocks are full of the calcium carbonate fossils of the ancient sea creatures, who, quite literally, are made from the atmospheres of the Earth's ancient past.

10.1 Exchanging Protons

The term acid comes from the Latin *acidus*, which means "sour." The sour taste of vinegar and citrus fruits is due to the presence of acids. Acids are essential in the chemical industry. For example, about 40 million tons of sulfuric acid is produced annually in the United States, making this the number-one manufactured chemical. Sulfuric acid is used to make fertilizers, detergents, paint dyes, plastics, pharmaceuticals, and storage batteries, as well as to produce iron and steel. It is so important in the manufacturing of goods that its production is considered a standard measure of a nation's industrial strength. **Figure 10.1** shows only a few of the acids we commonly encounter.

Figure 10.1 >

Examples of acids. (a) Citrus fruits contain many types of acids, including ascorbic acid, $C_6H_8O_6$, which is vitamin C. Vinegar contains acetic acid, $C_2H_4O_2$, and can be used to preserve foods. Many toilet bowl cleaners are formulated with hydrochloric acid, HCl. All carbonated beverages contain carbonic acid, H_2CO_3, while many also contain phosphoric acid, H_3PO_4.

Bases are characterized by their bitter taste and slippery feel. Interestingly, bases themselves are not slippery. Rather, they cause skin oils to transform into slippery solutions of soap. Most commercial preparations for unclogging drains contain sodium hydroxide, NaOH (also known as lye), which is extremely basic and hazardous when concentrated. Bases are also heavily used in industry. Each year in the United States, about 10 million tons of sodium hydroxide is manufactured for use in the production of various chemicals and in the pulp and paper industry. Solutions containing bases are often called *alkaline*, a term derived from the Arabic al-qali ("the ashes"). Ashes are slippery when wet because of the presence of bases such as potassium carbonate, K_2CO_3. **Figure 10.2** shows some familiar bases.

V Figure 10.2

Examples of bases. (a) Reactions involving sodium bicarbonate, $NaHCO_3$, cause baked goods to rise. (b) Ashes contain potassium carbonate, K_2CO_3. (c) Soap is made by reacting bases with animal or vegetable oils. The soap itself, then, is slightly alkaline. (d) Powerful bases, such as sodium hydroxide, NaOH, are used in drain cleaners.

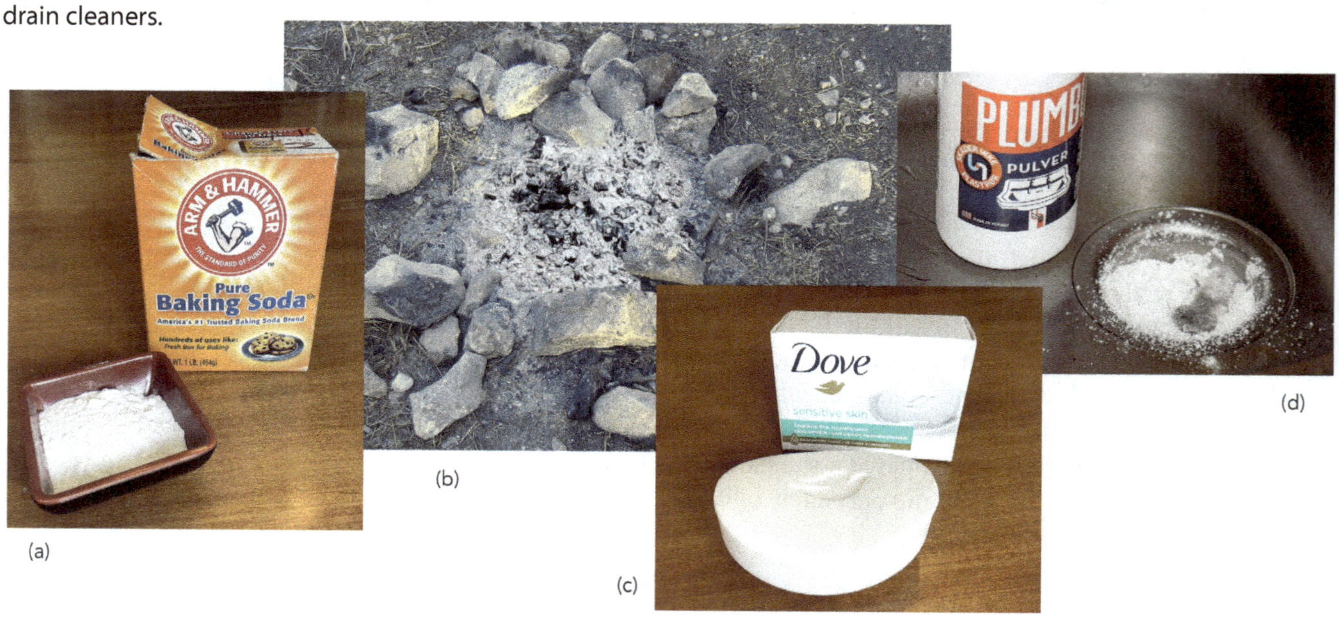

The Brønsted–Lowry Definition Focuses on Protons

Acids and bases may be defined in several ways. One of the earliest definitions proposed by Swedish chemist Svante Arrhenius (1859-1927) is that an acid is a chemical that increases the concentration of hydrogen ions, H^+, in water. Similarly, a base is a chemical that decreases the concentration of hydrogen ions. This idea was advanced further in the 1920s by the Danish chemist Johannes Brønsted (1879–1947) and the English chemist Thomas Lowry (1874–1936). Together, they pointed out that it was the very process of donating or accepting a hydrogen ion that was important, and this need not be restricted to water. In the Brønsted–Lowry definition, an **acid** is any chemical that donates hydrogen ions, H^+, and a **base** is any chemical that accepts hydrogen ions. Recall that a hydrogen atom consists of one electron surrounding a one-proton nucleus. A hydrogen ion, H^+, formed by the loss of an electron, therefore, is nothing more than a lone proton. Thus, it is also sometimes said that an acid is a chemical that donates a proton and a base is a chemical that accepts a proton.

READING CHECK

What is the Brønsted–Lowry definition of an acid and a base?

Consider what happens when hydrogen chloride is mixed into water:

$$HCl + H_2O \longrightarrow Cl^- + H_3O^+$$

H^+ donor (acid) H^+ acceptor (base)

Hydrogen chloride donates a hydrogen ion to one of the nonbonding electron pairs on a water molecule, resulting in a third hydrogen bonded to the oxygen. In this case, hydrogen chloride behaves as an acid (proton donor) and water behaves as a base (proton acceptor). The products of this reaction are a chloride ion and a **hydronium ion**, H_3O^+, which, as **Figure 10.3** shows, is made by adding a proton (hydrogen ion) to a water molecule.

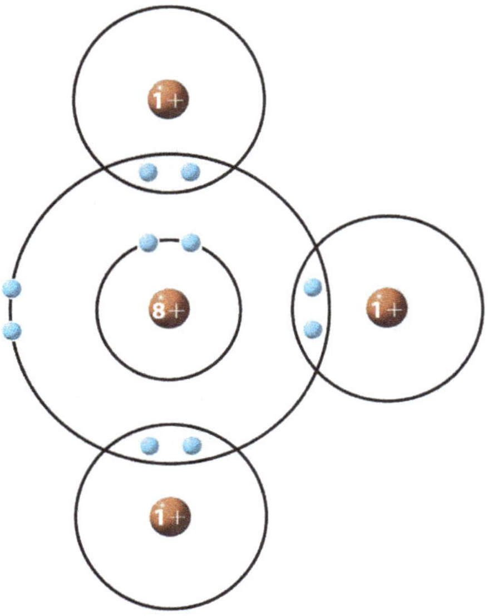

Electron-dot structure of hydronium ion

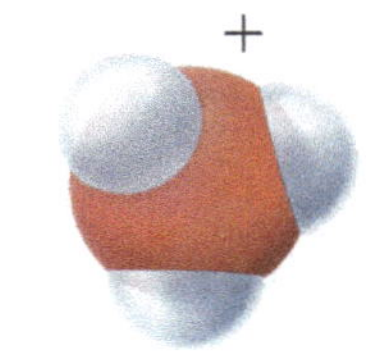

H_3O^+

Space-filling model of hydronium ion

Total protons	11+
Total electrons	10−
Net charge	1+

< Figure 10.3

The hydronium ion's positive charge is a consequence of the extra proton this molecule has acquired. Hydronium ions, which play a role in many acid-base reactions, are polyatomic ions, which, as mentioned in Section 6.2, are molecules that carry a net electric charge.

When added to water, ammonia behaves as a base, as its nonbonding electrons accept a hydrogen ion from water, which, in this case, behaves as an acid:

$$H_2O + NH_3 \longrightarrow OH^- + NH_4^+$$

H^+ donor (acid) — H^+ acceptor (base)

This reaction results in the formation of an ammonium ion and a **hydroxide ion**, which, as shown in Figure 10.4, is made by removing a proton (hydrogen ion) from a water molecule.

An important aspect of the Brønsted–Lowry definition is that it uses a *behavior* to define a substance as an acid or a base. We say, for example, that hydrogen chloride *behaves* as an acid when mixed with water, which *behaves* as a base. Similarly, ammonia *behaves* as a base when mixed with water, which, under this circumstance, *behaves* as an acid. Because labeling as an acid or a base depends on behavior, there is really no contradiction when a chemical such as water behaves as a base in one instance but as an acid in another instance. By analogy, consider yourself. You are who you are, but your behavior changes depending on whom you are with. Likewise, it is a chemical property of water to behave as a base (to accept H^+) when mixed with hydrogen chloride and as an acid (to donate H^+) when mixed with ammonia.

The products of an acid-base reaction can also behave as acids or as bases. An ammonium ion, for example, may donate a hydrogen ion back to a hydroxide ion to re-form ammonia and water:

$$H_2O + NH_3 \longleftarrow OH^- + NH_4^+$$

H^+ acceptor (base) — H^+ donor (acid)

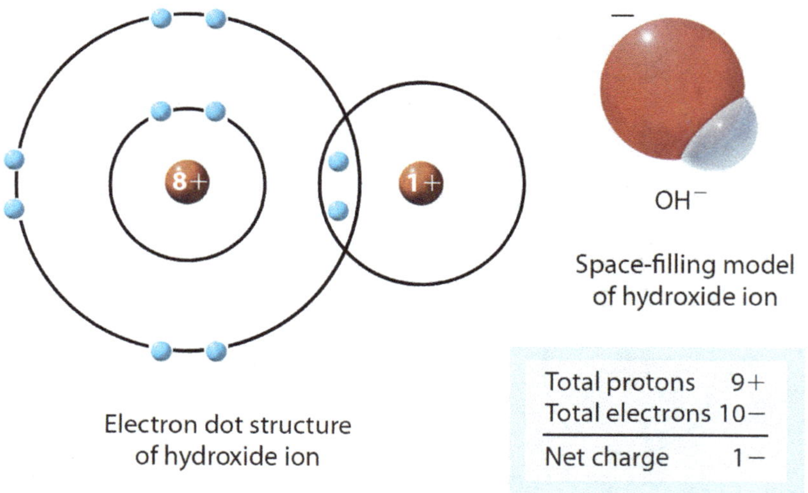

Figure 10.4 >

Hydroxide ions have a net negative charge, which is a consequence of having lost a proton. Like hydronium ions, they play a part in many acid-base reactions.

Forward and reverse acid-base reactions proceed simultaneously and can therefore be represented as occurring at the same time by using two oppositely facing arrows:

$$H_2O + NH_3 \rightleftarrows OH^- + NH_4^+$$

H^+ donor (acid)	H^+ acceptor (base)	H^+ acceptor (base)	H^+ donor (acid)

When the equation is viewed from left to right, the ammonia behaves as a base, because it accepts a hydrogen ion from the water, which therefore acts as an acid. Viewed in the reverse direction, the equation shows that the ammonium ion behaves as an acid, because it donates a hydrogen ion to the hydroxide ion, which therefore behaves as a base.

CONCEPT **CHECK**

Identify the behavior as an acid or a base of each participant in the reaction

$$H_2PO_4^- + H_3O^+ \rightleftarrows H_3PO_4 + H_2O$$

CHECK YOUR ANSWER

In the forward reaction (left to right), $H_2PO_4^-$ gains a hydrogen ion to become H_3PO_4. In accepting the hydrogen ion, $H_2PO_4^-$ is behaving as a base. It gets the hydrogen ion from the H_3O^+, which is behaving as an acid. In the reverse direction, H_3PO_4 loses a hydrogen ion to become $H_2PO_4^-$ and is thus behaving as an acid. The recipient of the hydrogen ion is the H_2O, which is behaving as a base as it transforms to H_3O^+.

The Lewis Definition Focuses on Lone Pairs

The Brønsted–Lowry definition of acids and bases is restricted to molecules that can donate or accept protons. A more general definition of acids and bases was proposed in the 1930s by Gilbert Lewis, the same chemist who introduced the idea of shells to explain the organization of the periodic table as well as chemical bonding (Chapters 4 and 6). According to the *Lewis definition*, a molecule behaves as a base when it donates a lone pair of electrons. Conversely, a molecule behaves as an acid when it accepts a lone pair. So, while the Brønsted–Lowry definition focuses on the action of a proton, the Lewis definition focuses on the action of a lone pair.

Looking back to the previous examples of bases, we see that a lone pair of each of these bases is being donated to a positively charged proton, as illustrated here. Note how the curved arrows indicate the movement of electrons:

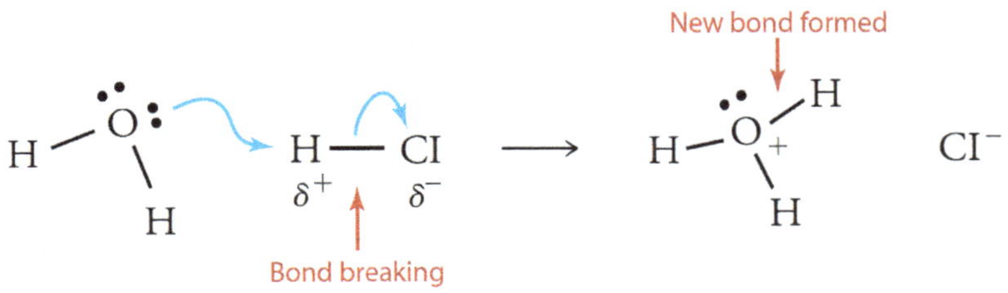

In this case, the water is behaving as a base, because its lone pair seeks out a positive charge (the proton), while the hydrochloric acid, HCl, behaves as an acid, because it accepts the lone pair.

Consider the formation of carbonic acid from water and carbon dioxide, as shown here:

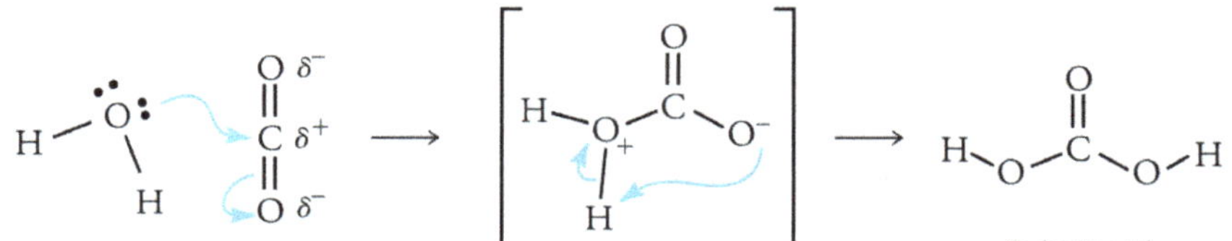

This reaction starts as a typical dipole–induced dipole attraction between the oxygen of the water and the carbon of the carbon dioxide. As the carbon accepts electrons from the lone pair of the water, however, it begins to lose electrons to one of its two adjacent oxygen atoms. This permits the formation of a covalent bond between the water and the carbon dioxide, which then become a larger molecule (shown in brackets) with both a positive and negative charge. This molecule exists only briefly before transforming into the more stable non-charged product, carbonic acid, which, as we described at the beginning of this chapter, is responsible for the natural acidity of rainwater.

CONCEPT **CHECK**

How is it possible for carbon dioxide, CO_2, to behave as an acid when it has no hydrogen ions to donate?

CHECK YOUR ANSWER

Carbon dioxide behaves as an acid when it accepts the lone pair on the oxygen atom of a water molecule.

Figure 10.5

"Lite" table salt substitutes contain potassium chloride in place of sodium chloride. Caution is advised in using these products, however, because excessive quantities of potassium salts can lead to serious illness. Furthermore, sodium ions are a vital component of our diet and should never be totally excluded. For a good balance of these two important ions, you might inquire about commercially available half-and-half mixtures of sodium chloride and potassium chloride, such as the one shown here.

A Salt Is the Ionic Product of an Acid-Base Reaction

In everyday language, the word *salt* implies sodium chloride, NaCl, table salt. In the language of chemistry, however, a **salt** is an ionic compound commonly formed from the reaction between an acid and a base. Hydrogen chloride and sodium hydroxide, for example, react to produce the salt sodium chloride and water:

HCl	+	NaOH	→	NaCl	+	H_2O
Hydrogen chloride (acid)		Sodium hydroxide (base)		Sodium chloride (salt)		Water

Similarly, the reaction between hydrogen chloride and potassium hydroxide yields the salt potassium chloride and water:

HCl	+	KOH	→	KCl	+	H_2O
Hydrogen chloride (acid)		Potassium hydroxide (base)		Potassium chloride (salt)		Water

Potassium chloride is the main ingredient in "lite" table salt, as noted in Figure 10.5.

Salts are generally far less corrosive than the acids and bases from which they are formed. A corrosive chemical has the power to disintegrate a material or wear away its surface. Hydrogen chloride is a remarkably corrosive acid, which makes it useful for cleaning toilet bowls and etching metal surfaces. Sodium hydroxide is a very corrosive base used for unclogging drains. Mixing hydrogen chloride and sodium hydroxide together in equal portions, however, produces an aqueous solution of sodium chloride—salt-water, which is not nearly as destructive as either starting material.

There are as many salts as there are acids and bases. Sodium cyanide, NaCN, is a deadly poison. "Saltpeter," which is potassium nitrate, KNO_3, is useful as a fertilizer and in the formulation of gunpowder. Calcium chloride, $CaCl_2$, is commonly used to de-ice walkways, and sodium fluoride, NaF, helps to prevent tooth decay. The acid-base reactions forming these salts are shown in **Table 10.1**.

The reaction between an acid and a base is called **neutralization**. As can be seen in the color-coding of the neutralization reactions in Table 10.1, the positive ion of a salt comes from the base and the negative ion comes from the acid. The remaining hydrogen and hydroxide ions join to form water.

Not all neutralization reactions result in the formation of water. In the presence of hydrogen chloride, for example, the drug pseudoephedrine behaves as a base by accepting H^+ from the hydrogen chloride. The negative Cl^- ions then join the pseudoephedrine–H^+ ions to form the salt pseudoephedrine hydrochloride, which is a common nasal decongestant, shown in **Figure 10.6**.

Because it is a salt, pseudoephedrine hydrochloride is much more water soluble than is the pseudoephedrine base, sometimes called the *free base* because it is "free" from ions. Because pseudoephedrine hydrochloride is water soluble, it is readily absorbed through the digestive system, which means it can be taken orally.

TABLE 10.1 Acid-Base Reactions and the Salts Formed

ACID		BASE		SALT		WATER
HCN Hydrogen cyanide	+	NaOH Sodium hydroxide	→	NaCN Sodium cyanide	+	H_2O
HNO_3 Nitric acid	+	KOH Potassium hydroxide	→	KNO_3 Potassium nitrate	+	H_2O
2 HCl Hydrogen chloride	+	$Ca(OH)_2$ Calcium hydroxide	→	$CaCl_2$ Calcium chloride	+	2 H_2O
HF Hydrogen fluoride	+	NaOH Sodium hydroxide	→	NaF Calcium fluoride	+	H_2O

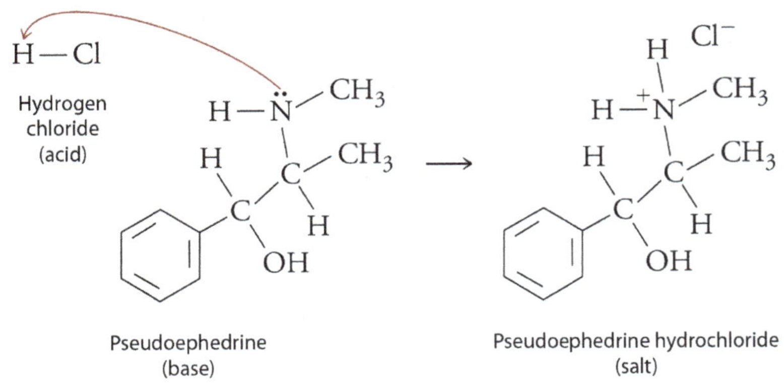

< Figure 10.6

Hydrogen chloride and pseudoephedrine react to form the salt *pseudoephedrine hydrochloride*, which, because of its solubility in water, is readily absorbed into the body. Most pharmaceuticals that can be taken orally are bases that have been converted to a salt form.

CONCEPT **CHECK**

Is a neutralization reaction best described as a physical change or a chemical change?

CHECK YOUR ANSWER

New chemicals are formed during a neutralization reaction, meaning the reaction is a chemical change.

10.2 Acid and Base Strength

READING CHECK

What is one way to assess the strength of an acid?

In general, the stronger an acid, the more readily it donates hydrogen ions. Likewise, the stronger a base, the more readily it accepts hydrogen ions. An example of a strong acid is hydrogen chloride, HCl, and an example of a strong base is sodium hydroxide, NaOH. The corrosiveness of these materials is a result of their strength.

One way to assess the strength of an acid or a base is to measure how much of it remains after it has been added to water. If little remains, the acid or base is strong. If a lot remains, the acid or base is weak. To illustrate this concept, consider what happens when the strong acid hydrogen chloride, HCl, is added to water and what happens when the weak acid acetic acid, $C_2H_4O_2$ (the active ingredient of vinegar), is added to water.

Being an acid, hydrogen chloride donates hydrogen ions to water, forming chloride ions and hydronium ions. Because HCl is such a strong acid, all of it is converted to these ions, as shown in **Figure 10.7**.

Because acetic acid is a weak acid, it has much less of a tendency to donate hydrogen ions to water. When this acid is dissolved in water, only a small portion of the acetic acid molecules are converted to ions, a process that occurs as the polar O—H bonds are broken. (The C—H bonds of acetic acid are unaffected by the water because of their nonpolarity.) The majority of acetic acid molecules remain intact in their original un-ionized form, as shown in **Figure 10.8**.

Figures 10.7 and 10.8 show the submicroscopic behavior of strong and weak acids in water. Because molecules and ions are too small to see, how, then, does a chemist measure the strength of an acid? One way is by measuring a solution's ability to conduct an electric current, as **Figure 10.9** illustrates. In pure water, there are practically no ions to conduct electricity. When a strong acid is dissolved in water, many ions are generated, as indicated within the circle of Figure 10.7. The presence of these ions allows the flow of a strong electric current. A weak acid dissolved in water generates only a few ions, as indicated within the circle of Figure 10.8. The presence of fewer ions means there can be only a weak electric current.

This same trend is seen with strong and weak bases. Strong bases, for example, tend to accept hydrogen ions more readily than weak bases. In solution, a strong base allows the flow of a strong electric current and a weak base allows the flow of a weak electric current.

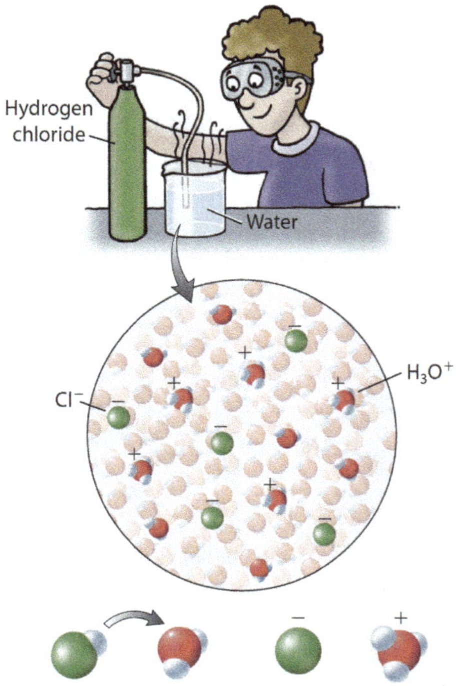

$$HCl + H_2O \longrightarrow Cl^- + H_3O^+$$

Hydrogen chloride | Water | Chloride ion | Hydronium ion

∧ Figure 10.7

Immediately after gaseous hydrogen chloride is added to water, it reacts with the water to form hydronium ions and chloride ions. That very little HCl remains (none shown here) lets us know that HCl acts as a strong acid.

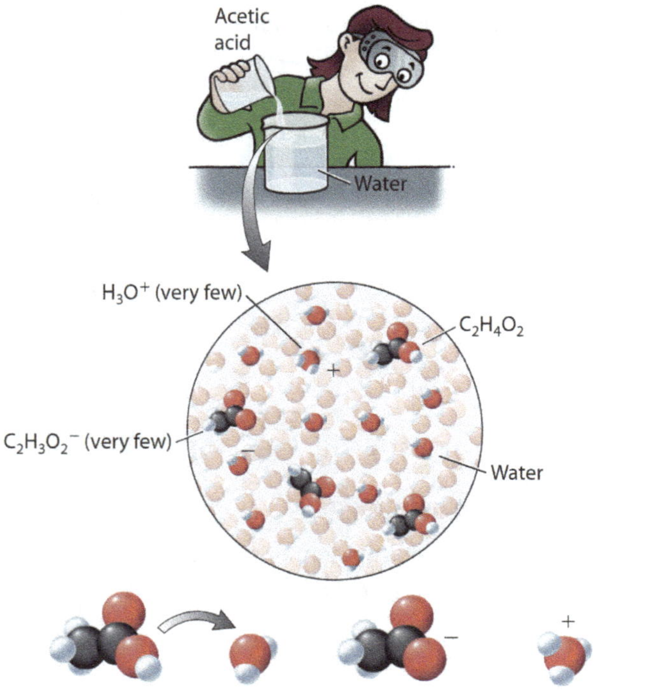

$$C_2H_4O_2 + H_2O \rightleftharpoons C_2H_3O_2^- + H_3O^+$$

Acetic acid | Water | Acetate ion | Hydronium ion

∧ Figure 10.8

When liquid acetic acid is added to water, only a few acetic acid molecules react with water to form ions. The majority of the acetic acid molecules remain in their un-ionized form, which implies that acetic acid is a weak acid

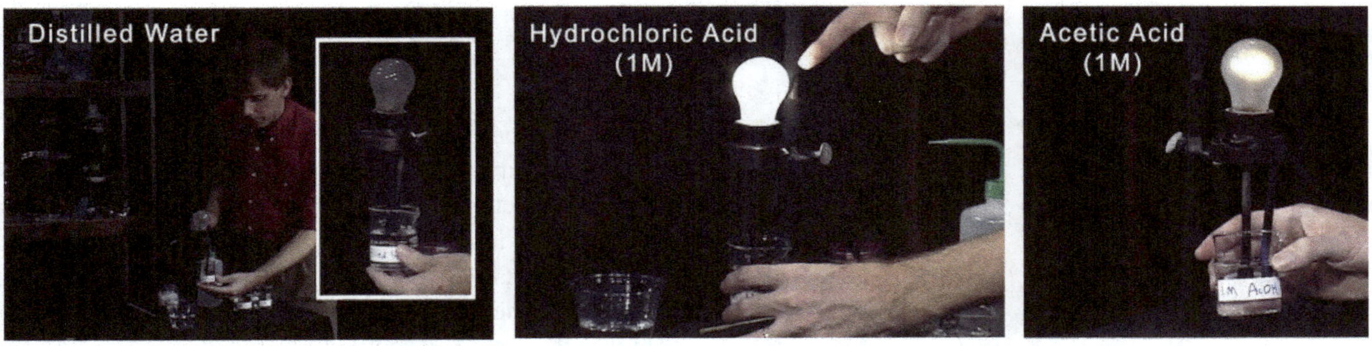

(a) (b) (c)

∧ Figure 10.9

(a) The distilled water in this electric circuit does not allow the bulb to light because the water contains practically no ions. The lightbulb in the circuit therefore remains unlit. (b) Because HCl is a strong acid, nearly all of its molecules break apart in water, yielding a high concentration of ions, which are able to conduct an electric current that lights the bulb. (c) Acetic acid, $C_2H_4O_2$, is a weak acid. In water, only a small portion of its molecules break up into ions. Because fewer ions are generated, only a weak current exists, and the bulb is therefore dimmer.

CONCEPT **CHECK**

According to the aqueous solutions illustrated here, which is the stronger base, NH_3 or NaOH?

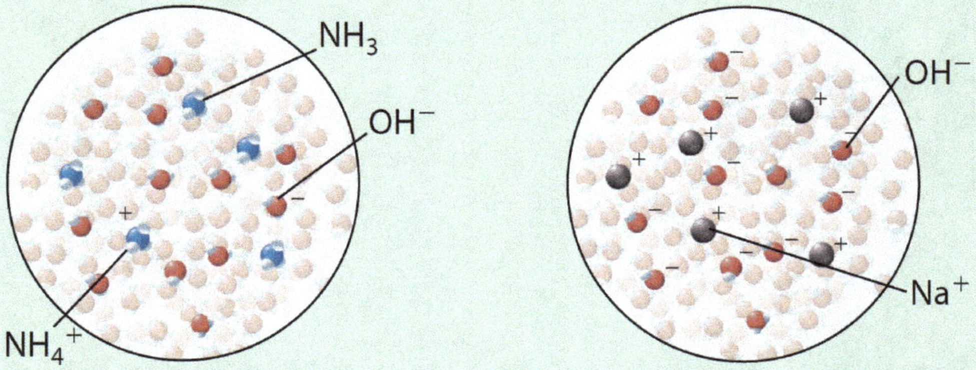

Aqueous solution of NH_3 Aqueous solution of NaOH

CHECK YOUR ANSWER

The solution on the right contains the greater number of ions, meaning sodium hydroxide, NaOH, is the stronger base. Ammonia, NH_3, is the weaker base, indicated by the relatively few ions in the solution on the left.

FOR YOUR INFORMATION

Aspirin is an acidic molecule, but not nearly as acidic as the hydrochloric acid, HCl, found in your stomach and used to digest food. So how is it that aspirin can cause damage to your stomach? Stomach acid is so strong that within this environment, aspirin is unable to donate its hydrogen ion, which means that it remains "un- ionized." This un-ionized aspirin is nonpolar and water insoluble, which helps it to penetrate through the largely nonpolar mucous membrane that lines the stomach. After passing through this membrane, the aspirin finds itself in a less acidic environment, where it can finally donate its hydrogen ion. This lowers the pH of the submucous inner wall the tissues and even cause bleeding. The remedy, of course, is to swallow specially coated aspirin tablets, which delay the release of the aspirin until after it passes through the stomach.

FOR YOUR INFORMATION

What makes one acid strong and another weak? One factor involves the stability of the negative ion that remains after the proton has been donated. Hydrogen chloride is a strong acid because the chloride ion is able to accommodate the negative charge rather well. Acetic acid, however, is a weaker acid because the resulting oxygen ion is less able to accommodate the negative charge.

Just because an acid or a base is strong doesn't mean a solution of that acid or base is corrosive. A *very* dilute solution of a strong acid or a strong base may have little corrosive action because in such solutions, there are only a few hydronium or hydroxide ions. Almost all the molecules of the strong acid or base break up into ions, but because the solution is dilute, there are only a few acid or base molecules to begin with. As a result, there are only a few hydronium or hydroxide ions. You shouldn't be too alarmed, therefore, when you discover that some toothpastes are formulated with small amounts of sodium hydroxide, one of the strongest bases known. The concentration of sodium hydroxide in these products is quite small.

On the other hand, a concentrated solution of a weak acid, such as the acetic acid in vinegar, may be just as corrosive as or even more corrosive than a dilute solution of a strong acid, such as hydrogen chloride. The relative strengths of two acids in solution or two bases in solution, therefore, can be compared only when the two solutions have the same concentration.

10.3 Acidic, Basic, or Neutral

A substance whose ability to behave as an acid is about the same as its ability to behave as a base is said to be **amphoteric**. Water is a good example. Because it is *amphoteric*, water has the ability to react with itself. In behaving as an acid, a water molecule donates a hydrogen ion to a neighboring water molecule, which, in accepting the hydrogen ion, is behaving as a base. This reaction produces a hydroxide ion and a hydronium ion, which can react together to reform the water molecule:

$$H_2O + H_2O \rightleftharpoons OH^- + H_3O^+$$

Water — Water — Hydroxide ion — Hydronium ion

When a water molecule gains a hydrogen ion, a second water molecule must lose a hydrogen ion. So, for every one hydronium ion formed, one hydroxide ion also forms. In pure water, therefore, the total number of hydronium ions must be the same as the total number of hydroxide ions. Experiments reveal that the concentration of hydronium and hydroxide ions in pure water is extremely low—about 0.00000010 *M* for each, where *M* stands for molarity, or moles per liter (Section 7.3). Water by itself, therefore, is a very weak acid as well as a very weak base, as evidenced by the unlit lightbulb in Figure 10.9a.

CHEMICAL CONNECTIONS

How is a piece of paper connected to a rock at the bottom of the ocean?

CONCEPT CHECK

Do water molecules react with one another?

CHECK YOUR ANSWER

Yes, but not to any large extent. When they do react, they form hydronium and hydroxide ions. (Note: make sure you understand this point, because it serves as a basis for most of the rest of the chapter.)

Further experiments reveal an interesting rule pertaining to the concentrations of hydronium and hydroxide ions in any solution that contains water. The concentration of hydronium ions in any aqueous solution multiplied by the concentration of the hydroxide ions in the solution always equals the constant K_w, which is a very, very small number:

$$\text{concentration } H_3O^+ \times \text{concentration } OH^- = K_w = 0.000000000000010$$

Concentration is given in molarity, which is indicated by abbreviating this equation using brackets:

$$[H_3O^+] \times [OH^-] = K_w = 0.000000000000010$$

FOR YOUR INFORMATION

K_w is an equilibrium constant, as described in Section 9.8. That K_w is such a small number tells us that the forward reaction of water reacting with water doesn't happen very often. By contrast, the reverse reaction of the hydronium ion reacting with the hydroxide ion happens quite readily.

The brackets mean this equation can be read as: "the molarity of H_3O^+ times the molarity of OH^- equals K_w." Writing in scientific notation, we have:

$$[H_3O^+][OH^-] = K_w = 1.0 \times 10^{-14}$$

For pure water, the value of K_w is the concentration of hydronium ions, 0.00000010 *M*, multiplied by the concentration of hydroxide ions, 0.00000010 *M*, which can be written in scientific notation as

$$[1.0 \times 10^{-7}][1.0 \times 10^{-7}] = K_w = 1.0 \times 10^{-14}$$

CONCEPT **CHECK**

1. In pure water, the hydroxide-ion concentration is 1.0×10^{-7} *M*. What is the hydronium-ion concentration?

2. What is the concentration of hydronium ions in a solution if the concentration of hydroxide ions is 1.0×10^{-3} *M*?

CHECK YOUR ANSWER

1. 1.0×10^{-7} *M*, because in pure water, $[H_3O^+] = [OH^-]$.
2. 1.0×10^{-11} *M*, because [H3O+][OH-] must equal $1.0 \times 10^{-14} = K_w$

FOR YOUR INFORMATION

The outer surface of hair is made of microscopic scale-like structures called cuticles that, like window shutters, are able to open and close. Alkaline solutions cause the cuticles to open up, which makes the hair "porous." Acidic solutions cause the cuticles to close down, which makes the hair "resistant." A beautician can control how long hair retains artificial coloring by modifying the pH of the hair-coloring solution. With an acidic solution, the cuticles close shut so that the dye binds only to the outside of each shaft of hair. This results in a temporary hair coloring, which may come off with the next hair washing. Using an alkaline solution permits the dye to penetrate through the cuticles into the hair for a more permanent effect. Congratulations for this deep understanding of cosmetology. If you've studied Chapter 5, then that means you've got a handle on cosmology as well, which is quite the spectrum.

READING CHECK
Why is pure water a neutral solution?

The constant value of K_w is quite significant because it means that no matter what is dissolved in the water, the product of the hydronium-ion and hydroxide-ion concentrations always equals 1.0×10^{-14}. So, if the concentration of H_3O^+ goes up, the concentration of OH^- must go down, and the product of the two remains 1.0×10^{-14}.

Any solution containing an equal number of hydronium and hydroxide ions is said to be **neutral**. Pure water is an example of a neutral solution—not because it contains so few hydronium or hydroxide ions, but because it contains *equal numbers* of these ions. A neutral solution is also obtained when equal quantities of acid and base are combined, which explains why acids and bases are said to *neutralize* each other.

The balance of hydronium and hydroxide ions in a neutral solution is upset by adding either an acid or a base. Add an acid and the water will react with that acid to produce more hydronium ions. Many of these additional hydronium ions neutralize hydroxide ions, which then become fewer. The final result is that the hydronium ion concentration is greater than the hydroxide-ion concentration. Such a solution is said to be **acidic**.

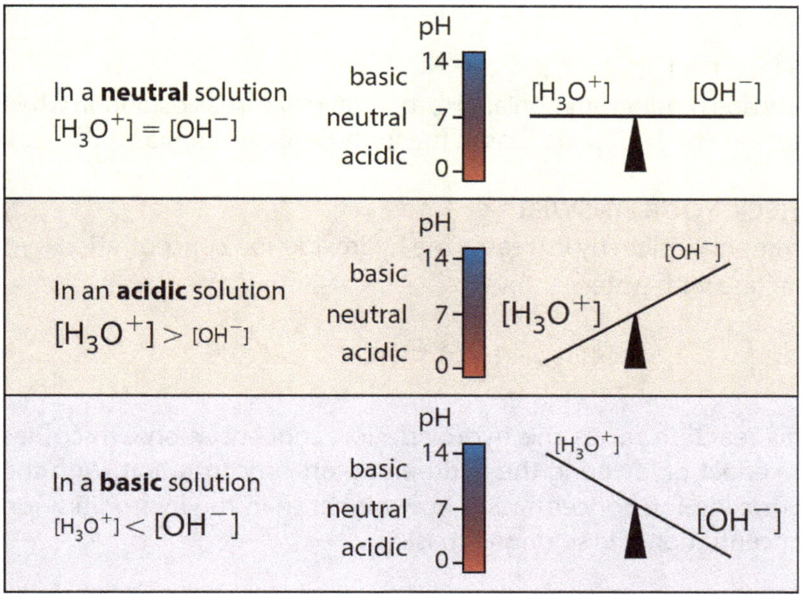

< Figure 10.10

The relative concentrations of hydronium and hydroxide ions determine whether a solution is neutral acidic, or basic. When these concentrations are equal, the solution is *neutral*. When the hydronium ion concentration increases, the hydroxide ion concentration necessarily decreases and the solution is *acidic*. Conversely, when the hydroxide ion concentration increases, the hydronium ion concentration necessarily decreases and the solution is *basic*.

Add a base to water and the reverse happens. The water will react with that base to produce more hydroxide ions. Many of these additional hydroxide ions will neutralize hydronium ions, which then become fewer. The final result is that the hydronium-ion concentration is less than the hydroxide-ion concentration. Such a solution is said to be **basic**, or sometimes *alkaline*. This is all summarized in **Figure 10.10.**

The pH scale Is used to Describe Acidity

The *pH scale* is a numeric scale used to express the acidity of a solution. Mathematically, **pH** is equal to the negative logarithm of the hydronium-ion concentration:

$$pH = -\log[H_3O^+]$$

Note again that brackets are used to represent molar concentrations, meaning $[H_3O^+]$ is read as: "the molar concentration of hydronium ions." To understand the logarithm function, see the Calculation Corner: Logarithms and pH.

Consider a neutral solution that has a hydronium-ion concentration of 1.0×10^{-7} *M*. To find the pH of this solution, we first take the logarithm of this value, which is -7 (see the Calculation Corner on logarithms). The pH is by definition the negative of this value, which means -(-7) = 7. Hence, in a neutral solution, where the hydronium-ion concentration equals 1.0×10^{-7} *M*, the pH is 7.

Acidic solutions have pH values less than 7. For an acidic solution in which the hydronium-ion concentration is 1.0×10^{-4} *M*, for example, $pH = -\log(1.0 \times 10^{-4}) = 4$. The more acidic a solution, the *greater* its hydronium-ion concentration and the lower its pH.

Basic solutions have pH values greater than 7. For a basic solution in which the hydronium-ion concentration is 1.0×10^{-8} *M*, for example, $pH = -\log(1.0 \times 10^{-8}) = 8$. The more basic a solution, the *smaller* its hydronium-ion concentration and the higher its pH. **Figure 10.11** shows typical pH values of some familiar solutions, and **Figure 10.12** shows two common ways of determining pH values.

fyi FOR YOUR INFORMATION

K_w is an equilibrium constant. As described in the previous chapter, an equilibrium constant equals the concentrations of the products over reactants. You'll notice that for this reaction the reactant is water. The concentration of water in water is about 55.5 M, which is huge compared to the super tiny concentrations of the hydronium and hydroxide ion products. Because water's concentration is so huge, any changes to it are insignificant, which is why, for K_w, it can be ignored. You may also recall that chemical equilibrium can be temperature dependent, as is the case for water reacting with water. Strictly speaking, $K_w = 1 \times 10^{-14}$ only at room temperature, 25°C. At higher temperatures, K_w gets larger. For example, at 40°C $K_w = 2.9 \times 10^{-14}$, which means the pH of neutral water is 6.7. Congratulations for this super deep understanding of acids and bases.

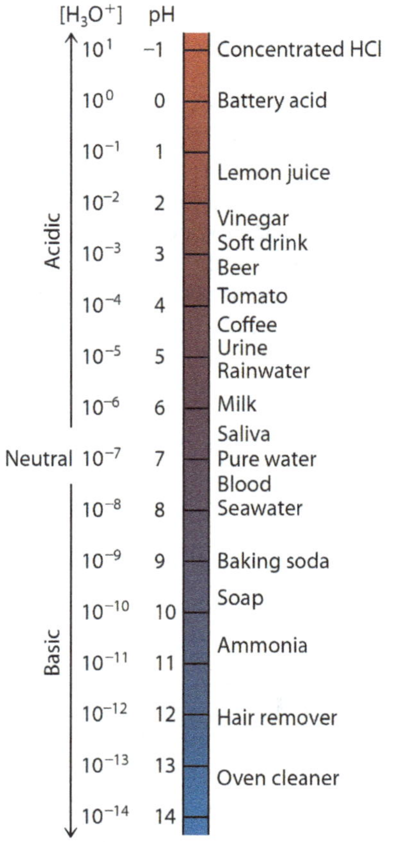

∧ Figure 10.11

The pH values of some common solutions.

CONCEPT **CHECK**

How does adding ammonia, NH_3, to water make a basic solution when there are no hydroxide ions in the formula for ammonia?

CHECK YOUR ANSWER

Ammonia indirectly increases the hydroxide-ion concentration by reacting with water:

$$NH_3 + H_2O \leftrightarrows NH_4^+ + OH^-$$

This reaction raises the hydroxide-ion concentration, which has the effect of lowering the hydronium-ion concentration. With the hydroxide-ion concentration now higher than the hydronium-ion concentration, the solution is basic.

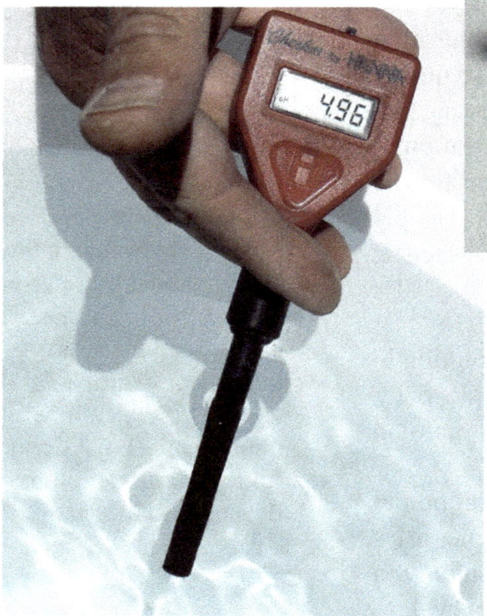

∧ Figure 10.12

(a) The pH of a solution can be measured electronically using a pH meter. (b) A rough estimate of the pH of a solution can be obtained with pH paper, which is coated with a dye that changes color with pH.

Calculation Corner: Logarithms and pH

Logarithms are super easy. When you ask for the logarithm, you are simply asking "to what power is 10 raised?" For example, the logarithm of 10^2 is 2, because that is the power to which 10 is raised. So, what is the logarithm of 10^3? Just look to what power 10 is raised and there is your answer: 3. If you know that 10^2 is equal to 100, then you'll understand that the logarithm of 100 is also 2. (And by no coincidence you'll note there are two zeros in 100.) Check for the logarithm of 100 using a [log] button–equipped calculator. Similarly, the logarithm of 1000 is 3, because 10 raised to the third power, 10^3, equals 1000. (And by no coincidence you'll note there are three zeros in 1000).

Any positive number, including a very small one, has a logarithm. The logarithm of 0.0001, which equals 10^{-4}, is -4 (the power to which 10 is raised to equal this number). As a review, study the pattern of the numbers shown below. Please keep in mind that no negative numbers are shown—all the numbers are greater than zero! The negative sign in the exponent merely means that the number is less than 1 (but greater than zero).

$10^3 = 1000$ (three zeros)
$10^2 = 100$ (two zeros)
$10^1 = 10$ (one zero)
$10^0 = 1$ (no zeros)
$10^{-1} = 0.1$ (one zero in the opposite direction)
$10^{-2} = 0.01$ (two zeros in the opposite direction)
$10^{-3} = 0.001$ (three zeros in the opposite direction)

EXAMPLE 1

What is the logarithm of 0.01?

ANSWER 1

The number 0.01 is 10^{-2}, the logarithm of which is -2 (the power to which 10 is raised).

The concentration of hydronium ions in most solutions is much less than 1 *M*. Recall, for example, that in neutral water, the hydronium-ion concentration is 0.0000001 *M* (10^{-7} *M*). The logarithm of any number less than 1 (but greater than zero) is a negative number. The definition of pH includes the minus sign so as to transform the logarithm of the hydronium-ion concentration to a positive number.

Note that when a solution has a hydronium-ion concentration of 1 *M*, the pH is 0 because 1 *M* = 10^0 *M*. A 10 *M* solution has a pH of -1 because 10 M = 10^1 *M*. So, yes, negative pH values are possible, but never do they exceed -2. Why not? Because a hydronium ion concentration of 100 *M* is chemically most difficult.

EXAMPLE 2

What is the pH of a solution that has a hydronium-ion concentration of 0.001 *M*?

ANSWER 2

The number 0.001 is 10^{-3}, so

$pH = -\log [H_3O^+]$

$= -\log (10^{-3})$

$= -(-3) = 3$

YOUR TURN

1. What is the logarithm of 10^5?
2. What is the logarithm of 100,000?
3. What is the pH of a solution having a hydronium-ion concentration of 10^{-9} *M*? Is this solution acidic, basic, or neutral?

10.4 Buffers Resist pH Changes

READING CHECK

How does a buffer solution work?

A **buffer solution** is any solution that resists large changes in pH. Buffer solutions work by containing two components. One component neutralizes any added acid, and the other neutralizes any added base. Buffer solutions can be prepared by mixing a weak acid with the salt of that weak acid, which, interestingly, is generally a weak base. An example is a solution of acetic acid, $C_2H_4O_2$ (a weak acid), and sodium acetate, $NaC_2H_3O_2$ (a weak base).

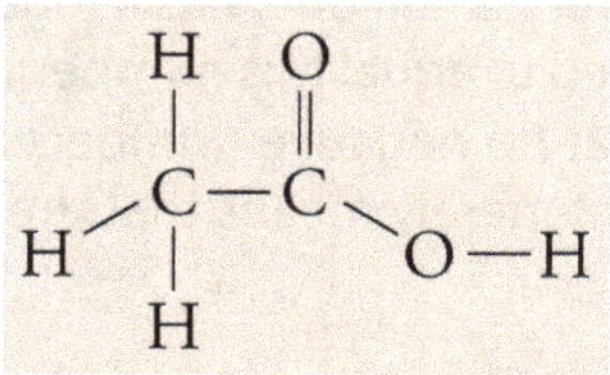

Acetic acid
(weak acid)

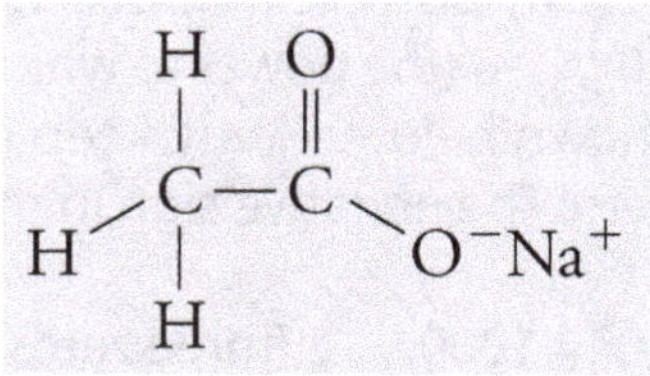

Sodium acetate
(salt of weak acid)

Any strong acid added to this particular buffer solution is neutralized by the sodium acetate, as shown in **Figure 10.13**. Similarly, any strong base added is neutralized by the acetic acid, as shown in **Figure 10.14**. So, a buffer works by containing both a weak acid and a weak base, each of which will neutralize any incoming material that might drastically alter the pH of the solution.

So, both strong acids and strong bases are neutralized by a buffer solution. This does not mean, however, that the solution's pH remains unchanged. When NaOH is added to the buffer system we are using as our example, sodium acetate is produced. Because sodium acetate behaves as a weak base (it accepts hydrogen ions but not very well), there is a slight increase in pH. When HCl is added, acetic acid is produced. Because acetic acid behaves as a weak acid, there is a slight decrease in pH. What makes a buffer solution special is its ability to resist *large* changes in pH.

CONCEPT **CHECK**

Why do most buffer solutions contain two dissolved components?

CHECK YOUR ANSWER

The weak base component neutralizes any incoming acid, and the weak acid component neutralizes any incoming base.

Many different buffer systems are useful for maintaining particular pH values. The acetic acid–sodium acetate system is good for maintaining a pH around 4.8. Buffer solutions containing equal mixtures of a weak base and a salt of that weak base maintain alkaline pH values. For example, a buffer solution of the weak base ammonia, NH_3, and its salt ammonium chloride, NH_4Cl, is useful for maintaining a pH of about 9.3.

Blood has several buffer systems that work together to maintain a narrow pH range between 7.35 and 7.45. A pH value above or below these levels can be lethal, primarily because it causes cellular proteins to become *denatured*, which is what happens to milk when vinegar is added to it.

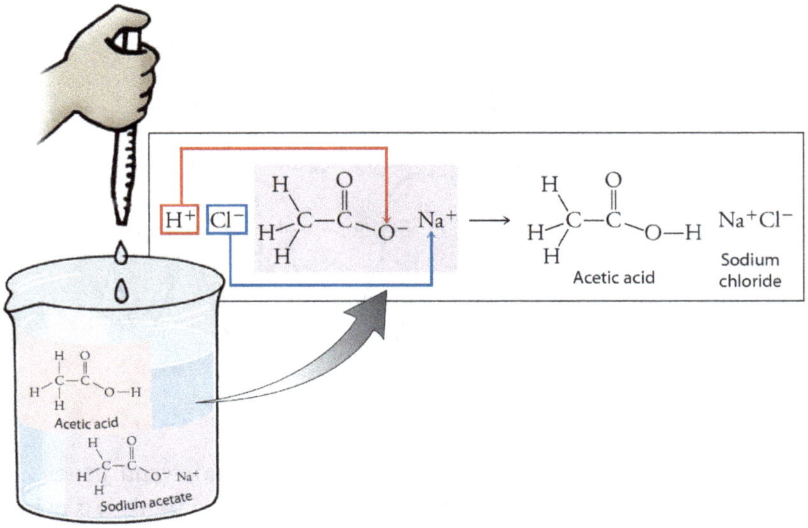

< Figure 10.13

Hydrochloric acid added to a solution containing acetic acid and sodium acetate is neutralized by the sodium acetate to form additional acetic acid and sodium chloride.

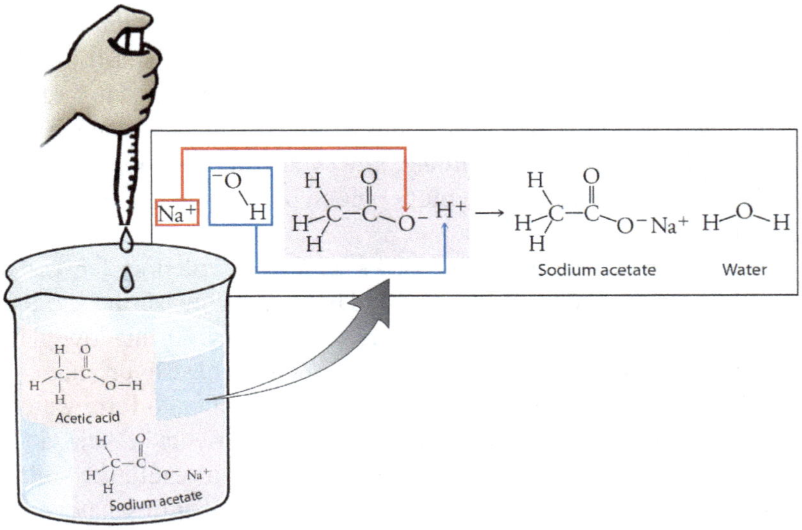

< Figure 10.14

Sodium hydroxide added to a solution containing acetic acid and sodium acetate is neutralized by the acetic acid to form additional sodium acetate and water.

The primary buffer system of our blood is a combination of carbonic acid and its salt sodium bicarbonate, shown in **Figure 10.15**. Any acid that builds up in the bloodstream is neutralized by the basic action of sodium bicarbonate, and any base that builds up is neutralized by the acidic action of carbonic acid.

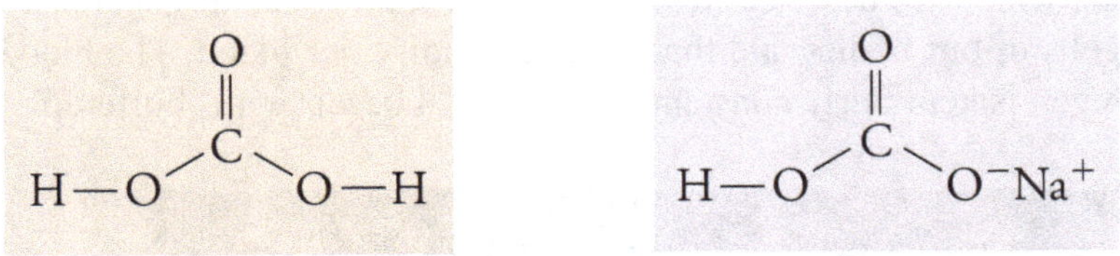

Carbonic acid (weak acid)

Sodium bicarbonate (salt)

< Figure 10.15
Carbonic acid and sodium bicarbonate.

The carbonic acid in your blood is formed as the carbon dioxide produced by your cells enters the bloodstream and reacts with water—this is the same reaction that occurs in a raindrop, as we discussed earlier. You fine-tune the levels of blood carbonic acid, and hence your blood pH, by your breathing rate, as **Figure 10.16** illustrates. Breathe too slowly or

Figure 10.16 >

(a) Hold your breath and CO_2 builds up in your bloodstream. This increases the amount of carbonic acid, which lowers your blood pH. (b) Hyperventilate and the amount of CO_2 in your bloodstream decreases. This decreases the amount of carbonic acid, which raises your blood pH.

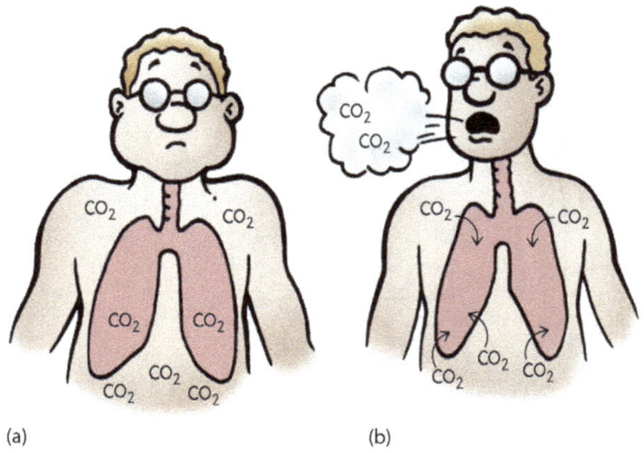

hold your breath and the amount of carbon dioxide (and hence carbonic acid) builds up, causing a slight but significant drop in pH. Hyperventilate and the carbonic acid level decreases by exhaling too much CO_2, causing a slight but significant increase in pH. Your body uses this mechanism to protect itself from changes in blood pH. One of the symptoms of a severe overdose of aspirin, for example, is hyperventilation. Aspirin, also known as acetylsalicylic acid, is an acidic chemical that, when taken in large amounts, can overwhelm the blood buffering system, causing a dangerous drop in blood pH. As you hyperventilate, however, your body loses carbonic acid, which helps to maintain the proper blood pH despite the overabundance of the acidic aspirin.

To summarize the acid/base concepts presented so far, consider the soil pH kit shown in **Figure 10.17**. Using a pH measuring kit, the farmer had found that the soil pH was unacceptably low. At this low pH, the soil contains an overabundance of hydronium ions. These hydronium ions react with alkaline *nutrients* of the soil, such as ammonia, to form water-soluble salts. Because of their water solubility, these nutrients in their salt form are readily washed away with the rainwater and as a result, the soil becomes nutrient-poor. The mechanism by which plants absorb whatever nutrients do remain in the soil is also disturbed by the soil's low pH.

As a result of all this, most plants do not grow well in acidic soil. To remedy this, the farmer spreads powdered limestone, a form of calcium carbonate, $CaCO_3$, which neutralizes the hydronium ions, thus raising the pH toward neutral.

Interestingly, the calcium carbonate reacts with the acidic soil to form carbon dioxide gas, which in the atmosphere helps to keep rainwater slightly acidic, which we describe next. This is the same gas that is generated by the cells of our bodies and that tends to acidify our blood. The blood pH, however, is kept fairly constant at around 7.4 because it is buffered.

Figure 10.17 >

Acidic soil tends to lose nutrients faster than alkaline soil, which is a reason for farmers to monitor the soil pH. Here the farmer has raised the pH by adding calcium carbonate in a process known as liming. Perhaps too much lime was added?

10.5 Rainwater Is Acidic

As previously mentioned, rainwater is naturally acidic. A main source of this acidity is carbon dioxide, the same gas that gives fizz to soda drinks. There are about 3,000 billion tons of CO_2 in the atmosphere, most of it from such natural sources as volcanoes and decaying organic matter but a growing amount (about 1000 billion tons) from human activities. In other words, about a third of the CO_2 in our present atmosphere is from ourselves.

Water in the atmosphere reacts with carbon dioxide to form *carbonic acid*:

$$\underset{\text{Carbon dioxide}}{CO_2(g)} + \underset{\text{Water}}{H_2O(l)} \leftrightarrows \underset{\text{Carbonic acid}}{H_2CO_3(aq)}$$

Carbonic acid, as its name implies, behaves as an acid and lowers the pH of water. The CO_2 in the atmosphere brings the pH of rainwater to about 5.6— noticeably below the neutral pH value of 7. Because of local fluctuations, the normal pH of rainwater varies between 5 and 7. This natural acidity of rainwater may accelerate the erosion of land and, under certain circumstances, can lead to the formation of underground caves.

By convention, *acid rain* is a term used for rain having a pH lower than 5. Acid rain is created when airborne pollutants, such as sulfur dioxide, are absorbed by atmospheric moisture. Sulfur dioxide is readily converted to sulfur trioxide, which reacts with water to form *sulfuric acid*:

$$\underset{\text{Sulfur dioxide}}{2\,SO_2(g)} + \underset{\text{Oxygen}}{O_2(g)} \rightarrow \underset{\text{Sulfur trioxide}}{2\,SO_3(g)}$$

$$\underset{\text{Sulfur trioxide}}{SO_3(g)} + \underset{\text{Water}}{H_2O(l)} \rightarrow \underset{\text{Sulfuric acid}}{H_2SO_4(aq)}$$

Most SO_2 is released into the atmosphere by the combustion of sulfur-containing coal and oil. Sulfuric acid is much stronger than carbonic acid, and as a result, rain laced with sulfuric acid and therefore higher H_3O^+ concentrations eventually corrodes metal, paint, and other exposed substances. This cumulative damage has cost billions of dollars. The cost to the environment is also high (**Figure 10.18**). Many rivers and lakes receiving acid rain become less capable of sustaining life. Much vegetation that receives acid rain doesn't survive. This is particularly evident in heavily industrialized regions.

fyi FOR YOUR INFORMATION

Acid rain remains a serious problem in many regions of the world. Significant progress, however, has been made toward fixing the problem. In the United States, for example, sulfur dioxide and nitrogen oxide emissions have been reduced by nearly half since 1980.

CONCEPT CHECK

When sulfuric acid, H_2SO_4, is added to water, what makes the resulting aqueous solution corrosive?

CHECK YOUR ANSWER

Because H_2SO_4 is a strong acid, it readily forms hydronium ions when dissolved in water. Hydronium ions are responsible for the corrosive action.

(a) (b) (c)

^ Figure 10.18

(a) and (b) These two photographs show the same obelisk in New York City's Central Park before and after the effects of acid rain. (c) Many forests downwind from heavily industrialized areas, such as in the eastern United States and in Europe, have been noticeably hard-hit by acid rain. The dead fir trees in this photo from the Great Smoky Mountains National Park are a result of both acid rain, as well as invasive beetles.

FOR YOUR INFORMATION

If the world is able to lower its SO_2 emissions, even with a growing population and greater economic activity, might we also be able to lower our CO_2 emissions? Collectively we release about 100 million tons of SO_2 each year. For perspective, we release about 40,000 million tons of CO_2 each year, which is about 400x as much. Carbon dioxide is also less chemically reactive than SO_2, which means that capturing it at the source is more difficult. Lowering our CO_2 emissions is many times more the challenge.

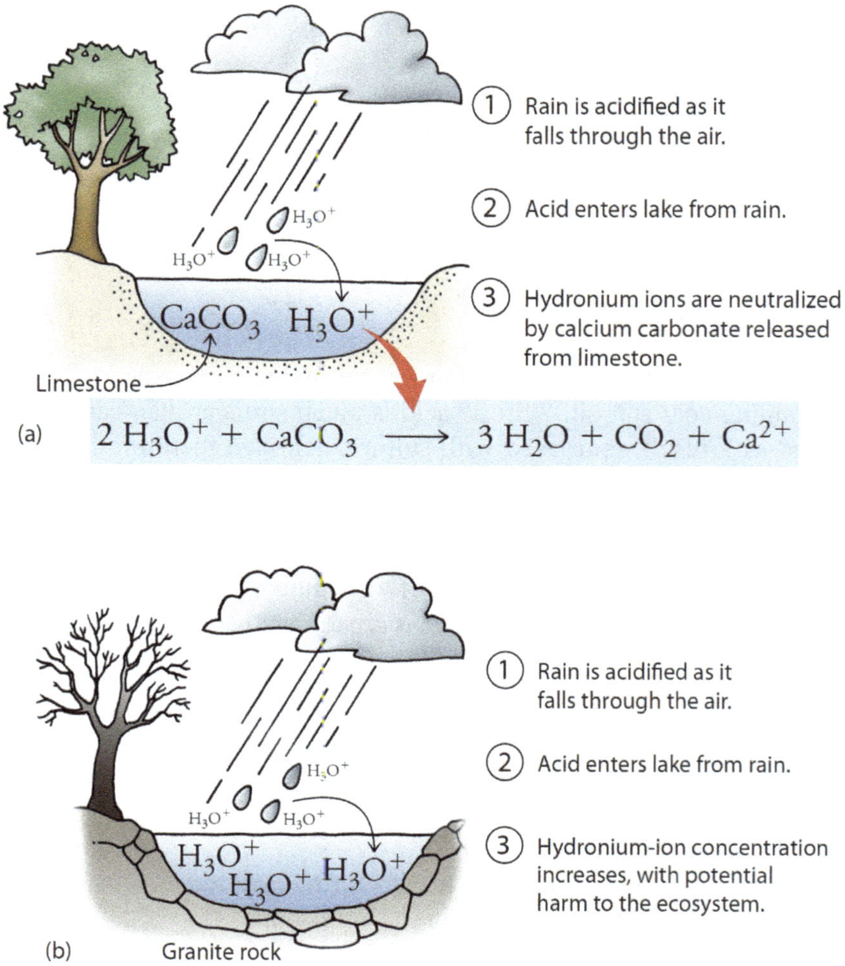

Figure 10.19 >

(a) The damaging effects of acid rain do not appear in bodies of fresh water lined with calcium carbonate, which neutralizes any acidity. (b) Lakes and rivers lined with inert materials are not protected.

The environmental impact of acid rain depends on local geology, as **Figure 10.19** illustrates. In certain regions, such as the midwestern United States, the ground contains significant quantities of the alkaline compound calcium carbonate (limestone), deposited when these lands were submerged under oceans, as has occurred several times over the past 500 million years. Acid rain pouring into these regions is often neutralized by the calcium carbonate before any damage is done. In the northeastern United States and many other regions, however, the ground contains very little calcium carbonate and is composed primarily of chemically less reactive materials, such as granite. In these regions, the effect of acid rain on lakes and rivers accumulates.

One demonstrated solution to this problem is to raise the pH of acidified lakes and rivers by adding calcium carbonate—a process known as liming. The cost of transporting the calcium carbonate, coupled with the need to monitor treated water systems closely, limits liming to only a small fraction of the vast number of water systems already affected. Furthermore, as acid rain continues to pour into these regions, the need to lime continues.

A longer-term solution to acid rain is to prevent most of the generated sulfur dioxide and other pollutants from entering the atmosphere in the first place. Toward this end, smokestacks have been designed or retrofitted to minimize the quantities of pollutants released. Although costly, the positive effects of these adjustments have been demonstrated as shown in **Figure 10.20**. An ultimate long-term solution, however, would be a shift from fossil fuels to cleaner energy sources, such as nuclear and solar energy.

READING CHECK
What is a longer-term solution to the problem of acid rain?

CONCEPT **CHECK**
What kind of lakes are most protected against acid rain?

CHECK YOUR ANSWER
Lakes that have a floor consisting of basic minerals, such as limestone, are more resistant to acid rain, because the chemicals of the limestone (mostly calcium carbonate, $CaCO_3$) neutralize any incoming acid.

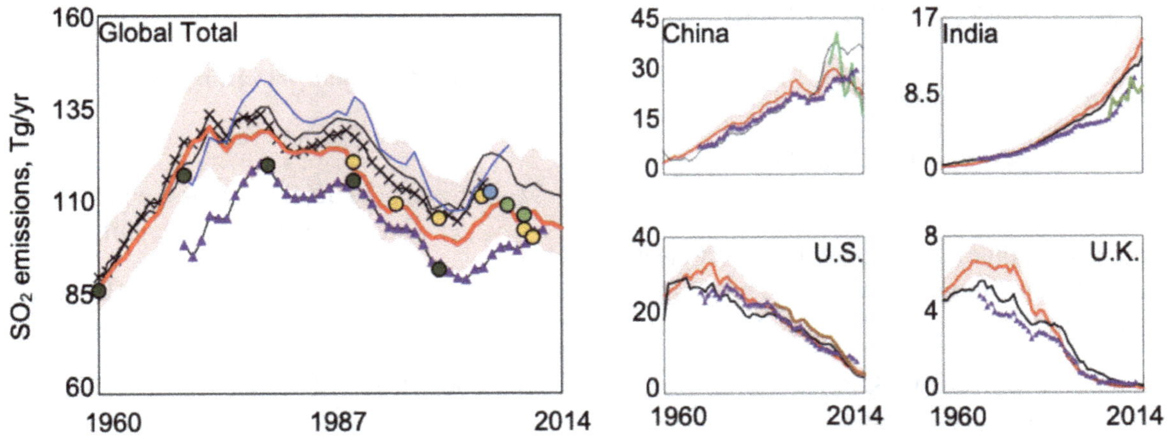

∧ Figure 10.20

Due in great part to public attention and economics, over the past few decades, developed nations have markedly decreased their SO_2 emissions, primarily by capturing the SO_2 rather than releasing it. The world leaders in SO_2 emissions are now collectively with developing nations such as India. (Source: Source: pubs.acs.org/doi/10.1021/acs.est.9b07696#)

Concept Review

Chapter 10

Summary of Terms

Acid A substance that donates hydrogen ions.

Acidic Said of a solution in which the hydronium ion concentration is higher than the hydroxide ion concentration.

Amphoteric A description of a substance that can behave as either an acid or a base.

Base A substance that accepts hydrogen ions.

Basic Said of a solution in which the hydroxide ion concentration is higher than the hydronium ion concentration. Also sometimes called alkaline.

Buffer solution A solution that resists large changes in pH, made from either a weak acid and one of its salts or a weak base and one of its salts.

Hydronium ion A polyatomic ion made by adding a proton (hydrogen ion) to a water molecule.

Hydroxide ion A polyatomic ion made by removing a proton (hydrogen ion) from a water molecule.

Neutral Said of a solution in which the hydronium ion concentration is equal to the hydroxide ion concentration.

Neutralization A reaction between an acid and a base.

pH A measure of the acidity of a solution, equal to the negative logarithm of the hydronium ion concentration.

Salt An ionic compound commonly formed from the reaction between an acid and a base.

Review Questions

10.1 Exchanging Protons

1. What are the Brønsted–Lowry definitions of acid and base?
2. When an acid is dissolved in water, what ion does the water form?
3. When a chemical loses a hydrogen ion, is it behaving as an acid or a base?

10.2 Acid and Base Strength

4. What does it mean to say that an acid is strong in aqueous solution?
5. Why does a solution of a strong acid conduct electricity better than a solution of a weak acid having the same concentration?
6. When can a solution of a weak base be more corrosive than a solution of a strong base?

10.3 Acidic, Basic, or Neutral

7. Is water a strong acid or a weak acid?
8. What is true about the relative concentrations of hydronium and hydroxide ions in an acidic solution? How about a basic solution? A neutral solution?
9. What does the pH of a solution indicate?
10. As the hydronium ion concentration of a solution increases, does the pH of the solution increase or decrease?

10.4 Buffers Resist pH Changes

11. What is a buffer solution?
12. A strong acid quickly drops the pH when added to water. Not so when added to a buffer solution. Why?
13. Why is it so important that the pH of our blood be maintained within a narrow range of values?

10.5 Rainwater Is Acidic

14. What is the product of the reaction between carbon dioxide and water?
15. What does sulfur dioxide have to do with acid rain?
16. How do humans generate the air pollutant sulfur dioxide?

Quantitative Questions

17. Show that the hydroxide ion concentration in an aqueous solution is 1×10^{-4} M when the hydronium ion concentration is 1×10^{-10} M? Recall that $10^a \times 10^b = 10^{(a+b)}$.

18. When the hydronium ion concentration of a solution is 1×10^{-10} M, what is the pH of the solution? Is the solution acidic or basic?

19. When the hydronium ion concentration of a solution is 1×10^{-4} M, what is the pH of the solution? Is the solution acidic or basic?

20. Show that an aqueous solution having a pH of 5 has a hydroxide ion concentration of 1×10^{-9} M.

21. When the pH of a solution is 1, the concentration of hydronium ions is 10^{-1} M = 0.1 M. Assume that the volume of this solution is 500 mL. What is the pH after 500 mL of water is added? You will need a calculator with a logarithm function to answer this question.

22. Show that the pH of a solution is –0.301 when its hydronium ion concentration equals 2 moles per liter. Is the solution acidic or basic?

23. Rank the following solutions in order of increasing concentration of hydronium ions, H_3O+.

a) Hydrogen Chloride, HCl (concentration = 2 M)

b) Acetic Acid, CH_3COOH (concentration = 2 M)

c) Ammonia, NH_3 (concentration = 2 M)

Solutions (Odd-Numbered)

1. The Brønsted-Lowry definition of an acid and base says that an acid is any chemical that donates a hydrogen ion and a base is any chemical that accepts a hydrogen ion.

3. A chemical that loses a hydrogen ion is behaving as an acid.

5. A solution of a strong acid has more ions in solution, which means it can conduct electricity better than a weak acid.

7. Water is a weak acid.

9. The pH of a solution indicates the acidity of the solution as judged by the concentration of hydronium ions.

11. A buffer solution is any solution that resists changes in pH.

13. A pH of blood that is too high or too low can be lethal.

15. Sulfur dioxide combines with oxygen and water in the air to make sulfuric acid.

17. The product of the hydroxide ion and hydronium ion concentration is always equal to 1×10^{-14}. So, if the hydroxide ion concentration equals 1×10^{-4} M, then the hydronium ion concentration must be equal to 1×10^{-10} M, because these two values multiply together to equal 1×10^{-14}.

19. The pH is of this solution is 4 and it is acidic.

21. The concentration of hydronium ions in the pH = 1 solution is 0.1M. Doubling the volume of solution with pure water means that its concentration is cut in half. The new concentration of hydronium ions after the addition of 500 mL of water, therefore, is 0.05 M. To calculate for pH:

$$pH = -\log [H_3O+] = -\log (0.05) = -(-1.3) = 1.3$$

23. All of these solutions have the same concentration. The difference between them is their acid strength. As discussed in the text, hydrogen chloride is a strong acid, which means nearly all of the hydrogen chloride molecules donate hydrogen ions to form hydronium ions. Acetic acid is a weak acid, which means only a few of the acetic acid molecules in solution donate hydrogen ions to form hydronium ions. The ammonia behaves better as a base than an acid, which means that it contributes very, very few hydrogen ions. The ammonia solution, therefore, has the lowest concentration of hydronium ions. The acetic acid solution has more hydronium ions, but not as many as are found in the hydrogen chloride solution.

Solutions to Chapter 10

Calculation Corner

Logarithms and pH

1. 5 (This is the exponent to which 10 is raised)

2. 5 (This is the number of zeros)

3. The pH is 9. This is a basic solution.

Chapter 9: Contextual Chemistry Essay

Ocean Acidification

The amount of carbon dioxide put into the atmosphere by human activities is growing. Surprisingly, however, the atmospheric concentration of CO_2 is not increasing proportionately. One explanation has to do with the oceans, as described in **Figure A**. When atmospheric CO_2 dissolves in any body of water—a raindrop, a lake, or an ocean—it forms carbonic acid, H_2CO_3. In fresh water, this carbonic acid transforms back to water and carbon dioxide, which is released back into the atmosphere. In the ocean, however, the carbonic acid is quickly neutralized by dissolved alkaline minerals. (The ocean is alkaline, pH = 8.1.) The products of this neutralization eventually end up on the ocean floor as insoluble solids. Thus, carbonic acid neutralization in the ocean prevents CO_2 from being released back into the atmosphere. The ocean, therefore, is an effective carbon dioxide sink—most of the CO_2 that goes in doesn't come out. Pushing more CO_2 into our atmosphere means pushing more of it into our vast oceans. So far, the oceans have been absorbing about one-third of our CO_2 emissions.

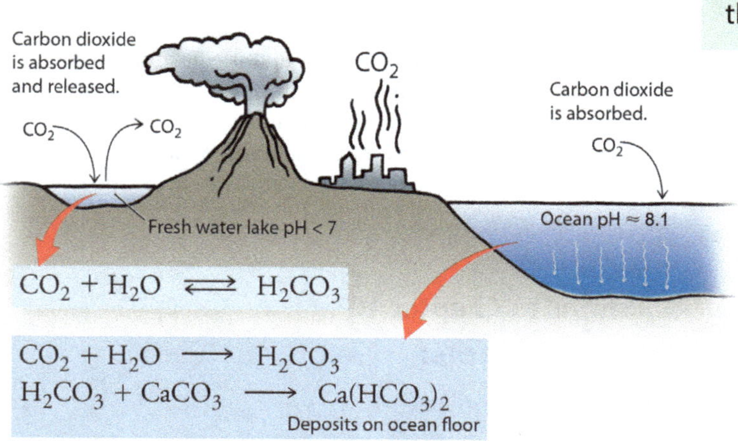

The movement of our CO_2 into the oceans, however, comes at a cost. Notice from Figure A that the CO_2 derived carbonic acid, H_2CO_3, reacts with calcium carbonate, $CaCO_3$. The addition of carbon dioxide into our oceans, therefore, has the effect of decreasing the amount of calcium carbonate in the ocean water, as well as other carbonates, such as magnesium carbonate, $MgCO_3$. Coral, shelled organisms, and many other marine species, however, use these carbonates to build and maintain their bodily structures. Importantly, they can only do this when the ocean water is saturated with these minerals, which is currently the case in most surface waters.

CONCEPT CHECK

How does adding CO_2 to the oceans cause a decrease in the concentration of carbonate ions, CO_3^{2-}?

CHECK YOUR ANSWER

As per Figure A, the CO_2 reacts with water, H_2O, to form carbonic acid, H_2CO_3, which effectively removes the carbonate ion, CO_3^{2-}, by reacting with it to form the bicarbonate ion, HCO_3^-.

The addition of CO_2 leads to ocean water *unsaturated* with carbonates. In regions where this occurs, the carbonate-based creatures begin to dissolve, which means they perish. This, in turn, can have a significant impact throughout the marine ecosystem, which would affect us as well. Consider this one example: pink salmon off the southern coast of Alaska live on a diet of sea snails, also known as pteropods, which are dependent on carbonate

Λ Figure A

Carbon dioxide forms carbonic acid upon entering any body of water. In fresh water, this reaction is reversible, and the carbon dioxide is released back into the atmosphere. In the alkaline ocean, the carbonic acid is neutralized to such compounds as calcium bicarbonate, $Ca(HCO_3)_2$, which precipitate to the ocean floor. As a result, As a result, the oceans are a net CO_2 absorber.

minerals. The destruction of this pteropod population by ocean acidification would also mean an end to the Alaskan pink salmon fishing industry. The immediate reality of this chemistry can be seen by pouring vinegar onto tropical sands, such as described in **Figure B.**

Is the effect of human-produced CO_2 on ocean pH measurable? Absolutely, and it is significant. Over the past 100 years, increases in atmospheric carbon dioxide—primarily due to the burning of fossil fuels—has lowered the average pH of the ocean by about 0.1 pH unit, which is about 1.26 times an increase in acidity. On a geologic timescale, this is lightning fast. The last comparable decrease in ocean pH occurred about 56 million years ago. At that time, major increases in atmospheric carbon dioxide caused the pH of the ocean to decrease by about 0.45 pH units. These changes, however, took place over 5000 years for an average decrease in ocean pH of about 0.01 pH units per century. The result was a huge die-off of marine organisms, which can be seen as a distinct layer of brown sediment within core samplings of the ocean floor (**Figure C**).

We are on track for surpassing the CO_2 event of 56 million years ago at a rate of about ten times as fast. This is indicated in **Figure D**, which shows the past concentration of atmospheric carbon dioxide. Of note, the most recent levels now exceed 400 ppm, which does not appear to have happened for over the past 800,000 years. You need to go back millions of years to find a comparable level. Our modern times where the human population has quickly eclipse 8 billion is most unique. We discuss many of the details concerning atmospheric carbon dioxide and climate in Chapter 16. For now, you should understand that our production of copious amounts of carbon dioxide is a two-fold problem. One is the potential for dramatic changes in global climate. The other is a very predictable change in ocean chemistry. Both need to be considered carefully.

READING CHECK

When was the last decrease in ocean pH that would have been comparable to what is taking place today?

∧ Figure B

Tropical beaches tend to be made of the crunched-up calcium carbonate containing bits and pieces of nearby coral reefs. The acetic acid in vinegar reacts with the calcium carbonate of these sands to produce gaseous carbon dioxide. Pour the vinegar onto these topical sands and you'll see much bubbling ensues. The oceans and creatures living within the oceans are most sensitive to pH.

Figure C >

Sediments that collect on the ocean floor are a physical record of current climate conditions. Because sediments build up over time, deeper sediments are a record of past climate conditions. Core samples from the ocean floor thus help in the study of ancient climates. Shown here is the ocean floor core repository of the Alfred Wegener Institute.

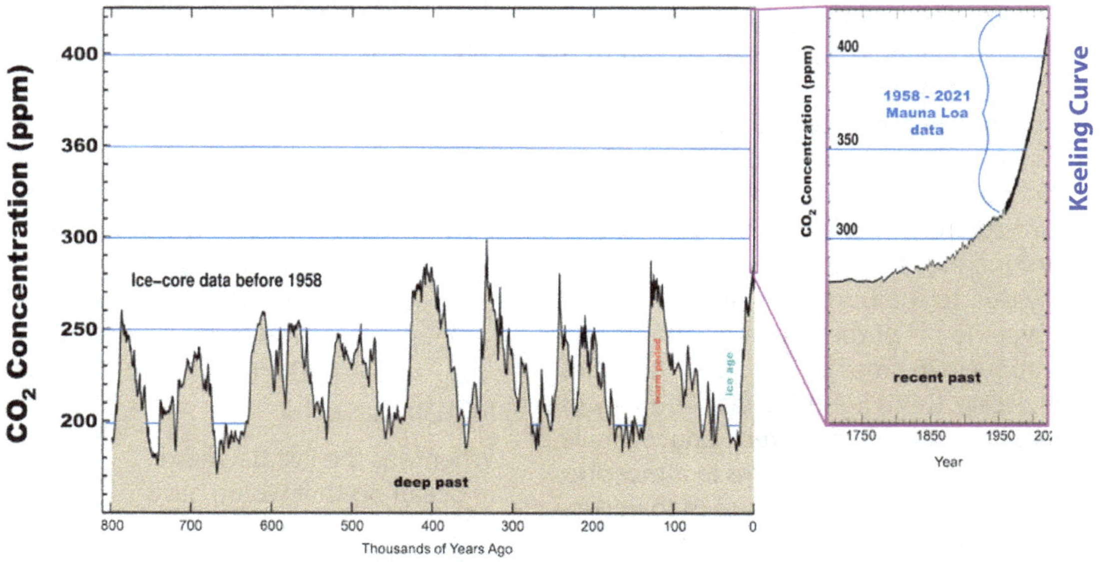

Λ Figure D
These graphs depict the levels of atmospheric carbon dioxide dating back 800,000 years ago as measured by ice core samples, and, more recently, via direct measurements, such as from the Mauna Loa Weather Observatory in Hawaii. Researchers at the Mauna Loa Weather Observatory have recorded increasing concentrations of atmospheric carbon dioxide since they began collecting data in 1958. Data from Mauna Loa forms a graph known as the Keeling curve, after the scientist Charles Keeling, who initiated this project and first noted the trends. (Source: keelingcurve.ucsd.edu)

Think and Discuss

1. Which is the greater problem: acid rain or ocean acidification?

2. How does burning fossil fuels lower the pH of the ocean?

3. How might warmer oceans have the effect of slowing down ocean acidification?

4. Scientists have experimented with ways of enhancing the ocean's ability to absorb atmospheric carbon dioxide. Adding powdered iron to a small plot of the ocean, they found, has the effect of fostering the growth of microorganisms that enhance the rate at which carbon dioxide is absorbed. Might this be a solution to the problem of global climate change? Why or why not?

CONCEPT **CHECK**

Why are atmospheric levels of carbon dioxide not rising as rapidly as might be expected based on the increased output of carbon dioxide resulting from human activities?

CHECK YOUR ANSWER

The ocean absorbs the CO_2 where it is neutralized and is not re-released.

Author Responses to Think and Discuss

1. Acid rain is localized to regions downwind of sulfur and nitrogen oxide producing industry. Ocean acidification, by contrast, results primarily from the release of carbon dioxide, which is produced in much greater quantities and across a much broader range of sources. Both acid rain and ocean acidification are significant problems. Ocean acidification, however, is much more global in its scope and therefore should be of greater concern.

2. When fossil fuels burn they react with the oxygen, O_2, in the air to form carbon dioxide, CO_2, and water vapor, H_2O. The carbon dioxide reacts with atmospheric moisture to form carbonic acid, which lowers the pH of rainwater. As the acidic rainwater precipitates into the ocean, it neutralizes alkaline ocean salts, such as calcium carbonate. This, in turn, lowers the pH of the ocean.

3. The warmer the ocean, the lower the solubility of any dissolved gases such as carbon dioxide, CO_2. Less CO_2 would be absorbed, which would slow the rate of acidification.

4. Adding iron to the ocean to enhance its carbon dioxide absorbing power my help to lower the amount of carbon dioxide in the atmosphere. At the same time, however, it might drastically alter the ocean's ecology. For example, the excess carbon dioxide would lower the pH of the ocean. Shelled marine organism would die, which could affect the whole food chain. We could end up trading one problem for another.

Chapter 11

Oxidations and Reductions

THE MAIN IDEA

Oxidation is the loss of electrons, and reduction is the gain of electrons.

∧ The chemical reactions going on in your body are quite similar to those going on within burning wood. In both cases, the products are carbon dioxide, water, and energy.

What do our bodies have in common with the burning of a campfire or the rusting of old farm equipment? Why does silver tarnish? How can aluminum restore tarnished silver? How are metals produced from minerals? How do batteries work, and what is their source of energy? What are fuel cells, and how do they generate electricity so efficiently? How do photovoltaic cells convert sunlight into electricity? Why is hydrogen such an environmentally friendly fuel but not itself a source of energy? The answers to all these questions involve a class of reactions called oxidation-reduction, sometimes "redox" for short. These reactions are similar to the acid-base reactions you learned about in the previous chapter in that they both involve the transfer of subatomic particles between reactants. For acid-base reactions, these subatomic particles are typically protons. Oxidation-reduction reactions, by contrast, involve the transfer of electrons.

11.1 Losing and Gaining Electrons

Oxidation is the process whereby a reactant loses one or more electrons. **Reduction** is the opposite process, whereby a reactant gains one or more electrons. Oxidation and reduction are complementary processes that occur at the same time. They always occur together; you cannot have one without the other. The electrons lost by one chemical in an oxidation reaction don't simply disappear; they are gained by another chemical in a reduction reaction.

READING CHECK
Is it possible for oxidation to occur without reduction?

An oxidation-reduction reaction occurs when sodium and chlorine react to form sodium chloride, as shown in **Figure 11.1**. The equation for this reaction is

$$2Na(s) + Cl_2(g) \rightarrow 2NaCl(s) + heat$$

(a) (b) (c)

Λ Figure 11.1

(a) Chlorine gas, Cl_2, is introduced to a small lump of sodium, Na, resting on sand within a reaction flask. (b) The two chemicals react explosively. (c) The end product is chemically stable sodium chloride, NaCl, which can be seen coating the inner surface of the flask. Through this exothermic reaction, sodium metal is *oxidized* by the chlorine gas and the chlorine gas is *reduced* by the sodium metal.

To see how electrons are transferred in this reaction, we can look at each reactant individually. Each electrically neutral sodium atom changes to a positively charged ion. At the same time, we can say that each atom loses an electron and is therefore oxidized:

$$2Na(s) \rightarrow 2Na^+ + 2e^- \quad \text{Oxidation}$$

Each electrically neutral chlorine molecule changes to two negatively charged ions. Each of these atoms gains an electron and is therefore reduced:

$$Cl_2(g) + 2e \rightarrow 2Cl^- \quad \text{Reduction}$$

The net result is that the two electrons lost by the sodium atoms are transferred to the chlorine atoms. Therefore, each of the two equations shown actually represents one-half of an entire process, which is why they are each a **half-reaction**. In other words, an electron won't be lost from a sodium atom without the presence of a chlorine atom available to pick up that electron. Both half-reactions are required to represent the *whole* oxidation-reduction process. Half-reactions are useful for showing which reactant loses electrons and which reactant gains them, which is why half-reactions are used throughout this chapter.

Because the sodium causes reduction of the chlorine, the sodium is acting as a *reducing agent*. A reducing agent is any reactant that causes another reactant to be reduced. Note that sodium is oxidized when it behaves as a reducing agent—it loses electrons. Conversely, the chlorine causes oxidation of the sodium and so is acting as an *oxidizing agent*. Because it gains electrons in the process, an oxidizing agent is reduced. Just remember that loss of electrons is oxidation and gain of electrons is reduction. Here is a helpful mnemonic: OIL RIG—**o**xidation **i**s **l**oss and **r**eduction **i**s **g**ain.

Different elements have different oxidation and reduction tendencies. For example, metals lose electrons readily, while most nonmetals tend to gain electrons, as **Figure 11.2** illustrates.

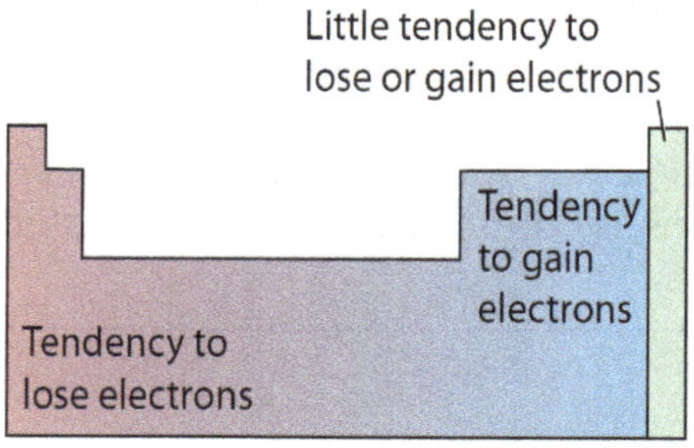

☐ More likely to behave as oxidizing agent (be reduced)

☐ More likely to behave as reducing agent (be oxidized)

☐ Noble gas elements resist both oxidation and reduction

Λ Figure 11.2

The ability of an atom to gain or lose electrons is indicated by its position in the periodic table. Those at the upper right (nonmetals) tend to gain electrons, and those at the lower left (metals) tend to lose them.

CONCEPT **CHECK**

True or false:

1. Reducing agents are oxidized in oxidation-reduction reactions.
2. Oxidizing agents are reduced in oxidation-reduction reactions.

CHECK YOUR ANSWER Both statements are true

Whether a reaction is classified as an oxidation-reduction reaction is not always immediately apparent. The chemical equation, however, can provide some important clues. First, look for changes in the ionic states of elements. Sodium metal, for example, consists of neutral sodium atoms. In the formation of sodium chloride, these atoms transform into positively charged sodium ions; this occurs as sodium atoms lose electrons (oxidation). A second way to identify a reaction as an oxidation-reduction reaction is to determine whether an element is gaining or losing oxygen atoms. As the element gains the oxygen atom, it is losing electrons to that oxygen, because of the oxygen's high electronegativity. The gain of oxygen, therefore, is oxidation (loss of electrons), while the loss of oxygen is reduction (gain of electrons). For example, hydrogen, H_2, reacts with oxygen, O_2, to form water, H_2O, as follows:

$$\text{H-H} + \text{H-H} + \text{O=O} \rightarrow \text{H-O-H} + \text{H-O-H}$$

Note that the element hydrogen becomes attached to an oxygen atom through this reaction. The hydrogen, therefore, is oxidized losing an electron to the oxygen with which it partnered.

A third way to identify a reaction as an oxidation-reduction reaction is to see whether an element is gaining or losing hydrogen atoms. The gain of hydrogen is reduction, while the loss of hydrogen is oxidation. For the formation of water shown previously, we see that the element oxygen is gaining hydrogen atoms, which means that the oxygen is being reduced—that is, the oxygen is gaining electrons from the hydrogen, which is why the oxygen atom within water is slightly negative, as discussed in Section 6.8. The three ways of identifying a reaction as an oxidation-reduction type of reaction are summarized in **Figure 11.3.**

Oxidation (Ionic state becomes more positive)	Reduction (Ionic state becomes more negative)
Loses electrons	Gains electrons
Gains oxygen	Loses oxygen
Loses hydrogen	Gains hydrogen

< Figure 11.3

Oxidation results in a greater positive charge, which can be achieved by losing electrons, gaining oxygen atoms, or losing hydrogen atoms. Reduction results in a greater negative charge, which can be achieved by gaining electrons, losing oxygen atoms, or gaining hydrogen atoms

CONCEPT **CHECK**

In the following equation, is carbon oxidized or reduced?

$$CH_4 + 2\,O_2 \rightarrow CO_2 + 2\,H_2O$$

CHECK YOUR ANSWER As the carbon of methane, CH_4, forms carbon dioxide, CO_2, it is losing hydrogen and gaining oxygen, which tells us that the carbon is being oxidized.

11.2 Harnessing the Energy

Electrochemistry is the study of the relationship between electrical energy and chemical change. It involves either the use of an oxidation-reduction reaction to produce an electric current or the use of an electric current to produce an oxidation-reduction reaction.

To understand how an oxidation-reduction reaction can generate an electric current, consider what happens when a reducing agent is placed in direct contact with an oxidizing agent: electrons flow from the reducing agent to the oxidizing agent. This flow of electrons is an electric current, which is a form of kinetic energy that can be harnessed for useful purposes.

Iron atoms, Fe, for example, are better reducing agents than are copper ions, Cu^{2+}. So, when a piece of iron metal and a solution containing copper ions are placed in contact with each other, electrons flow from the iron to the copper ions, as **Figure 11.4** illustrates. The result is the oxidation of iron atoms and the reduction of copper ions.

The elemental iron and copper ions need not be in physical contact for electrons to flow between them. If they are in separate containers but bridged by a conducting wire, the electrons can flow from the iron through the wire to the copper ions. The resulting electric current in the wire can be attached to some useful device, such as a lightbulb. But alas, an electric current is not sustained by this arrangement.

READING CHECK

Why won't electrons flow continuously through a wire connecting iron and copper in two containers?

The reason the electric current is not sustained is shown in **Figure 11.5**. An initial flow of electrons through the wire immediately results in a buildup of electric charge in both containers. The container on the left builds up positive charge as it accumulates Fe^{2+} ions from the nail. The container on the right builds up negative charge as electrons accumulate on this side. This situation prevents any further migration of electrons through the wire. Recall that electrons are negative, so they are repelled by the negative charge in the right container and attracted to the positive charge in the left container. The net result is that the electrons do not flow through the wire, and the bulb remains unlit. Put it this way: if you were an electron in a positively charged container, would YOU want to leave to go to another chamber that was negatively charged? Absolutely not. Remember, opposite signs attract and like signs repel.

V Figure 11.4

A nail made of iron placed in a solution of Cu^{2+} ions oxidizes to Fe^{2+} ions, which dissolve in the water. At the same time, copper ions are reduced to metallic copper, which coats the nail. (Negatively charged ions, such as chloride ions, Cl^-, must also be present to balance these positively charged ions in solution.)

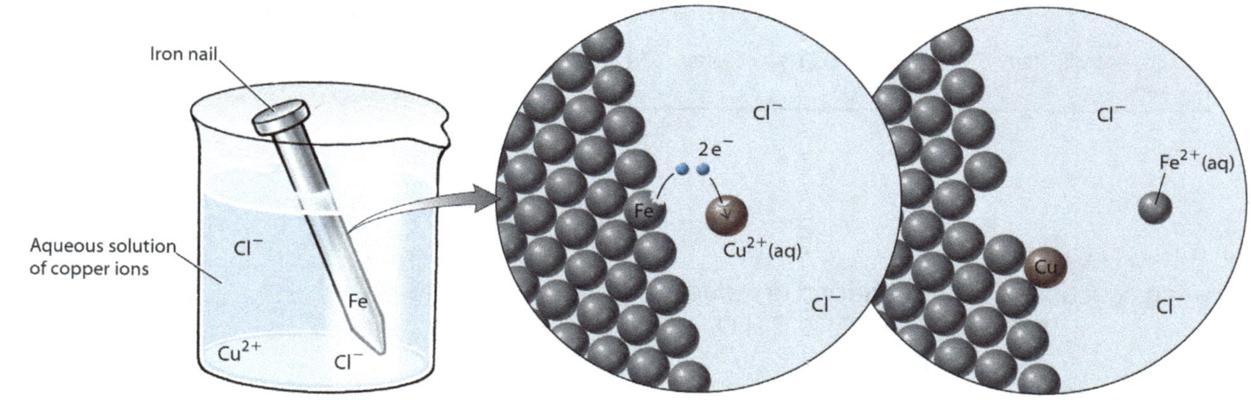

Oxidation $Fe \longrightarrow Fe^{2+} + 2\,e^-$

Reduction $Cu^{2+} + 2\,e^- \longrightarrow Cu$

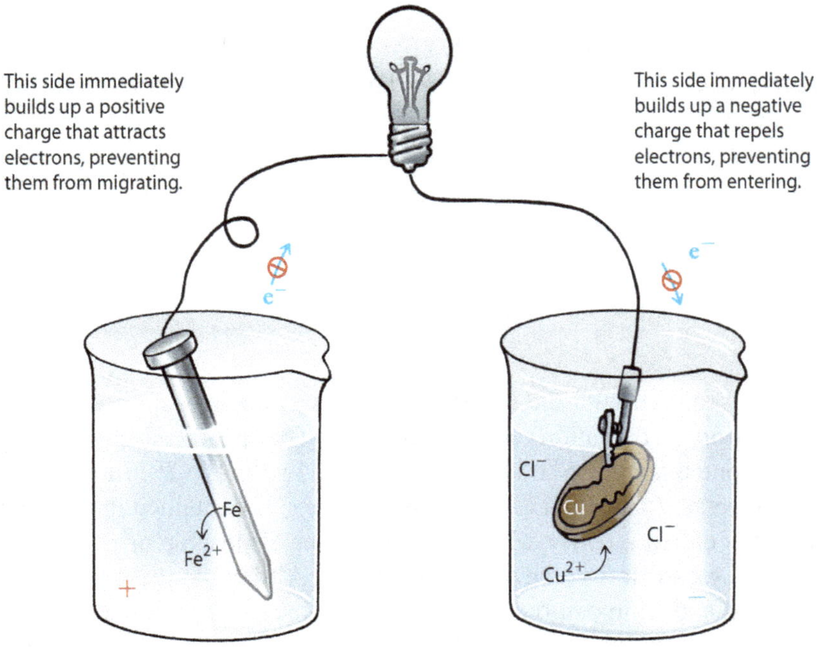

< Figure 11.5

An iron nail is placed in water and connected by a conducting wire to a solution of copper ions. Not much happens, because this arrangement results in a buildup of charge that prevents the further flow of electrons.

The solution to this problem is to allow ions to migrate between containers so that neither builds up any positive or negative charge. This is accomplished with a *salt bridge*, which may be as simple as a paper towel soaked in salt water stretching between both containers. Commonly, a salt bridge is made of a U-shaped tube filled with a paste of salts, such as sodium nitrate, $NaNO_3$, and closed with semi-porous plugs. **Figure 11.6** shows how such a salt bridge allows ions to enter either container as necessary. This creates a complete electric circuit, which permits the flow of electrons through the conducting wire as iron atoms yield their electrons to copper ions.

V Figure 11.6

The salt bridge completes the electric circuit. Electrons freed as the iron is oxidized pass through the wire to the container on the right. Nitrate ions, NO_3^-, from the salt bridge flow into the left container to balance the positive charges of the Fe^{2+} ions that form, thereby preventing any buildup of positive charge. Meanwhile, Na^+ ions from the salt bridge enter the right container to balance the Cl^- ions "abandoned" by the Cu^{2+} ions as the Cu^{2+} ions pick up electrons to become metallic copper.

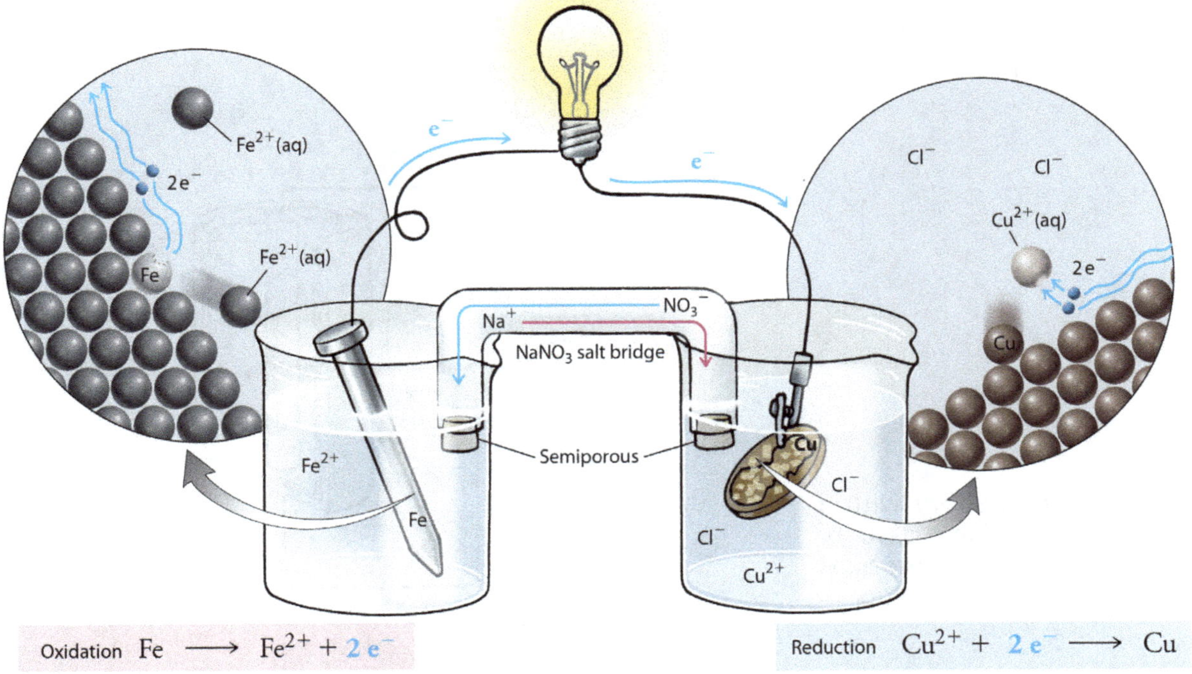

CONCEPT CHECK

Electrons are negatively charged. Does this mean they flow toward any negatively charged region?

CHECK YOUR ANSWER No! Recall that like signs repel. A negatively charged electron will be repelled by any region it finds to be also negatively charged. Conversely, electrons tend to flow toward regions that carry a positive charge.

11.3 Electricity from Batteries

READING CHECK

What is the principle by which any battery works?

We can see that, with the proper setup, it is possible to harness electrical energy from an oxidation-reduction reaction. The apparatus shown in Figure 11.6 is one example. Such devices are called *voltaic cells*. Instead of two containers, a voltaic cell can be an all-in-one self-contained unit, in which case it is called a *battery*. Batteries are either disposable or rechargeable, and here we explore some examples of each. Although the two types differ in design and composition, they function by the same principle: two materials that oxidize and reduce each other are connected by a medium through which ions travel to balance an external flow of electrons.

Let's look at disposable batteries first. The common *dry-cell battery*, which was invented in the 1860s, is still used today, and it is probably the cheapest disposable energy source for flashlights, toys, and the like. The basic design consists of a zinc cup filled with a thick paste of ammonium chloride, NH_4Cl, zinc chloride, $ZnCl_2$, and manganese dioxide, MnO_2. Immersed in this paste is a porous stick of graphite that projects to the top of the battery, as shown in **Figure 11.7.**

Graphite is a good conductor of electricity. Chemicals in the paste receive electrons at the graphite stick and so are reduced. The reaction for the ammonium ions is

CHEMICAL CONNECTIONS

How is a flashlight connected to a coal-fired power plant?

$$2\,NH_4^+(aq) + 2\,e^- \rightarrow 2\,NH_3(g) + H_2(g) \quad \text{Reduction}$$

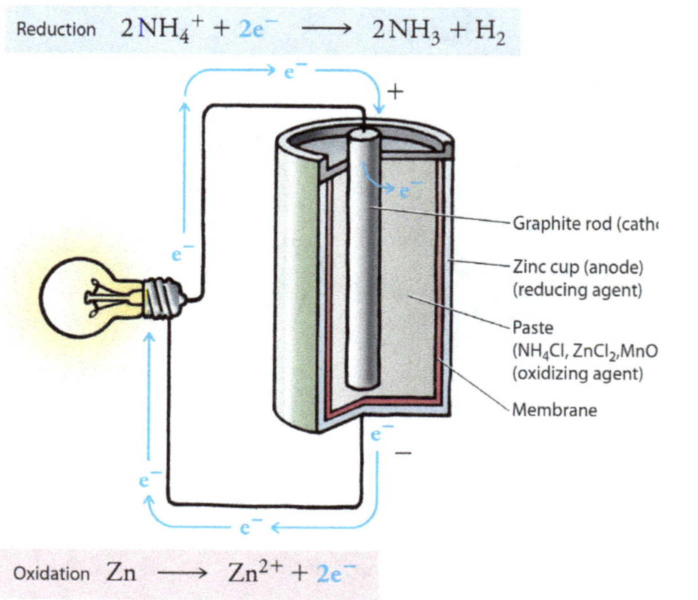

Figure 11.7 >

A common dry-cell battery with a graphite rod immersed in a paste of ammonium chloride, manganese dioxide, and zinc chloride in a zinc cup.

An **electrode** is any material that conducts electrons into or out of a medium in which electrochemical reactions are occurring. The electrode where chemicals are reduced is called a **cathode**. For any battery, such as the one shown in Figure 11.7, the cathode is always positive (+), which indicates that electrons are naturally attracted to this location. The electrons gained by chemicals at the cathode originate at the **anode**, which is the electrode where chemicals are oxidized. For any battery, the anode is always negative (–), which indicates that electrons are streaming away from this location (Remember, electrons are repelled by a negative charge). The anode in Figure 11.7 is the zinc cup, where zinc atoms lose electrons to form zinc ions:

$$Zn(s) \rightarrow Zn^{2+}(aq) + 2\,e^- \qquad \text{Oxidation}$$

The reduction of ammonium ions in a dry-cell battery produces two gases—ammonia, NH_3, and hydrogen, H_2—that need to be removed to avoid a pressure buildup and a potential explosion. Removal is accomplished by having the ammonia and hydrogen react with the zinc chloride and manganese dioxide:

$$ZnCl_2(aq) + 2\,NH_3(g) \rightarrow Zn(NH_3)_2Cl_2(s)$$

$$2\,MnO_2(s) + H_2(g) \rightarrow Mn_2O_3(s) + H_2O(l)$$

The life of a dry-cell battery is relatively short. Oxidation causes the zinc cup to deteriorate, and eventually, the contents leak out. Even while the battery is not operating, the zinc corrodes as it reacts with ammonium ions. This zinc corrosion can be inhibited by storing the battery in a refrigerator. As discussed in Section 9.6, chemical reactions slow down with decreasing temperature. Chilling a battery, therefore, slows down the rate at which the zinc corrodes, which increases the life of the battery.

Another type of disposable battery, the more expensive *alkaline battery*, shown in **Figure 11.8**, avoids many of the problems of dry-cell batteries by operating in a strongly alkaline paste. In the presence of hydroxide ions, the zinc is oxidized to insoluble zinc oxide:

Λ Figure 11.8

Alkaline batteries last a lot longer than dry-cell batteries and give a steadier voltage, but they are more expensive.

$$Zn(s) + 2\,OH^-(aq) \rightarrow ZnO(s) + H_2O(l) + 2\,e^- \qquad \text{Oxidation}$$

At the same time, manganese dioxide is reduced:

$$2\,MnO_2(s) + H_2O(l) + 2\,e^- \rightarrow Mn_2O_3(s) + 2\,H_2O(l) \qquad \text{Reduction}$$

Note how these two reactions avoid the use of the zinc-corroding ammonium ion (which means alkaline batteries last a lot longer than dry-cell batteries) and prevent formation of any gaseous products. Furthermore, these reactions are better suited to maintaining a given voltage during longer periods of operation.

The small mercury and lithium disposable batteries used for calculators and cameras are variations of the alkaline battery. In the mercury battery, mercuric oxide, HgO, rather than manganese dioxide, is reduced. Manufacturers are phasing out these batteries because of the environmental hazard posed by mercury, which is poisonous. In the lithium battery, lithium metal, rather than zinc, is used as the source of electrons. Lithium not only is able to maintain a higher voltage than zinc but also is about one-thirteenth as dense, which makes for a lighter battery.

Disposable batteries have relatively short lives, because electron-producing chemicals are consumed in their use. The main feature of *rechargeable* batteries is the reversibility of the oxidation and reduction reactions. A noteworthy example is the nickel metal hydride, NiMH, battery. Charging this battery causes the nickel metal to extract hydrogen from water to form the negatively charged *hydride ion*, shown in the equation below as H:, where the two dots represent two electrons.*

$$\underset{\text{water}}{H_2O} + \underset{\text{nickel metal}}{Ni} + e^- \rightarrow \underset{\text{nickel hydride}}{H{:}Ni} + \underset{\text{hydroxide ion}}{HO^-}$$

The role of the nickel is to stabilize the two electrons on the hydrogen, which because it contains an added electron is called the *hydride* ion, much the way chlorine with an extra electron is called the *chloride* ion, using the *-ide* suffix. A fully charged battery thus contains an abundance of nickel hydride. As the battery provides electricity, the hydride ion releases electrons, which allows it to join with the hydroxide ion to re-form water:

$$\underset{\text{nickel hydride}}{H{:}Ni} + \underset{\text{hydroxide ion}}{HO^-} \rightarrow \underset{\text{water}}{H_2O} + \underset{\text{nickel metal}}{Ni} + e^-$$

Recharging a rechargeable battery simply means regenerating the chemicals that can release electrons on demand. For the NiMH battery, this chemical is nickel hydride, H:Ni. For a traditional car battery, this chemical is lead, Pb, which transforms into lead sulfate, $PbSO_4$, as it releases electrons. As the car battery is recharged, the $PbSO_4$ is transformed back into lead, Pb. This is an endothermic reaction, which requires the input of energy that comes from an on-board generator, also known as the alternator. The alternator, in turn, gets its energy from the car's internal-combustion engine, which is driven by the energy of the fuel in the gas tank.

Rechargeable lithium-ion batteries have found a wide range of applications, from computer laptops to cell phones. Safer lithium phosphate ion batteries are used for today's hybrid and fully electric cars, which have grown in popularity as shown in **Figure 11.9.** Hybrids have improved gas mileage because as the car slows down, its kinetic energy is transformed into the electric potential energy of the battery rather than

FOR YOUR INFORMATION

As a battery provides electricity, electrons move from the negative anode to the positive cathode. When a rechargeable battery is being recharged, however, what was once the negative anode now becomes a negative cathode, which is where electrons are needed for reduction. Because electrons won't travel to a negative cathode on their own, they must be forced to do so. Hence, recharging is an energy-consuming process.

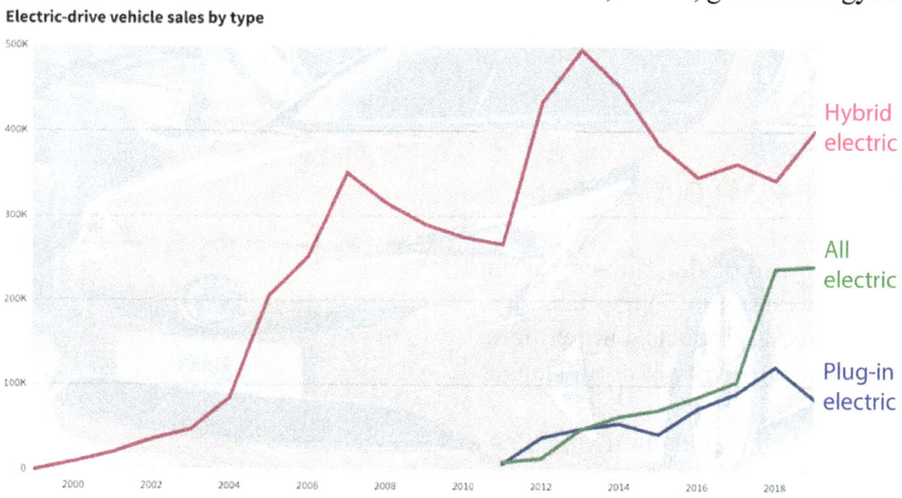

Figure 11.9

Sales of electric-drive vehicles in the United States. (Source: afdc.energy.gov/)

*The nickel in this reaction is actually an intermetallic compound of nickel and various rare-earth elements, such as lanthanum, La.

being wasted as heat from the car's brake pads. The captured electrical energy of the battery is subsequently used to assist the gas-powered engine to get the car moving. Also, the hybrid's battery system allows the engine to shut off when the car is merely idling or moving slowly, as occurs in heavy traffic. The fully electric cars are recognized for being environmentally friendly, though they still rely on fossil fuels due to the manufacture of the car itself, and potentially when the electricity powering the car comes from a fossil fuel dependent source.

New generation batteries are now being developed in which the anode is made of a 3-dimensional network of nanowires that cover a surface area 10,000 times greater than a traditional battery. These so-called "sponge" batteries hold more energy per unit mass and can be re-charged in a matter of minutes, rather than hours. For an example, visit PrietoBattery.com.

CONCEPT **CHECK**

What chemicals are produced as a nickel metal hydride battery is recharged?

CHECK YOUR ANSWER Nickel hydride, H:Ni, and hydroxide ions, HO^-.

11.4 Electricity from Fuel Cells

A *fuel cell* is a device that converts the energy of a fuel to electrical energy. Fuel cells are an efficient means of generating electricity. A hydrogen–oxygen fuel cell is shown in **Figure 11.10**. It has two compartments, one for entering hydrogen fuel and the other for entering oxygen, separated by a set of porous electrodes. Hydrogen is oxidized upon contact with hydroxide ions at the hydrogen-facing electrode (the anode). The electrons from this oxidation flow through an external circuit and provide electric power before meeting up with oxygen at the oxygen-facing electrode (the cathode). The oxygen readily picks up the electrons (in other words, the oxygen is reduced) and reacts with water to form hydroxide ions. To complete the circuit, these hydroxide ions migrate across the porous electrodes and through an ionic paste of potassium hydroxide, KOH, to join with hydrogen at the hydrogen-facing electrode.

READING CHECK

What is a fuel cell?

As the oxidation equation shown at the top of Figure 11.10 demonstrates, the hydrogen and hydroxide ions react to produce energetic water molecules that arise in the form of steam. This steam may be used for heating or for generating electricity in a steam turbine. Furthermore, the water that condenses from the steam is pure water, suitable for drinking.

Although fuel cells are similar to batteries, they will run continuously as long as fuel is supplied. The International Space Station uses hydrogen–oxygen fuel cells to meet its electricity needs. The cells also produce drinking water for the astronauts. Back on the Earth, researchers are developing fuel cells for buses and automobiles. As shown in **Figure 11.11**, experimental fuel-cell buses are already operating in many cities. These vehicles produce very few pollutants and can run much more efficiently than vehicles that burn fossil fuels.

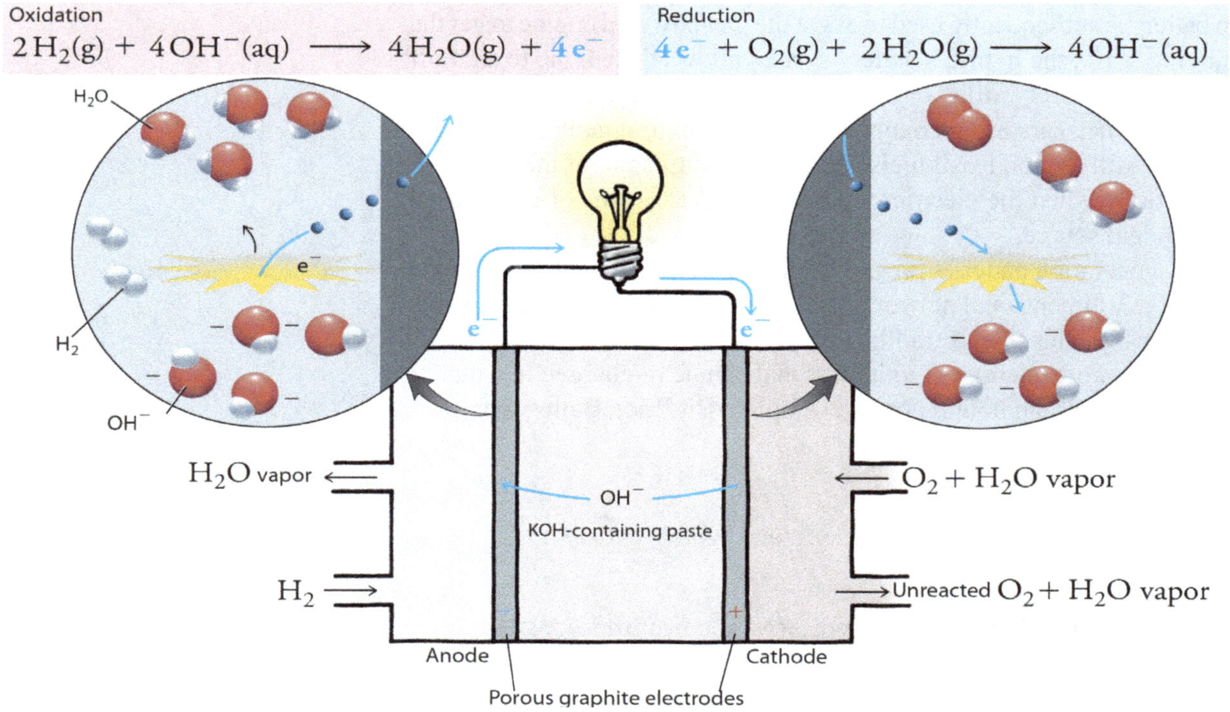

∧ Figure 11.10

In the hydrogen–oxygen fuel cell, hydrogen, H_2, combines with hydroxide ions, OH^-, at the anode to form water in addition to electrons that power an external circuit. The electrons then return to the cathode where oxygen, O_2, combines with water to form hydroxide ions that migrate back to the anode.

A number of problems are associated with using hydrogen as a fuel for fuel cells. Perhaps foremost is that hydrogen gas, H_2, is not a naturally abundant material. The hydrogen, therefore, must be manufactured, which is an energy-requiring process that defeats much of the utility of the fuel cell. For example, a most direct way of producing hydrogen gas is from the electrolysis of water. As described in Section 11.6, electrolysis requires a source of electrical energy. If this energy comes from a pollution-producing power plant, the environmental benefits of the fuel cell are lost. Alternatively, as discussed in Section 11.5, photovoltaic cells can produce the electricity needed for electrolysis. But today's photovoltaic cells are relatively expensive to produce, and they only work when the Sun is shining.

Hydrogen gas, however, can be produced from fossil fuels. The most efficient and environmentally friendly fossil fuel is methane, CH_4, also known as natural gas, which is now in great supply because of the advent of *fracking*, discussed in Chapter 17. When heated with steam in the presence of a catalyst, this methane can be converted to hydrogen and carbon dioxide:

$$\underset{\text{methane}}{CH_4\,(g)} + \underset{\text{steam}}{2\,H_2O\,(g)} \rightarrow \underset{\text{hydrogen}}{4\,H_2\,(g)} + \underset{\text{carbon dioxide}}{CO_2\,(g)}$$

∧ Figure 11.11

Because this bus is powered by a fuel cell, its tail pipe emits mostly water vapor.

A fuel cell now used on an industrial scale is the molten carbonate fuel cell (MCFC). These fuel cells have been engineered to work in tandem with the on-site production of hydrogen from methane. As shown in **Figure 11.12**, the resulting hydrogen is oxidized at the anode as it reacts with carbonate ions to produce carbon dioxide, water, and electrons. The electrons pass through an external circuit providing electric power. Meanwhile, the carbon dioxide is piped to the cathode where electrons are received by oxygen, which reacts with the carbon dioxide to form carbonate ions. The carbonate ions migrate through the electrolyte back to the anode for subsequent reactions.

The MCFC can be scaled to create a power plant from a few kilowatts to as many as 50 MW. For reference, a typical coal-fired power plant is about 500 MW. The coal-fired power plant, however, is centralized, producing

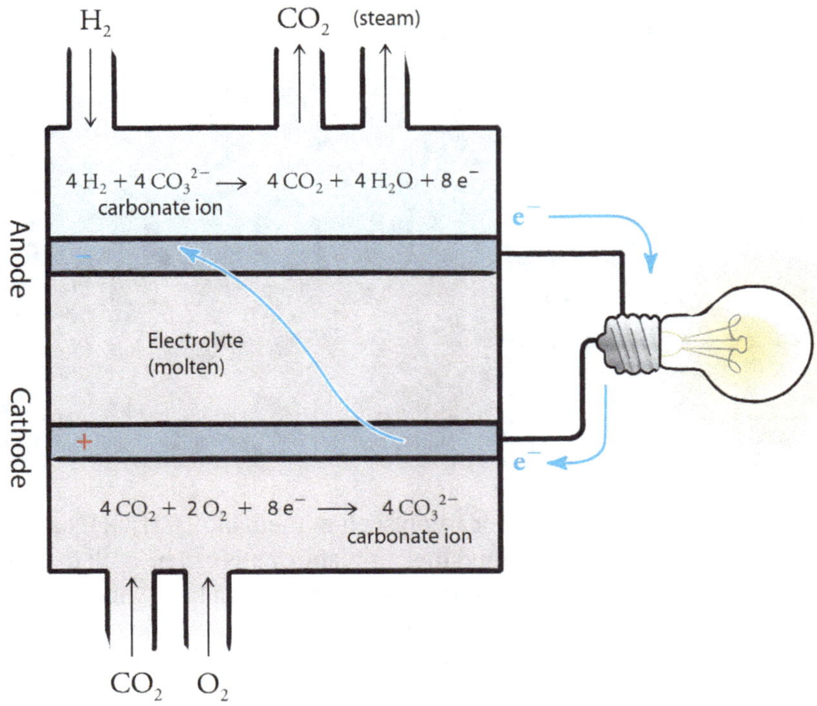

< Figure 11.12

In the molten carbonate fuel cell (MCFC), carbonate ions facilitate the oxidation of hydrogen, H2, at the anode and the reduction of oxygen, O2, at the cathode. While the oxygen comes from the air, the hydrogen can come from the transformation of a gaseous hydrocarbon, such as methane. Even compost gas, which is mostly methane, can be used as the fuel source.

electricity for an entire region. The fuel cell power plant, by contrast, is suitable for individual buildings, such as a hospital, factory, or skyscraper (**Figure 11.13**). It runs quietly, producing few exhausts from the basement or just outside the building it serves. Aside from electricity, the MCFC also produces much thermal energy, which can be used to heat water and ventilation systems, which boosts the overall efficiency.

The electricity from a fuel cell can be used to charge the batteries of an electric vehicle. But might the fuel cell also be placed on board to power the electric vehicle directly? While only 3 kilograms of hydrogen are needed to drive a car some 500 kilometers, the volume of this much hydrogen is about 36,000 liters. Compressing the gas into smaller volumes (as occurs in the bus shown in Figure 11.11) or chilling it to the liquid phase is an energy-intensive process, and this lowers the efficiencies. Researchers, however, are looking to porous materials, such as the carbon nanofibers shown in **Figure 11.14,** that can hold large amounts of hydrogen to their surfaces, behaving in effect like hydrogen "sponges." The hydrogen can be "squeezed" out of these porous materials on demand by controlling the temperature—the warmer the material, the more hydrogen is released.

< Figure 11.13

Molten Carbonate Fuels Cells offer relatively clean and affordable power that can be scaled to the needs of a single industrial complex or neighborhood, which bypasses the older model of centralized power. Such systems have the ability to operate on readily available natural gas or renewable biogas. (Source: fuelcellenergy.com)

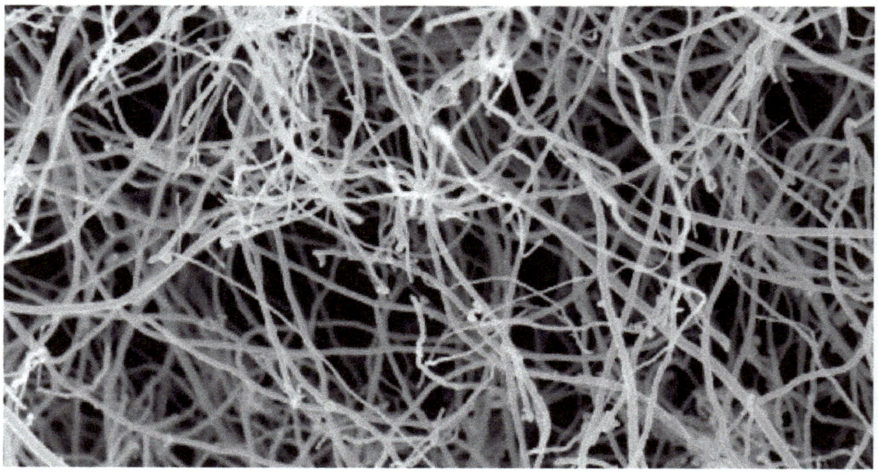

Figure 11.14 >

Carbon nanofibers consist of near-submicroscopic tubes of carbon atoms. They outclass almost all other known materials in their ability to absorb hydrogen molecules.

Alternatively, liquid hydrocarbon such as methanol, CH_3OH, can also be used for fuel cells. With a liquid fuel, miniature cells can be built to power portable electronic devices such as smartphones and tablets. Such devices could operate for long periods of time on a single ampoule of liquid methanol available at a local supermarket. The fuel cell industry holds much promise. Stay tuned for further developments.

CONCEPT CHECK

As long as fuel is available to it, a given fuel cell can supply electrical energy indefinitely. Why can't dry-cell or alkaline batteries do the same?

CHECK YOUR ANSWER Dry-cell and alkaline batteries generate electricity as the chemical reactants they contain are reduced and oxidized. Once these reactants are consumed, the battery can no longer generate electricity.

11.5 Energy from Photovoltaics

Photovoltaic cells are the most direct way of converting sunlight to electrical energy. Since their invention in the 1950s, photovoltaics have made remarkable strides. The first major application was in the 1960s with the U.S. space program—space satellites carried photovoltaic cells to power radios and other small electronic devices. Photovoltaics gained further momentum during the energy crisis of the mid-1970s.

The cost of receiving electricity through photovoltaics is primarily a function of the cost of purchasing the equipment and having it installed. But the system pays for itself in savings from utility bills within about a decade, or earlier in sunny climates. Also, payback time will become quicker as the cost of purchasing electricity from fossil fuel-based utility companies continues to rise. Further, keep in mind that the price of photovoltaic systems continues to drop as photovoltaic cell technology improves. Beyond economic reasons, many people are switching to photovoltaic systems simply because it is a clean source of energy and good for the environment. Today's average residential photovoltaic system has an estimated life span of 25 years.

Worldwide, sales of photovoltaics have grown exponentially over the last several decades. Photovoltaics are now a multi-billion-dollar

< Figure 11.15

Photovoltaic cells come in many sizes, from those in a handheld devices to those in a rooftop unit that provides electricity to a house. Each photovoltaic, however has a limited life span. To be considered is the environmental impact of photovoltaic e-waste from millions of future obsolete installments.

industry with a strong promise of continued growth. Because they need minimal maintenance and no water, they are well suited to remote or arid regions. Photovoltaics can operate on any scale and are now cost-competitive with electricity from fossil and nuclear fuels (**Figure 11.15**).

Conventional photovoltaic cells are made from thin slabs of ultrapure silicon. The four valence electrons in a silicon atom allow four single bonds to four silicon atoms, as shown in **Figure 11.16a.** This configuration can be changed by incorporating trace amounts of other elements that have either more or fewer than four valence electrons. Arsenic atoms, for example, have five valence electrons each, also shown in Figure 11.16a. Within the silicon lattice, four of these arsenic electrons participate in bonding with four silicon atoms, but the fifth electron remains free. This is *n-type silicon*, so called because of the presence of free negative charges (electrons) brought in by the arsenic.

Boron atoms, on the other hand, have only three valence electrons, represented in **Figure 11.16b**. Incorporating boron into a silicon crystal lattice creates "electron holes," which are bonding sites where electrons should be but are not. This is *p-type silicon*, so called because any passing electron is attracted to this hole as though the hole were a positive charge.

What happens when a slice of *n*-type silicon is pressed against a slice of *p*-type silicon? Remember that the *n*-type contains free electrons and the *p*-type contains electron holes just waiting to attract any available electrons.

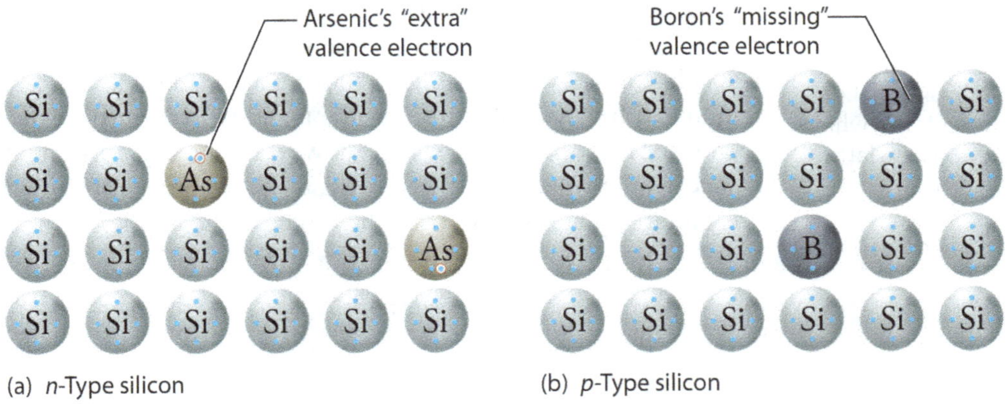

∧ Figure 11.16

(a) The four valence electrons in a silicon atom can form four bonds. The fifth valence electron of arsenic is unable to participate in bonding in the silicon lattice and so remains free. Silicon that contains trace amounts of arsenic (or any other element whose atoms have five valence electrons) is called *n-type silicon*. (b) Boron has only three valence electrons for bonding with four silicon atoms. One boron–silicon pair therefore lacks an electron for covalent bonding. Silicon containing trace amounts of boron (or any other element whose atoms have three valence electrons) is called *p-type silicon*.

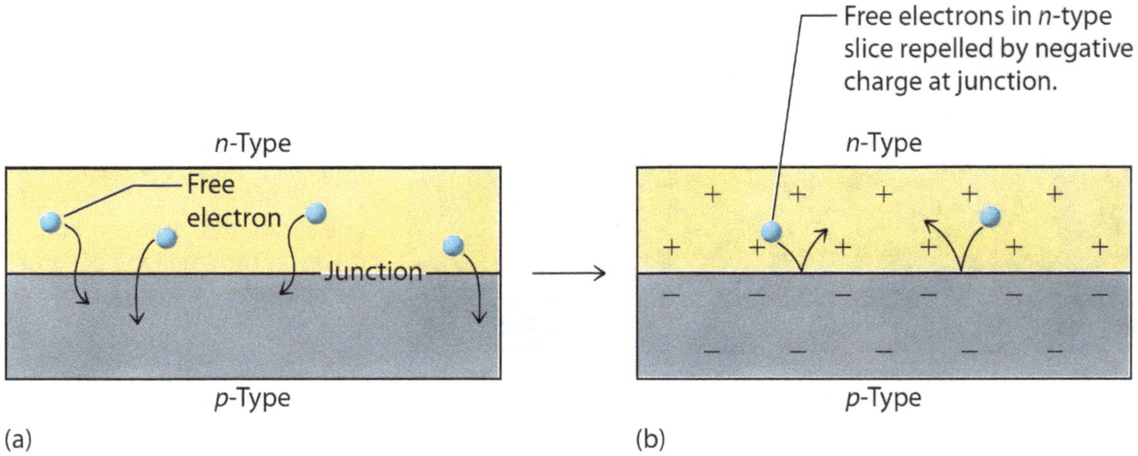

Λ Figure 11.17

(a) Initially, free electrons migrate from the *n*-type silicon to the *p*-type silicon. (b) Quickly, however, charge buildup at the junction prevents the continued flow of electrons.

READING CHECK

What happens to *n*-type silicon as it loses electrons to *p*-type silicon?

Sure enough, electrons migrate across the junction from the *n*-type to the *p*-type, as shown in **Figure 11.17a**. This occurs only up to a certain point, however, because losing or gaining electrons upsets the electron-to-proton balance. As the *n*-type silicon loses electrons, it develops a positive charge on its side of the junction. As the p-type silicon gains electrons, it develops a negative charge. This charge buildup at the junction serves as a barrier to continued electron migration, as **Figure 11.17b** illustrates.

Photovoltaic cells rely on the **photoelectric effect**, which is the ability of light to knock electrons away from the atoms in an object. In most materials, the electrons either are completely ejected from the object or simply fall back to their original positions. In a select number of materials, however, including silicon, dislodged electrons roam about randomly in and among neighboring atoms without being locked into any one place, as represented in **Figure 11.18**. However, random electron motion is not an electric current, because for every one electron moving to the left, there is another canceling that motion by moving to the right. Greater random motion simply means higher temperatures. This is why solar energy shining onto a piece of silicon produces nothing more than heat.

The junction barrier between adjacent slices of *n*-type and *p*-type silicon acts like a one-way valve. Electrons knocked loose in the *n*-type slice are inhibited from migrating across the junction to the *p*-type slice. Electrons knocked loose in the *p*-type slice, however, are readily drawn into the *n*-type slice, as shown in **Figure 11.19**. So, the flow of electrons across the barrier is unidirectional, always from the *p*-type to the *n*-type.

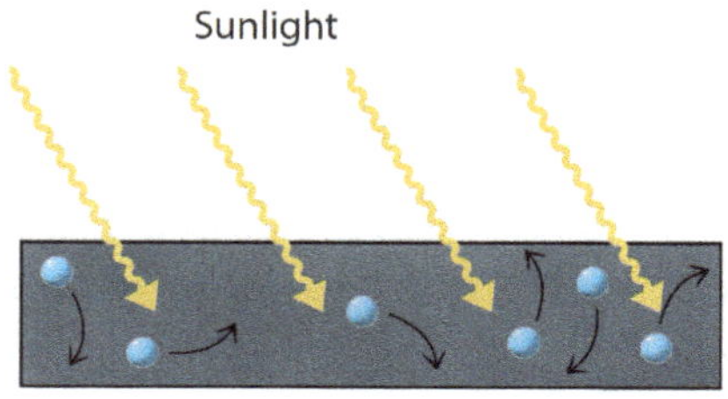

Λ Figure 11.18

The photoelectric effect in silicon. Light rays knock out bonding electrons, which are free to travel about the crystal lattice.

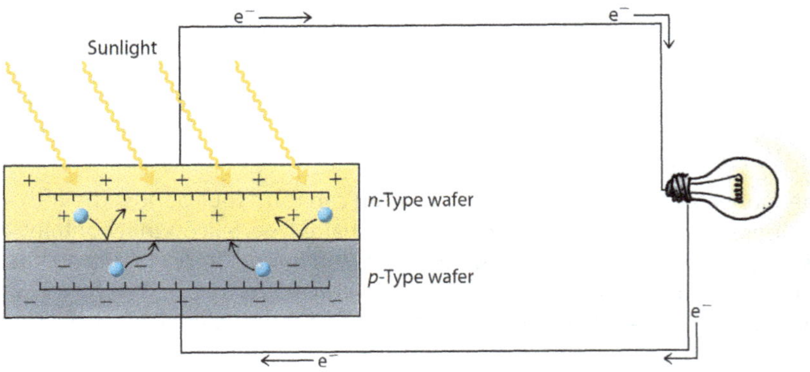

Figure 11.19 >

Free electrons in the *n*-type wafer are repelled by the negative charge of the *p*-type wafer. Free electrons in the p-type wafer, however, are attracted to the positive charge of the *n*-type wafer. The energy of sunlight gets these electrons moving to the point that they are pushed into an external circuit in a unidirectional fashion.

As Figure 11.19 shows, a complete electric circuit can be made by connecting a wire to the outside faces of the two slices of silicon. When light hits either slice, dislodged electrons are forced by the junction barrier to move in one direction: from the n-type slice through the wire. This, in turn, creates holes in the *p*-type slice which accepts electrons returning from the external circuit. Thus, rather than being transformed completely to heat, energy from sunlight is transformed to electricity.

CONCEPT **CHECK**

Why is the unidirectional nature of the *n*-type/*p*-type silicon junction so important?

CHECK YOUR ANSWER The unidirectional nature of this junction forces electrons knocked loose by the sunlight to flow in a specific direction, which is an electric current. Otherwise, the silicon would merely get warmer.

The goal of the photovoltaic industry is to create a cell that is highly efficient, inexpensive to manufacture, and easily mass-produced. Traditional photovoltaic cells manufactured from ultrapure crystalline silicon yield efficiencies as high as 15 percent, which is good. However, the high cost of producing ultrapure crystalline silicon prevents these cells from being cost-competitive. Although great strides have been made in the past 20 years, electricity from these cells still remains relatively expensive.

A promising area of photovoltaic research involves the second- and third-generation photovoltaic cells. These cells are not produced from expensive crystalline silicon. Rather, they are formed as vaporized silicon or some other photovoltaic material is deposited on a glass or plastic substrate. The resulting films are about 400 times thinner than traditional silicon wafers, which saves in material costs. Furthermore, the films are easy to mass-produce. The silicon "first-generation" cells, however, are well entrenched in the market, which means that it will still be a few years until these updated cells become commonly used. By that time, however, all these cells may be supplanted by even more efficient fourth-generation cells now being developed using nanotechnology.

Solar Fuels

Photovoltaic cells can be designed to produce chemical changes, such as the conversion of water into hydrogen, H_2, and oxygen, O_2. The hydrogen from this conversion can aptly be called a *solar fuel,* because it arises directly from solar energy. Any fuel is a form of chemical energy that can be stored for later use or transported for use elsewhere. In many ways, creating a solar fuel mimics the process of *photosynthesis* where a leaf uses sunlight to convert carbon dioxide and water into glucose and oxygen (See Chapter 13). A solar fuel cell is thus sometimes referred to as "artificial photosynthesis".

To be able to efficiently generate a fuel from sunlight is a major, if not ultimate, goal of present energy research. The technical challenges, however, are daunting. Consider that photosynthesis itself, which is a product of billions of years of evolution, has an energy efficiency of only about 4 percent. To be practical, a solar fuel cell would need a much higher level of efficiency. Unfortunately, the only known materials able to achieve this are very expensive. Most notably, this includes the element platinum. Another

Figure 11.20

Can photovoltaics be developed to create not only electricity, but fuels? If so, this would mimic the photosynthesis occurring within green leaves. This would allow us to leap frog over the burning of forests and fossil fuels for our energy needs.

Figure 11.21

The authors of Conceptual Chemistry with Professor Raj Rajeshwar (far left), a lead researcher in the area of solar fuels at the University of Texas, Arlington. This photo was taken after their podcast interview concerning recent solar fuel developments.

issue is that the materials of a solar fuel cell need to be durable, as well as non-toxic, without corroding over an extended period of time. Again, platinum fits this bill. But again, there is not much platinum on the planet, and all of it is very expensive.

At this time, researchers are focusing on various chemical element combinations. In a nutshell, the research involves experimenting with a bit of this element, bonded to a bit of that element in a very specific ratio, and then compounded with yet another element, and so on. Because there are only about 100 elements in the periodic table, the choice of elements is limited. But the various combinations is practically limitless. Progress is being made, but there is still much to learn in this important and exciting area of chemistry. Stay tuned. Efficient, non-toxic, and durable solar fuel cells would be revolutionary.

11.6 Electrolysis Produces Change

READING CHECK
What is electrolysis?

Electrolysis is the use of electrical energy to produce chemical change. The recharging of a car battery is an example of electrolysis. Another, shown in **Figure 11.22**, is passing an electric current through water, a process that breaks down the water into its elemental components:

$$\text{electrical energy} + 2\,H_2O\,(l) \rightarrow 2\,H_2\,(g) + O_2\,(g)$$

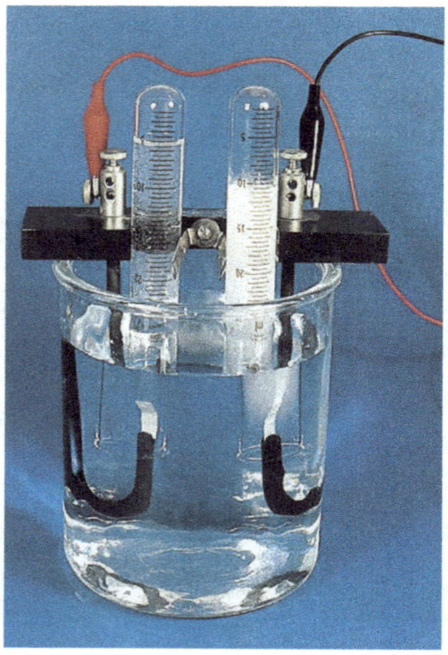

Figure 11.22

The electrolysis of water produces hydrogen gas and oxygen gas in a 2:1 ratio by volume, in accord with the chemical formula for water: H_2O. In order for this process to work, ions must be dissolved in the water so that electric charge can be conducted between the electrodes.

Electrolysis is used to purify metals from metal ores. An example is aluminum, the third most abundant element in the Earth's crust. Aluminum occurs naturally bonded to oxygen in an ore called *bauxite*. Aluminum metal wasn't known until about 1827, when it was prepared by reacting bauxite with hydrochloric acid. This reaction gave the aluminum ion, Al^{3+}, which was reduced to aluminum metal, with sodium metal acting as the reducing agent:

$$Al^{3+} + 3\,Na \rightarrow Al + 3\,Na^+$$

This chemical process was expensive, which is why the initial price of aluminum was well beyond the means of the average person. Aluminum was thus considered a rare and precious metal. In 1855, aluminum dinnerware and other items were exhibited in Paris with the crown jewels of France. Then, in 1886, two men working independently, Charles Hall (1863–1914) in the United States and Paul Heroult (1863–1914) in France, almost simultaneously discovered a process whereby aluminum could be produced from aluminum oxide, Al_2O_3, a main component of bauxite. In what is now known as the Hall–Heroult process, shown in **Figure 11.23**, a strong electric current is passed through a molten mixture of aluminum oxide and cryolite, Na_3AlF_6, a naturally occurring mineral. The fluoride ions of the cryolite react with the aluminum oxide to form various aluminum fluoride ions, such as $AlOF_3^{2-}$, which are then oxidized to the aluminum hexafluoride ion, AlF_6^{3-}. The Al^{3+} in this ion is then reduced to elemental aluminum, which collects at the bottom of the reaction chamber. This process, which is still in use by

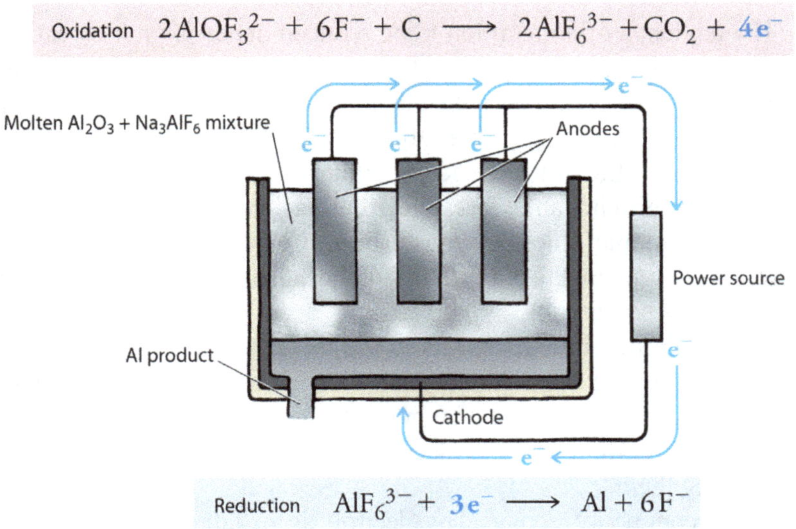

< Figure 11.23

The melting point of aluminum oxide (2030°C) is too high for efficiently electrolyzing it to aluminum metal. Aluminum oxide, however, dissolves in molten cryolite at a more reasonable 980°C. A strong electric current passed through the molten aluminum oxide–cryolite mixture generates aluminum metal at the cathode, where aluminum ions pick up electrons and are thus reduced to elemental aluminum.

manufacturers today, greatly facilitated mass production of aluminum metal, and by 1890, the price of aluminum had dropped to about \$2 per pound.

Today, worldwide production of aluminum is over 64 million metric tons annually. For each ton produced from ore, about 16,000 kilowatt-hours of electrical energy is required, as much as a typical American household consumes in 18 months. Processing recycled aluminum, on the other hand, consumes only about 700 kilowatt-hours for every ton. Thus, recycling aluminum not only reduces litter but also helps to reduce the load on power companies, which in turn reduces air pollution. Furthermore, reserves of high-quality aluminum oxide ores are already depleted in the United States. Recycling aluminum, therefore, also helps to minimize the need to develop new bauxite mines in foreign countries.

FOR YOUR INFORMATION

Charles Hall and Paul Heroult not only invented the cryolite process at about the same time, but they were born and died in the same years. Hall lived 51 years and 21 days. Heroult lived 51 years and 29 days. Both became rich and famous for their inventions and held no grudges against the other. Hall won the U.S. patent for the process in great part because of the careful records kept by his sister, Julia Brainerd Hall, who assisted him in his work. To be considered, if none of these individuals had been born, how much longer would it have taken for the cryolite method to be discovered? In the race to be the first, there is usually only one winner.

CONCEPT CHECK

Is the exothermic reaction in a hydrogen–oxygen fuel cell an example of electrolysis?

CHECK YOUR ANSWER No. During electrolysis, electrical energy is used to produce chemical change. In the hydrogen–oxygen fuel cell, chemical change is used to produce electrical energy.

11.7 Producing Metals

To convert a metal-containing compound to a metal requires an oxidation-reduction reaction. In the metal-containing compound, the metal exists as a positively charged ion, because it has lost one or more of its electrons to its bonding partner. To convert metal ions to neutral metal atoms requires that they gain electrons; that is, they must be reduced:

M^+	+	e^-	→	M^0
Metal ion		Electron		Metal atom

The tendency of a metal ion to be reduced depends on its location in the periodic table, as summarized in **Figure 11.24**. As discussed in Chapter 6, metals on the left in the periodic table readily lose electrons. This means it is relatively difficult to give electrons back to these metal ions—in other words, they are difficult to reduce. A sodium atom, for example, being on the left in the periodic table, loses electrons easily. Any ionic compound it forms, such as sodium chloride, tends to be very stable. Reducing the sodium ion, Na+, to sodium metal, Na^0, is difficult because doing so requires giving electrons to the sodium ion.

Metals on the left and especially the lower left of the periodic table therefore require the most energy-intensive methods of recovery, which include electrolysis. As was shown in the previous section, during electrolysis, an electric current supplies electrons to positively charged metal ions, thus reducing them.

Metals commonly recovered by electrolysis include the metals of groups 1–3, which occur most frequently as halides, carbonates, and phosphates. In addition, aluminum is commonly recovered by electrolysis, and other metals are obtained using electrolysis when very high purity is needed. The reactions involved when copper is produced this way are shown in **Figure 11.25**.

READING CHECK

Which metals are most difficult to reduce?

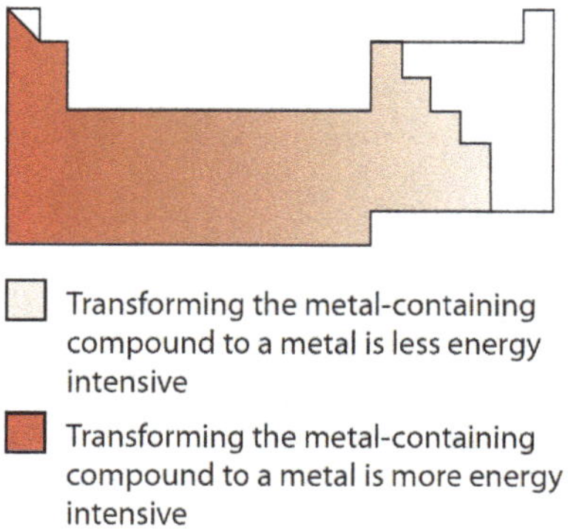

∧ Figure 11.24

The ions of metallic elements at the lower left of the periodic table are most difficult to reduce. For this reason, obtaining these elements from the metal-containing compounds they form is energy intensive. Metallic elements at the upper right of the periodic table tend to form compounds that require less energy to convert to metals.

CONCEPT CHECK

Why is it so difficult to obtain a group 1 metal from a compound containing ions of that metal?

CHECK YOUR ANSWER The metal ions do not readily accept electrons to form metal atoms.

Some Metals are Commonly Obtained from Metal Oxides

Ores containing metal oxides can be converted to metals fairly efficiently in a *blast furnace*. First, the ore is mixed with limestone and coke, which is a concentrated form of carbon obtained from coal. Then the mixture is dropped into the furnace, where the coke is ignited and used as a fuel. At

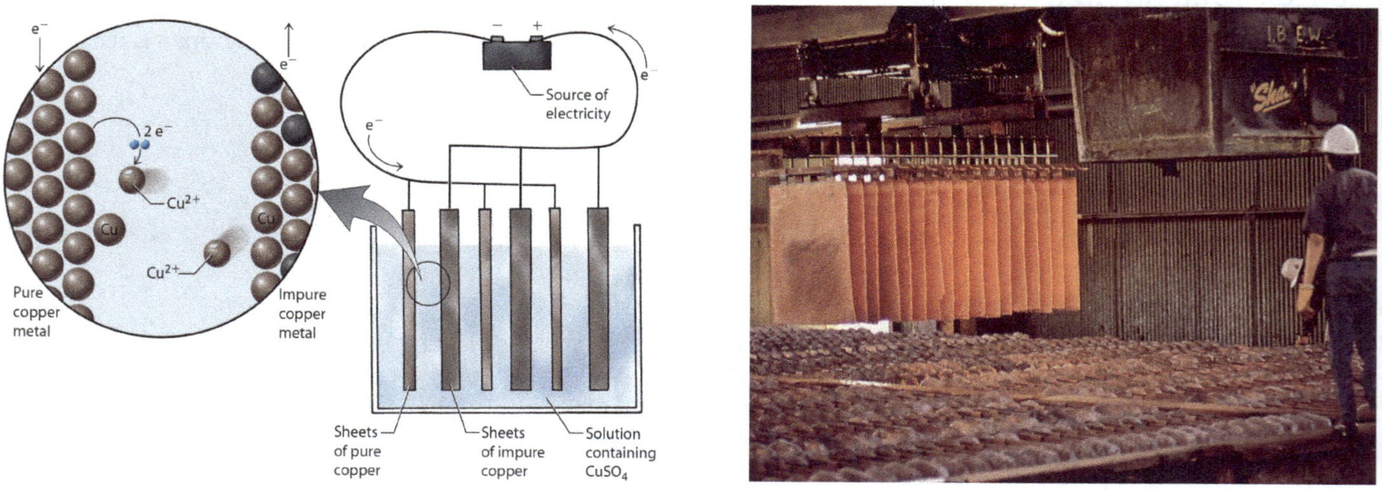

∧ Figure 11.25

High-purity copper is recovered by electrolysis. Pure copper metal deposits on the negative electrode as copper ions in solution gain electrons. The source of these copper ions is a positively charged electrode made of impure copper.

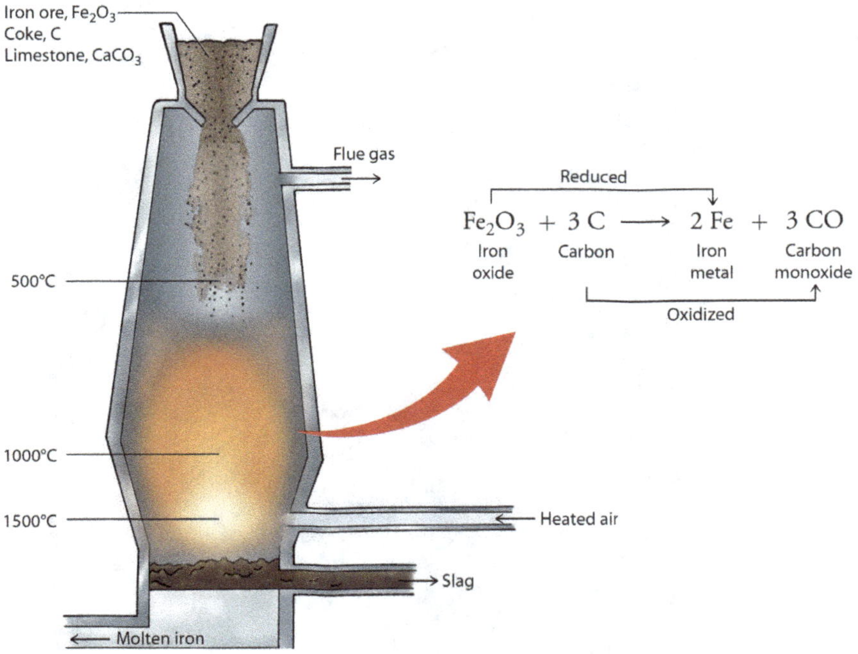

< Figure 11.26

A mixture of iron oxide ore, coke, and limestone is dropped into a blast furnace, where the iron ions in the oxide are reduced to metal atoms. Note how the carbon is oxidized as it gains an oxygen atom to form carbon monoxide.

high temperatures, the coke also behaves as a reducing agent, yielding electrons to the positively charged metal ions in the oxide and reducing them to metal atoms. **Figure 11.26** shows this method being used to produce iron, which is found in nature primarily as iron oxide.

In the furnace, the limestone reacts with ore impurities—predominantly silicon compounds—to form *slag*, which is primarily calcium silicate:

$SiO_2(s)$	+	$CaCO_3(s)$	→	$CaSiO_3(l)$	+	$CO_2(g)$
Silica sand (ore impurity)		Limestone		Molten slag (calcium silicate)		Carbon dioxide

Because of the high temperatures, both the metal and the slag are molten. They drain to the bottom of the blast furnace, where they collect in two layers, the less dense slag on top. The metal layer is then tapped through an opening at the bottom of the blast furnace.

Once cooled, the metal from a blast furnace is known as a *cast metal.* (Side note: With iron ore, the cast metal is known as *pig iron*.) A cast metal is brittle and relatively soft for a metal because it still contains impurities, such as phosphorus, sulfur, and carbon. To remove these impurities, oxygen is blown through the molten cast metal in a *basic oxygen furnace*, shown in **Figure 11.27**. The oxygen oxidizes the impurities to form additional slag, which floats to the surface and is skimmed off.

Most phosphorus and sulfur impurities are removed in a basic oxygen furnace, but the purified metal still contains about 3 percent carbon. For the production of iron, this carbon is desirable. Iron atoms are relatively large, and when they pack together, small voids are created between atoms, as shown in **Figure 11.28**. These voids tend to weaken the iron. Carbon atoms are small enough to fill the voids, and having the voids filled strengthens the iron substantially. Iron strengthened by small percentages of carbon is called **steel**. The tendency of steel to rust can be inhibited by alloying the steel with noncorroding metals, such as chromium or nickel. This yields the stainless steel used to manufacture eating utensils and countless other items.

fyi FOR YOUR INFORMATION

What is the difference between a ton and a metric ton? A ton is the United States Customary Standard unit for 2000 pounds. A metric ton is 1000 kilograms, which is equal to 2205 pounds. Which might you rather have: a ton of gold or a metric ton of gold?

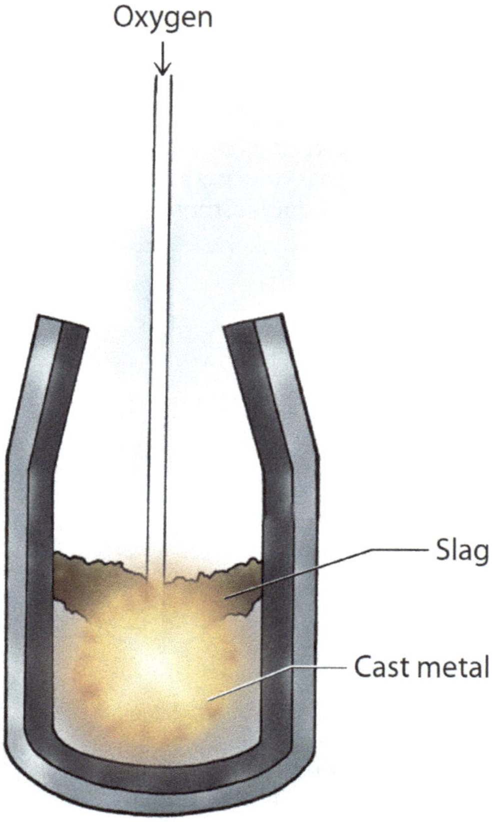

(a)

(b)

∧ Figure 11.27

(a) A flow of oxygen through a basic oxygen furnace oxidizes most of the impurities in a cast metal to form additional slag that may be skimmed away as it floats to the surface. (b) The basic oxygen furnace is hoisted and its purified contents can be poured into a reservoir used for molding iron pieces.

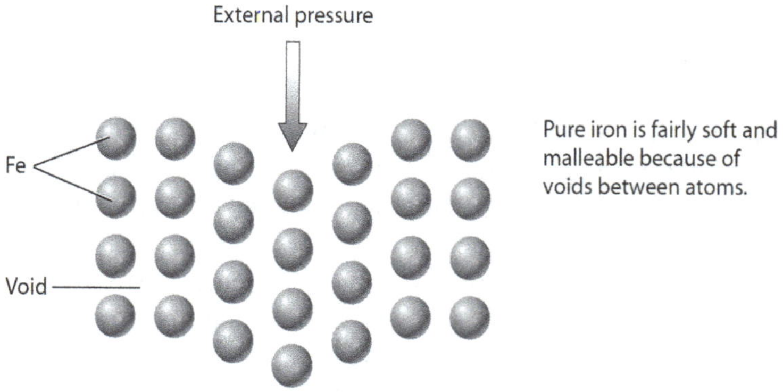

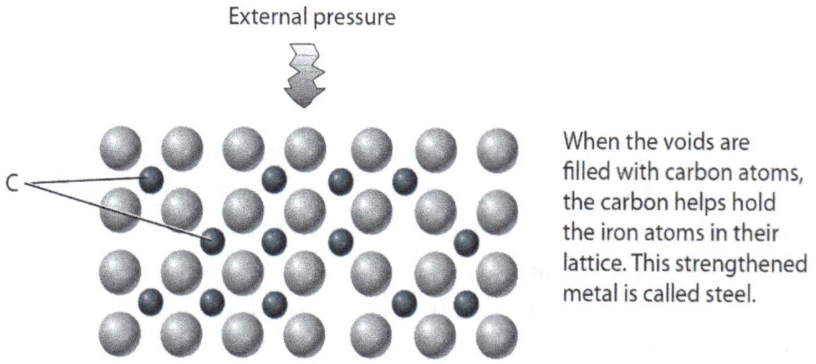

Figure 11.28 >

Steel is stronger than iron because of the small amounts of carbon it contains.

Other Metals are Commonly Obtained from Metal Sulfides

Metal sulfides can be purified by *flotation*, a technique that takes advantage of the fact that metal sulfides are relatively nonpolar and therefore attracted to oil. An ore containing a metal sulfide is first ground to a fine powder and then mixed vigorously with a lightweight oil and water. Compressed air is then forced up through the mixture. As air bubbles rise, they become coated with oil and metal sulfide particles. At the surface of the liquid, the coated bubbles form a floating froth, as shown in **Figure 11.29**. This froth, which is very rich in the metal sulfide, is then skimmed off.

The metal sulfides recovered from the froth are then roasted in the presence of oxygen. The net reaction is the oxidation of S^{2-} in the sulfide to S^{4+} in sulfur dioxide and the reduction of the metal ion to its elemental state:

∧ Figure 11.29

Air bubbles rising through a flotation container transport metal sulfide particles to the surface.

$$\underset{\text{Metal sulfide}}{MS(s)} + \underset{\text{Oxygen}}{O_2(g)} \rightarrow \underset{\text{Metal}}{M(l)} + \underset{\text{Sulfur dioxide}}{SO_2(g)}$$

The isolation of copper from its most common ore, chalcopyrite, $CuFeS_2$, requires several additional steps because of the presence of iron. First, the chalcopyrite is roasted in the presence of oxygen:

$$\underset{\text{Chalcopyrite}}{2\,CuFeS_2} + \underset{\text{Oxygen}}{O_2(g)} \rightarrow \underset{\text{Copper sulfide}}{CuS(s)} + \underset{\text{Iron oxide}}{2\,FeO(s)} + \underset{\text{Sulfur dioxide}}{2\,SO_2(g)}$$

The copper sulfide and iron oxide from this reaction are then mixed with limestone, $CaCO_3$, and sand, SiO_2, in a blast furnace, where CuS is converted to Cu_2S. The limestone and sand form molten slag, $CaSiO_3$, in which the iron oxide dissolves. The copper sulfide melts and sinks to the bottom of the furnace. The less dense iron-containing slag floats above the molten copper sulfide and is drained off. The isolated copper sulfide is then roasted to copper metal:

$$\underset{\text{Copper sulfide}}{Cu_2S(s)} + \underset{\text{Oxygen}}{O_2(g)} \rightarrow \underset{\text{Copper metal}}{2\,Cu(l)} \rightarrow \underset{\text{Sulfur dioxide}}{SO_2(g)}$$

Roasting metal sulfides requires a fair amount of energy. Furthermore, sulfur dioxide is a toxic gas that contributes to acid rain, so its emission must be minimized. Most companies comply with EPA emissions standards by converting the sulfur dioxide to marketable sulfuric acid, H_2SO_4.

CONCEPT **CHECK**

Of metal chlorides, metal oxides, and metal sulfides, which are most polar? Which are least polar?

CHECK YOUR ANSWER Metal chlorides, such as sodium chloride, tend to be most polar, while the metal sulfides, such as copper sulfide, tend to be least polar. The oxides, such as iron oxide, tend to be in between. This is an application of periodic trends as discussed in Chapter 6.

11.8 Corrosion and Combustion

If you look to the upper right of the periodic table, you will find one of the most common oxidizing agents—oxygen. In fact, if you haven't guessed already, the term *oxidation* is derived from the name of this element. Oxygen is able to pluck electrons from many other elements, especially those that lie at the lower left of the periodic table. Two common oxidation-reduction reactions involving oxygen as the oxidizing agent are *corrosion* and *combustion*.

READING CHECK

What is oxygen so good at doing?

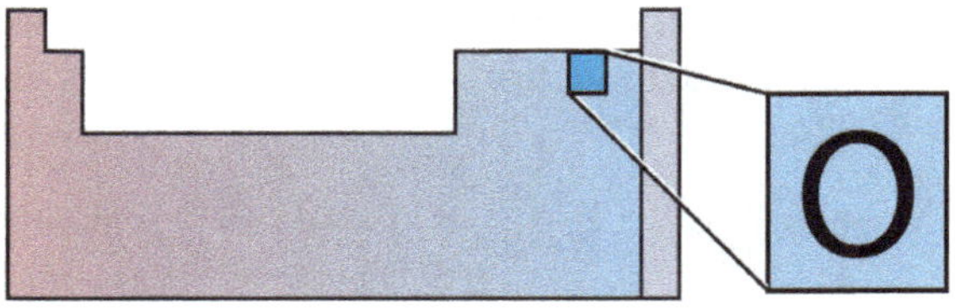

CONCEPT **CHECK**

Oxygen is a good oxidizing agent, but so is chlorine. What does this indicate about their relative positions in the periodic table?

CHECK YOUR ANSWER Chlorine and oxygen must lie in the same area of the periodic table. Both have strong effective nuclear charges and are strong oxidizing agents.

∧ Figure 11.30

Rust itself is not harmful to the iron structures on which it forms. It is the loss of metallic iron that ruins the structural integrity of these objects.

Corrosion

Corrosion is the process whereby a metal deteriorates. Corrosion caused by atmospheric oxygen is a widespread and costly problem. About one-quarter of the steel produced in the United States, for example, goes into replacing corroded iron at a cost of billions of dollars annually. Iron corrodes when it reacts with atmospheric oxygen and water to form hydrated iron oxide, which is the naturally occurring reddish-brown substance you know as rust, shown in **Figure 11.30**.

$$\underset{\text{Iron}}{4\,Fe} + \underset{\text{Oxygen}}{3\,O_2} + \underset{\text{Water}}{3\,H_2O} \rightarrow \underset{\text{Rust}}{2\,Fe_2O_3 \cdot 3\,H_2O}$$

∧ Figure 11.31

Galvanized nails are protected from rusting by the sacrificial oxidation of zinc.

Another common metal oxidized by oxygen is aluminum. The product of aluminum oxidation is aluminum oxide, Al_2O_3, which does not flake away from the aluminum. Rather it forms a protective coat that shields the metal from further oxidation. This coat is so thin that it's transparent, which is why aluminum maintains its metallic shine.

A protective, oxidized coat is the principle underlying a process called *galvanization*. Zinc has a slightly greater tendency to oxidize than does iron. For this reason, many iron objects, such as the nails pictured in **Figure 11.31**, are *galvanized* by coating them with a thin layer of zinc. The zinc oxidizes to zinc oxide, an inert, insoluble substance that protects the iron underneath it from rusting.

In a technique called *cathodic protection*, iron structures can be protected from oxidation by placing them in contact with certain metals, such as zinc or magnesium, that have a greater tendency to oxidize. This forces the iron to accept electrons, which means that it is behaving as a cathode. (Rusting occurs only where iron behaves as an anode.) Ocean tankers, for example, are protected from corrosion by strips of zinc affixed to their hulls, as shown in **Figure 11.32**. Similarly, outdoor steel pipes are protected by being connected to magnesium rods inserted into the ground.

Yet another way to protect iron and other metals from oxidation is to coat them with a corrosion-resistant metal, such as chromium, platinum, or gold. *Electroplating* is the operation of coating one metal with another by electrolysis, and it is illustrated in **Figure 11.33**. The object to be electroplated is connected to a negative battery terminal and then submerged in a solution containing ions of the metal to be used as the coating. The positive terminal of the battery is connected to an electrode made of the coating metal. The circuit is completed when this electrode is submerged in the solution. Dissolved metal ions are attracted to the negatively charged object, where they pick up electrons and are deposited as metal atoms. The ions in solution are replenished by the forced oxidation of the coating metal at the positive electrode.

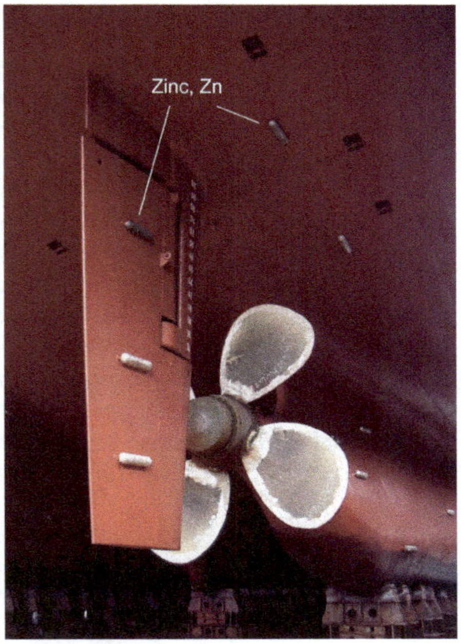

Figure 11.32

Zinc strips help to protect the iron hull of an oil tanker from oxidizing. The zinc strip shown here is attached to the hull's exterior surface.

Combustion

Combustion is a rapid oxidation-reduction reaction between a material and molecular oxygen. The burning of a campfire is a good example. Combustion reactions are characteristically exothermic (energy-releasing). A violent combustion reaction is the formation of water from hydrogen and oxygen. As discussed in Section 9.4, the energy from this reaction is used to power rockets into space. More common examples of combustion include the burning of wood and fossil fuels. The combustion of these and other carbon-based chemicals forms carbon dioxide and water. Consider, for example, the combustion of methane, the major component of natural gas:

$$\underset{\text{Methane}}{CH_4} + \underset{\text{Oxygen}}{2\,O_2} \rightarrow \underset{\text{Carbon dioxide}}{CO_2} + \underset{\text{Water}}{2\,H_2O} + \text{energy}$$

In combustion, electrons are transferred as polar covalent bonds are formed in place of nonpolar covalent bonds or vice versa. (This is in contrast with the other examples of oxidation-reduction reactions presented in this chapter, which involve the formation of ions from atoms or, conversely, atoms from ions.) This concept is illustrated in **Figure 11.34**, which compares the electronic structures of the combustion starting material, molecular oxygen, and the combustion product, water. Molecular oxygen is a nonpolar covalent compound. Although each oxygen atom in the molecule has fairly strong electronegativity, the four bonding electrons are pulled equally by both atoms and thus are unable to congregate on one side or the other. After combustion, however, the electrons are shared between the oxygen and hydrogen atoms in a water molecule and are pulled to the oxygen. This gives the oxygen a slight negative charge, which is another way of saying it has gained electrons and has thus been reduced. At the same time, the hydrogen atoms in the water molecule develop a slight positive charge, which is another way of saying they have lost electrons and have thus been oxidized. This gain of electrons by oxygen and loss of electrons by hydrogen

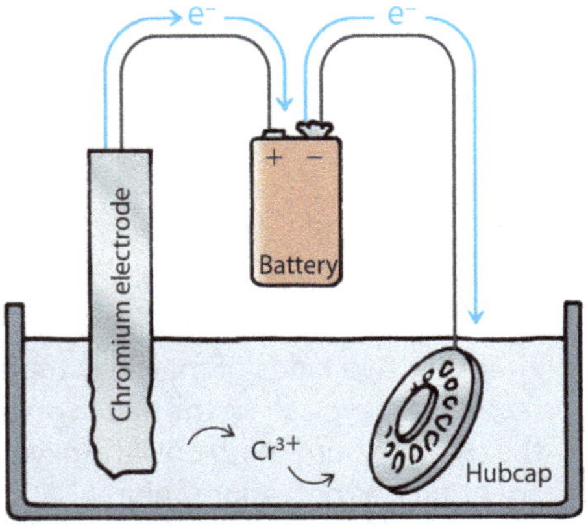

Figure 11.33

As electrons flow into the hubcap and give it a negative charge, positively charged chromium ions move from the solution to the hubcap and are reduced to chromium metal, which is deposited as a coating on the hubcap. The solution is supplied with ions as chromium atoms in the cathode are oxidized to Cr^{3+} ions.

Figure 11.34 >

(a) Neither atom in an oxygen molecule is able to preferentially attract the bonding electrons. (b) The oxygen atom of a water molecule pulls the bonding electrons away from the hydrogen atoms on the water molecule, making the oxygen slightly negative and the two hydrogens slightly positive. Consider: What then is the driving force of this reaction? Answer: Oxygen's "desire" to gain electrons (electronegativity). While an oxygen can't gain electrons from another oxygen atom, it can by attaching itself instead to a hydrogen atom.

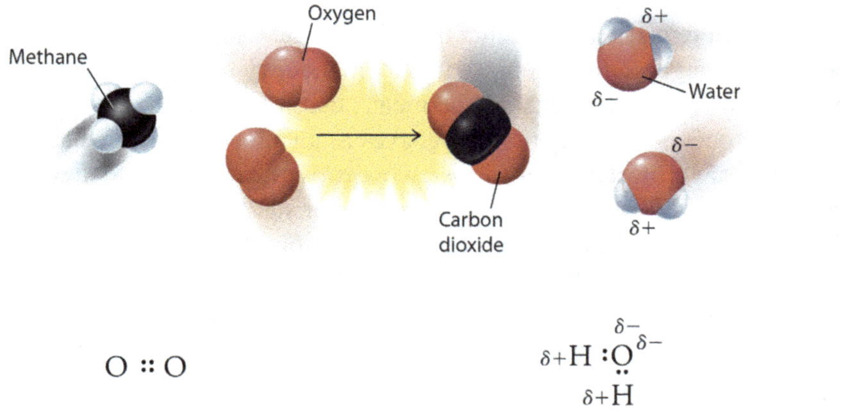

(a) Reactant oxygen atoms share electrons equally in O_2 molecules.

(b) Product oxygen atoms pull electrons away from H atoms in H_2O molecules and are reduced.

is an energy-releasing process. Typically, the energy is released either as molecular kinetic energy (heat) or as light (the flame).

Interestingly, oxidation-reduction reactions chemically related to combustion occur throughout your body. You can visualize a simplified model of your metabolism by reviewing Figure 11.34 and substituting a food molecule for the methane. Food molecules relinquish their electrons to the oxygen molecules you inhale. The products are carbon dioxide, water vapor, and energy. You exhale the carbon dioxide and water vapor, but much of the energy from the reaction is used to keep your body warm, and the rest is used to drive the many other biochemical reactions necessary for life.

CHEMICAL CONNECTIONS

How is your breath connected to a campfire?

FOR YOUR INFORMATION

There are two kinds of matches: the strike-anywhere type, usually having a "bull's-eye" tip, and the safety match, which requires that you strike the match on a provided strip. Both involve the burning of sulfur within the tip of the match. Getting the sulfur to burn using only the oxygen in the air, however, is difficult, which is why the sulfur is blended with an oxidizing agent such as potassium chlorate, $KClO_3$. for the strike-anywhere match, a third ingredient, red phosphorus, P_4, is included. The heat of friction causes the red phosphorus to convert into white phosphorus—an alternative form of phosphorus that burns rapidly in air. This initiates the redox reaction between the sulfur and the potassium chlorate, which in turn ignites the burning of the matchstick. Safety matches work the same way, except for the fact that the red phosphorus is embedded within the striking strip, which is the only place the match can be lit.

Concept Review

Chapter 11

Summary of Terms

Anode The electrode where chemicals are oxidized.

Cathode The electrode where chemicals are reduced.

Combustion The rapid exothermic oxidation–reduction reaction between a material and molecular oxygen.

Corrosion The deterioration of a metal, typically caused by atmospheric oxygen.

Electrochemistry The study of the relationship between electrical energy and chemical change.

Electrode Any material that conducts electrons into or out of a medium in which electrochemical reactions are occurring.

Electrolysis The use of electrical energy to produce chemical change.

Half reaction One half of an oxidation–reduction reaction, represented by an equation showing electrons as either reactants or products.

Oxidation The process whereby a reactant loses one or more electrons.

Photoelectric Effect The ability of light to knock electrons away from atoms.

Reduction The process whereby a reactant gains one or more electrons.

Steel Iron strengthened by small percentages of carbon.

Review Questions

11.1 Losing and Gaining Electrons

1. Which elements have the greatest tendency to behave as oxidizing agents?

2. What elements have the greatest tendency to behave as reducing agents?

3. Write an equation for the half reaction in which a potassium atom, K, is oxidized.

4. What happens to a reducing agent as it reduces?

11.2 Harnessing the Energy

5. What is electrochemistry?

6. What is the purpose of the salt bridge in a voltaic cell?

11.3 Electricity from Batteries

7. What type of reaction occurs at the cathode?

8. What type of reaction occurs at the anode?

9. Does lithium metal accept or donate electrons in a lithium battery?

11.4 Electricity from Fuel Cells

10. What is the prime difference between a battery and a fuel cell?

11. What else do fuel cells produce besides electricity?

11.5 Energy from Photovoltaics

12. How is *p*-type silicon made?

13. What happens when a slice of *n*-type silicon is joined to a slice of *p*-type silicon?

14. Light shining on a slab of pure silicon dislodges electrons from the silicon atoms. What happens to these electrons?

11.6 Electrolysis Produces Change

15. What is electrolysis, and how does it differ from what goes on inside a battery?

16. In what year was the efficient electrolysis of aluminum discovered?

11.7 Producing Metals

17. Which group of metals requires the most energy to be recovered from metal-containing compounds?

18. When iron ions are reduced to neutral iron atoms in a blast furnace, where do the electrons come from?

19. How are copper (II) sulfide, CuS, and iron (III) oxide, Fe_2O_3, separated from each other in a blast furnace in the preparation of copper metal?

11.8 Corrosion and Combustion

20. What metal coats a galvanized nail?

21. What do the oxidation of zinc and the oxidation of aluminum have in common?

22. What are some differences between corrosion and combustion?

23. What happens to the polarity of oxygen atoms as they transform from molecular oxygen, O_2, into water molecules, H_2O?

Quantitative Questions

24. Each year about 1.6×10^7 (16 million) metric tons (mt) of aluminum are produced. How many grams is this? (Recall, 1 mt is 1000 kg.)

25. Use the following balanced chemical equation to show that the production of 1.6×10^7 metric tons of aluminum, Al, through electrolysis each year produces about 2.0×10^7 metric tons of carbon dioxide, CO_2. Interestingly, the total mass of our atmosphere is only about 5×10^{15} metric tons.

$$4\ Al_2O_3 + 3\ C \rightarrow 4\ Al + 3\ CO_2$$

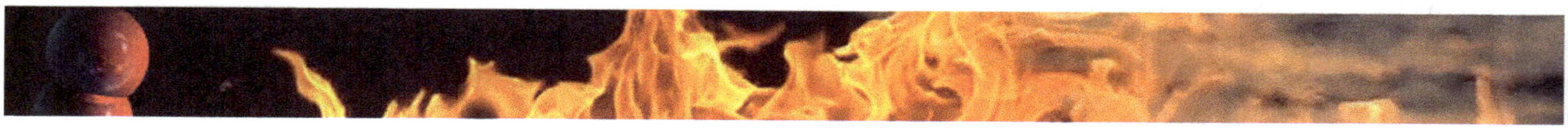

Solutions (Odd-Numbered)

1. The elements in the upper right of the periodic table (except for noble gases) have the greatest tendency to behave as oxidizing agents.

3. K → K+ + 1e-

5. Electrochemistry is the study of the relationship between electric energy and chemical change.

7. Reduction occurs at the cathode. Remember the "red cat."

9. The lithium donates electrons.

11. Fuel cells also produce the chemical products of the electricity generating chemical reactions. A hydrogen fuel cell, for example, produces clean water suitable for drinking.

13. Electrons migrate from *n*-type silicon to *p*-type silicon across the junction when these two slices of silicon are pressed close together.

15. Electrolysis uses electrical energy to produce a chemical change whereas a battery uses a spontaneous chemical change to produce electrical energy. Both involve oxidation-reduction reactions.

17. Metals located toward the lower left corner of the periodic table, which have the greatest tendency to lose electrons, are the most difficult to be recovered from metallic compounds.

19. Copper (II) sulfide and iron (III) oxide are mixed with limestone and sand and heated. The iron (III) oxide dissolves in the $CaSiO_3$ formed and the copper sulfide melts and goes to the bottom of the furnace where it is removed.

21. After oxidation, zinc and aluminum both form water-insoluble oxidized coats, impervious to penetration by air and moisture, that prevent further oxidation.

23. The oxygen atoms become somewhat negatively charged, which is to say they are gaining electron.

25. Use the techniques of Section 9.3 to solve for this problem. First convert metric tons of aluminum to grams of aluminum and then to moles of aluminum. Then follow the coefficients of the balanced equation to calculate the moles of carbon dioxide formed. Lastly, convert from moles of carbon dioxide to grams of carbon dioxide to metric tons of carbon dioxide:

    (1.6×10^{13} mt Al)(1000 kg/1 mt)(1000 g/1 kg)(1 mole Al/27.0 g Al)(3 moles CO_2/4 moles Al)(44 g CO_2/1 mole CO_2)(1 kg/1000 g)(1 mt/1000 kg) $= 2.0 \times 10^{13}$ mt of CO_2.

Chapter 11: Contextual Chemistry Essay

The Wonder Chemical, but . . .

What chemical holds the record for saving the most lives? Penicillin? Aspirin? Guess again. When it comes to saving lives, perhaps the most successful chemical is chlorine, which has saved many millions of us by disinfecting our drinking water. For this reason alone, chlorine ranks as a "wonder chemical," but wait, there's more! About 85 percent of all pharmaceuticals and about 95 percent of all crop protection chemicals are synthesized using chlorine. Chlorine is also an important component of many plastics, and it is used in the manufacture of a countless number of products. In all, about 45 percent of the U.S. gross domestic product is in some way rooted in chlorine chemistry. For this essay, we focus on some of the details of the wonders of chlorine, beginning with its role in disinfecting water and ending with a mindful look at some of chlorine's downsides.

Municipalities chlorinate both drinking water and wastewater by bubbling chlorine gas, Cl_2, through the water. The chlorine reacts with the water to produce hypochlorous acid, HOCl, and hydrogen and chloride ions. The hypochlorous acid is a weak acid, so it mostly stays in a molecular form. Notably, the hypochlorous acid molecule is able to penetrate the nonpolar microbial cell walls, which makes it an effective disinfectant.

$$\underset{\text{Chlorine gas}}{Cl_2(g)} + \underset{\text{Water}}{H_2O(l)} \rightarrow \underset{\text{Hypochlorous acid}}{HOCl(aq)} + \underset{\text{Hydrogen ion}}{H^+(aq)} + \underset{\text{Chloride ion}}{Cl^-(aq)}$$

Chlorine and hypochlorous acid, as well as water-disinfecting bleach solutions, can react with organic components of drinking and waste waters, creating chlorinated hydrocarbons, many of which are cancer-causing. Some municipalities have therefore switched to chlorine dioxide, ClO_2, which doesn't form chlorinated hydrocarbons so readily.

Most chlorine is manufactured from the electrolysis of salt water, as shown in the following equation. Notably, this reaction also produces two other valuable chemicals, sodium hydroxide, NaOH, and hydrogen, H_2.

$$2\,NaCl(aq) + 2\,H_2O(l) \xrightarrow{\text{electricity}} Cl_2(g) + 2\,NaOH(aq) + H_2(g)$$

Chlorine is useful for the manufacture of many chemical products, even though these products themselves contain no chlorine. To create titanium metal, for example, the mineral titanium dioxide is reacted with chlorine to form titanium tetrachloride, $TiCl_4$ (sometimes humorously called "tickle"), which is then reduced by magnesium metal, Mg, as shown in the following equations:

$$TiO_2 + 2\,Cl_2 + \text{carbon} \rightarrow TiCl_4 + CO_2$$

$$TiCl_4 + 2\,Mg \rightarrow 2\,MgCl_2 + Ti$$

Interestingly, titanium tetrachloride is the "ink" used by skywriters. When the pilot is ready to maneuver the airplane to spell out words, she releases a spray of $TiCl_4$, which reacts with atmospheric moisture to form visibly white titanium dioxide particles.

Chemists designing pharmaceuticals and crop protection chemicals often add chlorine atoms to the structure in order to modify their potency. The anti-anxiety agent Valium® is an example.

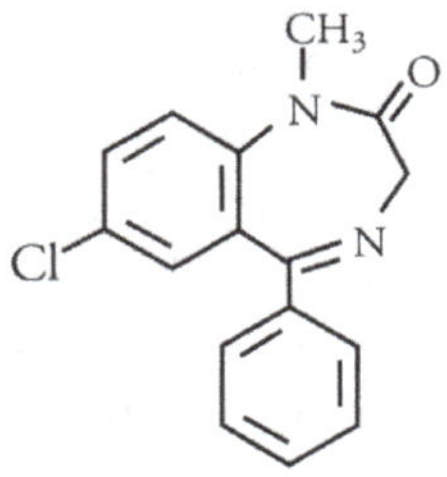

Valium

About one-third of the 40 million tons of chlorine produced annually goes to the manufacture of polyvinyl chloride, PVC, which is one of the most versatile of all plastics. PVC is ubiquitous, being used for pipes, flooring, electrical insulation, wallpaper, school supplies, swimming pools, and many other common products.

So, chlorine has become an integral part of modern life. However, we should beware of another important aspect of chlorine chemistry. Specifically, chlorine reacts with organic molecules to form a class of toxic molecules known as *persistent organic pollutants*, also known as POPs. A well-known group of POPs are the dioxins, which are represented by the compound 2, 3, 7, 8-tetrachloro-benzo-*p*-dioxin, also known as TCDD. These agents cause cancer and disrupt many bodily systems, especially those related to reproduction, immune responses, and hormones. Children are particularly susceptible because their bodies are still developing.

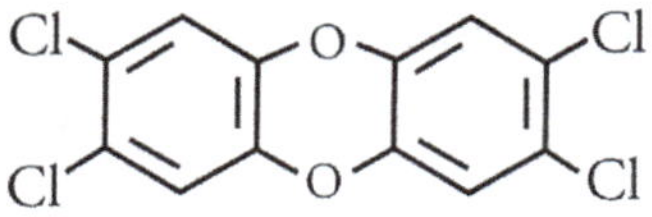

TCDD

All persistent organic pollutants share at least three qualities: (1) They are chemically stable, so they don't decompose in the environment—that is, they can "persist" for many years. (2) They are nonpolar toxins that tend to bioaccumulate within fatty tissues, especially in organisms higher in the food chain, such as humans. (3) They are semi-volatile, which enables them to evaporate into the atmosphere and be carried long distances by the wind, so they are found globally.

Some POPs, such as the insecticide DDT (Discussed in Chapter 15), are created on purpose. Many persistent organic pollutants, however, are created inadvertently, such as POPs that form upon the chlorine bleaching of paper. When possible, these POPs are sequestered as wastes. In the past, these wastes were not known to be harmful and they were often improperly buried, only to create later environmental hazards, as occurred at Love Canal, New York, in the late 1970s.

Wherever chlorine organic compounds are burned, persistent organic pollutants are formed, especially when there is a less-than-optimal quantity of oxygen available and the burning is incomplete. Since the 1970s, however, industrial emissions of POPs have declined over 90 percent, largely due to enforced regulations and technology that allows cleaner burning. Today, the most significant source of POPs, especially dioxins, is backyard trash burning, typically done in large metal barrels. Many municipalities now outlaw this hazardous practice.

Just as we are surrounded by the beneficial products of chlorine, so are we surrounded by chlorine's toxic by-products. The average American diet, for example, provides about 0.10 nanograms of dioxins per day. Even remote regions such as Arctic habitats have measurable quantities of POPs. In response, nations organized the Stockholm Convention on Persistent Organic Pollutants in 2001, during which a global protocol for containing POPs was created (chm.pops.int).

You are encouraged to do your own literature research. Many websites, such as ejnet.org/dioxin will tell you POPs are a problem of epic proportions. Others, such as AmericanChemistry.com will point to evidence showing that the concentration of POPs in humans has been reduced dramatically over the past 30 years, while the production of PVC and other chlorine compounds has more than tripled. Chlorine is indeed a wonder chemical, but as with any technology, we need to make sure that the benefits are well worth the risks. Becoming a well-informed citizen is a good place to start.

CONCEPT CHECK

Why should you never discard plastic wrap in a hot barbeque grill?

CHECK YOUR ANSWER Many plastic wraps contain chlorine. The combustion of this plastic puts forth a large dose of dioxin and other POPs directly onto your food and into the air you breathe.

Think and Discuss

1. We live in a time when over a billion people around the world lack access to safe drinking water. Chlorine powders, such as sodium hypochlorite, NaClO, however, can provide a family with safe drinking water for as little as 10 cents a day. What agencies or institutions should be involved in getting such water disinfectants to people who need them? What obstacles might be. encountered? How might they be overcome?

2. What would you do if you discovered that your neighborhood was built upon a leaking toxic waste dump? Might you be able to sell your house? What would you do if governmental agencies told you that there was no proof that the toxic wastes were causing you harm?

3. The compound tetraethyl lead, $Pb(CH_2CH_3)_4$, was first formulated with gasoline in the 1920s to help car engines run more smoothly. It was known to be poisonous to

humans, yet a ban on leaded gasoline in the United States didn't occur until the 1970s. In many countries, leaded gasoline is still in use. Explain how this does or doesn't relate to the chlorine industry.

4. Why are we not in danger of ever running out of chlorine? Does this mean that the cost of producing chlorine will remain relatively stable?

5. How much effort should really be put into controlling POPs when average life spans of humans continue to increase?

Author Responses to Think and Discuss

1. All agencies, be they governmental, private, religious, or whatever, ought to consider being involved in helping to provide water disinfectants to people who need them. According to the United Nations, about 6000 children die every day from diseases that can be prevented by improved water and sanitation. Over the past decade this has added up to the deaths of about 22 million children.

2. You should contact National Superfund Hotline of the Environmental Protection Agency at 1-800-424-9346. To report oil or chemical spills, call 1-800-424-8802. To brush up on your knowledge of the history of the Superfund Act, be sure to read the essay appearing with Chapter 6.

3. Tetraethyl lead, like dioxins and PCBs, were known to be causing harm to humans well before regulations were set in place. As presented in Chapter 15, the chlorinated persistent organic pollutant, DDT, was banned from use in the United States in the 1970s. This compound, however, is still widely used in other regions of the world, primarily to control the spread of malaria-causing mosquitos, which, to these nations, pose an even greater risk.

4. Chlorine is produced from sodium chloride, which is most abundant, especially in our oceans. The energy it takes to transform sodium chloride to chlorine, however, is significant. As the cost of energy increases, so will the cost of producing chlorine.

5. Quality of life counts as much if not more so than quantity of life. The world, however, is faced with so many issues. Ideally, each issue gains the attention it needs, but we would be wrong not to prioritize the issues. For example, many would argue that top priority should be given to helping developing nations to control birth rates, which happens naturally when the standard of living within a country is increased.

Chapter 12

Organic Compounds

THE MAIN IDEA

Carbon can form a limitless number of chemical structures.

Λ Organic molecules can be quite tasty, especially those that provide the flavor of our favorite ice cream.

Carbon atoms have the rare ability to bond with themselves repeatedly. This allows the building of large molecular structures. Add to this the fact that carbon atoms can also bond with atoms of other elements and you see the possibility of an endless number of different carbon-based molecules. Each molecule has its own unique set of physical, chemical, and biological properties. The flavor of vanilla, for example, is perceived when the compound *vanillin* is absorbed by the sensory organs in the nose. Vanillin is the essential ingredient in anything having the flavor of vanilla—without vanillin, there is no vanilla flavor. The flavor of chocolate, on the other hand, is generated when not just one, but a wide assortment of carbon-based molecules are absorbed in the nose. One of the more significant of these molecules is *tetramethylpyrazine*.

Life is based on carbon's ability to form diverse structures. Reflecting this fact, the branch of chemistry that is the study of carbon-containing chemical compounds has come to be known as **organic chemistry**. Because organic compounds are so closely tied to living organisms and because they have many applications—from flavorings to fuels, polymers, medicines, agriculture, and more—it is important to have a basic understanding of them.

12.1 Hydrocarbons

We begin with the simplest organic compounds—those consisting of only carbon and hydrogen. Organic compounds that contain only carbon and hydrogen are called **hydrocarbons**. The simplest hydrocarbon is methane, CH_4, with only one carbon per molecule as shown in **Figure 12.1** on the next page. Methane is the main component of natural gas. The hydrocarbon octane, C_8H_{18}, has eight carbons per molecule and is a component of gasoline. The hydrocarbon polyethylene contains hundreds of carbon and hydrogen atoms per molecule. Polyethylene is a plastic used to make many familiar items, such as milk containers and plastic bags.

READING CHECK
What types of atoms are found in hydrocarbons?

Hydrocarbons may differ in the way the carbon atoms connect to one another. **Figure 12.2** shows the three hydrocarbons pentane, isopentane, and neopentane. These hydrocarbons have the same molecular formula,

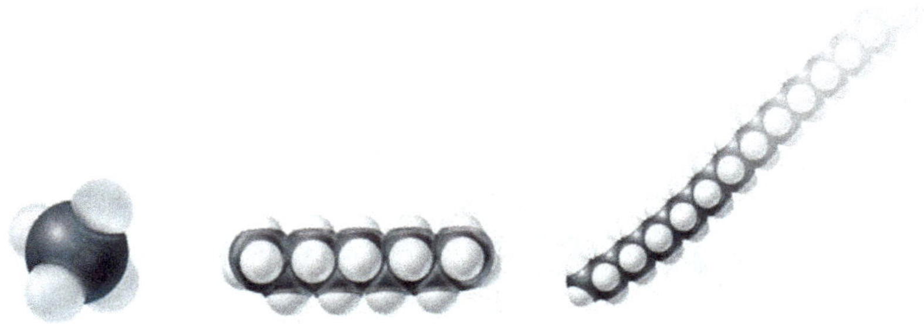

Methane, CH_4 Octane, C_8H_{18} Polyethylene

Figure 12.1 >
Hydrocarbons are made of hydrogen atoms (shown in white) and carbon atoms (shown in black). All hydrocarbons are non-polar. Light hydrocarbons, such as methane, tend to be gaseous. Medium-sized hydrocarbons, such as octane, are liquids. Heavier hydrocarbons, such as polyethylene, are solid.

C_5H_{12}, but they are structurally different from one another. The carbon framework of pentane is a chain of five carbon atoms. In isopentane, the carbon chain branches, so that the framework is a four-carbon chain branched at the second carbon. In neopentane, a central carbon atom is bonded to *four* surrounding carbon atoms.

We can see the different structural features of pentane, isopentane, and neopentane more clearly by drawing the molecules in two dimensions, as shown in the middle row of Figure 12.2. Alternatively, we can represent them by the *stick structures* (sometimes called line-angle drawings or skeletal structures) shown in the bottom row. A stick structure is a commonly used shorthand notation for representing an organic molecule. Each line (stick) represents a covalent bond, and carbon atoms are understood to exist at the end of any line or wherever two or more straight lines meet (unless another type of atom is drawn at the end of the line). Any hydrogen atoms bonded to the carbons also typically are not shown. Instead, their presence is assumed so that the focus can remain on the structure formed by the carbon atoms.

Molecules such as pentane, isopentane, and neopentane have the same molecular formula, which means they have the same number of the same kinds of atoms. The way these atoms are put together, however, is different. We say that each has its own *configuration*, where the term **configuration** refers to how the atoms are connected. Different configurations result in different chemical structures. Molecules with the same chemical formula but different configurations (and hence, different structures) are known as **structural isomers**. Structural isomers are

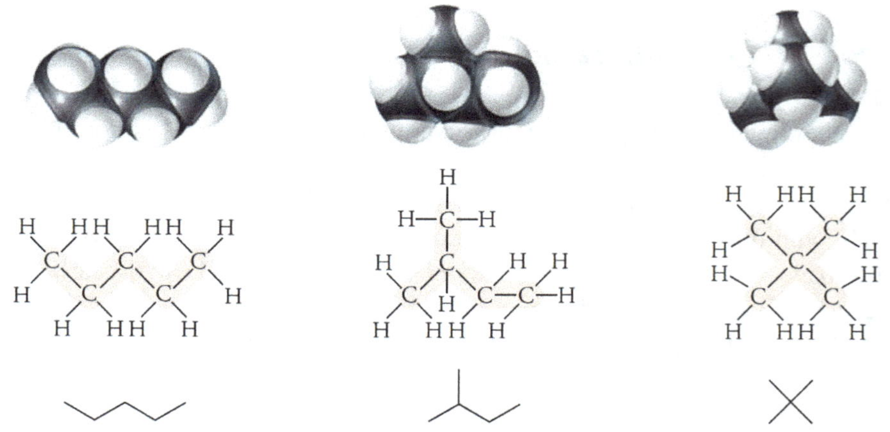

Pentane, C_5H_{12} Isopentane, C_5H_{12} Neopentane, C_5H_{12}

Figure 12.2 >
These three hydrocarbons all have the same molecular formula. We can see their different structural features by highlighting the carbon framework in two dimensions. Easy-to-draw stick structures that use lines for all carbon–carbon covalent bonds are also useful.

different from each other, having different physical and chemical properties. For example, pentane has a boiling point of 36°C, isopentane's boiling point is 30°C, and neopentane's is 10°C.

The number of possible structural isomers for a chemical formula increases rapidly as the number of carbon atoms increases. There are three structural isomers for compounds having the formula C_5H_{12}, 18 for C_8H_{18}, 75 for $C_{10}H_{22}$, and a whopping 366,319 for $C_{20}H_{42}$!

A carbon-based molecule can have different spatial orientations, called **conformations**. Flex your wrist, elbow, and shoulder joints and you'll find your arm passing through a range of conformations. Likewise, organic molecules can twist and turn about their carbon–carbon single bonds and thus have a range of *conformations*. The structures in **Figure 12.3**, for example, are different conformations of pentane. In the language of organic chemistry, we say that the *configuration* of a molecule such as pentane has a broad range of *conformations*. Change the configuration of pentane, however, and you no longer have pentane. Rather, you have a different structural isomer, such as isopentane, which has its own range of different conformations.

The number of carbon atoms within a hydrocarbon is indicated by the hydrocarbon's name, as shown in **Table 12.1**.

When the hydrocarbon is branched, the name of the hydrocarbon is based upon the longest carbon chain. Smaller branches off this longest chain are written as a prefix ending in *-yl*. As shown in Table 12.1, a one-carbon

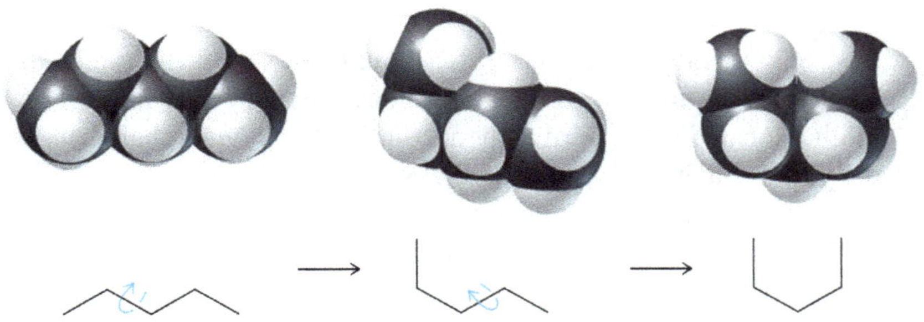

< Figure 12.3
Three conformations for a molecule of pentane. The conformation changes as bonds rotate as indicated by the blue arrows. The molecule looks different in each conformation, but the five-carbon framework is the same in all three conformations. In a sample of liquid pentane, the molecules are found in all conformations—not unlike a bucket of worms.

CONCEPT CHECK

Which carbon–carbon bond was rotated to go from the "before" conformation of isopentane to the "after" conformation?

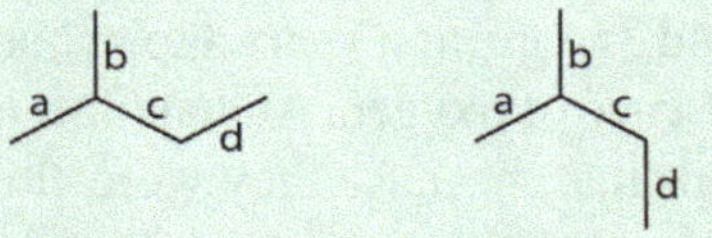

Before After

CHECK YOUR ANSWER Bond "c." This rotation is similar to that of the arm of an arm wrestler who, with the arm just above the table while on the brink of losing, suddenly gets a surge of strength and swings the opponent's arm through a half-circle arc and wins. The best way to understand the geometrical shapes of organic molecules is to get your hands on a set of molecular models.

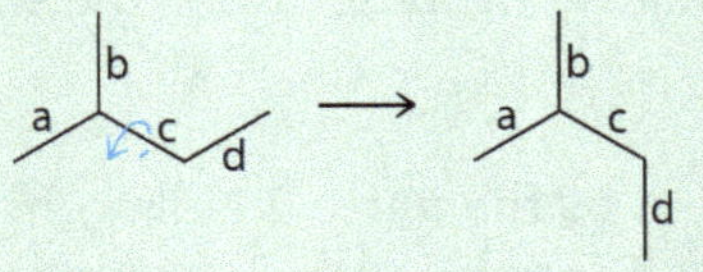

Before After

CHEMICAL CONNECTIONS

How is aspirin connected to petroleum?

TABLE 12.1 Names of Simple Straight-Chain Hydrocarbons

FORMULA	HYDROCARBON NAME	–YL PREFIX
CH_4	methane	methyl
C_2H_6	ethane	ethyl
C_3H_8	propane	propyl
C_4H_{10}	butane	butyl
C_5H_{12}	pentane	pentyl
C_6H_{14}	hexane	hexyl
C_7H_{16}	heptane	heptyl
C_8H_{18}	octane	octyl
C_9H_{20}	nonane	nonyl
$C_{10}H_{22}$	decane	decyl

branch would be indicated by the prefix methyl, where *meth-* means a single carbon. Also, the longest chain is numbered to indicate where the branching takes place. For example, the following compound is 3-methylhexane because it has a methyl group branching off the third carbon of hexane:

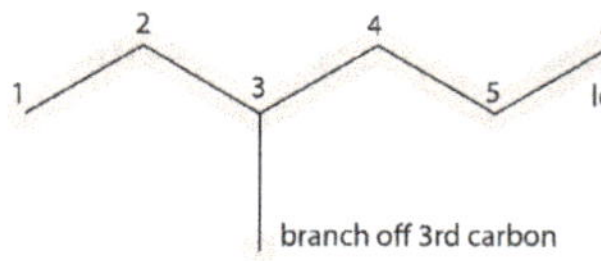

Below is the structure for 2,3-dimethylhexane, which tells us that there are two methyl groups—one located at the second carbon and another located at the third carbon. Number the longest chain backwards and you would have 4,5-dimethylhexane, which is the identical structure. The convention, however, is to always use the lowest numerals possible when naming an organic compound.

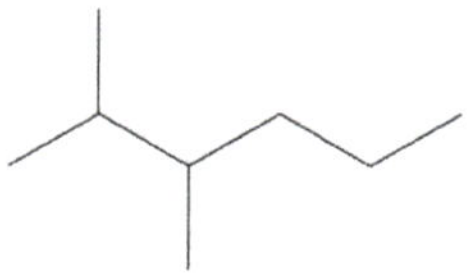

Hydrocarbons are obtained primarily from coal and petroleum. Most of the coal and petroleum that exist today were formed between 290 million and 354 million years ago, when plant and animal matter decayed in the absence of oxygen. At that time, the Earth was covered with extensive swamps that, because they were close to sea level, periodically became submerged. The organic matter of the swamps was buried beneath layers of marine sediments and was eventually transformed to either coal or petroleum.

CONCEPT **CHECK**

Neopentane, shown in Figure 12.2, has an alternate name based upon the above naming methodology. What is this alternate name for neopentane?

CHECK YOUR ANSWER Neopentane also goes by the name 2,2-dimethylpropane. The "di" is a prefix that means two, which in this case means two methyl's. The numerals tell us on what carbon these two methyl groups are attached. Note that in this case, the numeral 2 is used twice because each group needs its own number.

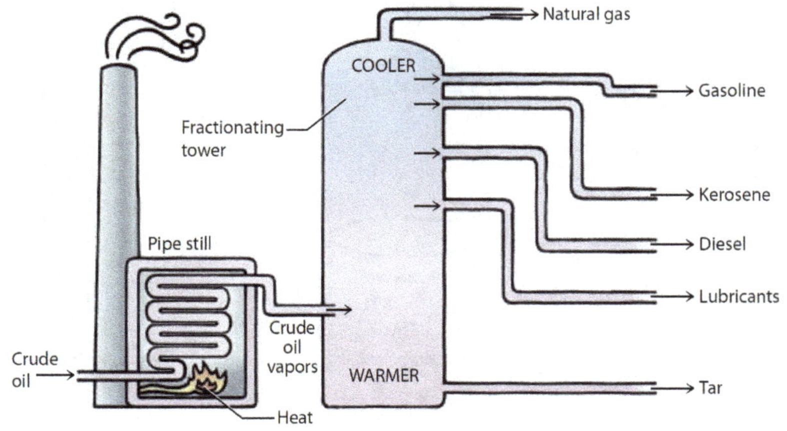

Figure 12.4
A schematic for the fractional distillation of petroleum into its useful hydrocarbon components.

FOR YOUR INFORMATION

We are placing unusually large amounts of carbon dioxide into the atmosphere by burning fossil fuels. The atmosphere, however, is not the only possible repository for the carbon dioxide we produce. The smokestacks of power plants, for example, can be modified to capture CO_2, which is then liquified and pumped kilometers deep into the ground. We discuss this technology further in Chapter 17.

Coal is a solid material containing many large, complex hydrocarbon molecules. Most of the coal mined today is used for the production of steel and for generating electricity at coal-burning power plants.

Petroleum, also called crude oil, is a liquid readily separated into its hydrocarbon components through a process known as *fractional distillation*, shown in **Figure 12.4**. The crude oil is heated to a temperature high enough to vaporize most of its components. The hot vapor flows into the bottom of a fractionating tower, which is warmer at the bottom than at the top. As the vapor rises in the tower and cools, the various components begin to condense. Hydrocarbons that have high boiling points, such as tar, condense first at warmer temperatures. Hydrocarbons that have low boiling points, such as gasoline, travel to the cooler regions at the top of the tower before condensing. Pipes drain the various liquid hydrocarbon fractions from the tower. Natural gas, which is primarily methane, does not condense. It remains a gas and is collected at the top of the tower.

CONCEPT CHECK

Do the heavier or the lighter molecules of crude oil rise to the top of the fractionation tower?

CHECK YOUR ANSWER The lighter molecules found in crude oil are the ones that rise highest within the fractionation tower. These are the molecules that have the lower boiling points. So, the lower the boiling point is, the higher the molecules rise.

Differences in the strength of molecular attractions explain why different hydrocarbons condense at different temperatures. As discussed in Section 7.1, in our comparison of induced dipole–induced dipole attractions in methane and octane, larger hydrocarbons experience many more of these attractions than smaller hydrocarbons do. For this reason, the larger hydrocarbons condense readily at high temperatures and so are found at the bottom of the tower. Smaller molecules, because they experience fewer attractions to neighbors, condense only at the cooler temperatures found at the top of the tower.

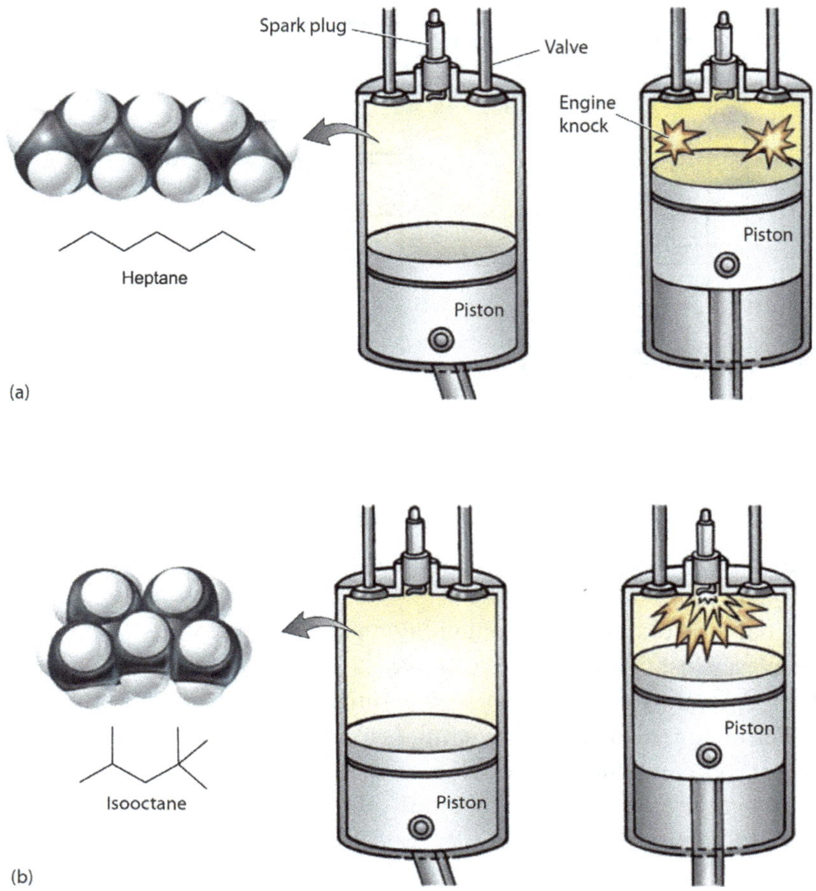

Figure 12.5 >
A straight-chain hydrocarbon, such as hexane, can be ignited from the heat generated as gasoline is compressed by the piston—before the spark plug fires. This upsets the timing of the engine cycle, giving rise to a knocking sound. Branched hydrocarbons, such as isooctane, burn less readily and are ignited not by compression alone, but only when the spark plug fires.

∧ Figure 12.6
Octane ratings are posted on gasoline pumps. Most modern car engines are designed to run optimally on regular octane gasoline. For such cars, higher-octane gasoline may provide poorer performance at greater cost.

The gasoline obtained from the fractional distillation of petroleum consists of a wide variety of hydrocarbons having similar boiling points. Some of these components burn more efficiently than others in a car engine. The straight-chain hydrocarbons, such as hexane, burn too quickly, causing what is called engine knock, as illustrated in **Figure 12.5.** Gasoline hydrocarbons that have more branching, such as isooctane, burn slowly and result in the engine's running more smoothly. The compounds heptane and isooctane are used as standards in assigning octane ratings to gasoline. An octane number of 100 is arbitrarily assigned to isooctane, and heptane is assigned an octane number of 0. The anti-knock performance of a particular gasoline is compared with those of various mixtures of isooctane and heptane, and an octane number is assigned. **Figure 12.6** shows the octane information appearing on a typical gasoline pump.

CONCEPT CHECK

Which structural isomer shown in Figure 12.2 should have the highest-octane rating?

CHECK YOUR ANSWER The structural isomer with the greatest amount of branching in the carbon framework will likely have the highest-octane rating, making neopentane the clear winner. For your information, the ratings are as follows:

Compound	Octane Rating
Pentane	61.7
Isopentane	92.3
Neopentane	116.0

12.2 Unsaturated Hydrocarbons

Recall from Section 6.1 that carbon has *four* unpaired valence electrons. As shown in **Figure 12.7,** each of these electrons is available for pairing with an electron from another atom, such as hydrogen, to form a covalent bond.

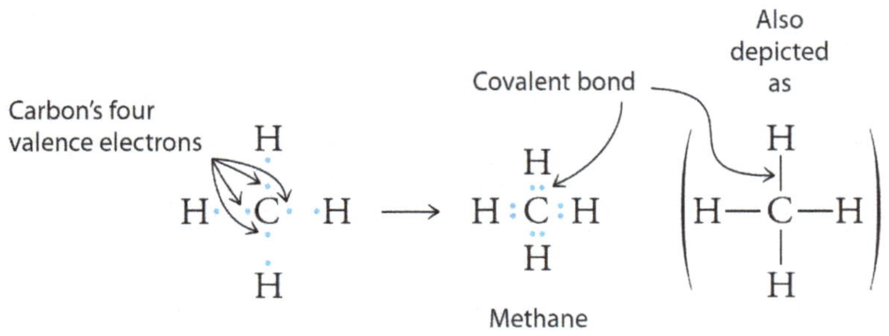

< Figure 12.7

Carbon has four valence electrons. Each electron pairs up with an electron from a hydrogen atom in the four covalent bonds of methane.

In all the hydrocarbons discussed so far, including the methane shown in Figure 12.7, each carbon atom is bonded to four neighboring atoms by four single covalent bonds. Such hydrocarbons are known as **saturated hydrocarbons**. The term *saturated* means that each carbon has as many atoms bonded to it as possible. We now explore cases where one or more carbon atoms in a hydrocarbon are bonded to *fewer* than four neighboring atoms. This occurs when at least one of the bonds between a carbon and a neighboring atom is a multiple bond. (See Section 6.5 for a review of multiple bonds.) A hydrocarbon that has a multiple carbon-carbon bond—either double or triple—is known as an **unsaturated hydrocarbon**. Because of the multiple bond, two of the carbons are bonded to *fewer* than four other atoms. These carbons are thus said to be *unsaturated.*

READING CHECK

What type of hydrocarbon has double or triple bonds?

Figure 12.8 compares the saturated hydrocarbon butane with the unsaturated hydrocarbon *cis*-2-butene. The number of atoms bonded to each of the two middle carbons of butane is four, whereas each of the two middle carbons of *cis*-2-butene is bonded to only three other atoms—a hydrogen and two carbons.

A generic term for a saturated hydrocarbon is *alkane*. All of the molecules discussed in Section 12.1 are examples of **alkanes**. This is indicated by the suffix *-ane*, as in *butane*. Unsaturated hydrocarbons containing one or more double bonds are called **alkenes**. This is indicated by the suffix *-ene*, as in *butene*.

An important aspect of the double bond of alkenes is that it cannot rotate more than a few degrees. By analogy, connect two gumdrops with a single toothpick and you'll be able to hold one gumdrop while rotating the

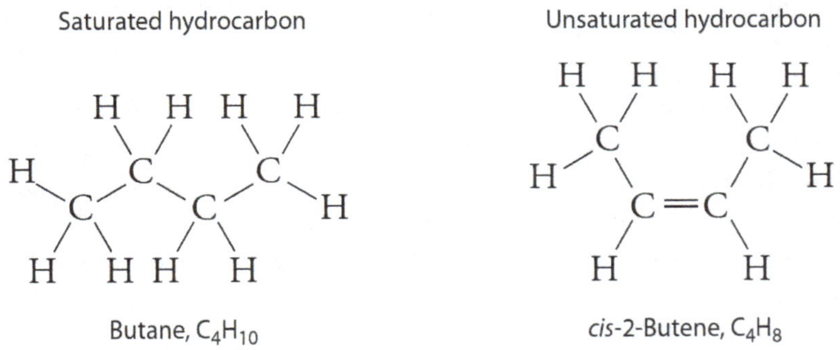

< Figure 12.8

The carbons of the hydrocarbon butane are saturated, each being bonded to four other atoms. Because of the double bond, two of the carbons of the hydrocarbon *cis*-2-butene are bonded to only three other atoms, which makes the molecule an unsaturated hydrocarbon.

other. Connect the two gumdrops with two toothpicks, however, and this rotational motion doesn't work because the two toothpicks can twist only so far. The significance of this for alkenes is what we call cis/trans isomerism. We use the prefix *cis* to indicate a structure in which the bulk of the carbons are on the same side of the length of the double bond, as shown in **Figure 12.9**. We use the prefix *trans* to indicate a structure in which the bulk of the carbons are on opposite sides of the length of the double bond. We'll be discussing the nature of "trans" fatty acids in our diet within Chapter 13.

Figure 12.9 >

The *cis*-isomer has the bulk of carbon atoms on the same side of the length of the double bond. The *trans*-isomer has these carbons on opposite sides. As a mnemonic, think "cis-same-side."

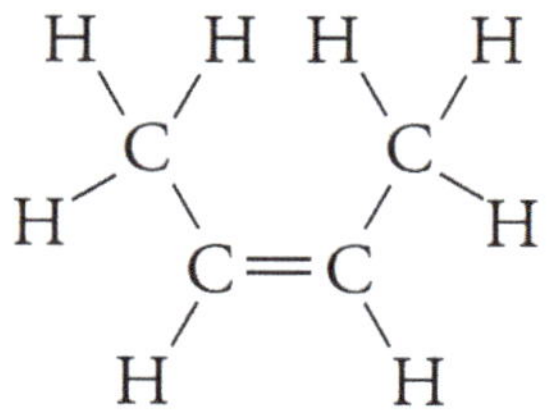

cis-2-butene

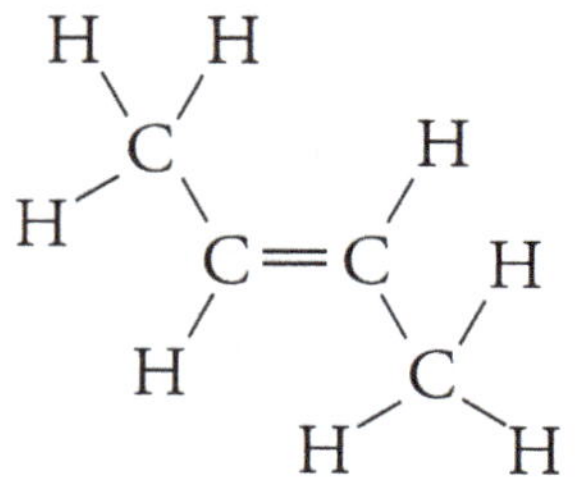

trans-2-butene

An important unsaturated hydrocarbon (alkene) is benzene, C_6H_6, which may be drawn as three double bonds contained within a flat hexagonal ring, as is shown in **Figure 12.10a**. Unlike the double-bond electrons in most other unsaturated hydrocarbons, the electrons of the double bonds in benzene are not fixed between any two carbon atoms. Instead, these electrons move freely around the ring. This is commonly represented by drawing a circle within the ring, as shown in **Figure 12.10b**, rather than showing individual double bonds.

Figure 12.10 >

The double bonds of benzene, C_6H_6, are able to migrate around the ring. For this reason, they are often represented by a circle within the ring.

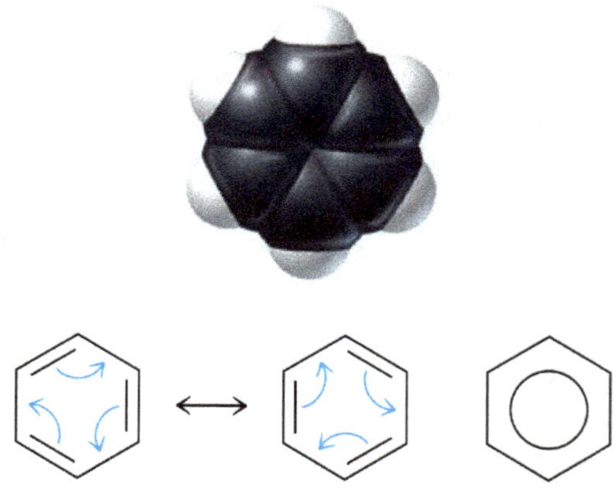

(a) (b)

Many organic compounds contain one or more benzene rings in their structure. Because many of these compounds are fragrant, any organic molecule containing a benzene ring is classified as an **aromatic** compound (even if it is not particularly fragrant). **Figure 12.11** shows a few examples. Toluene, a common solvent used as a paint thinner, is toxic and gives airplane glue its distinctive odor. Some aromatic compounds, such as naphthalene, contain two or more benzene rings fused together. At one time, mothballs were made of naphthalene. Most mothballs sold today, however, are made of the less toxic 1,4-dichlorobenzene.

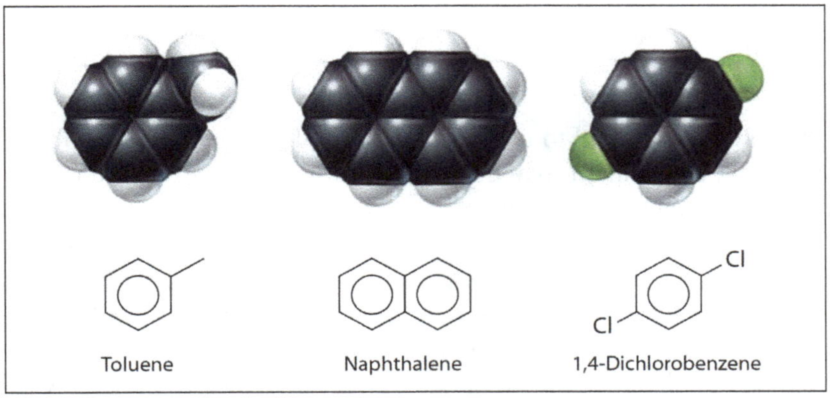

< Figure 12.11

The structures for three odoriferous organic compounds containing one or more benzene rings: toluene, naphthalene, and 1,4-dichlorobenzene.

Unsaturated hydrocarbons containing a triple bond are called **alkynes**. An example of an unsaturated hydrocarbon containing a triple bond is ethyne, which is more widely known as acetylene, C_2H_2. A confined flame of acetylene burning in oxygen is hot enough to melt iron, which makes acetylene a choice fuel for welding (**Figure 12.12**). Alkynes are generally not as common as alkanes or alkenes.

< Figure 12.12

The unsaturated alkyne acetylene, C_2H_2 (also known as ethyne), when burned in this torch, produces a flame hot enough to melt iron.

CONCEPT **CHECK**

Prolonged exposure to benzene increases the risk of developing certain cancers. The structure of aspirin contains a benzene ring. Does this indicate that prolonged exposure to aspirin will increase a person's risk of developing cancer?

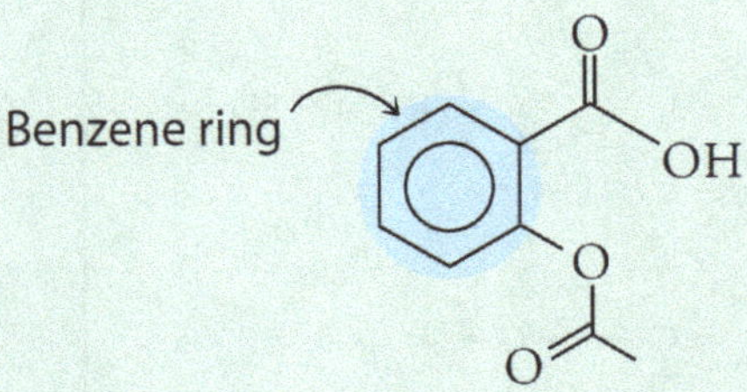

Aspirin

CHECK YOUR ANSWER No. Although benzene and aspirin both contain a benzene ring, these two molecules have different overall structures and quite different chemical properties. Each carbon-containing organic compound has its own set of unique physical, chemical, and biological properties. While benzene may cause cancer, aspirin works as a remedy for headaches.

12.3 Functional Groups

READING CHECK
What is a heteroatom?

Carbon atoms can bond to one another and to hydrogen atoms in many ways. This results in an incredibly large number of hydrocarbons. But carbon atoms can bond to atoms of other elements as well, further increasing the number of possible organic molecules. In organic chemistry, any atom other than carbon or hydrogen in an organic molecule is called a **heteroatom**, where *hetero-* means "different from either carbon or hydrogen."

A hydrocarbon structure can serve as a framework for the attachment of various heteroatoms. This is analogous to a Christmas tree serving as the scaffolding on which ornaments are hung. Just as the ornaments give character to the tree, heteroatoms give character to an organic molecule. Heteroatoms have profound effects on the properties of an organic molecule.

FOR YOUR INFORMATION

The chemistry of hydrocarbons is interesting, but start adding heteroatoms to these organic molecules and the chemistry becomes extraordinarily interesting. The organic chemicals of living organisms, for example, all contain heteroatoms.

Consider ethane, C_2H_6, and ethanol, C_2H_6O, which differ from each other by only a single oxygen atom. Ethane has a boiling point of -88°C, making it a gas at room temperature, and it does not dissolve in water very well. Ethanol, by contrast, has a boiling point of +78°C, making it a liquid at room temperature. It is infinitely soluble in water, and it is the active ingredient of alcoholic beverages.

Consider further ethylamine, C_2H_7N, which has a nitrogen atom on the same two-carbon framework. This compound is a corrosive, pungent, highly toxic gas—most unlike either ethane or ethanol. Why these vast differences? Answer: Because of the different heteroatoms they contain.

Organic molecules are classified according to the *functional groups* they contain. We define a **functional group** as a combination of atoms that behave as a single unit. Most functional groups are distinguished by the heteroatoms they contain, and some common groups are listed in Table 12.2. Notice that the structures shown in Table 12.2 are not the structures of complete molecules. The atoms highlighted in blue, however, are all the

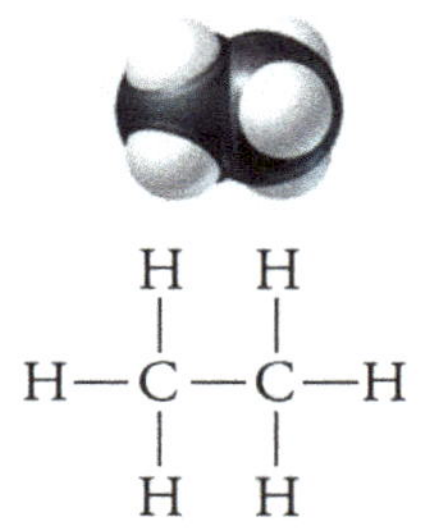
Ethane

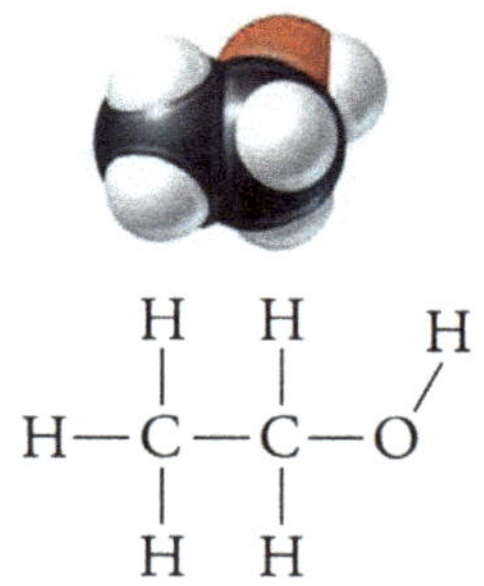
Ethanol

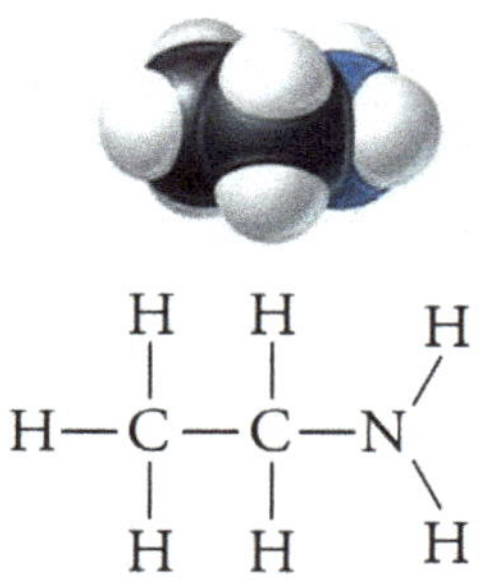
Ethylamine

TABLE 12.2 Common Functional Groups in Organic Molecules

GENERAL STRUCTURE	CLASS	GENERAL STRUCTURE	CLASS
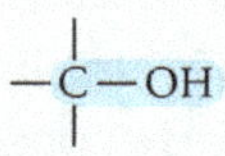	Alcohols	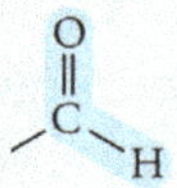	Aldehydes
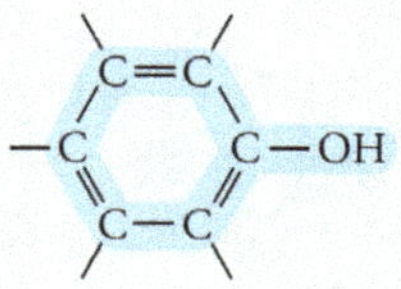	Phenols	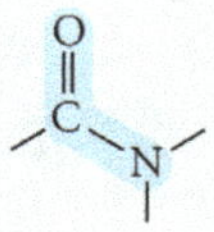	Amides
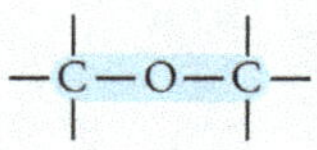	Ethers	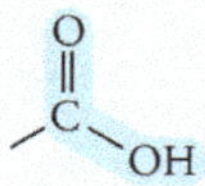	Carboxylic acid
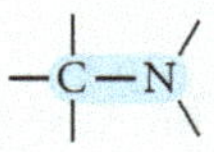	Amines	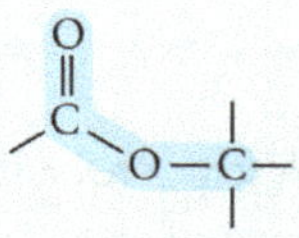	Esters
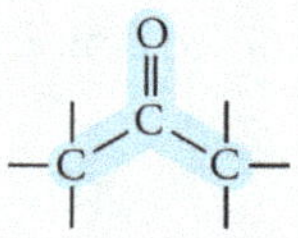	Ketones	**Functional Group:** A collection atoms within a molecule that behave as a single unit.	

atoms that go together to make a particular functional group. It is typical that when an organic chemist looks to an organic molecule, the first thing they look for are the various functional groups that organic molecule may contain. This helps them to quickly classify the molecule and to ascertain the various physical, chemical, and even biological properties it might have.

The remainder of this chapter introduces the classes of organic molecules shown in Table 12.2. The role heteroatoms play in determining the properties of each class is the underlying theme. As you study this material, focus on understanding the chemical and physical properties of the various classes of compounds, for doing so will give you a greater appreciation of the remarkable diversity of organic molecules and their many applications.

fyi FOR YOUR INFORMATION

It is common for organic chemists to use the letter R to indicate an unspecified hydrocarbon coming off a functional group. For example, an alcohol can be indicated as R—OH, while an ether can be indicated as R—O—R. The R could equal a methyl group, —CH_3, an ethyl group, —CH_2CH_3, or any other group of carbon atoms as shown in Table 12.1.

CONCEPT **CHECK**

What is the significance of heteroatoms in an organic molecule?

CHECK YOUR ANSWER Heteroatoms largely determine an organic molecule's physical and chemical properties.

12.4 Alcohols, Phenols, and Ethers

Alcohols are organic molecules in which a *hydroxyl group* is bonded to a saturated carbon. The hydroxyl group consists of an oxygen bonded to a hydrogen. Because of the polarity of the oxygen–hydrogen bond, low formula mass alcohols are often soluble in water, which is itself very polar. Some common simple alcohols and their melting and boiling points are listed in **Table 12.3**.

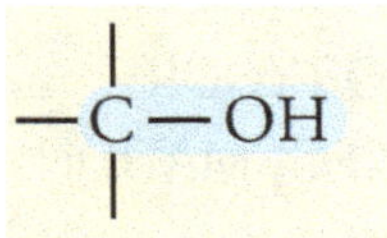

Billions of pounds of methanol, CH_3OH, are produced annually in the United States. Most of it is used for making formaldehyde and acetic acid, important starting materials in the production of plastics. In addition, methanol is used as a solvent, an octane booster, and an anti-icing agent in gasoline. Sometimes called wood alcohol because it can be obtained from

TABLE 12.3 Some Simple Alcohols

STRUCTURE	SCIENTIFIC NAME	COMMON NAME	MELTING POINT (°C)	BOILING POINT (°C)
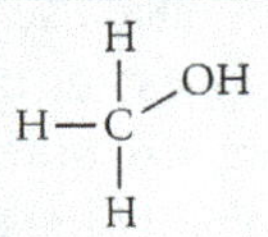	Methanol	Methyl alcohol	−97	65
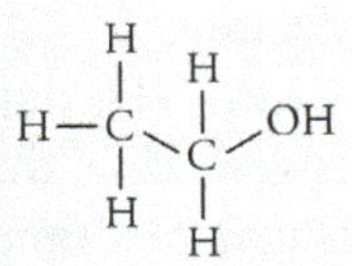	Ethanol	Ethyl alcohol	−115	78
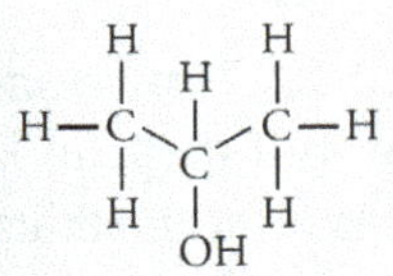	2-Propanol	Isopropyl alcohol	−89	83

> **FOR YOUR INFORMATION**
>
> We're classifying organic molecules based upon the functional groups they contain. As you will see shortly, organic molecules may contain more than one type of functional group. A single organic molecule, therefore, might be classified as both a phenol and an ether.

wood, methanol should never be ingested because in the body, it is metabolized to formaldehyde, H_2CO, and formic acid, HCO_2H. Formaldehyde is harmful to the eyes, can lead to blindness, and was once used to preserve dead biological specimens. Formic acid, the active ingredient in an ant bite, can lower the pH of the blood to dangerous levels. Ingesting only about 15 milliliters (about 3 tablespoons) of methanol may lead to blindness, and about 30 milliliters can cause death.

Ethanol, C_2H_5OH, on the other hand, is the "alcohol" of alcoholic beverages, and it is one of the oldest chemicals manufactured by humans. Ethanol is prepared by feeding the sugars of various plants to certain yeasts, which produce ethanol through a biological process known as *fermentation*. Ethanol is also widely used as an industrial solvent. For many years, ethanol intended for this purpose was made by fermentation, but today industrial-grade ethanol is more cheaply manufactured from petroleum by-products such as ethene, as **Figure 12.13** illustrates.

Figure 12.13 >

Ethanol can be synthesized from the unsaturated hydrocarbon ethene, with phosphoric acid as a catalyst.

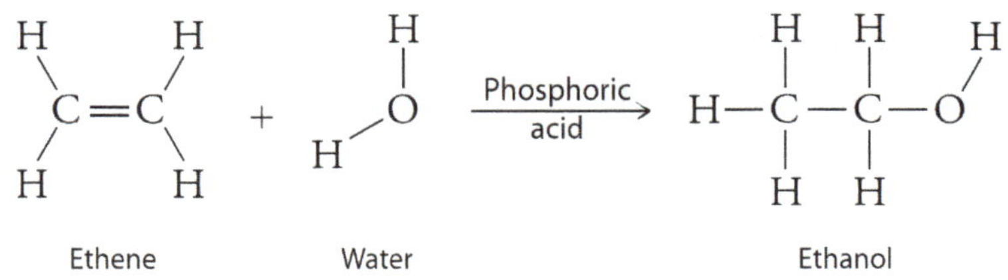

The solution produced by fermentation has an ethanol concentration no greater than about 12 percent, because at this concentration, the yeast cells begin to die. This is why most wines have an alcohol content of about 12 percent—they are produced solely by fermentation. To attain the higher ethanol concentrations found in such "hard" alcoholic beverages as gin and vodka, the fermented liquid must be distilled. In the United States, the ethanol content of distilled alcoholic beverages is measured as proof, which is twice the percentage of ethanol. An 86-proof whiskey, for example, is 43 percent ethanol by volume. The term *proof* evolved from a crude method once employed to test alcohol content. Gunpowder was wetted with a beverage of suspect alcohol content. If the beverage was primarily water, the powder would not ignite. If the beverage contained a significant amount of ethanol, the powder would burn, thus providing "proof" of the beverage's worth.

> **FOR YOUR INFORMATION**
>
> Just as the body metabolizes methanol into formaldehyde, H_2CO, it metabolizes ethanol into acetaldehyde, CH_3COH. Acetaldehyde won't cause blindness, but it does lead to some painful side effects, which people who drink too much experience as part of their hangover.

A third well-known alcohol is isopropyl alcohol, also called 2-propanol. This is the rubbing alcohol you buy at the drugstore. Although 2-propanol has a relatively high boiling point, it evaporates readily, leading to a pronounced cooling effect when it is applied to skin—an effect once used to reduce fevers. However, isopropyl alcohol is very toxic if ingested. In place of isopropyl alcohol, washcloths wetted with cold water are also quite effective in reducing fever, and they are far safer. You are probably most familiar with the use of isopropyl alcohol as a topical disinfectant.

> **READING CHECK**
>
> Is the phenol group slightly acidic or slightly basic?

Phenols contain a phenolic group, which consists of a hydroxyl group bonded to a benzene ring. Because of the presence of the benzene ring, the hydrogen of the hydroxyl group is readily lost in an acid-base reaction, which makes the phenolic group mildly acidic.

The reason for this acidity is illustrated in **Figure 12.14**. How readily an acid donates a hydrogen ion is a function of how well the acid is able to accommodate the resulting negative charge it gains after donating the hydrogen ion. After a phenol donates the hydrogen ion, it becomes a

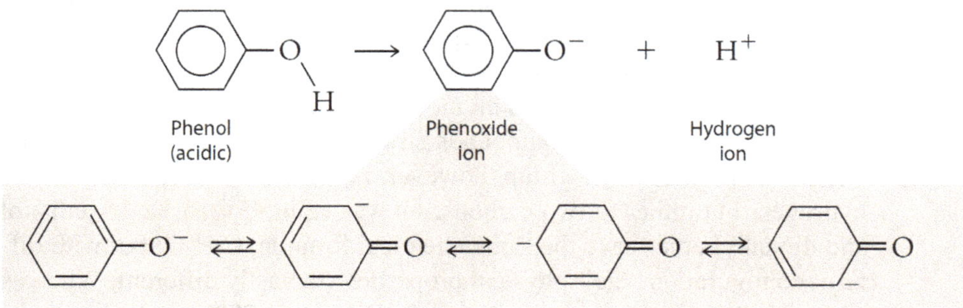

< Figure 12.14

The negative charge of the phenoxide ion is able to migrate on certain positions on the benzene ring. This mobility helps to accommodate the negative charge, which is why the phenolic group readily donates a hydrogen ion.

negatively charged phenoxide ion. The negative charge of the phenoxide ion, however, is not restricted to the oxygen atom. Recall that the electrons of the benzene ring are able to migrate around the ring. In a similar manner, the electrons responsible for the negative charge of the phenoxide ion are also able to migrate around the ring, as shown in Figure 12.14. Just as it is easy for several people to hold a hot potato by quickly passing it around, it is easy for the phenoxide ion to hold the negative charge because the charge gets passed around. Because the negative charge of the ion is so nicely accommodated, the phenolic group is more acidic than it would be otherwise.

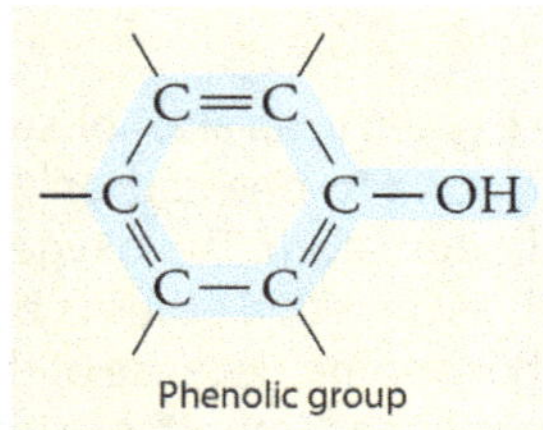

CONCEPT **CHECK**

Why are alcohols less acidic than phenols?

CHECK YOUR ANSWER An alcohol does not contain a benzene ring adjacent to the hydroxyl group. If the alcohol were to donate the hydroxyl hydrogen, the result would be a negative charge on the oxygen. Without an adjacent benzene ring, this negative charge has nowhere to go. As a result, an alcohol behaves only as a very weak acid, much like water.

The simplest phenol, shown in **Figure 12.15**, is called phenol. In 1867, Joseph Lister (1827–1912) discovered the antiseptic value of phenol, which, when applied to surgical instruments and incisions, greatly increased surgery survival rates. Phenol was the first purposefully used antibacterial solution, or *antiseptic*. Phenol damages healthy tissue, however, so a number of milder phenols have since been introduced. The phenol 4-hexylresorcinol,

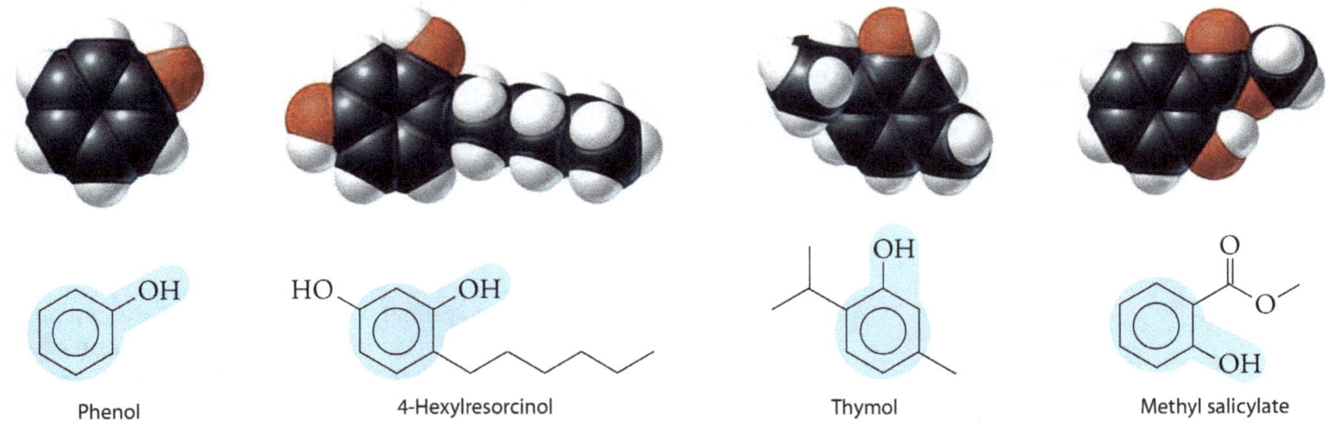

∧ Figure 12.15

Every phenol contains a phenolic group (highlighted in blue).

for example, is commonly used in throat lozenges and mouthwashes. This compound has even greater antiseptic properties than phenol, yet it does not damage tissue. Listerine® brand mouthwash (named after Joseph Lister) contains the antiseptic phenols thymol and methyl salicylate.

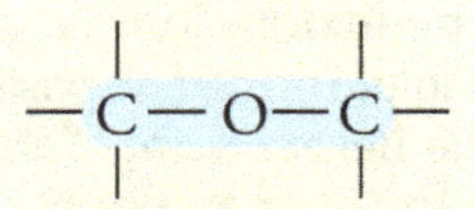

Ether group

Ethers are organic compounds structurally related to alcohols. The oxygen atom in an ether group, however, is bonded not to a carbon and a hydrogen, but rather to two carbons. As we see in **Figure 12.16**, ethanol and dimethyl ether have the same chemical formula, C_2H_6O; but with differnt configurations their physical properties are vastly different. Whereas ethanol is a liquid at room temperature (boiling point 78°C) and mixes quite well with water, dimethyl ether is a gas at room temperature (boiling point –25°C) and is much less soluble in water.

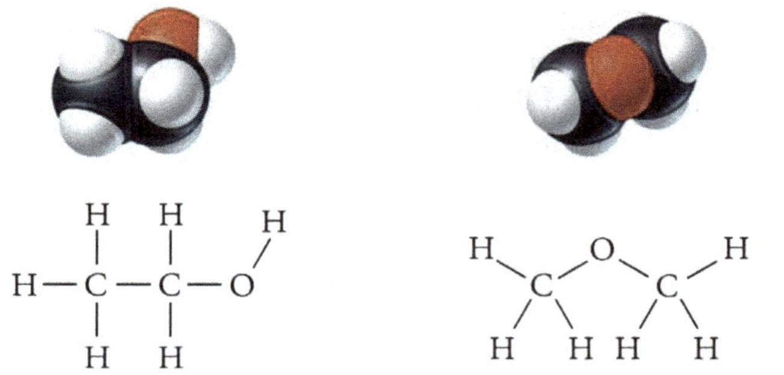

Ethanol: Soluble in water, boiling point 78°C

Dimethyl ether: Insoluble in water, boiling point –25°C

Figure 12.16 >

The oxygen in an alcohol, such as ethanol, is bonded to one carbon atom and one hydrogen atom. The oxygen in an ether, such as dimethyl ether, is bonded to two carbon atoms. Because of this difference, alcohols and ethers of similar molecular mass have vastly different physical properties.

Ethers are not very soluble in water because, without the hydroxyl group, they are unable to form strong hydrogen bonds with water (Section 7.1). Furthermore, without the polar hydroxyl group, the molecular attractions among ether molecules are relatively weak. As a result, little energy is required to separate ether molecules from one another. This is why low formula mass ethers have relatively low boiling points and evaporate so readily.

Diethyl ether, shown in **Figure 12.17**, was one of the first general anesthetics. The anesthetic properties of this compound were discovered in the early 1800s, and its use revolutionized the practice of surgery. Because of its high volatility at room temperature, inhaled diethyl ether rapidly enters the bloodstream. Because this ether has low solubility in water and high volatility, it quickly leaves the bloodstream once introduced. Because of these physical properties, a surgical patient can be brought into and out of *general anesthesia* (a state of unconsciousness) simply by regulating the gases breathed. Modern-day gaseous anesthetics have fewer side effects than diethyl ether, but they operate on the same principle.

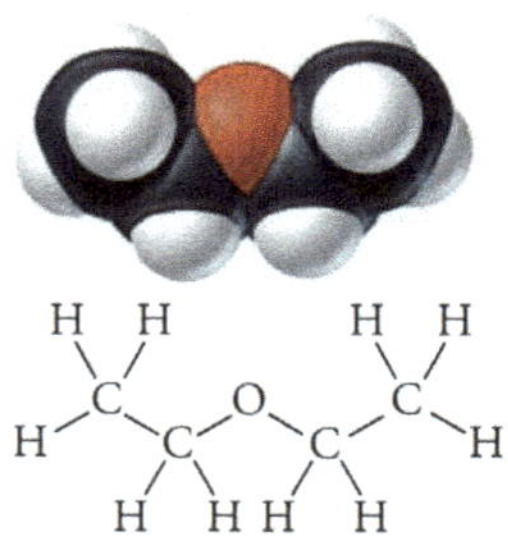

Diethyl ether, boiling point 35°C

Λ Figure 12.17

Diethyl ether is the systematic name for the "ether" historically used as an anesthetic.

fyi FOR YOUR INFORMATION

Alcohols can be classified as being primary, secondary, or tertiary. In a primary alcohol, the hydroxyl group is attached to the terminal carbon of a carbon chain. An example is ethanol, shown in Table 12.3. In a secondary alcohol, the hydroxyl group is attached to a central carbon, as seen with 2-propanol. For a tertiary alcohol, the hydroxyl group is attached to a carbon that is bonded to three other carbon atoms, as occurs in 2-methyl-2-propanol. Can you draw this structure?

12.5 Amines and Alkaloids

Amines are organic compounds that contain the amine group, which is a nitrogen atom bonded to one, two, or three saturated carbons. Amines are typically less soluble in water than are alcohols, because the nitrogen–hydrogen bond is not as polar as the oxygen–hydrogen bond. The lower polarity and weaker hydrogen bonding of amines also means their boiling points are typically lower than those of alcohols of similar formula mass. Table 12.4 lists three simple amines.

READING CHECK
What atoms are in an amine group?

Table 12.4 Three Simple Amines

STRUCTURE	NAME	MELTING POINT (°C)	BOILING POINT (°C)
	Ethylamine	−81	17
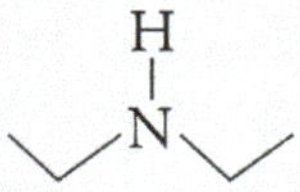	Diethylamine	−50	55
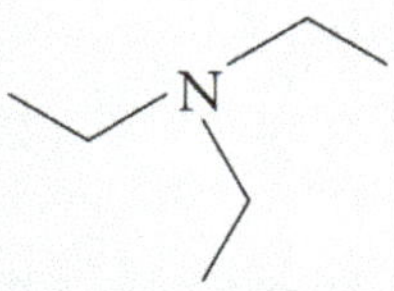	Triethylamine	−7	89

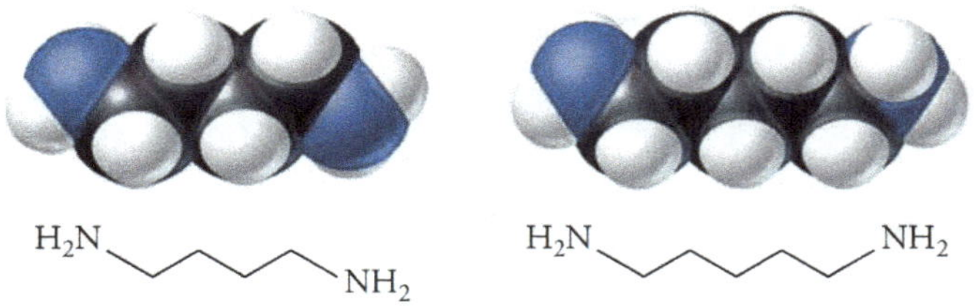

Putrescine (1,4-butanediamine) | Cadaverine (1,5-pentanediamine)

< Figure 12.18
Low-formula-mass amines such as these tend to have offensive odors.

One of the most notable physical properties of many low formula mass amines is their offensive odor. Figure 12.18 shows two appropriately named amines, putrescine and cadaverine, which are responsible, in part, for the odor of decaying flesh.

Amines are typically *alkaline*, because the nitrogen atom readily accepts a hydrogen ion, as Figure 12.19 illustrates. A class of alkaline amines found in nature are the *alkaloids*. Because many alkaloids have medicinal or other biological effects, there is great interest in isolating these compounds from

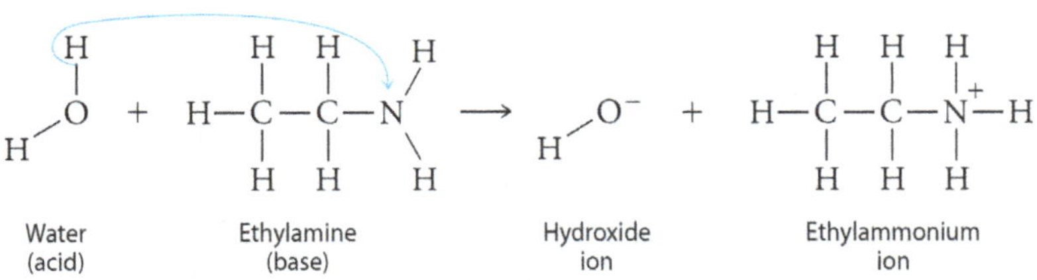

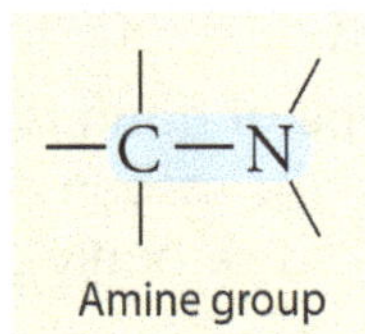

∧ Figure 12.19
Ethylamine acts as a weak base and accepts a hydrogen ion from water to become the ethyl ammonium ion. This reaction generates a small amount of hydroxide ions, which slightly increases the pH of the solution.

FOR YOUR INFORMATION

As we discuss in Chapter 14, nearly all pharmaceuticals that can be administered orally contain the nitrogen heteroatom in the water-soluble salt form.

the plants or marine organisms that contain them. As shown in **Figure 12.20,** an alkaloid, such as caffeine, reacts with an acid to form a salt that is usually quite soluble in water. This is in contrast to the unionized form of the alkaloid, known as a *free base*, which is typically insoluble in water.

Most alkaloids exist in nature not in their free-base form, but rather as the salts of naturally occurring acids known as *tannins*. The alkaloid salts of these acids are usually much more soluble in hot water than in cold water. The caffeine in coffee and tea exists in the form of the tannin salt, which is why coffee and tea are more effectively brewed in hot water. As **Figure 12.21** relates, tannins are also responsible for the stains caused by these beverages.

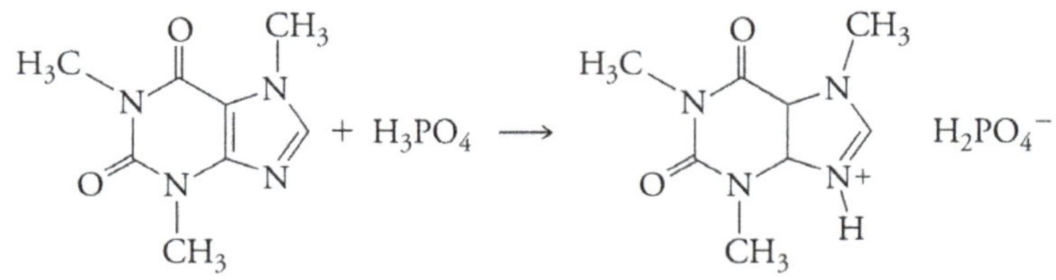

Caffeine, free-base form (water insoluble) | Phosphoric acid | Caffeine–phosphoric acid salt (water soluble)

Figure 12.20

All alkaloids are bases that react with acids to form salts. An example is the alkaloid caffeine, shown here reacting with phosphoric acid; both are common ingredients in soda beverages.

Figure 12.21

Tannins are responsible for the brown stains in coffee mugs or on a coffee drinker's teeth. Because tannins are acidic, they can be readily removed with an alkaline cleanser such as baking soda.

CONCEPT CHECK

Why do most caffeinated soft drinks also contain phosphoric acid?

CHECK YOUR ANSWER The phosphoric acid reacts with the caffeine to form the caffeine–phosphoric acid salt, which is much more soluble in cold water than the naturally occurring tannin salt. Interestingly, it also adds a desirable tingle to the tongue.

12.6 Carbonyl Compounds

READING CHECK

What type of bond is found between the carbon and the oxygen of a carbonly group?

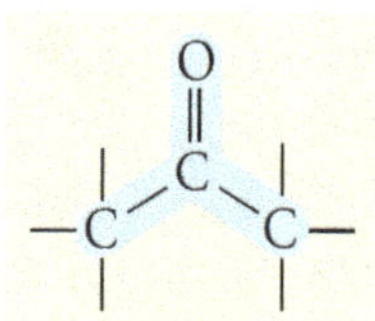

Ketone group

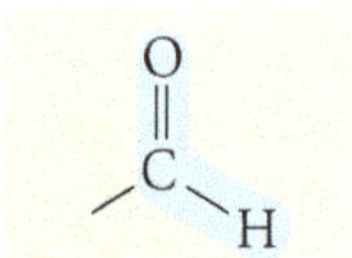

Aldehyde group

The **carbonyl group** consists of a carbon atom double-bonded to an oxygen atom. It occurs in the organic compounds known as *ketones*, *aldehydes*, *amides*, *carboxylic acids*, and *esters*.

A **ketone** is a carbonyl-containing organic molecule in which the carbonyl carbon is bonded to two carbon atoms. A familiar example of a ketone is *acetone*, which is often used in fingernail-polish remover and is shown in **Figure 12.22a**. In an **aldehyde**, the carbonyl carbon is bonded either to one carbon atom and one hydrogen atom, as in **Figure 12.22b**, or, in the case of formaldehyde, the simplest aldehyde, to two hydrogen atoms.

Many aldehydes are particularly fragrant. A number of flowers, for example, owe their pleasant odor to the presence of simple aldehydes. The smells of lemons, cinnamon, and almonds are due to the aldehydes citral, cinnamaldehyde, and benzaldehyde, respectively. The structures

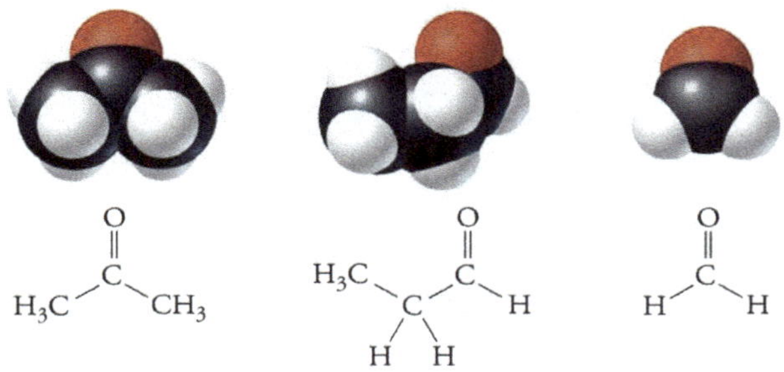

Acetone (a) | Propanal (b) | Formaldehyde (c)

< Figure 12.22
(a) When the carbon of a carbonyl group is bonded to two carbon atoms, the result is a ketone. An example is acetone. (b) When the carbon of a carbonyl group is bonded to at least one hydrogen atom, the result is an aldehyde. An example is propanal. The simplest aldehyde is formaldehyde, which has two hydrogens bonded to the carbon of the carbonyl.

of these three aldehydes are shown in **Figure 12.23**. Another aldehyde, vanillin, which was introduced at the beginning of this chapter, is the key flavoring molecule derived from seed pods of the vanilla orchid. You may have noticed that vanilla seed pods and vanilla extract are fairly expensive. Imitation vanilla flavoring is less expensive because it is merely a solution of the compound vanillin, which is economically synthesized from waste chemicals from the wood pulp industry. Imitation vanilla does not taste the same as natural vanilla extract, however, because in addition to vanillin, many other flavorful molecules contribute to the complex taste of natural vanilla. Many books manufactured in the days before acid-free paper smell of vanilla because of the vanillin formed and released as the paper ages, a process accelerated by the acids the paper contains.

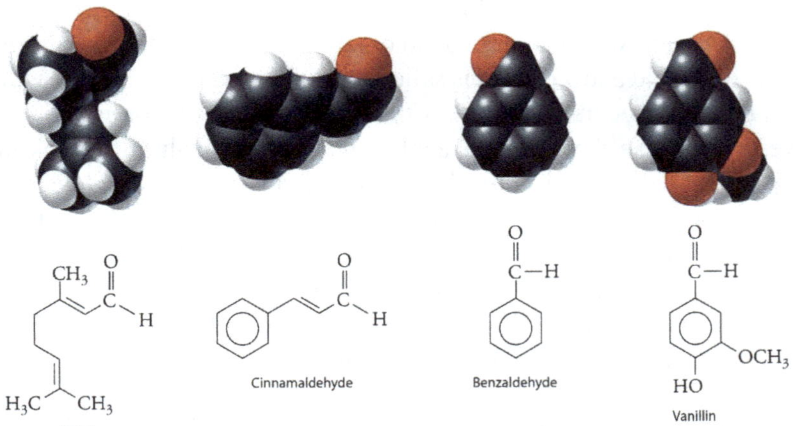

Citral | Cinnamaldehyde | Benzaldehyde | Vanillin

< Figure 12.23
Aldehydes are responsible for many familiar fragrances.

An **amide** is a carbonyl-containing organic molecule in which the carbonyl carbon is bonded to a nitrogen atom. The active ingredient of most mosquito repellents is an amide whose chemical name is N, N-diethyl-*m*-toluamide but is commercially known as DEET, shown in **Figure 12.24**. This compound is actually not an insecticide. Rather, it causes certain insects, especially mosquitoes, to lose their sense of direction, which effectively protects DEET wearers from being bitten.

In a **carboxylic acid** the carbonyl carbon is bonded to a hydroxyl group. As its name implies, this functional group is able to donate hydrogen ions. Organic molecules that contain it are therefore weakly acidic. An example is acetic acid, $C_2H_4O_2$, which, after water, is the main ingredient of vinegar—when you smell vinegar, you are smelling acetic acid. You may recall that this organic compound was used as an example of a weak acid back in Chapter 10.

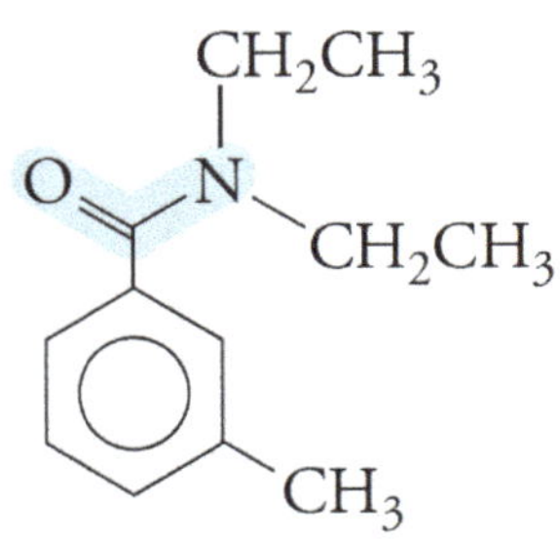

N,N-Diethyl-*m*-toluamide
(DEET)

∧ Figure 12.24
N,N-diethyl-*m*-toluamide is an example of an amide. Amides contain the amide group, shown highlighted in blue.

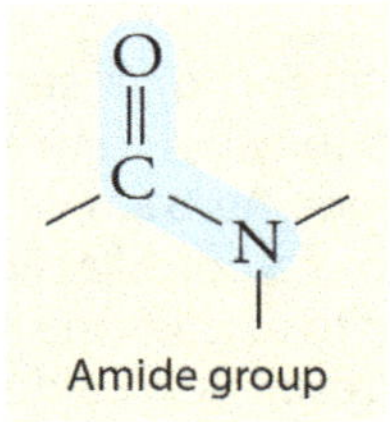

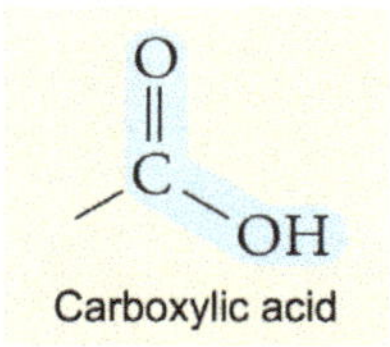

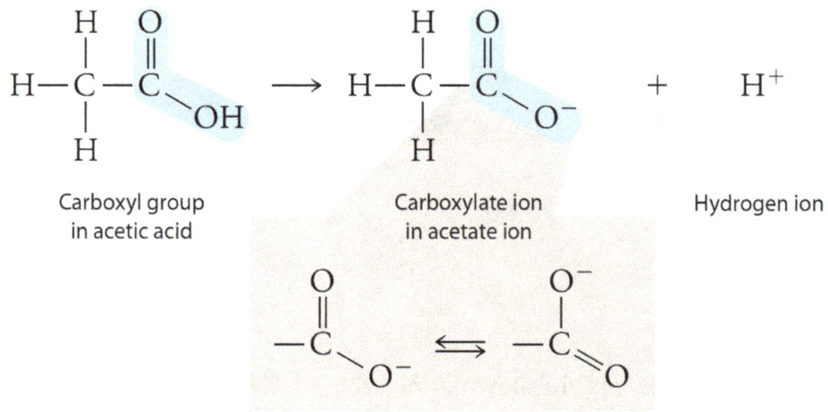

∧ Figure 12.25

The negative charge of the carboxylate ion is spread across the two oxygen atoms of the carboxyl group. This stabilizes the negative charge, which makes it easier for the carboxylic acid group to donate a hydrogen ion.

As with phenols, the acidity of a carboxylic acid results in part from the ability of the functional group to accommodate the negative charge of the ion that forms after the hydrogen ion has been donated. As shown in **Figure 12.25**, a carboxylic acid transforms to a carboxylate ion as it loses the hydrogen ion.

The negative charge of the carboxylate ion is then distributed between the two oxygens. This spreading out helps to accommodate the negative charge. An interesting example of an organic compound that contains both a carboxylic acid and a phenol is salicylic acid, found in the bark of willow trees and illustrated in **Figure 12.26a**. At one time brewed for its antipyretic (fever-reducing) effect, salicylic acid is an important analgesic (painkiller), but it causes nausea and stomach upset due to its relatively high acidity, a result of the presence of two acidic functional groups. In 1899, Friedrich Bayer and Company, in Germany, introduced a chemically modified version of this compound in which the acidic phenolic group was transformed into an *ester* functional group. The result was the less acidic and more tolerable drug called acetylsalicylic acid, the chemical name for aspirin, shown in **Figure 12.26b**.

An **ester** is an organic molecule similar to a carboxylic acid except that in the ester, the hydroxyl hydrogen is replaced by a carbon. Unlike

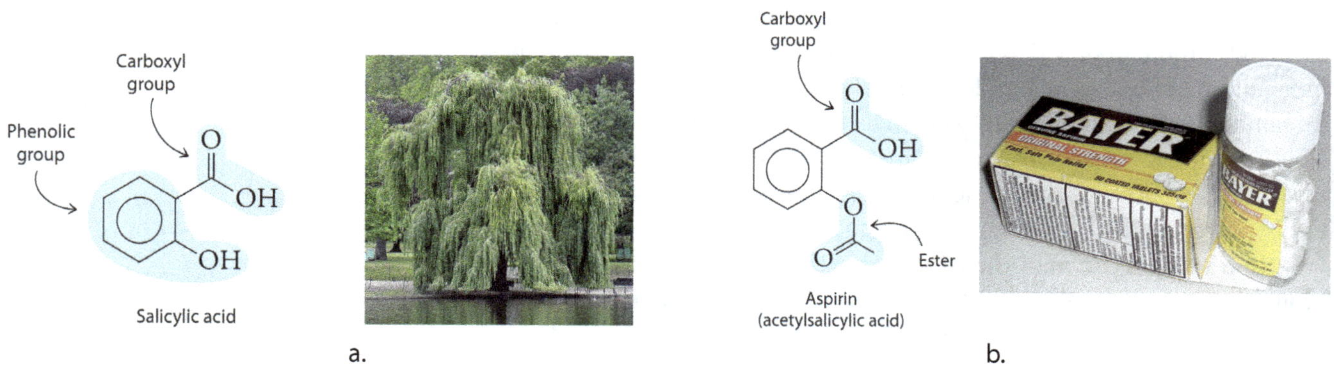

∧ Figure 12.26

(a) Salicylic acid, which is found in the bark of willow trees, is an example of a molecule containing both a carboxyl group and a phenolic group. (b) Aspirin, acetylsalicylic acid, is less acidic than salicylic acid because it no longer contains the acidic phenolic group, which has been converted to an ester.

TABLE 12.5 Some Esters and Their Flavors and Odors

STRUCTURE	NAME	FLAVOR/ODOR
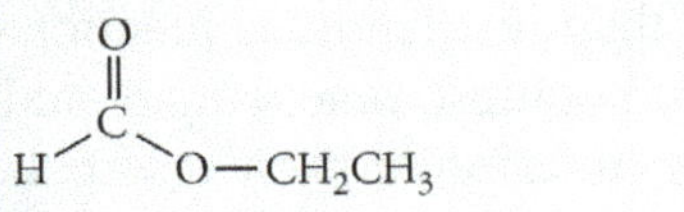	Ethyl formate	Rum
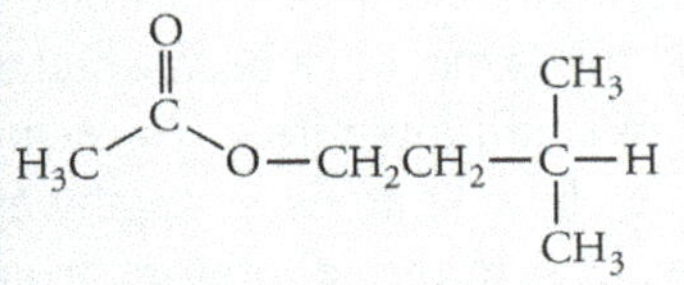	Isopentyl acetate	Banana
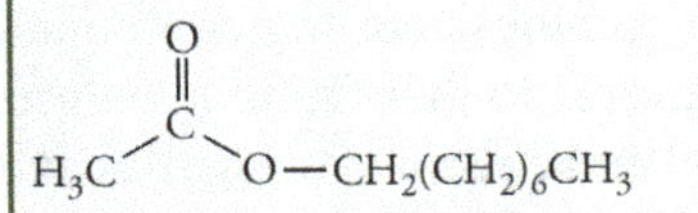	Octyl acetate	Orange
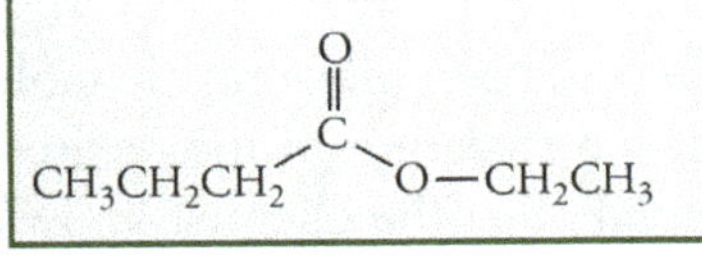	Ethyl butyrate	Pineapple

carboxylic acids, esters are not acidic because they lack the hydrogen of the hydroxyl group. Like aldehydes, many simple esters have notable fragrances and are often used as flavorings. Some familiar ones are listed in **Table 12.5.**

Esters are fairly easy to synthesize by dissolving a carboxylic acid in an alcohol and then bringing the mixture to a boil in the presence of a strong acid, such as sulfuric acid, H_2SO_4. Shown below is the synthesis of methyl salicylate from salicylic acid and methanol. Methyl salicylate is responsible for the smell of wintergreen and is a common ingredient of hard candies.

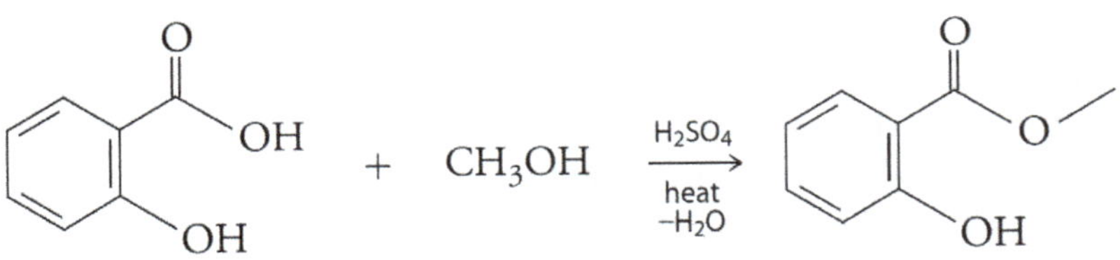

salicylic acid　　methanol　　methyl salicylate

> **fyi FOR YOUR INFORMATION**
>
> Felix Hoffmann was the chemist working at Bayer & Co, who in 1897, added the acetyl group to the phenol group of salicylic acid. According to Bayer, Hoffmann was inspired by his father, who had been complaining about the side effects of salicylic acid. To market the new drug, Bayer invented the name aspirin, where *a-* is for acetyl, *-spir-* is for the spirea flower, another natural source of salicylic acid, and *-in* is a common suffix for medications. After World War I, Bayer, a German company, lost the rights to use the name aspirin. Bayer didn't regain these rights until 1994, for a steep price of $1 billion.

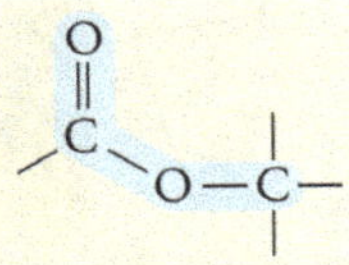

Ester group

CONCEPT CHECK

Identify all the functional groups in these two molecules (ignore the sulfur group in penicillin G).

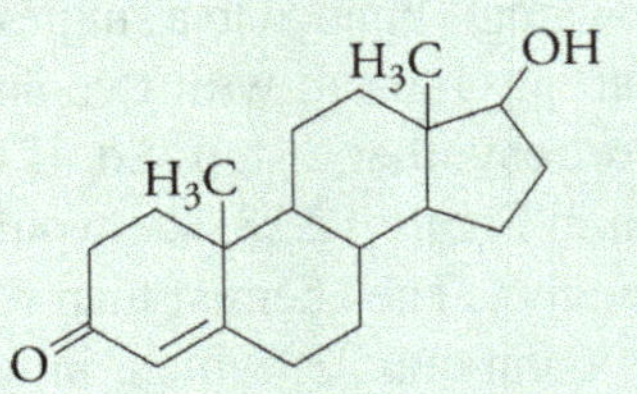

Testosterone

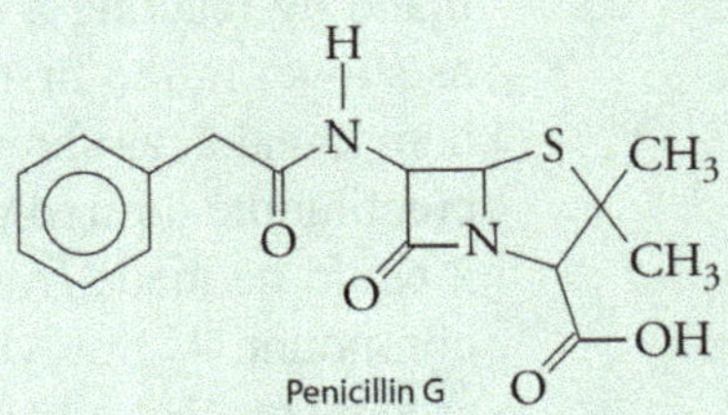

Penicillin G

CHECK YOUR ANSWER

Testosterone: alcohol and ketone. Penicillin G: amide (two amide groups) and carboxylic acid.

12.7 Organic Synthesis

There is much more to organic chemistry than just learning functional groups and their general properties. Many, if not most, practicing organic chemists dedicate much of their time to the synthesis of organic molecules that have practical applications, such as for agriculture or pharmaceuticals. Often these target molecules are organic compounds that have been isolated from nature, where they can be found only in small quantities. To create large amounts of these chemicals, the organic chemist devises a pathway through which the compound can be synthesized in the laboratory from readily available smaller compounds. Once synthesized, the compound produced in the laboratory is chemically identical to that found in nature. In other words, it will have the same physical and chemical properties and will have the same biological effects, if any.

READING CHECK

How might an organic chemist plan for the synthesis of a complex organic molecule?

How is it that an organic chemist synthesizes a complex organic molecule? One common approach is to look at the structure of the desired compound and to imagine cutting certain bonds that the chemist knows would be easy to re-form. The chemist, for example, might imagine cutting the structure into two halves. Each half is then divided into even simpler portions. In essence, the chemist is working backward, starting with the desired chemical product and arriving at a list of smaller reactant molecules that would be needed to build this product. This is called a *retrosynthesis* analysis. The chemist then goes to the laboratory to attempt the actual synthesis. Things rarely go exactly as planned, so the synthetic scheme is modified as necessary. Overall, the process requires a strong working knowledge of how to build various chemical bonds, along with a healthy dose of creativity, luck, and perseverance.

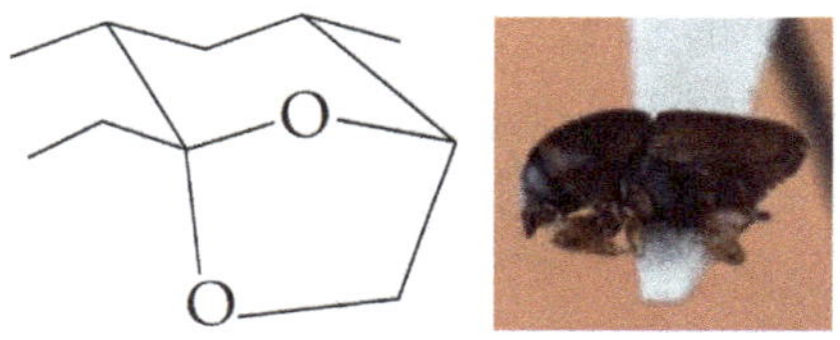

Multistriatin
(Elm bark beetle pheromone)

∧ Figure 12.27

Multistriatin is a pheromone produced only in small quantities by the female elm bark beetle

To illustrate how this is done, we present a retrosynthesis analysis of the compound multistriatin, shown in **Figure 12.27.** This compound is one of the pheromones of the elm bark beetle. It is a fragrant compound released by a virgin female beetle when she has found a good source of food, such as an elm tree. Male beetles follow this scent to the female beetle at the elm tree. The male beetle carries the fungus for Dutch elm disease, so the tree upon which it lands gets infected. Larger quantities of this pheromone, which is produced only in small quantities by the beetle, are useful for traps that capture the male beetles, thereby preventing the spread of the disease. This approach of targeting a specific pest with traps is more desirable than the application of a pesticide, many of which kill beneficial insects and have the potential to contaminate foodstuffs.

The chemical structure for mulitstriatin, **1**, as shown in **Figure 12.28**, may look complicated. The knowledgeable organic chemist, however, recognizes that two oxygen atoms bonded to a single carbon atom can be easily made by reacting a carbonyl group with two adjacent hydroxyl groups, as shown in the first retrosynthetic step. So, if the chemist could create compound **2**, as shown in Figure 12.28, she would be one step away from creating the desired product. The chemist then recognizes that compound **2** could be made from compound **3**, which, in turn, could be made from compound **4**. Studying compound **4** carefully, the experienced chemist recognizes that a bond two atoms away from a carbonyl group is easy to form. This leads her to break compound **4** evenly into compounds **5** and **6**. Compound **5** can be bought from a chemical supply company, while compound **6** requires only a few more steps to come to compound **8**, which is also commercially available.

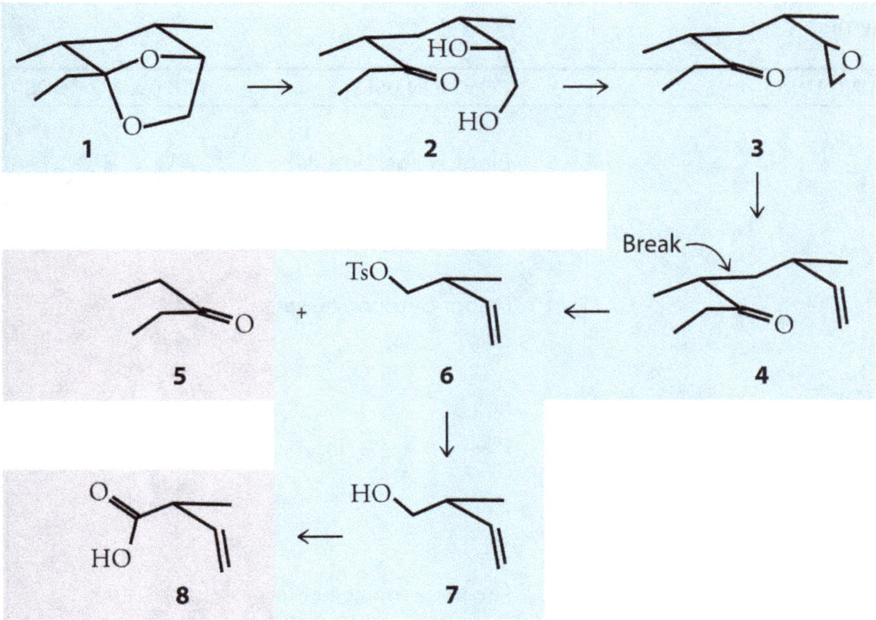

< **Figure 12.28**

A retrosynthetic analysis of the elm bark beetle pheromone multistriatin. The commercially available starting materials for the forward synthesis are in the purple shading.

After developing this retrosynthetic plan and seeking helpful input from her colleagues, the chemist heads to the laboratory to try to make some variation of the forward synthesis work.

Naturally, some organic molecules are more difficult to synthesize than others. Look ahead to Chapter 14 for the structure of the anticancer agent Taxol®, shown in Figure 14.4. This compound is very useful for the treatment of breast cancer, but unfortunately, it is produced in nature by the yew tree in only very small quantities. This prompted teams of organic chemists to synthesize this molecule from easily obtained starting materials. The first total synthesis of Taxol® by K. C. Nicolaou of the Scripps Research Institute in 1994 was a major accomplishment. Other researchers have since refined the synthesis of Taxol®, which has been of great benefit in the fight against cancer. Online, use the keywords *taxol total synthesis* to find the stunning synthetic pathways used by these practicing world-class chemists.

12.8 Polymer Chemistry

Polymers are exceedingly long molecules that consist of repeating molecular units called **monomers**, as **Figure 12.29** illustrates. Monomers have relatively simple structures, consisting of anywhere from 4 to 100 atoms per molecule. When monomers are chained together, they can form polymers consisting of hundreds of thousands of atoms per molecule. These large molecules are still too small to be seen with the unaided eye. They are, however, giants in the sub-microscopic world—if a typical polymer molecule were as thick as a kite string, it would be 1 kilometer long.

V **Figure 12.29**

A polymer is a long molecule consisting of many smaller monomer molecules linked together.

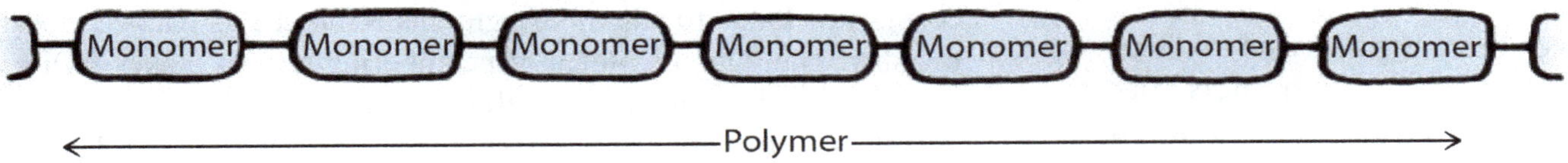

TABLE 12. 6 Addition and Condensation Polymers

POLYMERS	REPEATING UNIT	COMMON USES	RECYCLING CODE
Addition			
Polyethylene (PE)	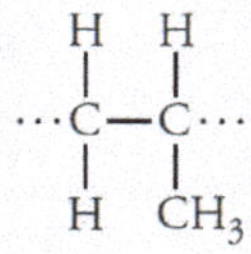	Plastic bags, bottles	2 HDPE, 4 LDPE
Polypropylene (PP)	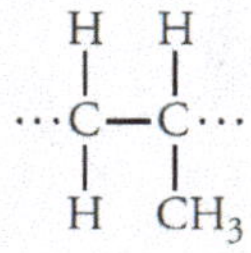	Indoor-outdoor carpets	5 PP
Polystyrene (PS)	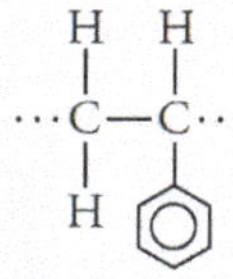	Plastic utensils, insulation	6 PS
Polyvinyl chloride (PVC)	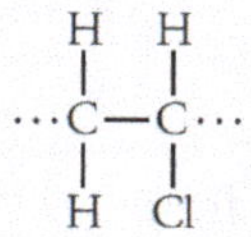	Shower curtains, tubing	3 V
Condensation			
Polyethylene terephthalate (PETE)	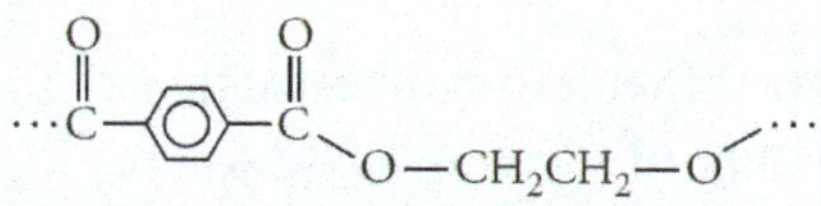	Clothing, plastic bottles	1 PET

Human-made polymers, also known as synthetic polymers, make up the class of materials commonly known as plastics. Two major types of synthetic polymers are used today—*addition polymers* and *condensation polymers*.

As shown in **Table 12.6**, addition and condensation polymers have a wide variety of uses. Solely the product of human design, these polymers pervade modern living. In the United States, for example, synthetic polymers have surpassed steel as the most widely used material.

fyi FOR YOUR INFORMATION

The amount of energy required to recycle plastic can be more than that required to create fresh plastic. For all the great utility of plastics, their environmental impact remains a significant drawback.

Addition Polymers

Addition polymers form simply by the joining together of monomer units. For this to happen, each monomer must contain at least one double bond. As shown in **Figure 12.30**, polymerization occurs when two of the electrons from each double bond split away from each other to form new covalent bonds with neighboring monomer molecules. During this process, no atoms are lost; so the total mass of an addition polymer is equal to the sum of the masses of all its monomers.

Over 22 million tons of polyethylene are produced annually in the United States; that's about 130 pounds per U.S. citizen. The monomer from which it is synthesized, ethylene, is an unsaturated hydrocarbon produced in large quantities from petroleum.

High-density polyethylene (HDPE), shown in **Figure 12.31a**, consists of long strands of straight-chain molecules packed closely together. The tight alignment of neighboring strands makes HDPE a relatively rigid, tough plastic useful for such things as bottles and milk jugs. Low-density polyethylene (LDPE), shown in **Figure 12.31b**, is made of strands with many branched chains, a pattern that prevents the

READING CHECK

Which has more mass: an addition polymer or the monomers that combined to make that addition polymer?

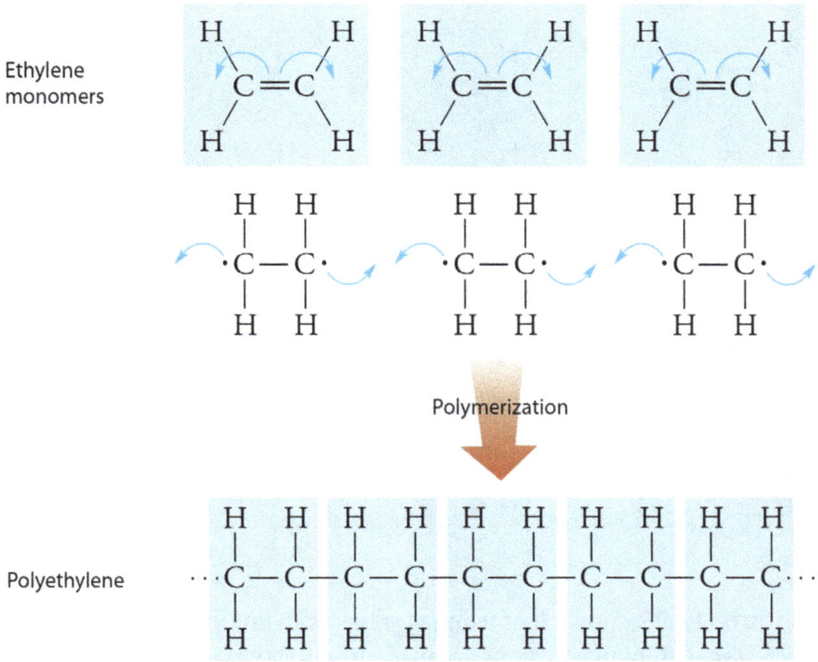

< Figure 12.30

The addition polymer polyethylene is formed as electrons from the double bonds of ethylene monomer molecules split away and become unpaired valence electrons. Each unpaired electron then joins with an unpaired electron of a neighboring carbon atom to form a new covalent bond that links two monomer units together.

strands from packing closely together. This makes LDPE more bendable than HDPE and gives it a lower melting point. While HDPE holds its shape in boiling water, LDPE deforms. It is most useful for such items as plastic bags, photographic film, and insulation for electric wire.

(a) Molecular strands of HDPE (b) Molecular strands of LDPE

< Figure 12.31
The polyethylene strands of HDPE are able to pack closely together, much like strands of uncooked spaghetti. The polyethylene strands of LDPE are branched, which prevents the strands from packing well.

Addition polymers are also created by using other monomers. The only requirement is that the monomer must contain a double (or triple) bond. The monomer propylene, for example, yields polypropylene, as shown in **Figure 12.32**. Polypropylene is a tough plastic material useful for pipes, hard-shell suitcases, and appliance parts. Fibers of polypropylene are used for upholstery, indoor-outdoor carpet, and even thermal underwear.

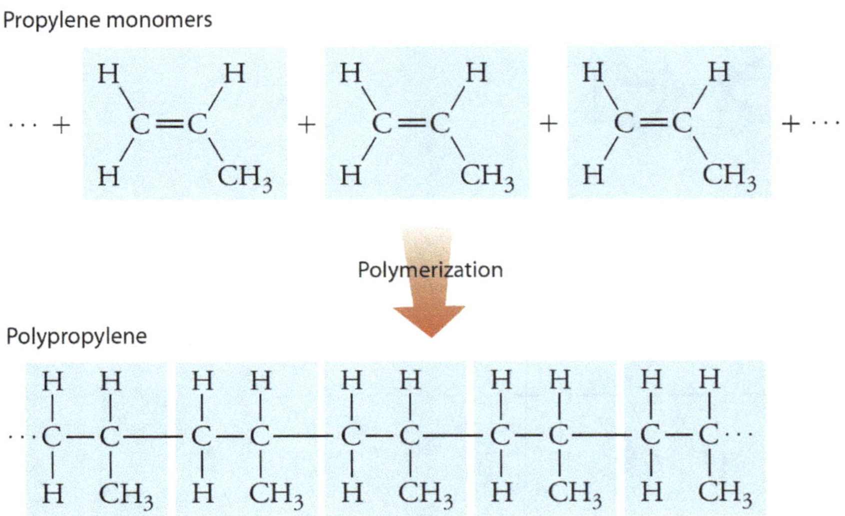

< Figure 12.32

Propylene monomers polymerize to form polypropylene.

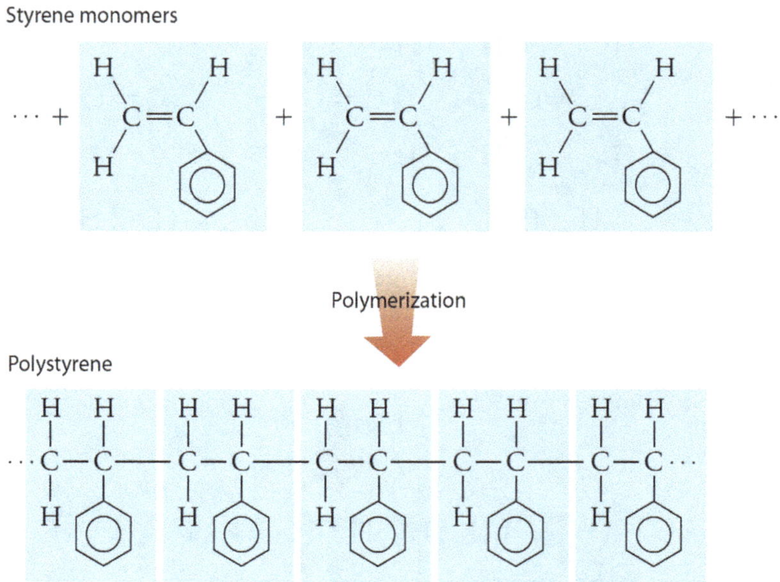

Figure 12.33 >

Styrene monomers polymerize to form polystyrene.

Figure 12.33 shows that using styrene as the monomer yields polystyrene. Transparent plastic cups are made of polystyrene, as are thousands of other household items. Blowing air into liquid polystyrene generates Styrofoam™, which is widely used for coffee cups, packing material, and insulation.

Another important addition polymer is polyvinyl chloride (PVC), which is tough and easily molded. Floor tiles, shower curtains, and pipes are most often made of PVC, shown in **Figure 12.34.**

Rigid polymers such as PVC can be made soft by incorporating small molecules called plasticizers. Pure PVC, for example, is a tough material great for making pipes. Mixed with a plasticizer, the PVC becomes soft and flexible and thus useful for making shower curtains, toys, and many other products now found in most households. One of the more commonly used types of plasticizers is the phthalates, some of which have been shown to disrupt the development of reproductive organs, especially in fetuses and in growing children. Governments and manufacturers are now working to phase out these plasticizers. But some phthalates, such as DINP, have been shown to be much less dangerous. Should all phthalates be banned or just the ones proven to be harmful? This is a social and political question that has yet to be resolved.

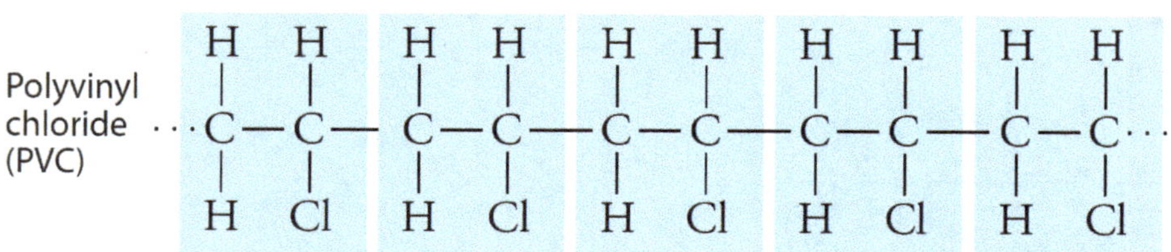

< Figure 12.34

PVC is tough and easily molded, which is why it often is used to fabricate many household items.

Polyvinylidene chloride (Saran)^{TM}

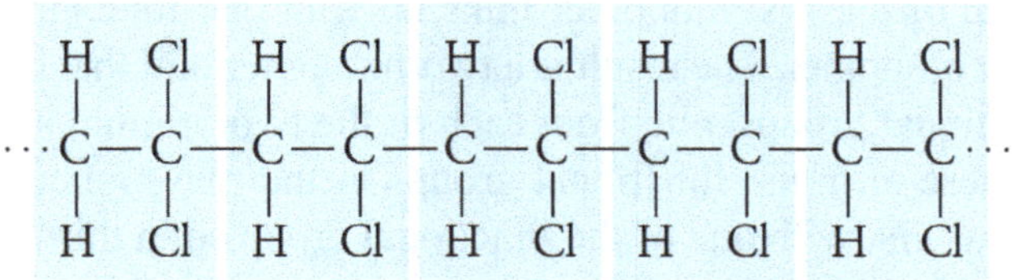

< **Figure 12.35**

The many large chlorine atoms in polyvinylidene chloride make this addition polymer sticky.

The addition polymer polyvinylidene chloride (trade name Saran™), shown in **Figure 12.35**, was once used to make plastic wrap for food. The large chlorine atoms in this polymer help it stick to such surfaces as glass by induced dipole attractions, as discussed in Section 7.1. Because of environmental concerns, however, Saran™ brand food wrap is now made from polyethylene. The original polyvinylidene chloride polymer contains twice the number of chlorine atoms as PVC and thus readily transforms into harmful dioxins upon burning. (See the Chapter 11 Essay)

The addition polymer polytetrafluoroethylene, shown in **Figure 12.36**, is what you know as Teflon®. In contrast to the chlorine-containing Saran, fluorine-containing Teflon® has a nonstick surface, because the fluorine atoms tend not to experience any molecular attractions. In addition, because carbon–fluorine bonds are unusually strong, Teflon® can be heated to high temperatures before decomposing. These properties make Teflon® an ideal coating for cooking surfaces. It is also relatively inert, which is why many corrosive chemicals are shipped or stored in Teflon® containers.

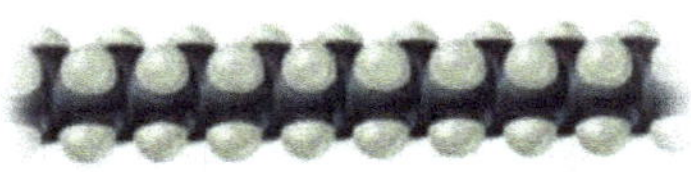

Polytetra-fluoroethylene (Teflon)®

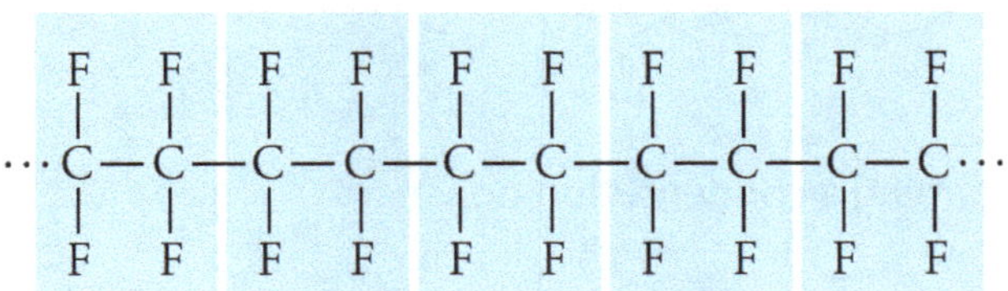

< **Figure 12.36**

The fluorine atoms in polytetrafluoroethylene tend not to experience molecular attractions, which is why this addition polymer is used as a nonstick coating and lubricant.

CONCEPT **CHECK**

What do all monomers that are used to make addition polymers have in common?

CHECK YOUR ANSWER A multiple covalent bond between two carbon atoms.

Condensation Polymers

A **condensation polymer** is formed when the joining of monomer units is accompanied by the loss of a small molecule, such as water or hydrochloric acid. Any monomer capable of becoming part of a condensation polymer must have a functional group on each end. When two such monomers come together to form a condensation polymer, one functional group of the first monomer links up with one functional group of the other monomer. The result is a two-monomer unit that has two terminal functional groups, one from each of the two original monomers. Each of these terminal functional groups in the two-monomer unit is now free to link with one of the functional groups of a third monomer, then a fourth, and so on. In this way, a polymer chain is built.

Figure 12.37 shows this process for the condensation polymer called Nylon®. This polymer is composed of two different monomers, as shown in Figure 12.37. One monomer is adipic acid, which contains two reactive end groups, both carboxyl groups. The second monomer is hexamethylenediamine, in which two amine groups are the reactive end groups. One end of an adipic acid molecule and one end of a hexamethylamine molecule can be made to react with each other, splitting off a water molecule in the process. After two monomers have joined, reactive ends still remain for further reactions, which leads to a growing polymer chain. Aside from its use in hosiery, Nylon® also finds important uses in the manufacture of ropes, parachutes, clothing, and carpets.

Another widely used condensation polymer is polyethylene terephthalate (PET or PETE), which is formed from the polymerization of ethylene glycol and terephthalic acid, as shown in **Figure 12.38.** Plastic soda bottles are made from this polymer. Also, PETE fibers are sold as Dacron™ polyester, a product used in clothing and stuffing for pillows and sleeping bags. Thin films of PETE, which are called Mylar®, can be coated with metal particles to make magnetic recording tape or those metallic-looking balloons you see for sale at grocery store checkout counters.

fyi FOR YOUR INFORMATION

Since their introduction, plastics have become ubiquitous. This includes the daily plastics we throw away as trash, as well as microscopic plastic fibers that measurably seep toward the oceans from our collective laundry machines and other human sources. The air we breathe, especially indoors, is also rife of microplastics. Wave some sticky tape through the air. Examine that sticky tape with a magnifying glass and you'll likely see many brightly colored plastic microfibers. Of course, the sticky tape itself is yet another form of plastic you'll likely discard. Interestingly, the root of the problem is not the plastics themselves, but the great number of humans using them.

CONCEPT CHECK

The structure of 6-aminohexanoic acid is the following:

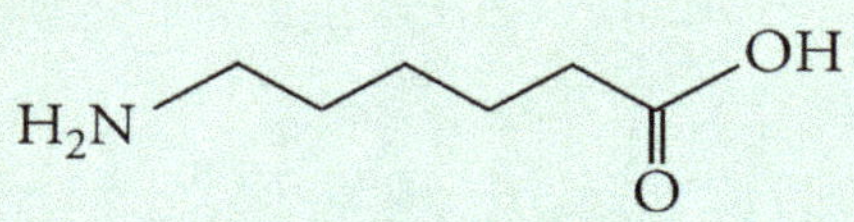

Is this compound a suitable monomer for forming a condensation polymer?

CHECK YOUR ANSWER Yes, because the molecule has two reactive ends. You know both ends are reactive because they are the ends shown in **Figure 12.37**. The only difference here is that both types of reactive ends are on the *same* molecule. Monomers of 6-aminohexanoic acid combine by splitting off water molecules to form the polymer known as nylon-6.

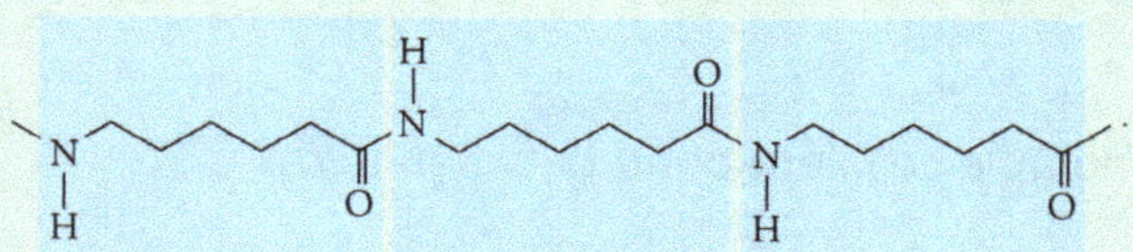

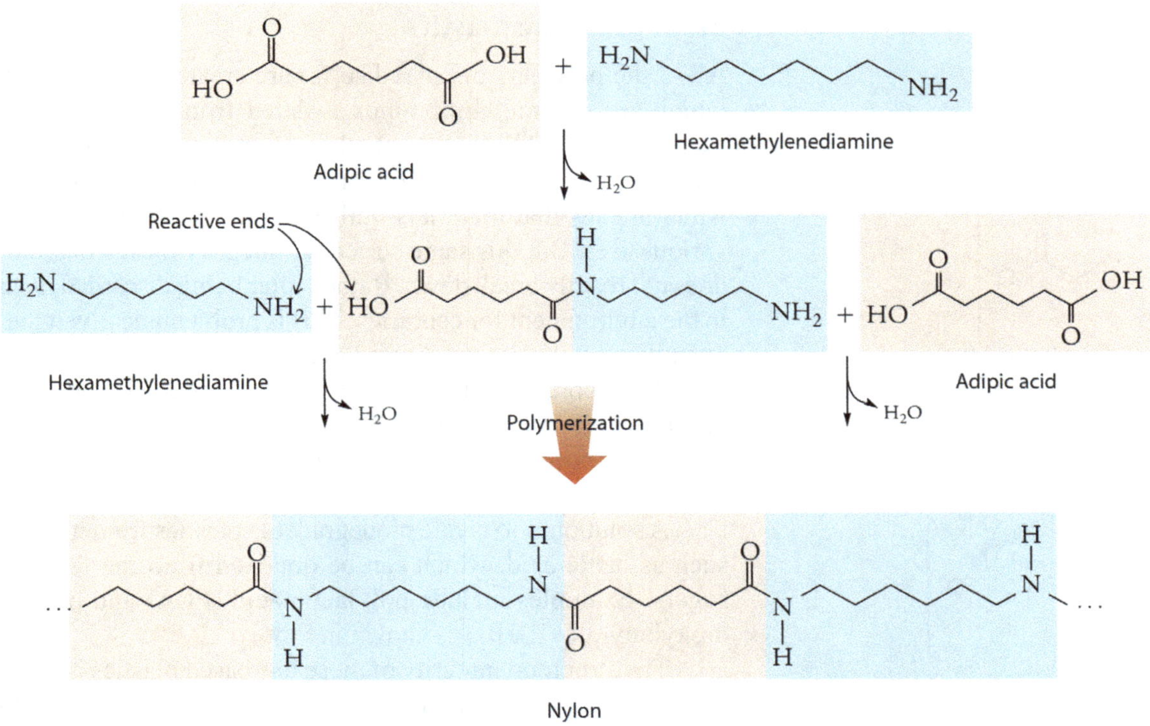

∧ Figure 12.37

Adipic acid and hexamethylenediamine polymerize to form the condensation polymer Nylon®.

Monomers that contain three reactive functional groups can also form polymer chains. These chains become interlocked in a rigid 3-dimensional network that lends considerable strength and durability to the polymer. Once formed, these condensation polymers cannot be remelted or reshaped, which makes them hard-set, or *thermoset*, polymers. Hard plastic dishes (Melmac™) and countertops (Formica™) are made of this material. A similar polymer, Bakelite™, made from formaldehyde and phenols containing multiple oxygen atoms, is used to bind plywood and particleboard. As discussed later in this chapter, Bakelite was synthesized in the early 1900s, and it was the first widely used polymer.

∨ Figure 12.38

Terephthalic acid and ethylene glycol polymerize to form the condensation polymer polyethylene terephthalate.

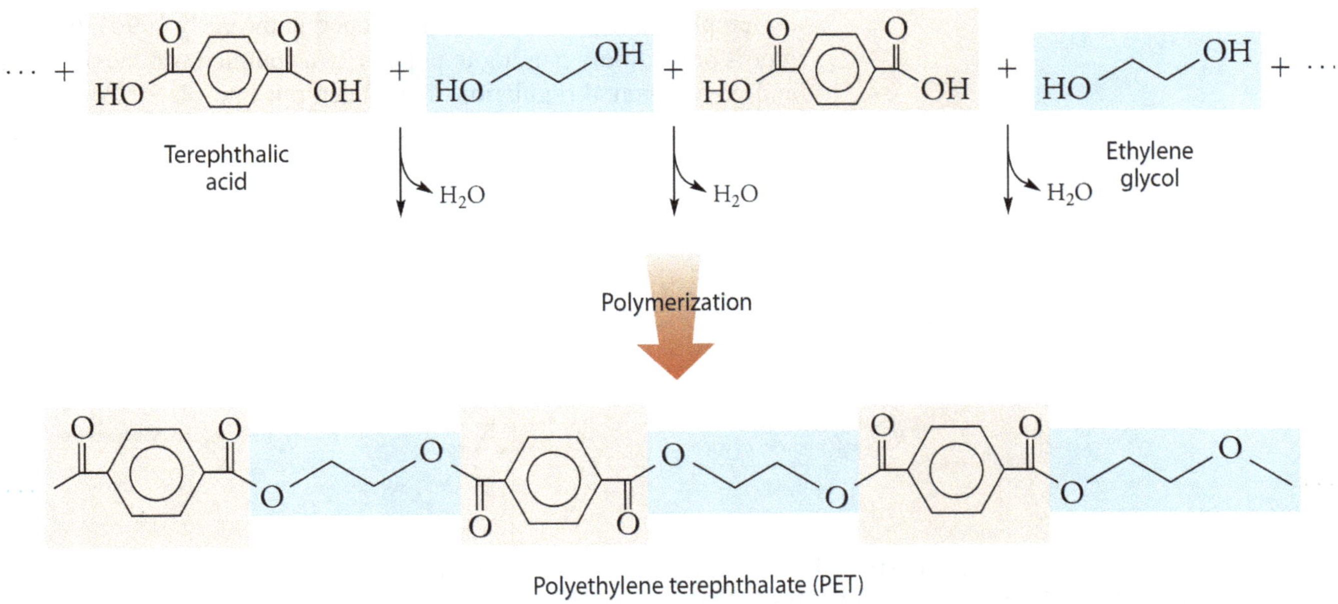

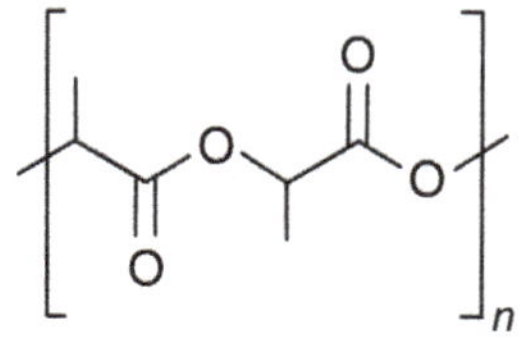

PLA

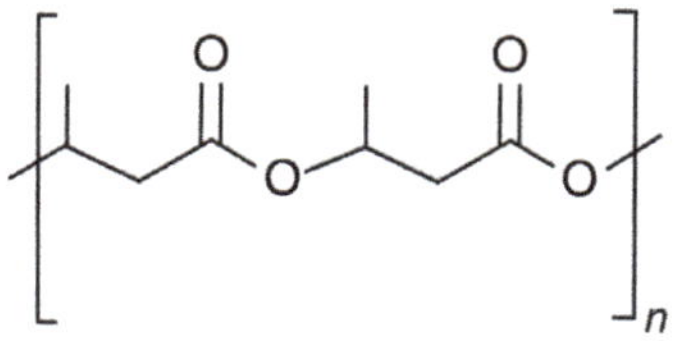

P3HB

Λ Figure 12.39

PLA and P3HB are two examples of polymers produced from natural sources through fermentation. Though they lack much of the structural integrity of petrochemical plastics, they readily bio-degrade.

Biodegradable Plastics

All of the polymers discussed so far are created from *petrochemicals*, which are organic compounds isolated from fossil fuels. With the exception of PETE, you'll note the polymer chain consists of repeated carbon-carbon bonds. This gives the polymer a great structural integrity, which means that materials built from the polymer hold up well for various uses. But this same structural integrity means that the polymer does not readily break down. Rather, after being discarded, it can persist in the environment for centuries. This is problematic given the growing quantities of plastics we discard every year.

An important solution is to recycle these plastics. But despite public education efforts, less than 10% of all plastics get recycled. The bulk still end up within the landfill. Furthermore, plastics, once recycled, do not hold up as well as fresh plastic. It's a problem.

A solution is to create biodegradable plastics from natural sources, such as lactic acid, which can be obtained from the fermentation of sugars. Examples include polylactic acid (PLA) and poly-(R)-3-hydroxybutyrate (P3HB) as shown in **Figure 12.39**.

The structural integrity of these bio-based plastics is not as strong. Ideally the environmentally-aware consumer can come to appreciate this as an advantage rather than disadvantage. The chemistry-aware consumer will recognize that the monomers of these polymers are held together by ester linkages, which are relatively easy to break apart. How so? By *hydrolysis*, which happens when the ester is attacked by a water molecule as shown in **Figure 12.40**.

Now look back to the chemical structure of PETE, which is perhaps the most successful of all petrochemically-based plastics for its strength, versatility, and recyclability. You'll see that it too consists of ester linkages. Might it then also be biodegradable? Unfortunately, the answer is no. Alongside those ester functional groups are multiple benzene rings. Because these benzene rings are most nonpolar they effectively repel water molecules, which greatly inhibits any hydrolysis.

A solution to this problem is to build a polymer with ester linkages that looks much like PETE, but has fewer benzene rings, as shown in **Figure 12.41**. This is exactly what we have with polybutylene adipate co-terephthalate (PBAT). First developed in the early 1990s, this polymer is now rapidly gaining in popularity as public awareness has risen and governmental regulations have been put into place. Consider the one example of laying down a mulch film to prevent the growing of

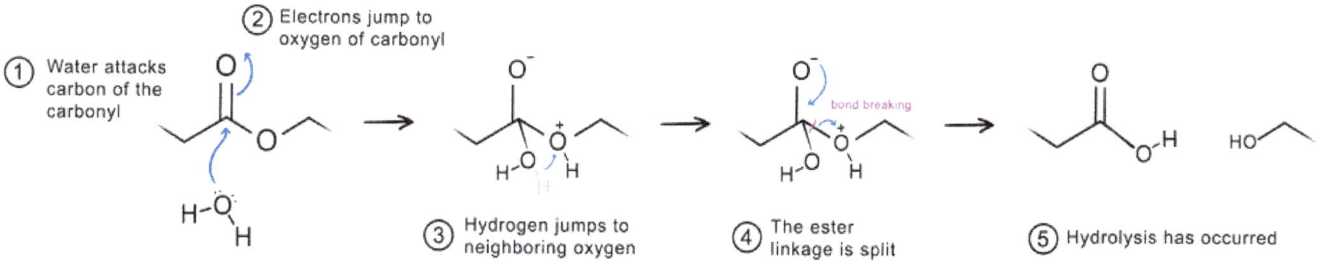

Λ Figure 12.40

The ester functional group is susceptible to hydrolysis, which is a process whereby a water molecule inserts itself in between the carbon-oxygen single bond. The end products are a carboxylic acid and an alcohol.

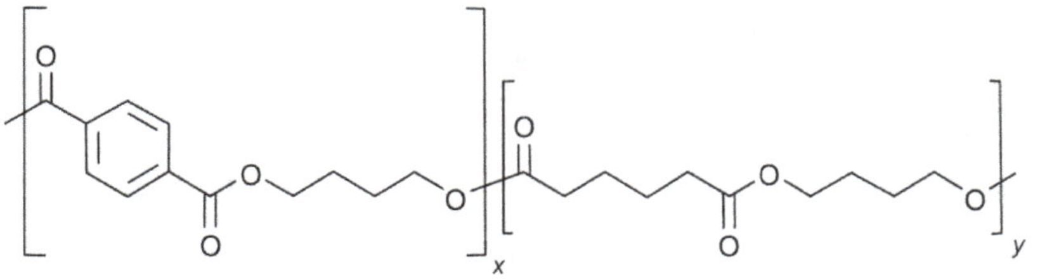

PBAT

< **Figure 12.41**

The PBAT polymer works well for the formulation of plastics with sufficient structural integrity. It can be produced from petrochemicals, and thereby readily generated in mass quantities. Importantly, because of the fewer benzene rings, this ester readily biodegrades.

weeds within an agricultural field. The bio-based polymer will fall apart too quickly. A traditional petrochemical polymer would have to be pulled-up and landfilled after use. The PBAT based film, by contrast, can be mulched into the soil where it degrades in a timely manner.

The synthetic polymer industry has grown remarkably. Annual production of polymers in the United States alone has grown from 3 billion pounds in 1950 to more than 100 billion pounds in 2020. Today, it is a challenge to find any consumer item that does not contain a plastic of one sort or another.

In the near future, watch for new kinds of polymers having a wide range of remarkable properties. One interesting application is shown in **Figure 12.42**. We already have polymers that conduct electricity, others that emit light, others that replace body parts, and still others that are stronger but much lighter than steel. Imagine synthetic polymers that mimic photosynthesis by transforming solar energy to chemical energy or that efficiently separate fresh water from ocean water. These are not dreams. They are realities that chemists have already been demonstrating in the laboratory.

∧ **Figure 12.42**

Flexible and flat video displays known as OLEDs (organic light-emitting diodes) are now fabricated from polymers.

12.9 A Brief History of Plastics

The search for a lightweight, non-breakable, moldable material began with the invention of vulcanized rubber. This material is derived from natural rubber, which is a semisolid, elastic, natural polymer. The fundamental chemical unit of natural rubber is polyisoprene, which plants produce from isoprene molecules, as shown in **Figure 12.43**. In the 1700s, natural rubber was noted for its ability to rub off pencil marks, which is the origin of the term *rubber*. Natural rubber has few other uses, however, because it turns gooey at warm temperatures and brittle at cold temperatures.

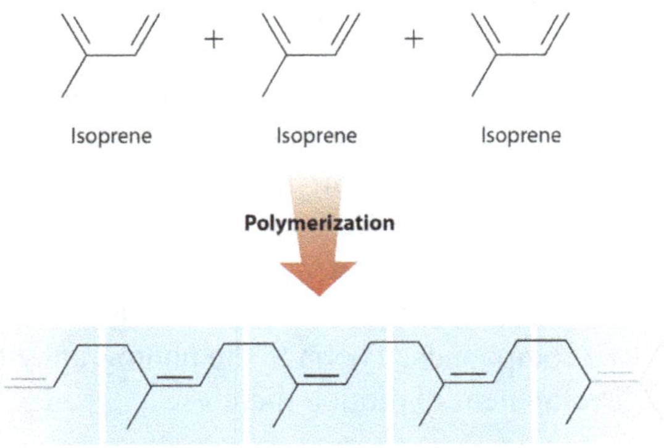

< **Figure 12.43**

Isoprene molecules react with one another to form polyisoprene. Isoprene is the fundamental chemical unit, or monomer, of natural rubber.

Figure 12.44 >

(a) When stretched, the individual polyisoprene strands in natural rubber slip past one another and the rubber stays stretched.

(b) When vulcanized rubber is stretched, the sulfur crosslinks hold the strands together, allowing the rubber to return to its original shape.

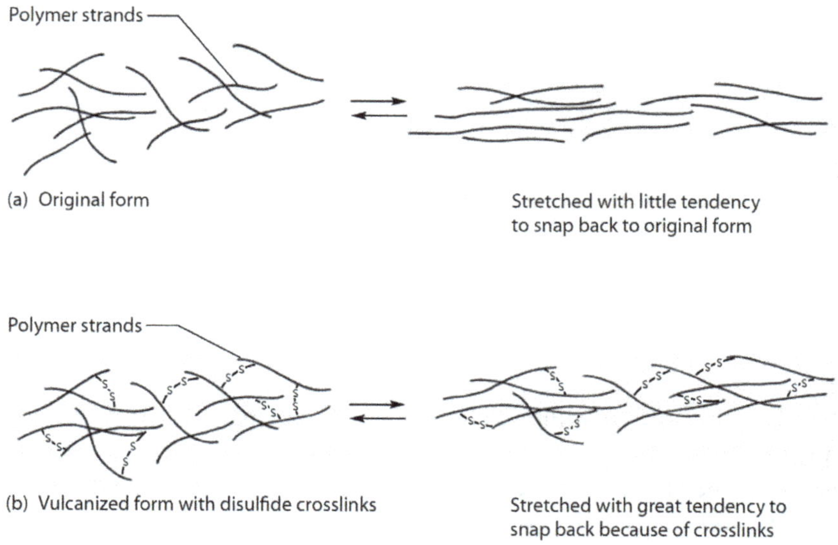

CHEMICAL CONNECTIONS

How is celluloid connected to the many exits now in movie theaters?

In 1839, an American inventor, Charles Goodyear, discovered *rubber vulcanization*, a process in which natural rubber and sulfur are heated together. The product, vulcanized rubber, is harder than natural rubber and retains its elastic properties over a wide range of temperatures. This is the result of *disulfide cross-linking* between polymer chains, as illustrated in **Figure 12.44.** To help quench our ever-growing thirst for vulcanized rubber, the polymer of rubber (polyisoprene) is now also produced from petroleum.

Charles Goodyear was the classic eccentric inventor. He lived most of his life in poverty, obsessed with transforming rubber into a useful material. Goodyear was a man of ill health who died in debt, yet he remained stubbornly optimistic. The present-day Goodyear Corporation was founded not by Goodyear, but by others who sought to pay tribute to his name 15 years after he died.

In 1845, as vulcanized rubber was becoming popular, the Swiss chemistry professor Christian Schönbein wiped up a spilled mixture of nitric and sulfuric acids with a cotton rag that he then hung up to dry. Within a few minutes, the rag burst into flames and then vanished, leaving only a tiny bit of ash. Schönbein had discovered nitrocellulose, in which most of the hydroxyl groups in cellulose are bonded to nitrate groups, as **Figure 12.45** illustrates. Schönbein's attempts to market nitrocellulose as a smokeless gunpowder (*guncotton*) were unsuccessful, mainly because of a number of lethal explosions at plants producing the material. Researchers in France discovered that solvents such as diethyl ether and alcohol transformed nitrocellulose to a gel that could be molded into various shapes. This workable nitrocellulose material was dubbed *collodion*, and its first application was as a medical dressing for cuts.

In 1870, John Hyatt, a young inventor from Albany, New York, discovered that collodion's moldable properties were vastly improved by using camphor as a solvent. This camphor-based nitrocellulose material was named *celluloid*, and it became the plastic of choice for the manufacture of many household items, such as combs and hair fasteners. In addition, thin transparent films of celluloid made excellent supports for photosensitive compounds, a boon to the photography industry and a first step in the development of motion pictures.

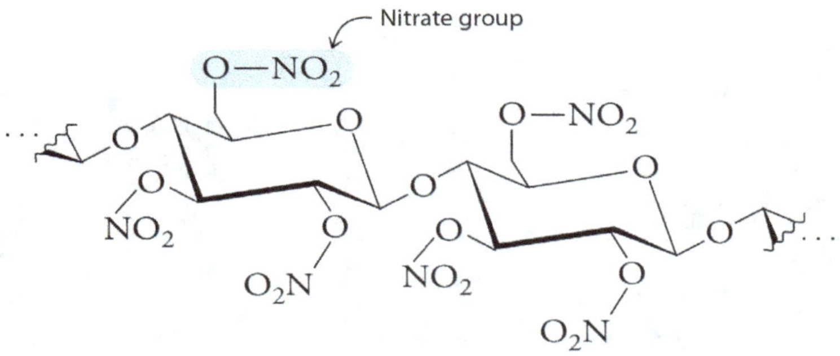

Nitrocellulose (cellulose nitrate)

< Figure 12.45

Nitrocellulose, also known as cellulose nitrate, is highly combustible because of its many nitrate groups, which facilitate oxidation.

As wonderful as celluloid was, it still had the major drawback of being highly flammable. Today, one of the few commercially available products made of celluloid is Ping-Pong balls, shown in **Figure 12.46**.

READING CHECK

What was a major drawback of celluloid?

Bakelite Was the First Widely Used Plastic

About 1899, Leo Baekeland, a chemist who had immigrated to the United States from Belgium, developed an emulsion for photographic paper that was exceptional in its sensitivity to light. He sold his invention to George Eastman, who had made a fortune selling celluloid-based photographic film along with his portable Kodak camera. Expecting no more than \$50,000 for his invention, Baekeland was shocked at Eastman's initial offer of \$750,000 (in today's dollars, that would be about \$25 million). Suddenly a very wealthy man, Baekeland was free to pursue his chemical interests.

Baekeland explored a tar-like solid once produced in the laboratories of Alfred von Baeyer, the German chemist who played a role in the development of aspirin. Whereas Baeyer had dismissed the solid as worthless, Baekeland saw it as a virtual gold mine. After several years, he produced a resin that, when poured into a mold and then heated under pressure, solidified into a transparent positive of the mold. Baekeland's resin was a mixture of formaldehyde and a phenol that polymerized into the complex network shown in **Figure 12.47**.

The solidified material, which he called *Bakelite*, was impervious to harsh acids or bases, wide temperature extremes, and just about any solvent. Bakelite quickly replaced celluloid as a molding medium, finding a wide variety of uses for several decades. It wasn't until the 1930s that alternative thermoset polymers began to challenge Bakelite's dominance in the evolving plastics industry.

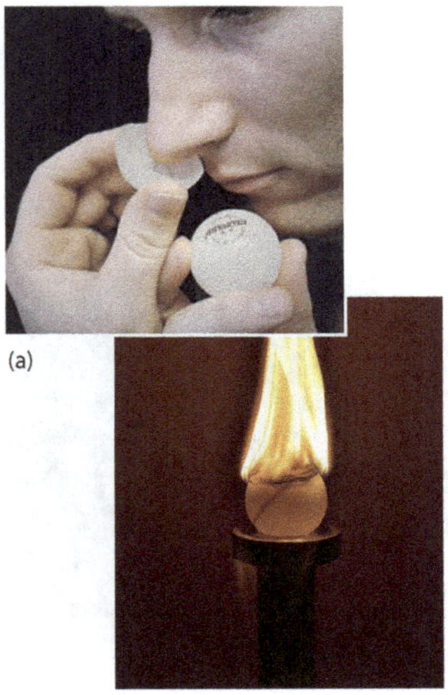

(a)

(b)

∧ Figure 12.46

(a) Smell a freshly cut Ping-Pong ball, and you will note the distinct odor of camphor, which is the same smell that arises from heat cream for sore muscles. This camphor comes from the celluloid from which the ball is made. (b) Ping-Pong balls burn rapidly because they are made of nitrocellulose.

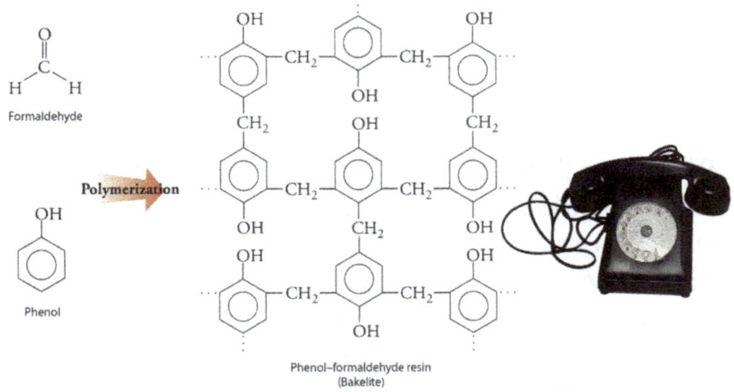

∧ Figure 12.47

The molecular network of Bakelite shown in two dimensions. (The actual structure projects in all three dimensions.) (b) The first handset telephones were made of Bakelite.

Figure 12.48 >

Viscose is still used today in the manufacture of fibers used to make the synthetic fabric called rayon. The fibers are formed as viscose and extruded through holes in metal dies. Shown here is a viscose factory from the early 1900s.

The First Plastic Wrap Was Cellophane

Cellophane had its beginnings in 1892, when Charles Cross and Edward Bevan of England found that treating cellulose with concentrated sodium hydroxide, NaOH, followed by carbon disulfide, CS_2, created a thick, syrupy yellow liquid they called *viscose*. Extruding the viscose into an acidic solution generated a tough cellulose filament that could be used to make a synthetic silky cloth today called *rayon* (**Figure 12.48**).

In 1904, Jacques Brandenberger, a Swiss textile chemist, observed restaurant workers discarding fine tablecloths that had only slight stains on them. Working with viscose at the time, he had the idea of extruding it not as a fiber, but as a thin, transparent sheet that might be adhered to tablecloths and provide an easy-to-clean surface. By 1913, Brandenberger had perfected the manufacture of a viscose-derived, thin, transparent sheet of cellulose, which he named *cellophane*.

Λ Figure 12.49

Colored cellophane samples. In the early 1900s, cellophane transformed the way foods and other items were marketed.

Within several years, the Du Pont Corporation bought the rights to cellophane. Hermetically sealed by cellophane, a product could be kept free of dust and germs. And unlike paper or tin foil—the alternatives of the day—cellophane was transparent and thus allowed the consumer to view the packaged contents. With properties such as these, cellophane played a great role in the success of supermarkets, which first appeared in the 1930s. Perhaps cellophane's greatest appeal to the consumer, however, was its shine. As marketing people soon discovered, nearly any product—soaps, canned goods, and golf balls—would sell faster when wrapped in *cellophane*.

Polymers Win in World War II

In the 1930s, more than 90 percent of the natural rubber used in the United States came from Malaysia. In the days after Pearl Harbor was attacked in December 1941 and the United States entered World War II, however, Japan captured Malaysia. As a result, the United States—the land with plenty of everything, except rubber—faced its first natural resource crisis. The military implications were devastating because without rubber for tires, military airplanes and jeeps were useless. Petroleum-based synthetic rubber had been developed in 1930 by Du Pont chemist Wallace Carothers but was not widely used because it was much more expensive than natural rubber. With Malaysian rubber impossible to get and a war on, however, cost was no longer an issue. Synthetic rubber factories were constructed across the nation, and within a few years, the annual production of synthetic rubber rose from 2000 tons to about 800,000 tons.

Also, in the 1930s, British scientists developed radar as a way to track thunderstorms. With war approaching, these scientists turned their attention to the idea that military planes could be equipped with radar. To make the radar equipment light enough to fly on an airplane, however, they needed the wires of the device coated with some sort of electrical insulator. Fortunately, the polymer polyethylene, recently developed in Britain, turned out to be an ideal electrical insulator. The radar-equipped planes were slow, but flying at night or in poor weather, they could detect, intercept, and destroy enemy aircraft.

Other polymers that had a significant impact on the outcome of World War II included Nylon®, Plexiglas®, PVC, Saran™, and Teflon®. Nylon® was invented in 1937 and, as discussed in the previous section, found great use in the manufacture of ropes, parachutes, and clothing. Interestingly, tear-resistant Nylon® stockings, which had just been introduced, were extremely popular during the war and often used as a commodity for favors or to repay debts among soldiers. Plexiglas®, shown in **Figure 12.50**, is a polymer known to chemists as poly(methyl methacrylate). This glass-like but moldable and lightweight material made excellent domes for the gunner's nests on fighter planes and bombers.

Polyvinyl chloride (PVC) had been developed by a number of chemical companies in the 1920s. The problem with this material, however, was that it lost resiliency when heated. In 1929, Waldo Semon, a chemist at BFGoodrich, found that PVC could be made into a workable material by the addition of a *plasticizer*. Semon got the idea of using plasticized PVC as a shower curtain when he observed his wife sewing together a shower curtain made of rubberized cotton. Other uses for PVC were slow to appear, however, and it wasn't until World War II that this material became recognized as an ideal waterproof material for tents and rain gear. After the war, PVC replaced Bakelite as the medium for making phonograph records.

Originally designed as a covering to protect theater seats from chewing gum, Saran™ found great use in World War II as a protective wrapping for artillery equipment during sea voyages. (Before Saran™, the standard operating procedure had been to disassemble and grease the artillery to avoid corrosion.) After the war, this polymer soon pushed cellophane aside to become the most popular food wrap of all time (**Figure 12.51**). Because of environmental issues associated with Saran™, polyethylene now holds the title of most widely used food wrap.

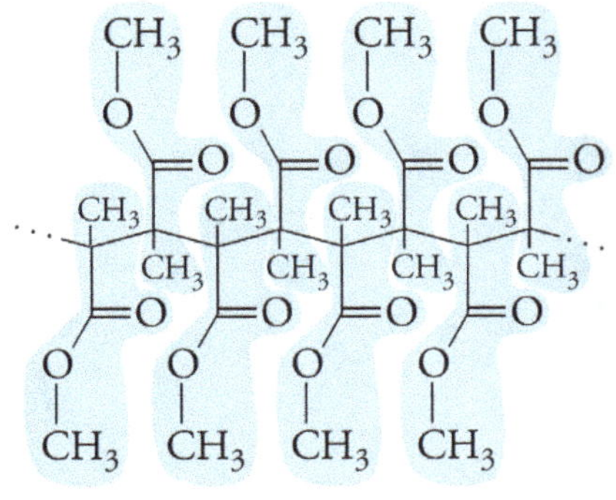

Poly(methyl methacrylate)

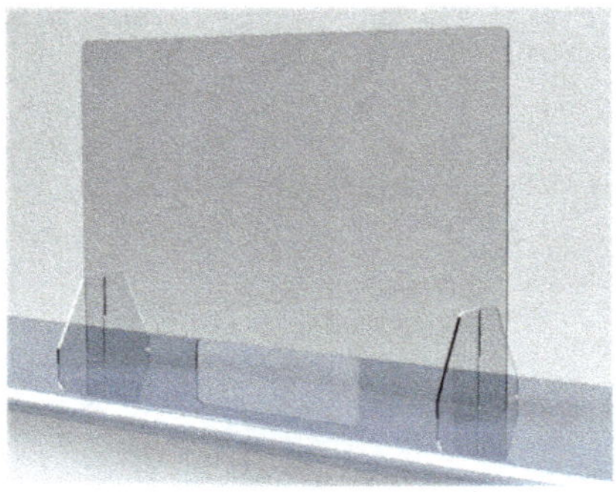

< Figure 12.50

The bulky side groups in poly(methyl methacrylate) prevent the polymer chains from aligning with one another. This makes it easy for light to pass through the material, which is tough, transparent, lightweight, and moldable, perfect for the creation of sneeze-guards that became popular during the Covid pandemic. (Plexiglas® is a registered trademark belonging to ATOFINA.)

∧ Figure 12.51

The now-familiar plastic food wrap carton with a cutting edge was introduced in 1953 by Dow Chemical for its Saran™ Wrap brand.

In the late 1930s, the discoverers of Teflon® were impressed by the long list of things this new material would not do. It would not burn, and it would not completely melt. Instead, at 620°F it congealed into a gel that could be conveniently molded. It would not conduct electricity, and it was impervious to attack by mold or fungus. No solvent, acid, or base could dissolve or corrode it. And most remarkably, nothing would stick to it, not even chewing gum.

Because of all the things Teflon® would not do, DuPont was not quite sure what to do with it. Then in 1944, the company was approached by governmental researchers in desperate need of a highly inert material to line the valves and ducts of an apparatus being built to isolate uranium-235 in the manufacture of the first nuclear bomb. Thus, Teflon® found its first application, and one year later, World War II came to a close with the nuclear bombing of Japan.

Attitudes about Plastics Have Changed

With a record of wartime successes, plastics were readily embraced in the postwar years. In the 1950s, Dacron polyester was introduced as a substitute for wool. Also, the 1950s were the decade during which the entrepreneur Earl Tupper created a line of polyethylene food containers, known as Tupperware®. By the 1960s, a decade of environmental awakening, many people began to recognize the negative attributes of plastics. Being inexpensive, disposable, and non-biodegradable, plastic readily accumulated as litter and as landfill. With petroleum so readily available and inexpensive, however, and with a growing population of plastic-dependent baby boomers, little stood in the way of an ever-expanding array of plastic consumer products. By 1977, plastics surpassed steel as the number-one material produced in the United States. Environmental concerns also continued to grow, and in the 1980s, plastics recycling programs began to appear. Some 40 years later, however, still less than 10 percent of plastics are recycled. Our attitudes toward plastics have certainly changed over the years. One might argue from excitement to fear to boredom. Given the nature of plastics themselves, one might also argue that each of these attitudes is warranted.

CONCEPT CHECK

Name at least four polymers that had a significant impact on the effectiveness of Allied forces in World War II.

CHECK YOUR ANSWER Here are six: synthetic rubber, polyethylene, Nylon, Plexiglas, PVC, Saran, and Teflon.

Concept Review

Chapter 12

Summary of Terms

Addition polymer A polymer formed by the joining together of monomer units with no atoms being lost as the polymer forms.

Alcohol An organic molecule that contains a hydroxyl group bonded to a saturated carbon.

Aldehyde An organic molecule containing a carbonyl group, the carbon of which is bonded either to one carbon atom and one hydrogen atom or to two hydrogen atoms.

Alkane A generic word for a saturated hydrocarbon.

Alkene An unsaturated hydrocarbon containing one or more double bonds.

Alkyne An unsaturated hydrocarbons containing a triple bond.

Amide An organic molecule containing a carbonyl group, the carbon of which is bonded to a nitrogen atom.

Amine An organic molecule containing a nitrogen atom bonded to one or more saturated carbon atoms.

Aromatic Said of any organic molecule containing a benzene ring.

Carbonyl group A carbon atom double-bonded to an oxygen atom; found in ketones, aldehydes, amides, carboxylic acids, and esters.

Carboxylic acid An organic molecule containing a carbonyl group, the carbon of which is bonded to a hydroxyl group.

Condensation polymer A polymer formed by the joining together of monomer units accompanied by the loss of small molecules, such as water.

Configuration A term used to describe how the atoms within a molecule are connected. For example, two structural isomers will consist of the same number and same kinds of atoms, but in different configurations.

Conformation One of a wide range of possible spatial orientations of a particular configuration.

Ester An organic molecule containing a carbonyl group, the carbon of which is bonded to one carbon atom and one oxygen atom bonded to another carbon atom.

Ether An organic molecule containing an oxygen atom bonded to two carbon atoms.

Functional group A specific combination of atoms that behaves as a unit in an organic molecule.

Heteroatom Any atom other than carbon or hydrogen in an organic molecule.

Hydrocarbon An organic compound containing only carbon and hydrogen atoms.

Ketone An organic molecule containing a carbonyl group, the carbon of which is bonded to two carbon atoms.

Monomers The small molecular units from which a polymer is formed.

Organic chemistry The study of carbon-containing compounds.

Phenol An organic molecule in which a hydroxyl group is bonded to a benzene ring.

Polymer A long organic molecule made of many repeating units.

Saturated hydrocarbon A hydrocarbon containing no multiple covalent bonds, with each carbon atom bonded to four other atoms.

Structural isomers Molecules that have the same chemical formula but different chemical structures.

Unsaturated hydrocarbon A hydrocarbon containing at least one multiple covalent bond.

Review Questions

12.1 Hydrocarbons

1. How do two structural isomers differ from each other?

2. How are two structural isomers similar to each other?

3. What physical property of hydrocarbons is used in fractional distillation?

4. What types of hydrocarbons are more abundant in higher-octane gasoline?

12.2 Unsaturated Hydrocarbons

5. What is the difference between a saturated hydrocarbon and an unsaturated hydrocarbon?

6. How many multiple bonds must a hydrocarbon have in order to be classified as unsaturated?

7. Aromatic compounds contain what kind of ring?

12.3 Functional Groups

8. What is a heteroatom?

9. Why do heteroatoms make such a difference in the physical and chemical properties of an organic molecule?

12.4 Alcohols, Phenols, and Ethers

10. Why are low-formula-mass alcohols soluble in water?

11. What distinguishes an alcohol from a phenol?

12. What distinguishes an alcohol from an ether?

12.5 Amines and Alkaloids

13. Which heteroatom is characteristic of an amine?

14. Do amines tend to be acidic, neutral, or basic?

15. Are alkaloids found in nature?

12.6 Carbonyl Compounds

16. Which elements make up the carbonyl group?

17. How are ketones and aldehydes related to each other? How are they different from each other?

18. How are amides and carboxylic acids related to each other? How are they different from each other?

12.7 Organic Synthesis

19. Why might an organic chemist want to synthesize a molecule that already occurs in nature?

20. What is a retrosynthesis analysis?

12.8 Polymer Chemistry

21. What happens to the double bond of a monomer participating in the formation of an addition polymer?

22. What is released in the formation of a condensation polymer?

23. Why is plastic wrap made of polyvinylidene chloride stickier than plastic wrap made of polyethylene?

12.9 A Brief History of Plastics

24. What is one of the major drawbacks of celluloid?

25. What role did Teflon play in World War II?

Quantitative Questions

26. Rank the following hydrocarbons in order of increasing number of hydrogen atoms:

Cyclobutane Butane trans-2-Butene

27. Rank the following hydrocarbons in order of increasing number of hydrogen atoms:

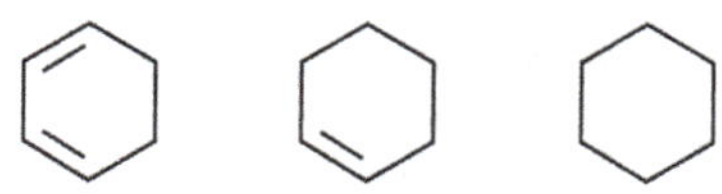

(a) (b) (c)

28. Rank the following hydrocarbons in order of smallest to largest carbon chain:

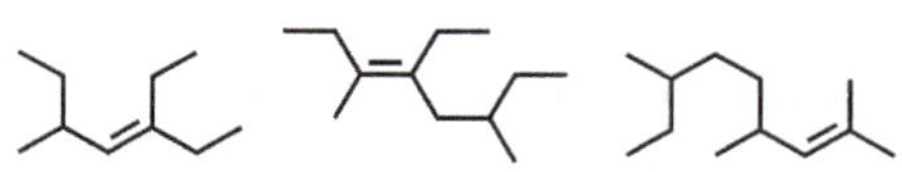

(a) (b) (c)

29. Rank the following organic molecules in order of increasing solubility in water:

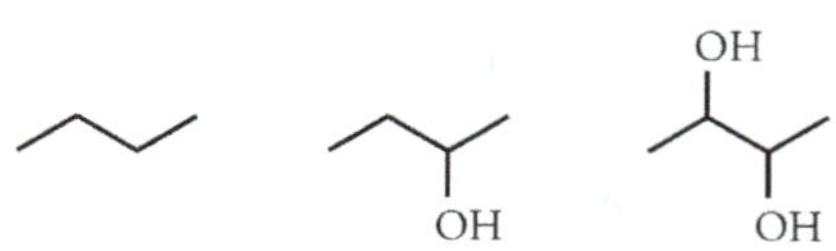

(a) (b) (c)

Solutions (Odd-Numbered)

1. Structural isomers have different arrangements of their carbon atoms, but the number of carbon atoms they each have is the same.

3. Boiling points of hydrocarbons are used for fractional distillation.

5. Saturated hydrocarbons have only single bonds. Unsaturated hydrocarbons have multiple bonds.

7. Aromatic compounds contain a 6-membered ring system called a benzene ring.

9. Heteroatoms give an organic molecule its "character". They are typically electronegative and affect the physical and electrical properties of the molecule.

11. An alcohol simply contains a hydroxyl functional group, but a phenol has a hydroxyl group attached to a benzene ring.

13. Nitrogen is found in all amines.

15. Alkaloids are alkaline amines found in nature, though they can also be synthesized in the laboratory.

17. Both contain carbonyl groups. Ketones have the carbonyl carbon bonded to two adjacent carbon atoms, but aldehydes have the carbonyl carbon bonded to either one hydrogen and one carbon or to two hydrogens.

19. The quantities of the molecule available in nature may be insufficient to allow for its use by many people.

21. One of the bonds in the multiple bond of a monomer opens to form a new bond with a neighboring monomer molecule.

23. The chlorine atoms of the polyvinylidene chloride are good at forming induced dipole-induced dipole molecular interactions as was discussed in Section 7.1.

25. Teflon was used to line the valves and ducts of an apparatus they were building to make nuclear weapons.

27. Compound c, C_6H_{12}, has more hydrogens than compound b, C_6H_{10}, which has more hydrogens than compound a, C_6H_8.

29. Recall that the hydroxyl group, -OH, is very polar. The more hydroxyl groups a compound has, the greater its polarity, which means the greater its solubility in water. Compound a is non-polar and insoluble in water. Compound b has a single hydroxyl group, which makes this compound somewhat soluble in water. Compound c has two hydroxyl groups, which makes it very soluble in water.

Chapter 12: Contextual Chemistry Essay

Hair and Skin Care

Through our modern lifestyles, we have become accustomed to chemical-based products that clean and protect our hair and skin. These include soaps, shampoos, conditioners, moisturizers, sunscreens, and many other products formulated to help us maintain a healthy look.

The essential ingredients of all these hair and skin products are part of a broad class of compounds called *surfactants*, which is short for "surface-active agents." All surfactant molecules are polar on one end and nonpolar on the other. Because of their nonpolar ends, they resist dissolving in water and instead cling to water's *surface*, where their nonpolar ends stick out of the water. Pour some oil onto the surfactant-containing water and the nonpolar tails will cling into the oil's surface. In this way, surfactants bind together the surfaces of water and a nonpolar material.

With a little agitation, the surfactant's surface-seeking behavior causes the formation of high-surface-area suds. With such an increased surface area, significant amounts of the oil are able to mix with the water, forming a single phase known as an *emulsion*. So, surfactants don't help nonpolar materials to "dissolve in water." Instead, they just help these two materials to become mixed.

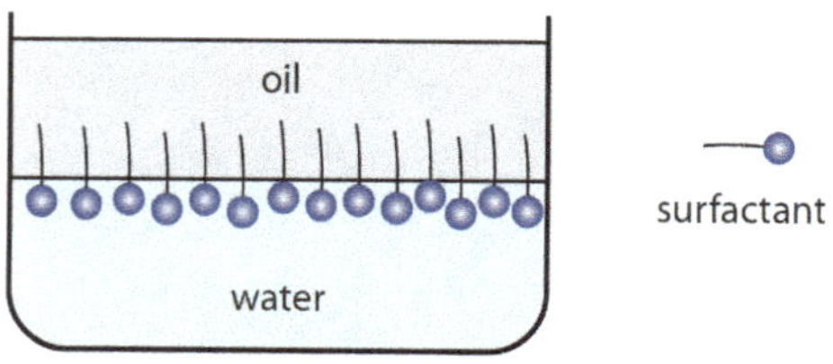

Sodium lauryl sulfate

The prototypical shampoo surfactant is the detergent sodium lauryl sulfate. A big advantage of detergents is that in hard water, they don't form scum (see Section 7.5). For a shampoo, this is particularly important because, otherwise, the scum would adhere to the hair. Rinsing hair first allows the removal of polar materials, such as salts, dirt, or any loose debris. Wetted hair also swells, and this opens the hair cuticles, which are scale-like structures lining the outside of the hair (**Figure A**).The main action of the detergent is to remove *sebum*, a skin oil that migrates up the hair and into the cuticles, making one's hair "greasy". Shampoo formulas vary in their cleansing strength. Those formulated for "oily" hair are generally stronger than those formulated for "normal" hair.

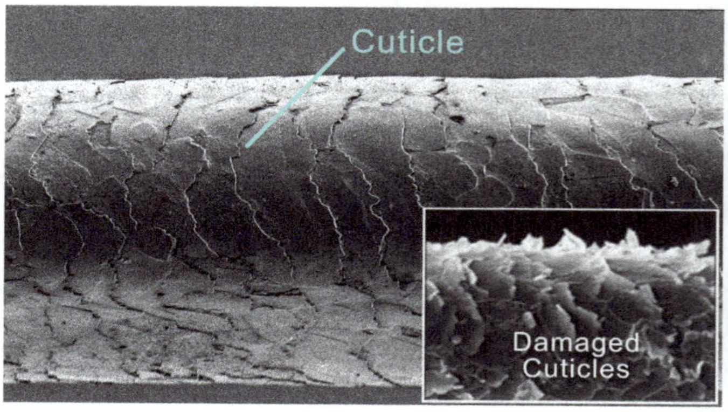

Λ Figure A
One thousand times magnification of a human hair shows a series of folds on the surface known as *cuticles*. These cuticles can open and shut depending upon the conditions. For damaged hair, these cuticles have been splintered.

Following the shampoo, many people apply a hair conditioner, which helps to make the hair feel soft and helps to minimize frizz due to the buildup of static charge. Notably, the polar end of a shampoo's surfactant is a negatively charged ion. Such a surfactant is called an anionic surfactant. The polar end of a hair conditioner's surfactant, however, is a positively charged ion. Such a surfactant is called a cationic surfactant. Shampoos and conditioners are generally not combined into a single formula because the attraction these oppositely charged surfactants have for each other would defeat their functions. Products that are "all-in-one" have the cationic surfactants microencapsulated.

The prototypical hair conditioning surfactant is cetyl trimethyl ammonium chloride, also known as cetrimonium chloride. Its positive charge helps it to bind with the hair so that much of it remains on the hair even after rinsing. The result is hair that feels smoother and is easier to manage. Laundry fabric softeners work by the same principle.

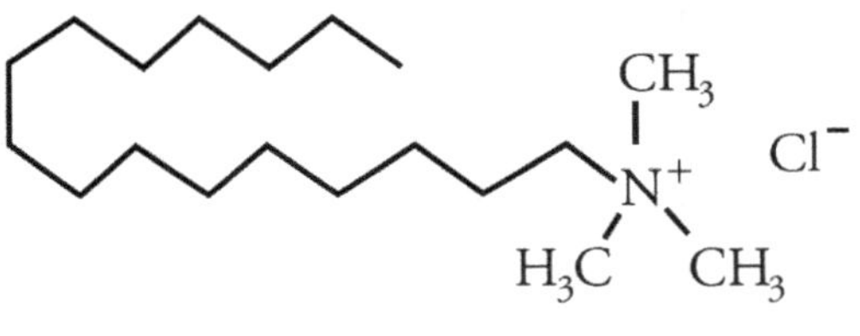

Cetyl trimethyl ammonium chloride

The trick to having comfortably moist skin is to maintain the moisture already found within the skin. The deeper parts of your skin are up to 80 percent water. Because skin has so much more water compared to the surrounding air, the general movement of moisture is from the skin outward. Skin oils help the outer layers of skin to retain much of this moisture. Your skin, however, will begin to feel dry when there is an increase in the rate at which your skin's moisture evaporates. This happens when the air is very dry or when your skin oils are removed by strong soaps or detergents. Rather than splashing water on your dry skin, a more practical solution is to coat your skin with a moisture-protecting nonpolar material that can enhance or mimic the action of skin oils. Commonly used materials include compounds such as propylene glycol, glycerin, and dimethicone. The best way to apply these compounds to the skin is with an emulsion made using nonionic surfactants, such as polyethylene glycol stearate, also known as PEG-100. Nonionic surfactants are used for moisturizers because they tend to be less irritating and less likely to remove skin oils.

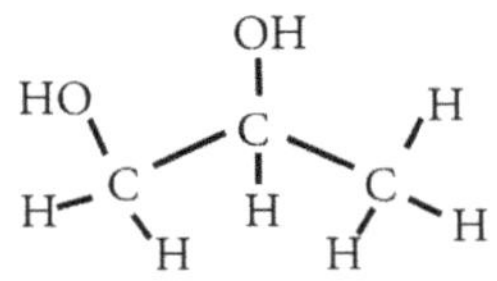

Propylene glycol

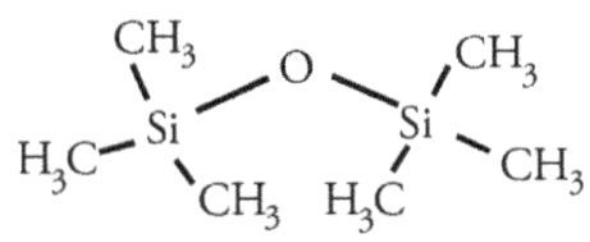

Dimethicone

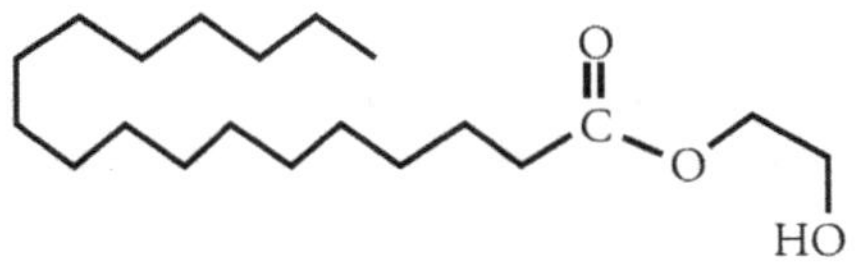

Polyethylene glycol stearate

Add UV-absorbing compounds to a skin moisturizer and you have sunblock lotion. Making a sunblock lotion water resistant, however, is problematic. Upon exposure to water, the surfactants within the lotion help to remove the lotion. Manufacturers generally recommend that you reapply the lotion after swimming. Lotions that minimize the use of surfactants have increased water resistance, but they tend to be oily or greasy. For these reasons, many water-resistant formulas include compounds that polymerize into a nonpolar, water-resistant film that helps to hold the sunscreen agent on the skin when wet.

A special consideration for sunscreens is the portion of the UV spectrum they are able to block. Nearly all sunscreens effectively block the region from 290 to 320 nm, known as UV-B. The traditional Sun Protection Factor (SPF) rating system indicates the effectiveness of a sunscreen lotion in blocking UV-B, which causes sunburn. Scientists, however, are finding that the range from 320 to 400 nm, known as UV-A, is more responsible for skin wrinkling as well as skin cancer. Sunscreens blocking both UV-A and UV-B are labeled "Broad Spectrum."

The most widely used UV-A blockers, avobenzone, is easily formulated into a lotion, but it quickly loses potency if not properly formulated with stabilizers. There is also the more stable UV-A blocker mexoryl SX. Two other UV-A blockers approved by the FDA are titanium dioxide and zinc oxide, both opaque minerals that are impractical to apply over large body-surface areas.

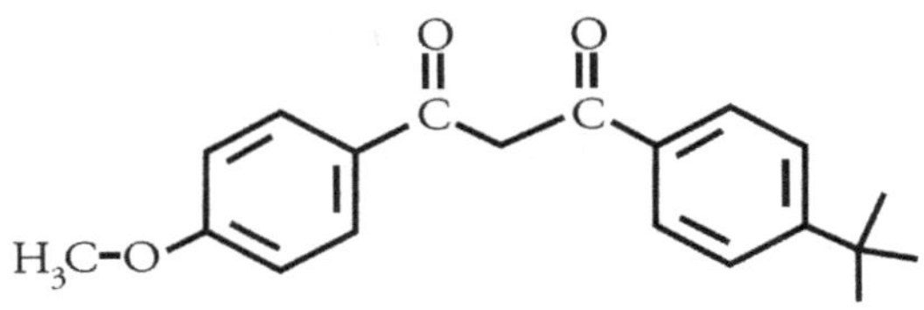

Avobenzone

CONCEPT **CHECK**

What do shampoos, hair conditioners, moisturizers, and sunscreen lotions all have in common?

CHECK YOUR ANSWER They all contain surfactants, which are molecules that have both polar and non-polar ends and permit the formation of emulsions

The lists of approved UV-B and UV-A sunscreens in countries such as Australia and Canada, and in European nations are much longer than in the U.S. These countries treat sunscreens as cosmetics, while the United States treats them as drugs, which are subject to more rigorous regulations.

Another way to help skin maintain a youthful look is to apply alpha-hydroxy acids, such as glycolic acid. These compounds serve as exfoliants, which means they help in the removal of dead outer skin. This stimulates the skin to generate new cells that help to hide thin-line wrinkles.

Discussed here are only the main chemicals used to formulate hair and skin-care products. As any casual look at an ingredients list tells you, many other chemicals serve other purposes.

Looking to the immediate future, there are some areas where skin care may soon make some significant advances. As we grow older, our skin wrinkles, because it has lost much of its elasticity. One of the reasons for this loss of elasticity is the formation of chemical bonds, called crosslinks, between adjacent strands of collagen, which is the fundamental fiber skin cells create to form skin. The search is now underway to find agents that effectively inhibit and even reverse these collagen crosslinks.

Think and Discuss

1. In an effort to speed up the drug approval process, the U.S. FDA created a special "fast-track" system for agents that show unusual promise or for agents that have been shown to be generally safe after years of use in other countries. Why are there so few sunscreen agents currently within this fast-track system? Should sunscreen agents continue to be classified as drugs?

2. If a collagen cross-linking agent is determined to be safe by the FDA, should this agent be available only by prescription, or should it be made available to consumers over the counter? If this agent noticeably helped to reduce skin wrinkling, how popular might it be? Assuming you were in your 50's, how much would you be willing to pay for a year's supply?

3. What criteria do you use in deciding what shampoo to buy? List the following in order of importance: the price, the brand name, the ingredients named on the front label, those listed on the back label, the fragrance, commercial advertisements, recommendation from a friend, and your experience with the shampoo.

Author Responses to Think and Discuss

1. Why are so few sunscreen agents currently within the FDA fast track system? There are likely several reasons, such as bureaucratic inertia or preference given to medicines. This would make a good research project for a team of students. Should sunscreen agents be classified as drugs even if that slows down the approval process?

2. If a collagen cross-linking agent is found to be safe, then the FDA might be tempted to initially regulate its formulation and distribution to assure quality control. Clinically shown to eliminate wrinkles? This would be the makings of a new billion-dollar industry.

3. Answers will vary, but remember: Brand is the hook that frees the mind from having to think much when shopping.

Index

Photo Credits

Chapter 1

Figure 1.2	William Harwood
Figure 1.3	Jim Baker
Figure 1.4a	Russ Hopcroft
Figure 1.4b	Jim Baker
Figure 1.9	Lisa Fabro
Figure 1.11a	Daniel Sone
Figure 1.11b	Famartin
Figure 1.11c	Wes Agresta
Figure 1.11d	Mike Voorhies
Figure 1.11e	John O'Meara

Chapter 2

Figure 2.1	Luca Galuzzi
Figure 2.8	ORNL
Figure 2.15	Frank Schulenburg
Figure 2.17	Pyrogrimace
Figure 2.32	Pep.per de Ré
Page 52 UNA01	Matthew Rader
Page 58 UNA01	Conall
Page 58 UNA02	Kelisi
Page 59 UNA01	Warren Gretz

Chapter 3

Figure 3.1a	Szaaman
Figure 3.1b	Catherine Lewis
Figure 3.1c	Derek Jensen
Figure 3.2a	Randi Hausken
Figure 3.2b	Jacek Halicki
Figure 3.3a	Ivan Radic
Figure 3.3c	William Haslet
Figure 3.6	Mobilus In Mobili
Figure 3.11 inset	Gaius Cornelius
Figure 13.12	Auckland Museum
Figure 13.12	Deutsche Fotothek
Figure 13.12	Ludvig14
Figure 13.12	Peellden
Figure 13.12	Daler Boi
Figure 13.12	Wilco Oelen
Figure 3.17	Davy Image
Figure 3.21	Nefronus
Figure 3.24	Carolos Eduardo Joos
Figure 3.27a	Baminnick
Figure 3.27b	James Yardley
Figure 3.28	Dicklyon
Figure 3.30a	B. Domangue
Figure 3.30b	Alzinous
Figure 3.30c	Dale Cruse
Figure 3.30d	Bob Haarmans
Figure 3.30e	Vyacheslav Argenberg
Figure 3.30f	Bence Fördős
Figure 3.31	Alan Light
Figure 3.34	NanoscaleRanger
Page 94	Murray Foubister

Chapter 4

Figure 4.4a	Tony Webster
Figure 4.6	Sigot
Figure 4.11	Nick Kellar
Figure 4.18a	盧樂
Figure 4.18b	Hegelrast
Figure 4.18c	G43
Figure 4.23a	George Webster
Page 130 UNA1	Tor Svensson
Page 130 UNA2	Japanexperterna.se

Chapter 5

CAC05CO	Martin Mergili
Figure 5.4	Kyle McDonald
Figure 5.7	BLK Cyberkife Hospital
Figure 5.16	Haruhiko Okumura
Figure 5.26	Photorush
Figure 15.32	ORNL/ITER
Page 166 Figure A1	IAEA Imagebank
Page 166 Figure A2	IAEA Imagebank

Chapter 6

Figure 6.9	Robert Lavinsky
Figure 6.10a	Parent Géry
Figure 6.10b	Robert Lavinsky
Figure 6.11	Edal Anton Lefterov
Figure 6.14	Raman Patel
Figure 6.15	Toby Hudson
Figure 6.16	Urbain Kinet
Figure 6.21b	Lisa Fern
Figure 6.22	Catherine Lewis
Figure 6.38	Toby Hudson
Page 200 UNA1	Adam Keck
Page 201 UNA1	Basil Morin

Chapter 7

CAC07CO	Antoine Taveneaux
Figure 7.6	Tomwsulcer
Figure 7.10	Irhanz
Figure 7.11a	Dnn87
Figure 7.12	Guista Pentol
Figure 7.14a	Michelle Jo
Figure 7.14b	Gee KongLawm
Figure 7.14c	Renato Ganoza
Page 212 UNA1	Evan-Amos
Page 212 UNA2	Gee KongLawm
Figure 7.20	Acabashi
Page 224 UNA1	Der Kreole
Figure 7.26	Hustvedt
Figure 7.31	Gary Beutler
Figure 7.32	Vmenkov
Figure 7.33	SolAqua

Chapter 8

CAC08CO	Kucheryash021
Figure 8.1	Tomwsulcer
Figure 8.5	David W. Carmichael
Figure 8.14	Kaldari
Figure 8.15	Tony Alter
Figure 8.18	Thirun Raveendran
Figure 8.23	Ashley Pomeroy
Figure 8.24	Bob Jones
Figure 8.26	Schorle
Figure 8.29d	GRAN
Figure 8.31b	Paxson Woelber
Figure 8.33	Skatebiker
Figure 8.43	Andreas Weith
Page 274 Figure D	Rob Rose
Page 275 Figure F	Kelvinsong

Chapter 9

CA09CO	jonwestra
Figure 9.4	Nefronus
Figure 9.6a	Alan Light
Figure 9.10	Soph556
Figure 9.11	Jarek Tuszyński
Figure 9.24	Kim2480
Figure 9.25	V1213
Figure 9.26	Silverish Lily
Figure 9.29	Eframgoldberg
Figure 9.30a	Inbal Weisman
Figure 9.30b	National Archives
Page 317 Figure A1	Wilco Oelen
Page 318 Figure D	Bill Morrow

Chapter 10

Figure 10.1d	M0tty
Figure 10.2b	Havelock
Figure 10.2d	Kjetil Ree
Figure 10.12a	Ildar Sagdejev
Figure 10.12b	Kiyok
Figure 10.17	CSIRO
Figure 10.18	Wallyg
Page 345 Figure C	Hannes Grobe

Chapter 11

CAC11CO	Dogora Sun
Figure 11.9	Rutger van der Maar
Figure 11.11	Jason Lawrence
Figure 11.13	FuelCellEnergy
Figure 11.14	Liberata Guadagno
Figure 11.15	ChristofferRiemer
Figure 11.20	Hazelares
Figure 11.25	Cornelius Keyes
Figure 11.27	LeRoy Woodson
Figure 11.28	Andreslan
Figure 11.29	Raysonho
Figure 11.32	Brosen
Page 373 UNA1	Maksym Kozlenko

Chapter 12

Figure 12.6	Anthony Inswasty
Figure 12.26a	Jdforrester
Figure 12.26b	Mosesofmason
Figure 12.27	Gerald J. Lenhard
Figure 12.35	Foster Bamford
Figure 12.36	Guista Pentola
Figure 12.43	Vis M
Figure 12.46b	Harsha K R
Figure 12.47	Rama
Figure 12.49	Itzuvit
Page 414 Figure A	Priscilla Du Preez
Page 414 Figure B	Jae Hong Ji

Chapter 13

Figure 13.1	RJ Raawat
Figure 13.2	Peter Shanks
Figure 13.3	Marsalo17
Figure 13.4	Santeri Viinamäki
Figure 13.5a	Klarkamatej
Figure 13.5b	Rushenb
Figure 13.6	Philippa Uwins

Figure 13.7	Marie-Lan Nguyen
Figure 13.10	Katja Schulz
Figure 13.12	Budd Christman
Figure 13.13a	MCB
Figure 13.13a	Achilles2019
Figure 13.18b	MayStudio
Figure 13.20	OpenStax College
Figure 13.28	Abogomazova
Figure 13.28	Thomas Shafee
Figure 13.28	Natalierussell77
Figure 13.33	Raysonho
Figure 13.36	Peter Van den Bossche
Figure 13.38	Squirrel Cottage
Page 455 Figure A	MayStudio
Page 457 Figure D	Chad Veal

Chapter 14

CAC14CO	Duke of W4
Figure 14.3	Werner Hammer
Figure 14.4	Jason Hollinger
Figure 14.8	Dominic Lockyer
Figure 14.14	Evilonan
Figure 14.16	Ввласенко
Figure 14.18	Hoffmann-LaRoche Ltd
Figure 14.26	Kristi Denby
Figure 14.29	Filo gèn'
Figure 14.31a	Arnoud Maaswinkel
Figure 14.31b	Alexander Klink
Figure 14.31c	Mattes
Figure 14.31d	Paolo Neo
Figure 14.34	Dav Hir
Figure 14.35	Daniel Oberhaus
Figure 14.36	Josh Berglund
Figure 14.46	Hiobson
Page 495 Figure A	Patrick Mallahan III
Figure 497 Figure B	Stokkete
Figure 497 Figure B	Eungchopan

Chapter 15

Figure 15.2	Bukky658
Figure 15.5c	Alandmanson
Figure 15.7	Nigel Chadwick
Figure 15.13	David McClenaghan
Figure 15.18	ForestWander
Figure 15.28a	Carolina Latorre Canet
Figure 15.28b	AlvarovegaCLP
Figure 15.29a	Acabashi
Figure 15.29b	normanack
Figure 15.31	James Brey
Figure 15.32b	bdk
Figure 15.33	Marjorie Hoy
Figure 15.34a	Ian Alexander
Figure 15.34b	Albert Herring
Figure 15.35a	Bluemoose
Figure 15.35b	Dehaan
Figure 15.36	Ryan Thompson
Page 527 Figure B	Jeffery Martin

Chapter 16

Figure 16.3	Ingo Mehling
Figure 16.6a	Bill Gillette
Figure 16.6b	Joe Mabel
Figure 16.7	ReunionVacationHomes
Figure 16.11	Smaack
Figure 16.22	Mike Mozart
Figure 16.26	Laurent Augustin
Figure 16.26	CSIRO
Figure 16.30a	Thomas Fuhrmann
Figure 16.30b	Ilya Grigorik
Figure 16.31	Jean-Pascal Quod
Page 559 Figure A	Andrei Porzhezhinskii

Chapter 17

Figure 17.5a	Amcyrus2012
Figure 17.5b	Nefronus
Figure 17.8	Aldo116
Figure 17.10	Calistemon
Figure 17.11	Iketani Daisei
Figure 17.17b	Carl Montgomery
Figure 17.17c	Firef7y
Figure 17.18	IAEA
Page 579 Figure A	Arne Müseler
Page 579 Figure B	Hansueli Krapf
Page 579 Figure C	TomaszSwatowski
Page 579 Figure D	御嶽木郎
Figure 17.22	Allen Watkin
Figure 17.23	Vitafougue
Figure 17.26	Moyan Brenn
Figure 17.27	Sw271149
Figure 17.28	Mariordo Ortiz
Figure 17.29	Marco Esparz
Figure 17.30	Paul Gipe
Figure 17.33	Kjkolb
Figure 17.34	kallerna
Figure 17.38	Acciona Energia
Figure 17.42	Jakub
Page 599 Figure A	Battenbrook

www.ingramcontent.com/pod-product-compliance
Lightning Source LLC
LaVergne TN
LVHW082109200926
848484LV00008B/1932